Lecture Notes in Computer Science 16513

The series Lecture Notes in Computer Science (LNCS), including its subseries Lecture Notes in Artificial Intelligence (LNAI) and Lecture Notes in Bioinformatics (LNBI), has established itself as a medium for the publication of new developments in computer science and information technology research, teaching, and education.

LNCS enjoys close cooperation with the computer science R & D community, the series counts many renowned academics among its volume editors and paper authors, and collaborates with prestigious societies. Its mission is to serve this international community by providing an invaluable service, mainly focused on the publication of conference and workshop proceedings and postproceedings. LNCS commenced publication in 1973.

Mitra Baratchi · Siegfried Nijssen ·
Jan N. van Rijn
Editors

Advances in Intelligent Data Analysis XXIV

24th International Symposium on Intelligent Data Analysis, IDA 2026
Leiden, The Netherlands, April 22–24, 2026
Proceedings

Springer

Editors
Mitra Baratchi
Universiteit Leiden
Leiden, The Netherlands

Siegfried Nijssen
KU Leuven
Leuven, Belgium

Jan N. van Rijn
Universiteit Leiden
Leiden, The Netherlands

ISSN 0302-9743 ISSN 1611-3349 (electronic)
Lecture Notes in Computer Science
ISBN 978-3-032-23832-0 ISBN 978-3-032-23833-7 (eBook)
https://doi.org/10.1007/978-3-032-23833-7

This Springer imprint is published by the registered company Springer Nature Switzerland AG
The registered company address is: Gewerbestrasse 11, 6330 Cham, Switzerland

Preface

This volume contains the proceedings of the 24th International Symposium on Intelligent Data Analysis, held from April 22–24, 2026, in Leiden, The Netherlands. After being founded in 1995, the symposium was organised biennially until 2009. Starting from 2010, the symposium was organised annually and shifted its focus towards submissions that present ground-breaking and innovative ideas, even if they might not be fully developed yet. This edition maintains this tradition, encouraging submissions that have the potential to significantly advance research in a certain field.

The IDA Symposium welcomes a broad range of modelling and analysis methodologies from all disciplines, aiming to be an interdisciplinary forum that fosters discussions on intelligent data analysis that span various fields. Continuing this tradition, IDA 2026 extended its support for data analysis beyond the conventional algorithmic solutions typically discussed in the academic literature, also accepting submissions that integrate established technologies within intelligent data analysis frameworks or apply such technologies in innovative ways to the analysis and/or modelling of complex systems.

Similarly to previous years, to encourage innovative research, reviewers and senior PC members were encouraged to favour innovative ideas. Additionally, the senior PC members had the mandate to promote such papers to the PC chairs, even when none of the reviewers championed them. The outcome was a diverse and interesting program, as reflected in this volume. The main research track received 102 submissions, of which 16 received a desk rejection for various reasons, and 86 underwent a rigorous single-blind review process by three members of the program committee and one senior member. Eventually, 33 papers were accepted.

We were pleased to have the following distinguished invited speakers at IDA 2026:

- Fredrik Heintz, of Linköping University (Sweden)
- Ana Lucic, of the University of Amsterdam (the Netherlands)
- Isabel Valera, of Saarland University (Germany)
- Geoff Webb, of Monash University (Australia)

The symposium was held at the Academy Building of Leiden University in Leiden, the Netherlands.

We wish to express our gratitude to all authors of submitted papers for their intellectual contributions; to the program committee members and emergency reviewers for their help in ensuring the quality of the program. In particular, we wish to express our gratitude to the chairs who played an important role in organising the symposium. We wish to thank the advisory chairs Élisa Fromont, Allan Tucker and Peter van der Putten for their various contributions, including the coordination of the Frontier Prize. The PhD Forum track was chaired by Christine Sinoquet. The local organisation was coordinated by Roy de Kleijn as local chair. Online communication was in the hands of Laurens Arp, Mischa Hautvast, Andreas Paraskeva and Julia Wąsala. The proceedings were organised by Matthijs van Leeuwen and Slawomir Nowaczyk.

Finally, we would like to express our gratitude to our sponsors and partners: Leiden&Partners (the city marketing organisation of the city of Leiden), Springer, the Netherlands Research School for Information and Knowledge Systems (SIKS), Leiden Institute of Advanced Computer Science (LIACS), and Society, Artificial Intelligence and Life Sciences (SAILS) at Leiden University. We thank the City of Leiden for welcoming us at Leiden City Hall and hosting the welcome reception there; finally, we thank Leiden University for offering its historic Academy Building and adjacent locations free of charge.

April 2026

Mitra Baratchi
Siegfried Nijssen
Jan N. van Rijn

Organization

General Chair

Jan N. van Rijn	Leiden University, the Netherlands

Program Committee Chairs

Mitra Baratchi	Leiden University, the Netherlands
Siegfried Nijssen	KU Leuven, Belgium

Senior Program Committee

Hendrik Blockeel	KU Leuven, Belgium
Henrik Boström	KTH Royal Institute of Technology, Sweden
Peggy Cellier	INSA Rennes, IRISA, France
Bruno Crémilleux	Université de Caen Normandie, France
Thi-Bich-Hanh Dao	University of Orléans, France
Wouter Duivesteijn	Technische Universiteit Eindhoven, the Netherlands
Elisa Fromont	Université Rennes, IRISA/Inria Rennes, France
João Gama	INESC TEC – LIAAD, Portugal
Riccardo Guidotti	University of Pisa, Italy
Barbara Hammer	CITEC, Bielefeld University, Germany
Frank Höppner	Ostfalia University of Applied Science, Germany
Arno Knobbe	Leiden University, the Netherlands
Jefrey Lijffijt	Ghent University, Belgium
Ioanna Miliou	Stockholm University, Sweden
Kai Puolamäki	University of Helsinki, Finland
Celine Robardet	INSA Lyon, France
Arno Siebes	Universiteit Utrecht, the Netherlands
Maryam Tavakol	TU Eindhoven, the Netherlands
Allan Tucker	Brunel University London, UK
Peter van der Putten	Leiden University, the Netherlands
Matthijs van Leeuwen	Leiden University, the Netherlands
David Weston	Birkbeck, University of London, UK
Albrecht Zimmermann	Université de Caen Normandie, France

Program Committee

Shahrooz Abghari	Blekinge Institute of Technology, Sweden
Pedro Henriques Abreu	CISUC, Portugal
Thiago Andrade	INESC TEC – LIAAD, Portugal
Fabrizio Angiulli	University of Calabria, Italy
Ali Ayadi	University of Strasbourg, France
Paulo Azevedo	Universidade do Minho, Portugal
José Luis Balcázar	Universitat Politècnica de Catalunya, Spain
Guillaume Bied	Ghent University, Belgium
Anton Björklund	Oxford University, UK
Veselka Boeva	Blekinge Institute of Technology, Sweden
Giacomo Boracchi	Politecnico di Milano, Italy
José Borges	FEUP/INESC-TEC, Portugal
Ali Braytee	University of Technology Sydney, Australia
Paula Brito	Universidade do Porto, LIAAD-INESC TEC, Portugal
David Broneske	German Centre for Higher Education Research and Science Studies, Germany
Dariusz Brzezinski	Poznań University of Technology, Poland
Mirko Bunse	TU Dortmund University, Germany
Vitor Cerqueira	University of Porto, Portugal
Paulo Cortez	University of Minho, Portugal
Joana Cristo Santos	University of Coimbra, Portugal
Bertrand Cuissart	University of Caen Normandy, France
Moises dos Santos	University of Porto, Portugal
Nuno Escudeiro	Instituto Politécnico do Porto, Portugal
Brigida Mónica Faria	Polytechnic Institute of Porto and LIACC, Portugal
Ad Feelders	Universiteit Utrecht, the Netherlands
Françoise Fessant	Orange, France
Johannes Fürnkranz	Johannes Kepler University Linz, Austria
Mikel Galar	Universidad Pública de Navarra, Spain
Rui Gomes	Universidade Portucalense Infante D. Henrique, Portugal
João Gonçalves	Universidade Católica Portuguesa, Portugal
Maciej Grzenda	Warsaw University of Technology, Poland
Alberto Fernandez Hilario	University of Granada, Spain
Martin Holena	Institute of Computer Science, Czech Academy of Sciences, Czech Republic
Tomáš Horváth	Edinburgh Napier University, UK
Dino Ienco	Irstea Institute, France

Szymon Jaroszewicz	Polish Academy of Sciences, Poland
Baptiste Jeudy	Université de Saint-Étienne, France
Ulf Johansson	Jönköping University, Sweden
Andrej Kastrin	University of Ljubljana, Slovenia
Jiri Klema	Czech Technical University in Prague, Czech Republic
Georg Krempl	Utrecht University, the Netherlands
Alejandro Kuratomi	Stockholm University, Sweden
Lazhar Labiod	Université Paris Cité, France
Christine Largeron	LabHC, University of Lyon, France
Mark Last	Ben-Gurion University of the Negev, Israel
Etienne Lehembre	LIFO, France
Tuwe Löfström	Jönköping University, Sweden
Marco Loog	Radboud University, the Netherlands
Jens Lundström	Halmstad University, Sweden
Luis Miguel Matos	University of Minho, Portugal
Joao Mendes-Moreira	University of Porto, Portugal
Mohamed Nadif	Université Paris Cité, France
Aurélie Névéol	CNRS, France
Ana Rita Nogueira	INESCTEC, Portugal
Abdelkader Ouali	University of Caen Normandy, France
Panagiotis Papapetrou	Stockholm University, Sweden
Kaustubh Patil	Research Centre Jülich, Germany
Ruggero Pensa	University of Turin, Italy
Nico Piatkowski	Fraunhofer IAIS, Germany
Marc Plantevit	EPITA, France
Senja Pollak	Jožef Stefan Institute, Slovenia
Luboš Popelínský	Masaryk University, Czech Republic
Filipe Portela	University of Minho, Portugal
Miguel Prada	Universidad de León, Spain
Rita Ribeiro	University of Porto, INESC TEC, Portugal
Duncan Ruiz	Pontifícia Universidade Católica do Rio Grande do Sul, Brazil
Petr Ryšavý	Czech Technical University in Prague, Czech Republic
Miriam Santos	University of Porto, Portugal
Jörg Schlötterer	University of Marburg, Germany
Johannes Schneider	University of Liechtenstein, Liechtenstein
Mikel Sesma-Sara	Universidad Pública de Navarra, Spain
Paula Silva	INESC TEC - LIAAD, FEUP, Portugal
Chitsutha Soomlek	Khon Kaen University, Thailand
Amina Souag	Canterbury Christ Church University, UK

Arnaud Soulet	University of Tours, France
Niclas Ståhl	Jönköping University, Sweden
Jerzy Stefanowski	Poznań University of Technology, Poland
Shazia Tabassum	INESCTEC, Portugal
Maguelonne Teisseire	INRAE - UMR Tetis, France
Sónia Teixeira	INESC TEC, Portugal
Alicia Troncoso	Pablo de Olavide University, Spain
Elena Tsiporkova	Sirris, Belgium
Cor Veenman	TNO/Leiden University, the Netherlands
Bruno Veloso	INESC TEC & FEP-UP, Portugal
Annelies Vroman	Sirris, Belgium
Pascal Welke	Lancaster University Leipzig, Germany
Guy Wolf, Mila	Université de Montréal, Canada
Paweł Zyblewski	Wrocław University of Science and Technology, Poland

Contents

Section: Data mining

Section: Temporal Data

Section: Applications of AI/ML

Foundations of AI/ML

Adaptive Local Kernel for Efficient Active Pairwise Constraint Clustering

Vincent Blase[1,3](✉), Julien Aligon[2], Moncef Garouani[2], Isabelle Ader[3], and Olivier Teste[1]

[1] IRIT, UMR 5505 CNRS, Université de Toulouse, UT2J, Toulouse, France
[2] IRIT, UMR 5505 CNRS, Université Toulouse Capitole, Toulouse, France
[3] RESTORE Research Center, Université de Toulouse, INSERM 1301, CNRS 5070, EFS, ENVT, Toulouse, France
vincent.blase@irit.fr

Abstract. Active constraint-based clustering iteratively incorporates user knowledge into the clustering process through pairwise constraints, specifying whether two elements should be linked (Must-Link) or separated (Cannot-Link). Existing active approaches often overlook the underlying data distribution, particularly across regions with varying scales, which can lead to uninformed exploration of the data space. We introduce a novel approach that employs a locally adaptive kernel, automatically adjusting its bandwidth based on local data density. This local adaptation identifies regions that deviate from uniform or Gaussian assumptions, prioritizing them for constraint queries and thus minimizing user effort while improving clustering quality. Unlike global kernels, which require extensive hyperparameter tuning incompatible with interactive learning, our method dynamically selects the appropriate local bandwidth, avoiding manual configuration. Experiments on 29 real-world datasets demonstrate that the proposed approach consistently outperforms state-of-the-art methods and global kernels in terms of query efficiency, clustering accuracy.

Keywords: Semi-supervised clustering · Pairwise clustering · Active clustering · Kernel representation

1 Introduction

Constraint-based clustering [1] has emerged as an effective approach for handling data with presence of complex structures, non-convex shapes or noisy conditions. It relies on supervision from human experts, providing pairwise Must-Link (ML) and Cannot-Link (CL) constraints to enforce that specific pairs of data points should or should not, belong to the same cluster [2]. Active Constraint-based Clustering (ACC) techniques select the most informative constraints iteratively,

I. Ader and O. Teste—Co-last author.

M. Baratchi et al. (Eds.): IDA 2026, LNCS 16513, pp. 3–17, 2026.
https://doi.org/10.1007/978-3-032-23833-7_1

optimizing the trade-off between clustering performance and the number of user interactions required [3].

Current ACC methods aim to leverage neighborhoods or super-instances [4,6], which aggregate nearby points that are assumed to belong to the same cluster, acting as higher-level representatives. These representatives enable constraints to be queried between groups rather than individual data points, thereby significantly reducing the number of user interactions required. Nevertheless, these approaches are usually based on graph-based or matrix methods, which assume Euclidean proximity. These assumptions often fail in datasets with complex manifolds, non-convex clusters, or multiple scales. This results in over-generalised or fragmented representations that do not accurately reflect the real data structure. Therefore, these methods either fail to explore significant portions of the data space, struggle with varying density, and often produce representatives that misrepresents the underlying distribution which delays refinement.

To address this gap, kernel representations provide an alternative by embedding the data in a Reproducing Kernel Hilbert Space (RKHS), enabling linear analysis of non-linear structures [7]. Yet global kernels suffer from density and bandwidth biases: small bandwidths emphasize tight clusters, while large ones oversmooth sparse regions. Our idea is to define a locally adaptive kernel as a general principle for resolving such biases. This principle alleviates these issues by adjusting their scale to local density [8], making them suitable for identifying representative regions and areas where the distribution is poorly captured.

We present a novel active constraint-based clustering method that learns an explicit functional representation of the data distribution through a locally adaptive kernel function. This representation enables identifying representative super-instances that capture the global structure of the data, while simultaneously detecting regions that are poorly represented or on a different scale. Constraints are then generated from these elements to guide user queries and minimize user interactions. The main contributions of the work are as follows:

1. We introduce a novel clustering method based on a *locally adaptive kernel* that captures variations in data density and constructs informative *super-instances.*
2. We propose a refinement strategy that detects ambiguous regions using a *witness function*, enforces balanced constraint selection through a *diversity regularizer*, and improves clustering by subdividing super-instances in these regions.
3. We present an *active clustering framework* that outperforms existing approaches in terms of clustering quality, query efficiency, and robustness. An open-source implementation is available on GitHub[1].

The remainder of the paper is organized as follows: Sect. 2 reviews related work on active clustering. Section 3 provides preliminaries on kernel-based methods, showing that locally adaptive kernels are more effective than global kernels, and describes the construction of super-instances along with the refinement and

[1] https://github.com/VincentBlase/LAK-for-active-constraint-clustering.

merging strategies. Section 4 presents an overview of the proposed method. Sections 5 and 6 report the empirical evaluation and discuss the results, demonstrating that the proposed approach outperforms existing methods. Finally, Sect. 7 concludes the paper.

2 Related Work

Active constraint-based clustering aims to incorporate human knowledge into unsupervised learning by selectively querying pairwise ML or CL relations [3]. The main challenge lies in identifying, at each iteration, the constraints that provide the highest information gain and guide the clustering process most effectively.

There are two major heuristics for selecting those queries: uncertainty [9] and representativeness [10,11]. The former assumes that querying points near decision boundaries yields the most information, while the latter seeks points that best summarize the dataset or capture its structural diversity. Some approaches attempt to combine these two heuristics by selecting points that are both uncertain and representative [4]. Although intuitive, these strategies often ignore the global distribution of the data and its multi-scale structure. As a result, they may oversample dense regions, fail to explore sparse ones, and ultimately provide a biased picture of the clustering space.

Super-instances and neighborhood sets are widely used as auxiliary structures that reduce annotation cost by grouping points believed to belong to the same cluster [4–6,12]. Querying at the super-instance level enables one interaction to inform multiple data points. However, these structures are typically derived through heuristics such as Euclidean-based clustering, graph construction, or matrix factorization. Such heuristics do not always reflect the true geometry of the data, leading to poorly aligned representatives, especially in datasets with heterogeneous densities, non-convex shapes, or manifold structure.

Several refinement strategies have been proposed to mitigate this limitation. Ensemble-based methods attempt to infer constraints from multiple clustering solutions [13,14], improving robustness but often at the cost of significant computational overhead. Other works refine super-instances by splitting those containing low-confidence points or those with uncertain cluster membership [10]. While effective in some settings, these approaches rely on strong prior assumptions about cluster shape or density that frequently break down in practice. Other techniques refine super-instances using clustering-based or graph-based procedures. For instance, some methods recursively split the largest super-instances [6], assuming that largest clusters should be refined first. Others rely on diffusion graphs to capture global structure and identify representative nodes for splitting [4]. Yet size-based heuristics risk overlooking small or sparse clusters, and diffusion-based approaches, despite improving structural fidelity, still struggle to account for multi-scale distributions. Consequently, these methods often require a large number of queries before adequately capturing the full complexity of the data.

3 Preliminary

Let $X = \{x_i\}_{i=1}^n$, with $x_i \in \mathbb{R}^d$, denote the dataset to be partitioned into clusters that approximate an unknown ground truth $Y = \{y_i\}_{i=1}^n$. In active constraint-based clustering, supervision is provided in the form of pairwise constraints. A *must-link* (respectively *cannot-link*) constraint indicates that two samples belong to the same (respectively different) cluster in Y.

Given a limited query budget b, the objective is to learn a clustering function $f : X \to \hat{Y}$ that approximates Y using as few pairwise queries as possible. The efficiency of an active clustering algorithm thus depends on how effectively it selects (i) representative subsets of data on which to query, and (ii) informative pairs that most improve the underlying cluster structure.

3.1 Kernel Representation

Kernel representation maps data into a high-dimensional feature space in which nonlinear structures become easier to model or even linearly separable. This implicit embedding enables kernel methods to capture both the geometry of the data distribution and the similarity between points, making them particularly attractive for tasks involving complex or nonconvex structures.

Given a positive semi-definite kernel $k : \mathcal{X} \times \mathcal{X} \to \mathbb{R}$, kernel-based methods typically operate through the Gram matrix $\mathcal{K}$. A widely used choice is the Gaussian RBF kernel:

$$\mathcal{K}(x_i, x_j) = \exp(-\frac{\|x_i - x_j\|^2}{2\sigma^2}). \tag{1}$$

where $\sigma > 0$ is the bandwidth controlling the notion of similarity [8].

The performance of kernel methods is highly sensitive to the calibration of σ. Classical approaches tune σ by evaluating a clustering method over multiple bandwidths and selecting the one yielding the best output. This is incompatible with active constraint-based clustering, where labels are unavailable and only provided incrementally via user queries. One might consider running kernel k-means [17], prior to active querying, but doing so implicitly assumes access to labels or ground truth feedback, contradicting the unsupervised nature of the task.

A common alternative is the median heuristic, [18], which sets

$$\sigma = \sqrt{H_n}, \quad H_n = \text{Med}\{|x_i - x_j|_2 : 1 \leq i < j \leq n\}, \tag{2}$$

where Med denotes the empirical median of pairwise distances. While simple and often effective, the median heuristic is known to be biased, especially in datasets with heterogeneous densities or multiple scales. A single global bandwidth generally cannot account for tight and sparse regions simultaneously, leading to distorted similarity estimates [7].

To address this limitation, we introduce a *locally adaptive kernel*, following the self-tuning approach of [8]. Instead of a global σ, each point x_i is assigned its own bandwidth σ_i. Therefore, the kernel matrix is computed with entries as:

$$\mathcal{K}(x_i, x_j) = \exp(-\frac{\|x_i - x_j\|^2}{2\sigma_i\sigma_j}). \tag{3}$$

This formulation naturally adapts to local density and scale: similarities within dense regions are sharpened, whereas similarities in sparse regions are smoothed. To leverage neighborhood scale and density, a common heuristic is to compute distance between a point and its neighborhood such as:

$$\sigma_i = d(x_i, x_K) \tag{4}$$

where x_K is the K'th neighbor of x_i. The selection of K remains independent of scale. In all our experiments, K is set as a percentage of the number of instances in our data. $K = 1\%$ has proven best results in our empirical study (see Sect. 6.3 for details).

Because σ_i can be sensitive to noise and outliers, we apply smoothing and clipping: each σ_i is replaced by the median of its neighbors and clipped to the 15th–85th percentiles of $\{\sigma_i\}_{i=1}^n$, stabilizing the kernel while preserving local adaptivity. Figure 1 illustrates the effect of the locally adaptive kernel compared with the global RBF kernel on a 2D-synthetic dataset containing clusters of varying density and scale (See Fig. 2 for a visualization of this dataset). The adaptive kernel successfully captures fine- and large-scale structures simultaneously, suppressing noise and improving similarity estimation. Best global σ value was found using cross validation and show similar result as locally adaptive kernel. In contrast, the median heuristic fails to recover multi-scale structure, and even the best global σ struggles to match the flexibility of the locally adaptive formulation.

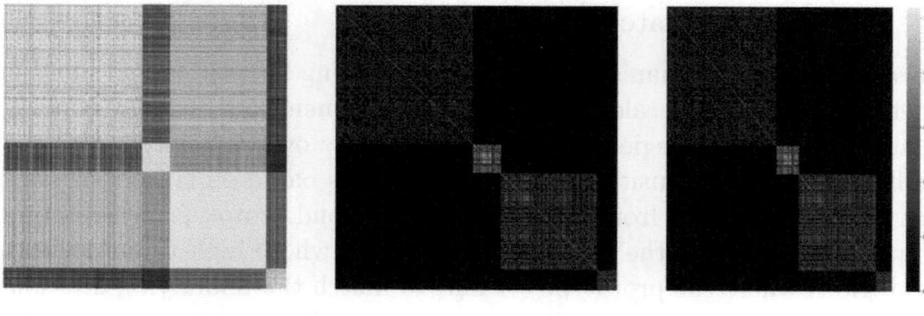

(a) Median global kernel (b) Locally adaptive kernel (c) Best global kernel

Fig. 1. Illustration of three similarity matrices obtained from the kernel representation with different σ values on a 2D synthetic dataset presented in Fig. 2.

3.2 Super-Instance

The Kernel representation provides more than pairwise similarities: it induces a representation of the data in a reproducing kernel Hilbert space (RKHS) where averages, variances, and other distributional properties take a simple linear form. This is important in our context because super-instances are intended to summarize regions of the dataset, and summarizing a region requires comparing distributions rather than individual points. The *Maximum Mean Discrepancy* (MMD) [15] offers a principled way to measure how well a subset of points preserves the kernel-defined data distribution. Unlike heuristic choices of representatives (e.g., cluster centroids or random samples), MMD explicitly quantifies the mismatch between the full dataset and a candidate prototype set under the geometry induced by the adaptive kernel.

Let $X_P = \{x_j^p\}_{j=1}^m \subset X$ denote the set of prototypes, drawn from the feature space $\mathcal{X}$. The discrepancy between the empirical distributions of X and X_P is:

$$\text{MMD}^2(\mathcal{K}, X, X_P) = \frac{1}{n^2}\sum_{i,j} k(x_i, x_j) - \frac{2}{nm}\sum_{i,j} k(x_i, x_j^p) + \frac{1}{m^2}\sum_{i,j} k(x_i^p, x_j^p), \quad (5)$$

where $n = |X|$ and $m = |X_P|$.

Maximizing the negation of this expression with an additive bias to cancel the constant terms yields prototypes that best reproduce the distribution encoded by the adaptive kernel, ensuring that both dense and sparse regions are proportionally represented. Each data point is then assigned to its closest prototype in the input space, forming a collection of disjoint *super-instances* $S = \{S_i\}_{i=1}^m$ each represented by its prototype X_P^i. These groups are compact, coherent, and aligned with the multi-scale structure revealed by the adaptive kernel. Figure 2 illustrates this behavior: prototypes learned from the adaptive kernel follow the true underlying geometry, whereas median-heuristic kernels concentrate representatives in dense areas and fail to capture structure at larger scales.

3.3 Refinement Strategy

A local refinement mechanism is crucial for revealing structures that are difficult to detect at the global scale. Without such refinement, queries may concentrate on large or uncertain regions while systematically overlooking clusters that lie at different scales or densities. Classical heuristics based on cluster size or local uncertainty often suffer from this limitation. Beyond prototype selection, MMD also provides access to the *witness function* [16], whose high-value points highlight regions where the prototype set fails to match the underlying distribution. These points, referred to as *criticisms*, indicate discrepancies between the full dataset and its prototype-based approximation, and therefore naturally serve as candidates for refinement. The cost function used to select criticisms X_{Cr} is:

$$L(Cr) = \sum_{l \in Cr} \left(\frac{1}{n} \sum_i k(x_i, x_l) - \frac{1}{m} \sum_j k(x_j^p, x_l) \right), \quad (6)$$

where Cr are the index of the given criticisms. To enforces diversity in the selection of criticisms, a log-determinant regularizer [19] $r(\mathcal{K}, Cr)$ is added. The criticisms are therefore selected as:

$$\max_{Cr \subseteq [n] \setminus P, |Cr| \leq m+1} L(Cr) + r(\mathcal{K}, Cr) \tag{7}$$

We identify the super-instance containing the highest concentration of criticisms and refine it by introducing additional prototypes within its region. This process ensures that refinement is directed toward areas where the kernel-based representation exhibits the largest mismatch, rather than being driven by global heuristics. Figure 2 illustrates this behavior. Under the locally adaptive kernel, criticisms cluster within a sparse structure that the initial prototypes could not capture, prompting refinement exactly where needed. In contrast, criticisms derived from the global median-bandwidth kernel are widely dispersed, with many falling on outliers rather than meaningful structures, leading to ineffective refinement.

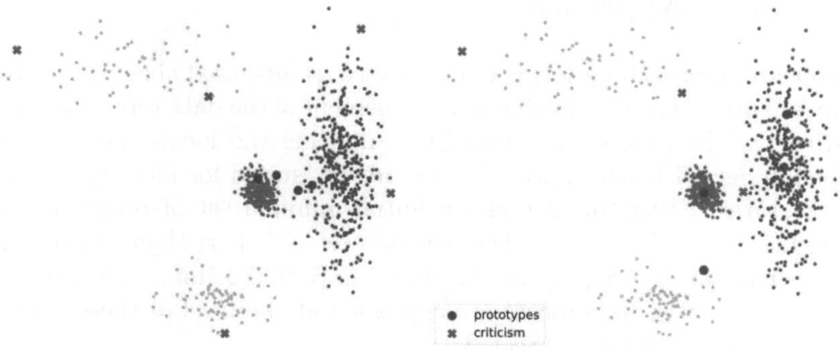

(a) Median global kernel (b) Locally adaptive kernel

Fig. 2. Distribution of prototypes and criticisms on a 2D synthetic dataset, comparing the median heuristic kernel and the locally adaptive kernel. Blue circle are prototypes and red cross are criticisms. (Color figure online)

3.4 Merging Super-Instance

To obtain a clustering solution $C = \{C_1, \ldots, C_t\}$, where t is the desired number of clusters, such that $\bigcup_i C_i = X$ and $C_i \cap C_j = \emptyset$ for all $i \neq j$, super-instances must be merged (or not) according to pairwise constraints. A *must-link* (ML) constraint merges two super-instances into the same cluster, whereas a *cannot-link* (CL) constraint prevents such a merge. To this end, whenever new super-instances are generated, queries are submitted to an oracle to determine the type of pairwise constraint between the representative prototypes of a pair of super-instances. Identifying ML constraints is particularly important, as they indicate that a super-instance should be assigned to a specific cluster. Consequently, an ordering of clusters is required. For each new prototype X_P^i, we compute the mean of the prototypes already assigned to each cluster C_j, and rank the clusters according to their distance to X_P^i. The distance between a prototype X_P^i and a cluster C_j is computed as:

$$d(X_P^i, C_j) = \left\| X_P^i - \frac{1}{|C_j|} \sum_{X_P^l \in C_j} X_P^l \right\|_2 \tag{8}$$

We then iterate over the sorted list of clusters, querying the user to determine whether the new super-instance should be merged with each cluster. The process stops as soon as a *must-link* constraint is obtained, indicating that the super-instance belongs to the queried cluster. The super-instance is then assigned to the selected cluster, and the procedure is repeated for the next prototype. If no must-link constraint is identified, a new cluster is created.

4 Proposed Approach

The proposed approach performs iterative constraint-based clustering driven by a locally adaptive kernel and a dynamic refinement of the data representation (see Algorithm 1). The procedure begins by computing the locally adaptive kernel described in Sect. 3.1, which provides a geometry suited for identifying coherent local structures. Using this kernel, an initial minimal set of prototypes X_P^i is generated (typically $|P| = 2$) . Then, the dataset X is partitioned into an initial set of super-instances $\{S_i\}_{i=1}^2$ as introduced in Sect. 3.2 (line 1–3). All pairwise queries issued during the process are expressed at the level of these prototypes rather than the raw samples (Sect. 3.4).

The objective is to cluster the super-instances into disjoint clusters $C = \{C_1, \ldots, C_t\}$. Each cluster C_i is defined as $C_i = \{S_q \mid S_q \subseteq C_i\}$ (line 4).

At each iteration, the method identifies regions that are poorly captured by the current prototypes. To this end, a set of *criticisms* $X_{\text{Cr}} = \{X_{\text{Cr}}^1, \ldots, X_{\text{Cr}}^{|P|+1}\}$ is computed from the locally adaptive kernel (line 6). Criticisms highlight samples whose local statistics are not well reproduced by the existing prototypes and therefore indicate where refinement is most beneficial. The super-instance containing the largest number of criticisms is selected for refinement; ties are resolved in favor of the largest super-instance.

The selected super-instance is then split by computing $|\{X_{\text{Cr}} \subset S_{\text{split}}\}| + 1$ new prototypes directly in its local kernel-induced space. These new representatives, denoted X_P^{new}, are added to the global prototype set. To maintain coherence across the entire dataset, all super-instances are recomputed by assigning each data point to its nearest prototype in Euclidean space (line 7–9), ensuring that the refinement of a single region propagates consistently.

Once the updated super-instances are obtained, the constraint-based clustering step is repeated on the newly added prototypes (line 10), and the query budget is updated accordingly. The refinement clustering loop continues until

Algorithm 1. Locally adaptive kernel for active constraint clustering

Require: X: a data set, b: a budget limit of queries
Ensure: C: a clustering of X
1: *budget* $\leftarrow 0$
2: $\mathcal{K} \leftarrow$ *Compute locally adaptive kernel* (Sect. 3.1)
3: $S \leftarrow \left\{S_i = \{x \in X \mid X_P^i = \arg\min_{j \in P} \|x - X_P^j\|_2\}\right\}_{i=1}^{2}$, X_P *computed from* $\mathcal{K}$(Eq. 5)
4: C, *budget* $\leftarrow$ *Update clustering and budget using constraints on* X_P (Sect. 3.4)
5: **while** *budget* $< b$ **do**
6: $X_{\mathrm{Cr}} \leftarrow$ *Compute criticisms from* $\mathcal{K}$ (Eq. 8)
7: $X_P^{new} \leftarrow$ *Compute new prototypes on :* $\arg\max_{S_i} |S_i \cap X_{Cr}|$ *using* $\mathcal{K}$ (Eq. 5)
8: $X_P \leftarrow X_P \cup X_P^{new}$
9: $S \leftarrow \left\{S_i = \{x \in X \mid X_P^i = \arg\min_{j \in P} \|x - X_P^j\|_2\}\right\}_{i=1}^{m}$
10: C, *budget* $\leftarrow$ *Update clustering and budget using constraints on* X_P^{new} (Sect. 3.4)
11: **end while**
12: **return** C

the budget is exhausted or the user is satisfied with the resulting clustering structure (line 5–11).

5 Experimental Study

5.1 Datasets

We evaluate the proposed approach on 29 real-world datasets from the UCI Machine Learning Repository, previously used in constraint-based active clustering benchmarks [4,6]. Instances containing missing and duplicate values were removed to ensure data quality. All features were scaled to [0,1]. Table 1 summarizes the selected datasets.

5.2 Implementations

All experiments were executed on a dedicated Linux server equipped with dual Intel Xeon Gold 6136 processors (12 cores, 3 GHz) and 192 GB RAM. Clustering quality was evaluated using the Adjusted Rand Index (ARI) [20]. To compare performance across heterogeneous datasets, we additionally report the *Aligned Rank (AR)* [21], a rank-based metric widely used for meta-evaluation; lower AR values indicate more consistently strong performance. Each run was capped at 72 h of wall-clock time for reproducibility and feasibility. Methods exceeding this limit were terminated and no results were reported.

To assess the effectiveness of our approach, we compare it against representative baselines covering the main paradigms of active and constraint-based clustering. These include state-of-the-art clustering and active methods :

- COBRAS [6]: An active clustering algorithm that employs a dynamic refinement of super-instances initialized via K-means. It is widely recognized for

Table 1. Description of the datasets used in the experiments.

Dataset	Number of Instances (m)	features (d)	Classes (Y)	Dataset	Number of Instances (m)	features (d)	Classes (Y)
Tae	106	5	3	Iris	147	4	3
Wine	178	13	3	Parkinsons	195	22	2
Seeds	198	7	3	Sonar	208	60	2
Segmentation	210	19	7	Glass	213	9	7
Soybean	263	35	19	Haberman	283	3	2
Columns_2C	310	6	2	Ecoli	336	7	8
Ionosphere	350	34	2	Dermatology	358	34	6
Musk	476	166	2	BreastCancerW.	569	30	2
HCV	589	12	2	Balance	625	4	5
Pima	767	8	2	Vehicle	845	18	4
Banknote	1348	4	2	Yeast	1453	8	9
Optdigits389	1688	64	3	mfeat_karhunen	1997	64	10
mfeat_zernike	1997	47	10	Spambase	4207	57	2
Waveform	5000	21	3	EEG	14980	14	2
Letter	18668	16	26				

achieving a strong balance between clustering quality, query efficiency, and computational cost.

- ACDM [4]: A graph-based active clustering method that leverages diffusion processes on sparse similarity graphs. It has demonstrated strong performance, particularly on large datasets as well as on some smaller datasets.
- SPACE [14]: An ensemble-based active clustering approach combining self-paced learning and consensus mechanisms to progressively integrate complex samples.
- MPCKMeans-NPU [2,23] A common constraint clustering method using Kmeans with NPU active process that repeatedly selects the most informative instance, and queries its neighborhood membership by means of pairwise constraints. This method is included as baseline for active method.
- Kmeans++ [22]: A widely used clustering method, included as a lower non-active baseline for comparison.

All methods were configured using the hyperparameters recommended in their original publications. Note that MPCKMeans-NPU need to know the desired number of clusters K prior to clustering. In our experiments, the true K (as indicated by the class labels) is given to these algorithms. We evaluate our method in two configurations: one using the locally adaptive kernel, and one using the median global kernel. For comparison across methods, we used a perfect oracle that provides pairwise constraints based on the ground-truth class labels available for each dataset.

6 Experimental Results

6.1 Clustering Quality (effectiveness)

Figure 3a reports the average ARI obtained by each method across all datasets. The proposed Kernel-based variants consistently outperform non-kernel baselines for all query budgets, yielding higher ARI scores throughout the entire query range. A Wilcoxon signed-rank test confirms that these improvements are statistically significant ($p < 0.01$). Among kernel variants, the locally adaptive kernel achieves the strongest and most stable performance. It delivers superior clustering quality across all query budgets, demonstrating that adapting bandwidths to local density leads to more informative queries and a markedly more efficient use of the available budget. The Aligned Rank analysis (Fig. 3b) confirms this trend. Across datasets and budgets, the locally adaptive kernel consistently appears among the top-ranked methods, indicating stable behavior even on heterogeneous data.

Table 2 provides detailed results for all query budgets. The locally adaptive kernel achieves the best ARI on 35% of the datasets (versus 11% for COBRAS and 26% for ACDM), and ranks first or second on 61% of them (compared to 27% and 35% for COBRAS and ACDM, respectively). These results highlight both the effectiveness and the robustness of the proposed approach across heterogeneous datasets.

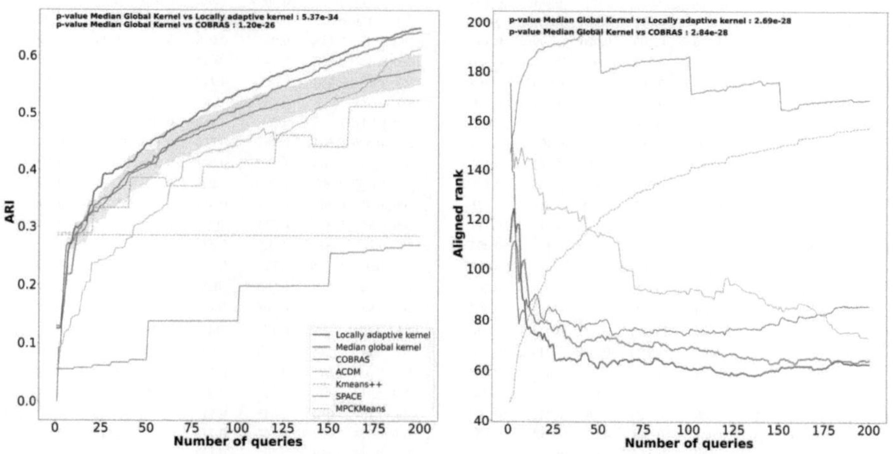

(a) Average ARI (b) Aligned rank

Fig. 3. Clustering quality across all datasets for varying query numbers. Lower Aligned Rank values indicate better performance.

6.2 Computational Efficiency

We evaluate the computational cost of the proposed approach using various datasets sizes. Figure 4a shows that its runtime is comparable to that of the median global kernel, indicating that the adaptive bandwidth estimation adds little overhead. As expected, kernel-based approaches are slower than non-kernel baselines, mainly due to the computation of full similarity matrices (Sect. 3.1).

Table 2. Clustering quality (ARI) for each model across datasets and query budgets (25, 50, 100, 200 for datasets < 1000 instances; 50, 100, 200, 500 for larger datasets). Highlighted and underlined values indicate the highest and second-highest ARI, respectively. "-" denotes runtimes exceeding 72 h. #Q: number of queries, LAK: Locally Adaptive Kernel, MGK: Median Global Kernel.

DS	#Q	LAK	MGK	COBRAS	ACDM	SPACE	MPC	Kmeans++
Tae	25	0.015	0.058	0.007	0.002	0.04	0.02	0.046
	50	0.168	0.096	0.019	0.028	0.04	0.225	0.046
	100	0.384	0.326	0.107	0.205	0.057	0.225	0.046
	200	1.0	1.0	0.546	1.0	0.098	0.916	0.046
Wine	25	0.817	0.801	0.897	0.582	0.014	0.917	0.869
	50	0.833	0.818	0.953	0.967	0.014	0.899	0.869
	100	0.933	0.897	0.967	1.0	0.052	0.983	0.869
	200	1.0	1.0	0.983	1.0	0.194	1.0	0.869
Seeds	25	0.79	0.721	0.778	0.561	0.249	0.603	0.727
	50	0.83	0.779	0.767	0.899	0.249	0.808	0.727
	100	0.871	0.842	0.827	0.985	0.39	0.955	0.727
	200	0.985	0.985	0.863	1.0	0.39	1.0	0.727
Segment.	25	0.503	0.515	0.344	0.106	0.223	0.438	0.403
	50	0.6	0.573	0.421	0.488	0.223	0.309	0.403
	100	0.793	0.73	0.645	0.562	0.223	0.59	0.403
	200	0.896	0.925	0.785	0.989	0.223	0.786	0.403
Soybean	25	0.275	0.148	0.365	0.018	0.026	0.416	0.406
	50	0.362	0.322	0.365	0.047	0.026	0.478	0.406
	100	0.436	0.454	0.456	0.105	0.026	0.399	0.406
	200	0.645	0.618	0.583	0.283	0.026	0.48	0.406
Clmn-2c	25	0.254	0.083	0.306	0.163	0.042	0.053	0.086
	50	0.241	0.228	0.281	0.205	0.042	0.195	0.086
	100	0.334	0.252	0.274	0.508	0.169	-0.015	0.086
	200	0.505	0.581	0.408	0.583	0.722	0.466	0.086
Ionosphere	25	0.45	0.43	0.563	0.456	0.009	0.181	0.171
	50	0.507	0.497	0.589	0.587	0.009	0.238	0.171
	100	0.652	0.578	0.615	0.623	0.036	0.287	0.171
	200	0.691	0.73	0.744	0.74	0.143	-0.01	0.171
Musk	25	0.17	0.075	-0.013	-0.013	-	-	0.003
	50	0.163	0.091	0.039	-0.013	-	-	0.003
	100	0.278	0.31	0.25	-0.002	-	-	0.003
	200	0.468	0.485	0.399	0.18	-	-	0.003
Hcv	25	0.0	0.322	0.0	0.488	0.004	0.033	0.046
	50	0.495	0.446	0.0	0.506	0.004	0.048	0.046
	100	0.591	0.579	0.524	0.716	0.015	0.06	0.046
	200	0.704	0.711	0.589	0.78	0.063	0.1	0.046
Pima	25	0.079	0.11	0.036	0.128	0.0	0.011	0.102
	50	0.135	0.15	0.054	0.144	0.002	0.209	0.102
	100	0.14	0.198	0.178	0.203	0.011	0.147	0.102
	200	0.221	0.251	0.201	0.277	0.046	0.286	0.102
Banknote	50	0.89	0.592	0.885	0.675	0.003	0.366	0.024
	100	0.956	0.896	0.979	0.8	0.011	0.039	0.024
	250	0.979	0.976	0.982	0.997	0.068	0.319	0.024
	500	0.994	0.994	0.997	1.0	0.272	0.319	0.024
Optdigits	50	0.818	0.831	0.773	0.89	0.0	0.824	0.785
	100	0.878	0.878	0.786	0.972	0.001	0.842	0.785
	250	0.934	0.94	0.871	0.996	0.009	0.897	0.785
	500	0.958	0.954	0.896	1.0	0.035	0.897	0.785
mfeat	50	0.291	0.271	0.397	0.0	0.012	0.291	0.446
	100	0.418	0.452	0.421	0.558	0.045	0.331	0.446
	250	0.583	0.589	0.475	0.576	0.168	0.332	0.446
	500	0.678	0.679	0.6	0.627	0.168	0.332	0.446
Waveform	50	0.282	0.342	0.426	0.0	-	-	0.254
	100	0.352	0.358	0.432	0.339	-	-	0.254
	250	0.428	0.38	0.419	0.364	-	-	0.254
	500	0.47	0.435	0.459	0.421	-	-	0.254
Letter	50	0.068	0.053	0.044	0.0	-	-	0.12
	100	0.07	0.052	0.044	0.0	-	-	0.12
	250	0.093	0.094	0.044	0.0	-	-	0.12
	500	0.128	0.147	0.088	0.0	-	-	0.12
# first		42	20	14	31	6	14	1
# second		31	37	19	11	3	7	5

DS	#Q	LAK	MGK	COBRAS	ACDM	SPACE	MPC	Kmeans++
Iris	25	0.901	0.846	0.722	0.959	0.169	0.901	0.706
	50	0.864	0.901	0.85	1.0	0.387	0.959	0.706
	100	0.939	0.939	0.909	1.0	0.387	1.0	0.706
	200	1.0	1.0	0.967	1.0	0.729	1.0	0.706
Parkinsons	25	0.329	0.552	0.268	0.473	0.04	-0.071	-0.068
	50	0.559	0.647	0.515	0.617	0.04	0.143	-0.068
	100	0.894	0.795	0.757	0.658	0.196	0.134	-0.068
	200	0.978	0.978	0.856	1.0	0.877	0.544	-0.068
Sonar	25	0.183	0.116	0.108	0.033	0.119	-0.004	0.009
	50	0.319	0.266	0.209	0.073	0.119	0.006	0.009
	100	0.546	0.62	0.497	0.246	0.433	0.053	0.009
	200	0.887	0.851	0.683	0.834	0.433	0.437	0.009
Glass	25	0.316	0.307	0.245	0.187	0.292	0.182	0.185
	50	0.32	0.274	0.318	0.215	0.292	0.232	0.185
	100	0.432	0.4	0.366	0.342	0.498	0.224	0.185
	200	0.605	0.511	0.534	0.464	0.498	0.381	0.185
Haberman	25	0.046	0.04	0.006	-0.002	0.029	0.11	-0.001
	50	0.156	0.029	0.026	0.012	0.029	0.032	-0.001
	100	0.17	0.118	0.074	0.086	0.096	0.29	-0.001
	200	0.337	0.287	0.244	0.56	0.399	0.105	-0.001
Ecoli	25	0.654	0.653	0.613	0.416	0.118	0.368	0.415
	50	0.674	0.71	0.686	0.43	0.118	0.485	0.415
	100	0.741	0.767	0.71	0.469	0.421	0.38	0.415
	200	0.826	0.804	0.767	0.776	0.746	0.397	0.415
Dermat.	25	0.84	0.766	0.903	0.271	0.008	0.721	0.592
	50	0.903	0.87	0.929	0.293	0.008	0.861	0.592
	100	0.938	0.909	0.931	0.517	0.043	0.938	0.592
	200	0.958	0.958	0.941	0.99	0.139	1.0	0.592
BreastCan.	25	0.707	0.811	0.73	0.477	0.01	0.665	0.73
	50	0.818	0.818	0.777	0.617	0.01	0.792	0.73
	100	0.799	0.713	0.786	0.93	0.023	0.883	0.73
	200	0.831	0.85	0.803	0.986	0.059	0.958	0.73
Balance	25	0.524	0.257	0.173	0.08	0.0	0.163	0.125
	50	0.595	0.473	0.263	0.094	0.0	0.16	0.125
	100	0.591	0.523	0.418	0.123	0.0	0.15	0.125
	200	0.613	0.626	0.563	0.225	0.002	0.307	0.125
Vehicle	25	0.102	0.075	0.048	0.117	-	0.124	0.076
	50	0.186	0.092	0.066	0.119	-	0.08	0.076
	100	0.307	0.189	0.127	0.122	-	0.125	0.076
	200	0.386	0.348	0.284	0.158	-	0.103	0.076
Yeast	50	0.172	0.13	0.17	0.06	0.001	0.125	0.155
	100	0.196	0.165	0.181	0.062	0.003	0.107	0.155
	250	0.179	0.184	0.201	0.064	0.018	0.149	0.155
	500	0.235	0.211	0.236	0.068	0.031	0.149	0.155
mfeat	50	0.272	0.3	0.559	0.0	0.012	0.485	0.51
	100	0.508	0.483	0.607	0.725	0.045	0.589	0.51
	250	0.761	0.764	0.742	0.757	0.168	0.616	0.51
	500	0.844	0.865	0.825	0.812	0.168	0.616	0.51
Spambase	50	0.377	0.246	0.353	0.161	-	-	0.342
	100	0.426	0.285	0.378	0.171	-	-	0.342
	250	0.452	0.337	0.331	0.241	-	-	0.342
	500	0.523	0.404	0.386	0.297	-	-	0.342
Eeg	50	0.014	0.003	-0.0	0.0	-	-	0.0
	100	0.02	0.009	0.0	0.0	-	-	0.0
	250	0.07	0.014	0.022	0.019	-	-	0.0
	500	0.14	0.059	0.053	0.023	-	-	0.0

The dominant cost in our method comes from the MMD computations, which scale as $O(N^2d)$ [15]. Although this makes our approach slower than ACDM in absolute runtime, active constraint-based clustering is ultimately limited by the number of user interactions rather than raw computational time. In this setting, the locally adaptive kernel achieves a favorable balance: it significantly reduces the query budget while keeping runtimes within practical limits.

6.3 Sensitivity Analysis of K

The locally adaptive kernel depends on a key parameter K, which defines the size of the local neighborhood used to compute each σ_i (Sect. 3.1). To assess its impact, we tested K values ranging from 1% to 10% of the dataset size across all 29 datasets. Figure 4b reports the average ARI over a query range from 0 to 200 for each K. The results indicate minimal differences between values, with $K = 1\%$ yielding slightly better performance. Based on this observation, we adopt $K = 1\%$ in all experiments.

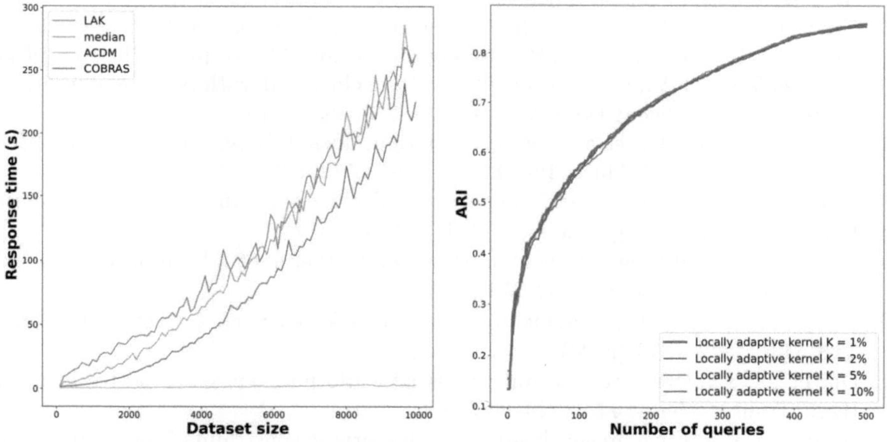

(a) Response Time vs Dataset size (b) Average ARI for different K %

Fig. 4. Runtime and hyperparameter analysis. LAK = Locally adaptive kernel.

7 Conclusion

We have introduced a novel active constraint-based clustering method that leverages a locally adaptive kernel to handle data with varying scales and densities. The experimental evaluation demonstrates its effectiveness, achieving superior clustering performance while minimizing the number of required queries. These results highlight the practical advantage of locally adaptive kernels, particularly in reducing sensitivity to hyperparameter selection. Despite these strengths, scalability remains a limitation for very large datasets. Future work will focus on improving computational efficiency, aiming to maintain the benefits of the adaptive kernel while extending its applicability to larger-scale problems.

Acknowledgment. This work is supported by the *European Regional Development Fund* (ERDF) and the *Occitanie Region, France* as part of the Emergence project OCC004676 AIBIOPRETRAS.

References

1. Gonzalez-Almagro, G., et al. Semi-supervised constrained clustering: an in-depth overview, ranked taxonomy and future research directions. AI Rev. **58** (2025)
2. Wagstaff, et al.: Constrained k-means clustering with background knowledge. In: Proceedings of the 18th ICML, pp. 577–584
3. Settles, B. Active learning literature survey. CS Technical Report, 1648 (2009)
4. Fu, et al.: Acdm: an effective and scalable active clustering with pairwise constraint. In: Proceedings of the 33rd acm CIKM, pp. 643–652 (2024)
5. Craenendonck, T.V., et al.: Cobra: a fast and simple method for active clustering with pairwise constraints. In: Proceedings of the 26th IJCAI, pp. 2871–2877 (2017)
6. Craenendonck, T.V., et al. Cobras: Interactive clustering with pairwise queries. In: Advances in Intelligent Data Analysis, pp. 353–366 (2018)
7. Marin, D., Tang, M., et al.: Kernel clustering: density biases and solutions. IEEE Trans. Pattern Anal. Mach. Int. **41**(1), 136–147 (2019)
8. Zelnik-manor, L., et al.: Self-tuning spectral clustering. In: Advances in Neural Information Processing Systems, vol. 17 (2004)
9. Abin, A.A.: Clustering with side information: further efforts to improve efficiency. Pattern Recogn. Lett. **84**, 252–258 (2016)
10. Deng, Y., et al.: Query-augmented active metric learning. J. Am. Stat. Assoc. **118**(543), 1862–1875 (2023)
11. Du, et al.: Exploring representativeness and informativeness for active learning. IEEE Trans. Cybern. **47**(1), 14–26 (2017)
12. Biswas, et al.: Active image clustering with pairwise constraints from humans. Int. J. Comput. Vis. **108**(1–2), 133–147 (2014)
13. Shen, et al.: Semi-supervised hierarchical ensemble clustering based on an innovative distance metric and constraint information. Eng. Appl. Artif. Intell. **124**(C), (2023)
14. Zhou, et al.: Active clustering ensemble with self-paced learning. IEEE Trans. Neural Networks Learn. Syst. **35**(9), 12186–12200 (2024)
15. Lloyd, et al.: Statistical model criticism using kernel two sample tests. In: Cortes, C., Lawrence, N., Lee, D., Sugiyama, M., Garnett, R. (eds.), Advances in Neural Information Processing Systems, vol. 28. Curran Associates, Inc. (2015)
16. Kim, et al.: Examples are not enough, learn to criticize! Criticism for Interpretability. In: NIPS'16: Proceedings of the 30th International Conference on Neural Information Processing Systems (2016)
17. Scholkopf, B., et al.: Nonlinear component analysis as a kernel eigenvalue problem. Neural Comput. **10**(5), 1299–1319 (1998)

18. Garreau, D., Jitkrittum, W., Kanagawa, M.: Large sample analysis of the median heuristic (2018)
19. Krause, et al.: Near-optimal sensor placements in gaussian processes: theory, efficient algorithms and empirical studies. J. Mach. Learn. Res. **9**, 235–284 (2008)
20. Hubert, L., Arabie, P.: Comparing partitions. J. Classif. **2**(1), 193–218 (1985)
21. Garcıa, S., et al.: Advanced nonparametric tests for multiple comparisons in the design of experiments in computational intelligence and data mining: Experimental analysis of power. Inf. Sci. **180**(10), 2044–2064 (2010)
22. Arthur, et al.: k-means++: the advantages of careful seeding. In: Proceedings of the Eighteenth Annual acm-siam Symposium on Discrete Algorithms (2007)
23. Xiong, et al.: Active learning of constraints for semi-supervised clustering. In: TKDE 26, pp. 43–54 (2014)

Evidential Deep Learning is not Evidential Learning: A Clear Distinction

Arthur Hoarau[1,2](✉)

[1] CentraleSupélec, Université Paris-Saclay, Metz, France
arthur.hoarau@centralesupelec.fr
[2] Loria, CNRS, Université de Lorraine, Nancy, France

Abstract. A large portion of the recent literature on uncertainty quantification has focused on second-order risk minimization, renaming it Evidential Deep Learning. We argue that this naming is misleading, as the naming of the evidential framework has already been established by the Mathematical Theory of Evidence, which has been applied to numerous problems for decades and deep evidential learning was defined long before second-order risk minimization. However, this naming overlap has given rise to particularly interesting ideas for (evidential learning in the sense of) second-order risk minimization, which is applied here both theoretically and empirically to (evidential learning in the sense of) Dempster-Shafer Theory, establishing a rule of monotonicity for epistemic uncertainty as essential to evidential learning. This rule is proven to be true in this paper for Dempster-Shafer Theory, whereas it has been proven impossible for second-order risk minimization. As a result, our work clarifies the distinctions between these approaches while contributing new insights to the broader discussion on uncertainty quantification.

Keywords: Machine Learning · Uncertainty Quantification

1 Introduction

The intersection between machine learning and uncertainty quantification has been gaining momentum for some time. Being able to assign a reliability (or an uncertainty) to a model's prediction is indeed a valuable objective. Two types of uncertainties are widely distinguished in the literature [18]: aleatoric and epistemic. Aleatoric uncertainty stems from the stochastic nature of the event (*i.e.*, the data-generating process) and is therefore irreducible within the same learning boundaries, whereas epistemic uncertainty corresponds to a lack of knowledge and can be reduced by acquiring more observations. This distinction has profound implications for the design of models and methods in machine learning, as it dictates how uncertainty is quantified and addressed in practical applications.

In 2018, a new method is introduced by [30], which allows for the estimation of epistemic uncertainty in the context of deep learning, addressing the issue of the model's ability to represent its ignorance rather than experiencing wrong confidence. The authors elaborate on the Mathematical Theory of

M. Baratchi et al. (Eds.): IDA 2026, LNCS 16513, pp. 18–30, 2026.
https://doi.org/10.1007/978-3-032-23833-7_2

Evidence, from which they draw inspiration, to propose a second-order risk minimization strategy. This approach aims to minimize both the error of the model and its epistemic uncertainty, which is represented by a second-order probability. This work was subsequently widely adopted by several authors [1,2,6,25,28,35], and the name of this application domain became recognized as Evidential Deep Learning.

Two decades earlier, the evidential framework [9,31], also known as Dempster-Shafer theory or the theory of belief functions, was applied to machine learning [11] using a K-Nearest Neighbors architecture based on Dempster's combination rule. This approach treated observations as sources of evidence and was applied to a wide range of real-world applications [7,15,27]. Five years later, the same framework was extended to neural networks [10], eventually leading to a proliferation of evidential and deep evidential models based on Dempster-Shafer theory [5,8,14,17,26,36,37]. It should also be noted that the application of the evidential framework to the field of artificial intelligence dates back to before 1995 [32], as retraced by the author of the theory himself, 40 years later.

However, the issue of overlapping nomenclature between these two domains is not solely due to its anachronism. The confusion was initially induced by the fact that a belief function can be interpreted as a Dirichlet distribution [19]. Therefore, second-order risk minimization can theoretically lead to a belief function. However, unlike credal sets, evidential framework, second-order probabilities (which include second-order risk minimization), and possibility theory, all of which are frameworks for representing uncertainty, second-order risk minimization is a method (and not a theory) for quantifying a second-order probability, fundamentally antagonistic to the theory of evidence. This frequentist approach does not align with Shafer's view that "the idea of chance and the idea of belief have been brought together under the same term of probability" [31] proposing the belief function to exclusively represent epistemic uncertainty in a more Bayesian than frequentist interpretation. Moreover, this method does not utilize any rule of combination such as Dempster's rule [9], which is at the core of the evidential framework.

This specific naming has led survey papers to use the term "Evidential Deep Learning", which can give the impression (especially to newcomers) that evidential and deep evidential learning did not exist before 2018. Our contribution includes highlighting the risks of such a naming convention, and therefore, we advocate for renaming the so-called approach (when epistemic regularization is incorporated directly into the loss) simply to *second-order risk minimization*, following the formalism of [20], as this is what it truly represents.

Nevertheless, this overlap in nomenclature has led to a plethora of criticisms regarding second-order risk minimization. Valuable insights can be drawn from these critiques and applied to Dempster-Shafer Theory. As part of this critique, the authors in [3] propose to formalize appropriate properties for second-order risk minimization with the following two rules when obtained through loss minimization: (i) The epistemic uncertainty of a second-order distribution should be monotonic with respect to an increase in sample size. (ii) The second-order distribution should converge to a Dirac delta function as the sample size tends to infinity. Several works have proven that this expected behavior is impossible to achieve [3,4,20,29], with the general conclusion being that second-order risk minimization is unsuitable for epistemic uncertainty quantification[1].

The mistake to avoid is generalizing these results to (evidential learning in the sense of) Dempster-Shafer Theory (DST). We therefore propose in this paper to generalize these two particularly relevant properties to DST and to demonstrate, both theoretically and experimentally, that they are indeed respected. More importantly, it would be relevant in machine learning and uncertainty quantification to define a proper evidential model as one that adheres to these rules. We will also show that there are several models that satisfy these properties and are already used in practice, highlighting a framework that truly allows for properly modeling epistemic uncertainty. It should be noted that this paper is not a critique of second-order risk minimization, but rather aims to highlight that some of the criticisms directed at this family are not (theoretically and practically) applicable to DST.

The paper is structured as follows: Sect. 2 introduces DST and presents the difference with second-order risk minimization, while also listing the risks associated with improper nomenclature concerning these two topics. Section 3 theoretically extends the expected rules to the current DST-evidential framework through a proof of the monotonicity of epistemic uncertainty. Section 4 experimentally demonstrates that there are already existing models in use that comply with these rules; we show application cases in both machine learning and deep learning. Finally, Sect. 5 concludes the article.

2 Evidential Learning and Deep Evidential Learning

2.1 Evidential Framework - Dempster-Shafer Theory

The theory of belief functions, also known as Dempster-Shafer theory, was introduced by Dempster [9] then was further developed and renamed as the "Theory of Evidence" by Shafer [31]. This theory extends the theory of probabilities by focusing on the notion of belief rather than chance.

The frame of discernment, denoted by Ω, is the set of C exclusive hypotheses such that $\Omega = \{\omega_1, \ldots, \omega_C\}$. The power set of Ω, denoted by 2^Ω, represents the set of all subsets of Ω: $2^\Omega = \{\emptyset, \{\omega_1\}, \{\omega_2\}, \{\omega_1, \omega_2\}, \ldots, \Omega\}$. A source s can then assign a belief to any element of 2^Ω.

The function $m : 2^\Omega \to [0, 1]$, which assigns belief (or mass) to each element of 2^Ω, is called the mass function and always satisfies the following normalization condition: $\sum_{A \in 2^\Omega} m(A) = 1$. Non-zero mass elements are called focal elements, and the union of focal elements forms the core. A vacuous mass function assigns a mass of 1 to the frame Ω, representing complete ignorance. A mass function

[1] Recent work [34] has proposed addressing some of these issues within the framework of second-order risk minimization. However, this lies outside the scope of our paper, which focuses on evidential learning in the sense of DST.

with only one focal element is referred to as a categorical mass function. A mass function with all its focal elements concentrated on singletons is referred to as a Bayesian mass function (special case of a probability function). A quasi-Bayesian mass function, on the other hand, has its focal elements on singletons as well as on the frame Ω (*i.e.*, partial ignorance).

Example 1. Consider the toss of a coin that may or not be fair. The two events h: "the coin lands on heads" and t: "the coin lands on tails" form a partition of the frame $\Omega = \{h, t\}$. Two sources s_1 and s_2 express their beliefs about this coin. Source s_1 states "I don't know whether the coin is biased or not" with $m_1 : m_1(\{h\}) = 0, m_1(\{t\}) = 0, m_1(\{h,t\}) = 1$, while source s_2 expresses "I know that the coin is not biased" with $m_2 : m_2(\{h\}) = 0.5, m_2(\{t\}) = 0.5, m_2(\{h,t\}) = 0$.

The concepts of upper and lower probabilities, referred to as plausibility and belief respectively, serve to bound the belief of a source in an event. The belief function, denoted as Bel, represents the minimal belief of the source and is calculated as follows for any $A \subseteq \Omega$: $Bel(A) = \sum_{B \subseteq A} m(B)$. It is the sum of the elements of information that imply A and represents the total belief in A.

The plausibility function Pl is the maximum belief that could be attributed to A, defined as: $Pl(A) = \sum_{A \cap B \neq \emptyset} m(B)$. It is the sum of elements of information that are non-contradictory with A. One can relate the plausibility of an element A to the belief of its complement $\bar{A}$ with $Pl(A) = 1 - Bel(\bar{A})$.

Example 2. Consider example 1, the belief of s_1 that h : "the coin lands on heads" can be bounded by the functions Bel and Pl, where $Bel_{s_1}(\{h\}) = 0$ and $Pl_{s_1}(\{h\}) = 1$. Therefore, its belief in the event is bounded by $[0, 1]$. Similarly, the belief of s_2 is bounded by $[0.5, 0.5]$, since this source knows that the coin is fair, $Bel_{s_2}(\{h\}) = Pl_{s_2}(\{h\}) = 0.5$.

To capture epistemic uncertainty within the evidential framework, the Hartley entropy was extended to the evidential framework by Dubois and Prade [13]. The Non-specificity NS of a mass function m is defined as:

$$NS(m) = \textstyle\sum_{A \subseteq \Omega} m(A) \log_2(|A|), \tag{1}$$

with $|A|$ the cardinality of the set A.

Example 3. Consider example 1, the Non-specificity of m_1 is given by $NS(m_1) = 1\log_2(2) = 1$. This represents maximal epistemic uncertainty, as s_1 is in a state of complete ignorance. The Non-specificity of m_2 is $NS(m_2) = 0.5\log_2(1) + 0.5\log_2(1) = 0$, meaning the epistemic uncertainty is zero, since m_2 corresponds to a purely aleatoric event, of a fair coin toss.

When multiple sources express themselves, it is possible to combine two or more mass functions defined on the same frame of discernment. Several methods exist, each with its own advantages and disadvantages depending on the fusion context (*cf.* [24]). The conjunctive combination (Dempster's rule) of two mass functions m_1 and m_2, defined on the same frame of discernment Ω, is denoted by $m_1 \oplus m_2$ and is defined as follows:

$$m_1 \oplus m_2(A) = \frac{1}{1-\kappa} \sum_{B \cap C = A} m_1(B) m_2(C), \quad \forall A \subseteq \Omega, \tag{2}$$

where κ is the degree of conflict defined as:

$$\kappa = \sum_{B \cap C = \emptyset} m_1(B) m_2(C), \tag{3}$$

where B and C are subsets of Ω. This rule combines the information from the two mass functions by aggregating the contributions of all intersecting subsets B and C that form A. In the case of more than two masses, the combination of multiple mass functions $m_i \in M$ is denoted by $\bigoplus_{m_i \in M_j} m_i(A)$.

Example 4. Let's consider three suspects: Alice, Bob, and Eve, forming the frame of discernment $\Omega = \{\text{Alice, Bob, Eve}\}$. During a trial, two reliable sources, s_1 and s_2, express the following beliefs: "the culprit's name has three letters", represented by the mass function $m_1 : m_1(\{\text{Bob, Eve}\}) = 1$, and "the culprit is a woman" represented by the mass function $m_2 : m_2(\{\text{Alice, Eve}\}) = 1$. These two pieces of evidence can be combined into m such that $m = m_1 \oplus m_2$, resulting in $m(\{\text{Eve}\}) = 1$, meaning "the culprit is Eve". It is also noteworthy that the Non-specificity of m_1 and m_2 are equivalent $NS(m_1) = NS(m_2) = 1$, while the Non-specificity of m is zero. Epistemic uncertainty has been reduced to zero.

By adding a few assumptions and using the multiple mathematical tools of the evidential framework, it is possible to learn from data, resulting in evidential learning, and, in the case of deep architectures, in deep evidential learning. The following section demonstrates that it would be problematic to have two frameworks coexisting under the same name and suggests drawing inspiration from second-order risk minimization approaches to make evidential learning even more robust.

2.2 An Overview of the Issue

A second-order risk minimization technique was introduced by [30]. It has been further extensively revisited by the community [1,2,6,25,28,35], and adopted under the name of Evidential Deep Learning. The actual Mathematical Theory of Evidence, nearly half a century older [9,31], at its heart is more about what we model than how we get to that representation. The evidential and deep evidential learning (in the sense of DST) partially address this second point, using various hypotheses. The authors of [30] clearly phrase that it could be useful if a model could *simply say "I do not know"*. Evidential and deep evidential models have inherently and transparently addressed this assertion (as shown in Sect. 2.1). However, it seems that these recent second-order risk minimization approaches do not inherently respond to it. Therefore, the two nomenclatures struggle to coexist.

The second problem is that the conclusions drawn regarding second-order risk minimization and epistemic uncertainty are disappointing [3,4,20,29,34] and sometimes mistakenly categorize evidential deep learning as inadequate. This risks tarnishing the reputation of DST, which satisfactorily meet the same criteria (as shown theoretically in Sect. 3 and empirically in Sect. 4). Therefore, it is important to be careful to critique second-order risk minimization and not generalize to evidential and deep evidential learning.

The last but most relevant problem concerns nomenclature. It pertains to the acceptance of this second-order risk minimization method as evidential deep learning by the scientific community. This is not evidential learning; any of the combinations in the evidence theory such as Dempster's rule are even utilized in these methods. It is scientifically questionable to use the name of Shannon's entropy from its Mathematical Theory of Communication [33] to refer to a different object. We advocate renaming these methods simply as "Second-Order Risk Minimization" (a term proposed by [20]) instead of "Evidential Deep Learning" for greater transparency and scientific integrity. This would avoid misleading conclusions such as "evidential deep learning methods do not represent epistemic uncertainty in a faithful manner", where the authors really mean "second-order risk minimization methods do not represent epistemic uncertainty in a faithful manner" supported by their strong theoretical claims and remarkable work.

There remains a positive aspect related to this overlap in nomenclature. The flourishing literature on second-order risk minimization has led to very relevant conclusions that can be applied to any evidential and deep evidential learning.

2.3 Second-Order Expected Properties

Two particularly interesting properties from the literature on second-order risk minimization, as defined by [3], are presented below. We believe that these properties should also apply to DST's evidential and deep evidential learning. Notably, they have been proven to be impossible [4,20] within second-order risk minimization framework, and we will demonstrate in this paper that these properties are validated both theoretically (Sect. 3) and experimentally (Sect. 4) in the framework of evidential and deep evidential learning.

Property 1 (Monotonicity). The epistemic uncertainty of a second-order probability distribution estimate should be monotonic as the sample size increases.

Property 2 (Convergence). A second-order probability distribution estimate should converge to a Dirac delta function when the sample size grows to infinity.

3 Extension of the Monotonicity Rule

In this section, the monotonicity rule is extended to the evidential framework in the case of quasi-Bayesian mass functions (see Sect. 2.1). Only this first rule will be studied in detail here and in the experiments, as it appears to be the most important and should be respected by any evidential and deep evidential model. Regarding the second rule, it is obvious that when available evidence becomes infinite, its combination will result in a categorical mass function, which is the discrete equivalent of a Dirac delta function (convergence rule).

Non-specificity has been advocated [21] as the best measure of DST for capturing epistemic uncertainty. Therefore, this measure will be used in this document to prove its monotonicity when combining multiple sources of information. For this demonstration, we consider the case of quasi-Bayesian mass functions (a common application *scenario*, such as in the widely used evidential learning model EK-NN [11]).

Lemma 1. *The Non-specificity (epistemic uncertainty) of two quasi-Bayesian pieces of evidence m_1 and m_2 is monotonic with regard to their combination.*

Given the Non-Specificity and the conjunctive combination defined respectively by equations (1) and (2), after combining the two mass functions m_1 and m_2, the Non-Specificity NS of the resulting mass $m_1 \oplus m_2$ is given as follows:

$$NS(m_1 \oplus m_2) = \sum_{A \subseteq \Omega} m_1 \oplus m_2(A) \log_2(|A|). \tag{4}$$

This can be trivially addressed in the general case of null conflict $\kappa = 0$, but in practical applications, there is often a positive conflict $\kappa \geq 0$ (*e.g.*, the conflict is null in the example 4, but if we combine two pieces of evidence, such as "I think the culprit is Bob" and "I think the culprit is Alice", conflict may arise).

Proof. We therefore address the case of non-null conflict in the combination of quasi-Bayesian mass functions. Let a quasi-Bayesian mass function be defined as follows:

$$\begin{aligned} m(\{\omega_i\}) &= \alpha_i, \ \ \forall \omega_i \in \Omega, \\ m(A) &= 0, \ \ \forall A \subset \Omega, |A| \neq 1, \\ m(\Omega) &= 1 - \textstyle\sum_{i=1}^{|\Omega|} \alpha_i, \end{aligned} \tag{5}$$

with $\alpha_i \in [0,1]$. Consider two quasi-Bayesian mass functions m_1 and m_2 such that:

$$\begin{aligned} m_1(\{\omega_i\}) = \alpha_i, \ \ m_1(\Omega) = 1 - \textstyle\sum_{i=1}^{|\Omega|} \alpha_i, \\ m_2(\{\omega_i\}) = \beta_i, \ \ m_2(\Omega) = 1 - \textstyle\sum_{i=1}^{|\Omega|} \beta_i. \end{aligned} \tag{6}$$

The Non-specificity of m_1, given by equation (1), is:

$$NS(m_1) = (1 - \textstyle\sum_{i=1}^{|\Omega|} \alpha_i) \log_2(|\Omega|). \tag{7}$$

By combination of m_1 and m_2, the conflict κ given by equation (3), is:

$$\kappa = \textstyle\sum_{i=1}^{|\Omega|} \alpha_i \sum_{i=1}^{|\Omega|} \beta_i - \sum_{i=1}^{|\Omega|} \alpha_i \beta_i. \tag{8}$$

During the combination, the mass on the ignorance in then adjusted by the conflict κ such that:

$$\begin{aligned} m_1 \oplus m_2(\Omega) &= \tfrac{m_1(\Omega) m_2(\Omega)}{1-\kappa}, \\ m_1 \oplus m_2(\Omega) &= \frac{(1 - \sum_{i=1}^{|\Omega|} \alpha_i)(1 - \sum_{i=1}^{|\Omega|} \beta_i)}{1 - \sum_{i=1}^{|\Omega|} \alpha_i \sum_{i=1}^{|\Omega|} \beta_i + \sum_{i=1}^{|\Omega|} \alpha_i \beta_i}. \end{aligned} \tag{9}$$

The Non-specificity of $m_1 \oplus m_2$ is given as follows:

$$NS(m_1 \oplus m_2) = \frac{(1 - \sum_{i=1}^{|\Omega|} \alpha_i)(1 - \sum_{i=1}^{|\Omega|} \beta_i)}{1 - \sum_{i=1}^{|\Omega|} \alpha_i \sum_{i=1}^{|\Omega|} \beta_i + \sum_{i=1}^{|\Omega|} \alpha_i \beta_i} \log_2(|\Omega|). \tag{10}$$

We can now compare left-hand terms of $NS(m_1)$ and $NS(m_1 \oplus m_2)$. For simplicity, we pose $\Gamma_\alpha = \sum_{i=1}^{|\Omega|} \alpha_i$ and $\Gamma_\beta = \sum_{i=1}^{|\Omega|} \beta_i$. Using arithmetic we can prove the following statement by assuming $\Gamma_\beta(1-\Gamma_\alpha)^2 + (1-\Gamma_\alpha)\sum_{i=1}^{|\Omega|} \alpha_i\beta_i \geq 0$, since all elements are positive:

$$\begin{aligned}
\Gamma_\beta(1-\Gamma_\alpha)^2 + (1-\Gamma_\alpha)\textstyle\sum_{i=1}^{|\Omega|} \alpha_i\beta_i &\geq 0,\\
\Gamma_\beta - 2\Gamma_\alpha\Gamma_\beta + \Gamma_\alpha^2\Gamma_\beta + (1-\Gamma_\alpha)\textstyle\sum_{i=1}^{|\Omega|} \alpha_i\beta_i &\geq 0,\\
-\Gamma_\alpha\Gamma_\beta + \Gamma_\alpha^2\Gamma_\beta + (1-\Gamma_\alpha)\textstyle\sum_{i=1}^{|\Omega|} \alpha_i\beta_i &\geq -\Gamma_\beta + \Gamma_\alpha\Gamma_\beta,\\
1 - \Gamma_\alpha\Gamma_\beta - \Gamma_\alpha + \Gamma_\alpha^2\Gamma_\beta + (1-\Gamma_\alpha)\textstyle\sum_{i=1}^{|\Omega|} \alpha_i\beta_i &\geq 1 - \Gamma_\alpha - \Gamma_\beta + \Gamma_\alpha\Gamma_\beta,\\
(1-\Gamma_\alpha)(1 - \Gamma_\alpha\Gamma_\beta + \textstyle\sum_{i=1}^{|\Omega|} \alpha_i\beta_i) &\geq (1-\Gamma_\alpha)(1-\Gamma_\beta),\\
(1-\Gamma_\alpha) &\geq \frac{(1-\Gamma_\alpha)(1-\Gamma_\beta)}{1 - \Gamma_\alpha\Gamma_\beta + \sum_{i=1}^{|\Omega|} \alpha_i\beta_i}.
\end{aligned} \tag{11}$$

The Non-specificity $NS(m_1 \oplus m_2)$ is therefore less than or equal to $NS(m_1)$. For any quasi-Bayesian mass functions m_1 and m_2, the monotonicity of the Non-specificity with respect to the combination of two pieces of evidence can thus be expressed as:

$$NS(m_1) \geq NS(m_1 \oplus m_2). \tag{12}$$

This demonstrates the monotonicity of Non-specificity for two quasi-Bayesian mass functions.

Theorem 1 (Extended monotonicity rule). *The Non-specificity of any given quasi-Bayesian set of evidence and a larger set is monotonic with regard to their combination.*

The combination of two quasi-Bayesian mass functions results in another quasi-Bayesian mass function, and since Dempster's rule (2) is both commutative and associative, the equation can be extended to any set of evidence. This establishes the extended rule of monotonicity. In other words, epistemic uncertainty decreases as the sample size increases. It may also be useful to note the following propositions derived from the properties of belief functions.

Proposition 1 (Zero property). *Combining any mass function* m_1 *to a Bayesian mass function* m_{α_i} *(first order probability) such that* $m_{\alpha_i}(\{\omega_i\}) = \alpha_i$, $\forall \omega_i \in \Omega$, *and* $\sum \alpha_i = 1$, *gives a zero Non-specificity:* $NS(m_1 \oplus m_{\alpha_i}) = 0$.

Simply put, adding specific knowledge (purely aleatoric) reduces epistemic uncertainty to zero (*e.g.*, adding the information "the coin is not biased" removes all epistemic uncertainty about the toss and degenerates into the probability of getting heads = probability of getting tails = 0.5).

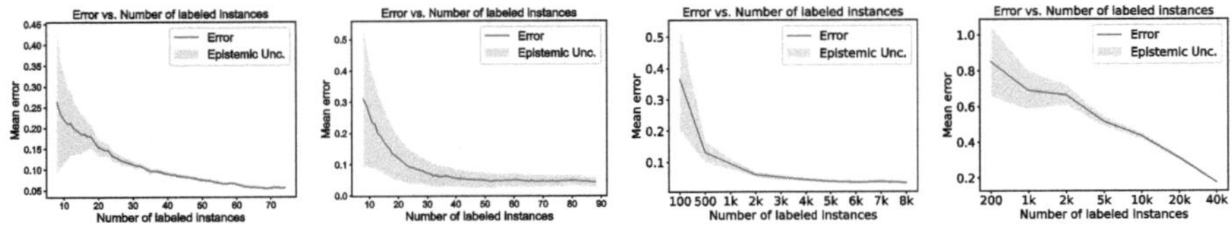

(a) Iris (b) Wine (c) LeNet / MNIST (d) ResNet18 / CIFAR10

Fig. 1. Epistemic uncertainty *vs.* knowledge increase.

Proposition 2 (Identity property). *Combining any mass function* m_1 *to a vacuous mass function* m_Ω *such that* $m_\Omega(\Omega) = 1$ *and* $m_\Omega(A) = 0, \forall A \in 2^\Omega \backslash \Omega$, *gives a Non-specificity identical to that of* m_1*:* $NS(m_1 \oplus m_\Omega) = NS(m_1)$.

In other words, adding completely ignorant information does not affect epistemic uncertainty (*e.g.*, adding the information "I don't know whether the coin is biased or not" does not change the prior epistemic uncertainty of the problem).

The monotonicity rule, particularly relevant in machine learning, has been extended to the evidential framework in this section. Although it has just been proven to be true, machine learning and deep learning models relying on evidence theory can make many assumptions to estimate such mass functions and can thus deviate from the theory. In the following section, we demonstrate that several models indeed adhere to this rule and, more importantly, that a proper evidential or deep evidential model must respect this rule.

4 Experiments

In this section, the extension of the monotonicity rule (clearly recalled as "the epistemic uncertainty of an evidential model should be monotonic when the sample size increases") is empirically demonstrated for a case of machine learning and deep learning. Initially, this behavior is demonstrated on the most popular evidential machine learning model, the Evidential K-nearest neighbors [11]. Then, a deep evidential learning version is also applied to show the continuity of the results within the deep architecture.

To simplify reproducibility, the uncertainty quantification framework is available to the scientific community at https://github.com/ArthurHoa/deep-eknn. The framework is compatible with pre-trained models and can be seamlessly integrated at the penultimate layer of a deep neural network.

4.1 DST's Evidential Learning

An evidential version of K-NN is introduced by [11], this Evidential K-Nearest Neighbors (EK-NN) uses belief functions to assign a label to a new incoming instance. Given a machine learning *scenario*, predictions then can be made. Let an instance space $\mathcal{X} = \mathbb{R}^P$ be a collection of P random variables and an output space $\Omega = \{\omega_1, \ldots, \omega_C\}$ be a set of C possible outcomes. Let $\mathbf{x}_i = \{x_i^1, \ldots, x_i^P\} \in \mathcal{X}$ be an instance associated with a class $\omega \in \Omega$ and the distance $d_{i,j}$ be the l_2-norm between two instances x_i and x_j. Given the K nearest neighbors of x_i according to d, one can assign a mass function to each neighbors $x_j, 1 \leq j \leq K$ of class $\omega^j \in \Omega$ discounted according to the distance from x_i. For further details, the reader is referred to [11] due to space constraints. Note that in this setup, we only consider quasi-Bayesian mass functions (*c.f.*, Sect. 2.1).

According to the theoretical proof outlined in Sect. 3, the number of neighbors should ideally be set to the size of the training set. However, we demonstrate that such complexity is unnecessary and that simpler parameters can yield satisfactory monotonicity. Therefore, for this experiment, we arbitrarily selected 7 as the number K of neighbors. The masses of all neighbors are then aggregated into the predicted belief function, enabling the Non-specificity (1) to be computed directly as the epistemic uncertainty.

Figures 1a and 1b illustrate the monotonicity of epistemic uncertainty in the context of machine learning. Two datasets from the UCI repository [12] are used, and the model's performance is shown as a function of the number of samples in the training set. The datasets are split into training and testing sets using a 70/30 ratio. This process is repeated 100 times, and the average model error is plotted. Along with the error, epistemic uncertainty (Non-specificity) is presented as an interval around the error. The results clearly show that Non-specificity decreases as the number of labeled instances increases. In other words, epistemic uncertainty decreases with regard to a sample size increase.

4.2 DST's Deep Evidential Learning

In this section, we use a Deep EK-NN version. This framework enables the quantification of epistemic uncertainty for most deep learning architectures and is compatible with pre-trained models. It can be seamlessly applied to the penultimate layer of a neural network (whether convolutional, residual, transformer, encoder, *etc.*) to perform uncertainty quantification. EK-NN [11] is used to predict the membership of new observations in the form of a belief function. For the experiments, observations are encoded using the penultimate layer of a neural network, and EK-NN is applied to this reduced feature space, generating a belief function. Epistemic uncertainty is computed according to Equation (1).

LeNet [23] is trained on the MNIST dataset with varying dataset sizes, and a ResNet18 [16] is pre-trained on CIFAR-10 [22]. The EK-NN framework is then applied to estimate epistemic uncertainty after each model's training. The model's error is plotted along with its epistemic uncertainty, showing that non-specificity decreases as the number of labeled instances increases. Figures 1c and 1d illustrate the monotonicity of epistemic uncertainty in deep learning, where a small value of $K = 10$ is sufficient for monotonicity. Again, the results show that non-specificity decreases with more labeled instances, confirming that epistemic uncertainty remains monotonic with respect to sample size in deep learning. This valuable property demonstrates that, both theoretically and empirically, the model becomes less ignorant as it gains access to more knowledge.

5 Conclusion

The main takeaway of this article is that second-order risk minimization is not Dempster-Shafer Theory but both are known under the same name of evidential (deep) learning. It has been showed that modeling a Dirichlet distribution is not sufficient for a model to be considered evidential. Furthermore, recent critiques of second-order risk minimization emphasize the difficulty in modeling epistemic uncertainty, whereas Dempster-Shafer Theory offers a built-in, reliable approach to modeling this type of uncertainty. A mathematical proof demonstrated the monotonicity of the learner's epistemic uncertainty in the application context of the evidential framework. Practical experiments also showed a satisfactory decreasing of epistemic uncertainty both in machine learning and deep learning, leveraging a proper framework to model the ignorance state of the learner. This property is particularly important, as it guarantees that the learner's ignorance decreases as it acquires more information. Although reliable and efficient in our *scenario*, evidential learning is a broad domain, and some models considered evidential do not adhere to this axiomatic behavior. Thus, the insights gained from second-order risk minimization allow us to extend valuable properties to the DST case, enabling the definition of an evidential model that properly estimates epistemic uncertainty, such as those presented in our experimental setup.

Disclosure of Interests. The authors have no competing interests to declare that are relevant to the content of this article.

References

1. Amini, A., Schwarting, W., Soleimany, A., Rus, D.: Deep evidential regression. In: NIPS, vol. 33. Curran Associates, Inc. (2020)
2. Bao, W., Yu, Q., Kong, Y.: Evidential deep learning for open set action recognition. In: 2021 ICCV, IEEE Computer Society (2021)
3. Bengs, V., Hüllermeier, E., Waegeman, W.: Pitfalls of epistemic uncertainty quantification through loss minimisation. In: NIPS. Curran Associates, Inc. (2022)
4. Bengs, V., Hüllermeier, E., Waegeman, W.: On second-order scoring rules for epistemic uncertainty quantification. In: ICML, JMLR.org (2023)
5. Bi, Y., Guan, J., Bell, D.: The combination of multiple classifiers using an evidential reasoning approach. Artif. Intell. **172**(15), 1731–1751 (2008)
6. Charpentier, B., Zügner, D., Günnemann, S.: Posterior network: Uncertainty estimation without ood samples via density-based pseudo-counts. In: NIPS (2020)
7. Chen, X., Wang, P., Hao, Y., Zhao, M.: Eknn-based condition monitoring and early warning method with applications in power plant. Neurocomputing (2018)
8. Côme, E., Oukhellou, L., Denœux, T., Aknin, P.: Learning from partially supervised data using mixture models and belief functions. Pattern Recogn. (2009)

9. Dempster, A.P.: Upper and lower probabilities induced by a multivalued mapping. Ann. Math. Stat. **38**(2), 325–339 (1967)
10. Denœux, T.: A neural network classifier based on dempster-shafer theory. IEEE Trans. Syst. Man Cybern. (2000)
11. Denœux, T.: A k-nearest neighbor classification rule based on dempster-shafer theory. Syst. Man Cybern. IEEE Trans. **219** (1995)
12. Dua, D., Graff, C.: UCI ML repository (2017), https://archive.ics.uci.edu/
13. Dubois, D., Prade, H.: Properties of measures of information in evidence and possibility theories. Fuzzy Sets Syst. **24**(2), 161–182 (1987)
14. Elouedi, Z., Mellouli, K., Smets, P.: Belief decision trees: theoretical foundations. Int. J. Approximate Reasoning **28**(2), 91–124 (2001)
15. Guettari, N., Capelle-Laizé, A.S., Carré, P.: Blind image steganalysis based on evidential k-nearest neighbors. In: ICIP (2016)
16. He, K., Zhang, X., Ren, S., Sun, J.: Deep residual learning for image recognition. In: CVPR (2016)
17. Hoarau, A., Martin, A., Dubois, J.C., Le Gall, Y.: Evidential random forests. Expert Syst. Appl. **230** (2023)
18. Hora, S.C.: Aleatory and epistemic uncertainty in probability elicitation with an example from hazardous waste management. Rel. Eng. Syst. Saf. (1996)
19. Jøsang, A., Elouedi, Z.: Interpreting belief functions as dirichlet distributions. In: Symbolic and Quantitative Approaches to Reasoning with Uncertainty (2007)
20. Jürgens, M., Meinert, N., Bengs, V., Hüllermeier, E., Waegeman, W.: Is epistemic uncertainty faithfully represented by evidential deep learning methods? In: ICML (2024)
21. Klir, G.J., Wierman, M.J.: Uncertainty-based information: elements of generalized information theory. In: Springer-Verlag (1998)
22. Krizhevsky, A.: Learning multiple layers of features from tiny images. In: University of Toronto (2009), https://api.semanticscholar.org/CorpusID:18268744
23. Lecun, Y., Bottou, L., Bengio, Y., Haffner, P.: Gradient-based learning applied to document recognition. Proc. IEEE **86**(11), 2278–2324 (1998)
24. Lefevre, E., Colot, O., Vannoorenberghe, P.: Belief function combination and conflict management. Inf. Fusion **3**(2), 149–162 (2002)
25. Liu, Z., Amini, A., Zhu, S., Karaman, S., Han, S., Rus, D.: Efficient and robust lidar-based end-to-end navigation. In: ICRA (2021)
26. Liu, Z., Pan, Q., Dezert, J.: Evidential classifier for imprecise data based on belief functions. Knowl.-Based Syst. **52**, 246–257 (2013)
27. Liu, Z., Pan, Q., Dezert, J.: A new belief-based k-nearest neighbor classification method. Pattern Recogn. **46**(3), 834–844 (2013)
28. Malinin, A., Gales, M.J.F.: Predictive uncertainty estimation via prior networks. In: Neural Information Processing Systems (2018)
29. Meinert, N., Gawlikowski, J., Lavin, A.: The unreasonable effectiveness of deep evidential regression. In: AAAI Conference on Artificial Intelligence (2022)
30. Sensoy, M., Kaplan, L., Kandemir, M.: Evidential deep learning to quantify classification uncertainty. In: NIPS. Curran Associates, Inc. (2018)
31. Shafer, G.: A Mathematical Theory of Evidence. Princeton University Press (1976)
32. Shafer, G.: A mathematical theory of evidence turns 40. Int. J. Approximate Reasoning (2016)
33. Shannon, C.E.: A mathematical theory of communication. Bell Syst. Tech. J. **27**(3), 379–423 (1948)
34. Shen, M., et al.: Are uncertainty quantification capabilities of evidential deep learning a mirage? In: NIPS (2024)

35. Soleimany, A.P., Amini, A., Goldman, S., Rus, D., Bhatia, S.N., Coley, C.W.: Evidential deep learning for guided molecular property prediction and discovery. ACS Central Sci. (2021)
36. Trabelsi, A., Elouedi, Z., Lefevre, E.: Decision tree classifiers for evidential attribute values and class labels. Fuzzy Sets Syst. **366**, 46–62 (2019)
37. Xu, P., Davoine, F., Zha, H., Denœux, T.: Evidential calibration of binary svm classifiers. In: IJAR (2016)

Enabling Context-Aware Data Reductions

Vlada Stegarescu[1,2](✉), Franck Ravat[1], Jiefu Song[1], Leonidas Papastamatis[2], and Benoit Baurens[2]

[1] IRIT, CNRS (UMR 5505), Université Toulouse Capitole, Toulouse, France
{vlada.stegarescu,franck.ravat,jiefu.song}@irit.fr
[2] Akkodis Research, Toulouse, France
{vlada.stegarescu,leonidas.papastamatis,benoit.baurens}@akkodis.com

Abstract. Data reduction is essential for managing the growing scale and heterogeneity, and computational cost of modern analytical pipelines. Selecting an appropriate reduction technique remains challenging due to the diversity of existing methods and the complex interplay between dataset properties, analytical constraints, and user-defined priorities. Existing taxonomies describe families of reduction strategies but do not provide solutions for choosing a method that is compatible with a given context. We address this gap by introducing a context-aware, paradigm-agnostic decision-support system that formalizes reduction-method selection as a multi-objective optimization problem. The system generalizes observed method behaviours via clustering and identifies Pareto-efficient trade-offs before ranking alternative techniques according to a preference-consistent operator. Experiments on seven real-world datasets from diverse domains demonstrate that the system consistently recommends methods that align with dataset characteristics and user priorities.

Keywords: Data Reduction · Decision Support · Multi-Objective Selection

1 Introduction

The rapid growth of data acquisition technologies, coupled with increasingly heterogeneous analytical environments, has made data reduction a central component of modern machine learning pipelines. Reducing the dimensionality or volume of data is often required to meet constraints related to scalability, storage, computational cost, learning-time efficiency, and energy consumption. Data reduction can also improve model robustness, mitigate noise, and enhance interpretability. Three main paradigms dominate the landscape of reduction strategies: (i) vertical methods that operate on the feature space, (ii) horizontal methods that act on observations, and (iii) hybrid approaches combining both.

While each paradigm offers distinct advantages, reduction techniques are not interchangeable. Based on the motivations articulated across existing studies on

M. Baratchi et al. (Eds.): IDA 2026, LNCS 16513, pp. 31–44, 2026.
https://doi.org/10.1007/978-3-032-23833-7_3

data reduction, we consider that their suitability depends on an intricate combination of factors: (i) dataset statistical properties, as certain methods only operate on specific data types or distributions [9]; (ii) analytical task requirements, such as particular data structures 0 [14]; and (iii) user-defined priorities [5] in terms of reduction objectives, including analytical performance preservation, interpretability, reduction level, or computational efficiency, which are often antagonistic. As a result, choosing an appropriate reduction technique is far from trivial.

Prior surveys catalogue reduction techniques and describe the characteristics of the three paradigms. They do not offer a systematic mechanism to link dataset characteristics, task requirements, user priorities, and reduction paradigms within a unified selection framework. More importantly, the selection of reduction techniques has not been formalized as a decision problem, which limits the ability to reason explicitly about trade-offs and context suitability.

To address this gap, we propose (i) the formalisation of data reduction selection as a multi-objective optimisation framework that jointly models the statistical properties of the dataset, the analytical task to be performed, and user preferences in terms of reduction objectives; (ii) a generic Decision Support System implementing this framework to empirically characterise the behavioural patterns of reduction methods from the three paradigms within these defined contexts; and (iii) an experimental study assessing the robustness and pertinence of the recommended strategies across the full spectrum of contextual constraints considered. These three elements of our contribution are the subjects of the three following sections.

2 State of the Art

Across the three reduction paradigms, reduction methods are organised into distinct families: (i) feature selection [9](e.g., mutual information, RFE) or dimensionality reduction techniques [4] (e.g., PCA, LDA) for vertical reduction; (ii) sampling strategies [11] (e.g., random, stratified, or cluster-based sampling) for horizontal reduction; (iii) and hybrid approaches combine methods drawn from both vertical and horizontal families [11]. In the literature, these families are generally examined independently, limiting any comparative understanding of how they behave across similar conditions and obscuring the interplay between their assumptions and empirical performance. When vertical and horizontal methods are jointly considered, analyses typically restrict themselves to technical requirements—principally dataset statistical properties—and often do so within big-data contexts [11]. As a result, existing surveys offer little guidance on how to match reduction families to datasets with diverse structural characteristics.

In addition, most studies evaluate reduction techniques with respect to a single analytical task, most often predictive analysis [5]. This task-specific evaluation prevents a unified analysis of how task requirements influence the suitability or behaviour of different reduction methods. In works that do consider task-related factors alongside statistical properties, a unique analytical task is considered and usually within narrow domain-specific settings—such as predictive modelling in medical imaging [6]—which limits the transferability of their conclusions to other analytical tasks or domains.

Most existing research further treats data reduction as a single-objective, accuracy-driven optimisation problem, overlooking user-driven trade-offs between objectives such as reduction magnitude, performance preservation, post-reduction interpretability, and computational cost. Although multi-objective formulations have been explored in evolutionary feature selection [13], these efforts remain limited in scope—typically balancing only accuracy and the number of selected features—and do not generalise to the wider landscape of vertical, horizontal, and hybrid methods. Consequently, current approaches offer limited support for incorporating user preferences into reduction-method selection in a broad and systematic manner.

Meanwhile, Automated Machine Learning (AutoML) constitutes a broader research effort focused on automating the full machine learning pipeline. Recent surveys describe AutoML as optimising end-to-end machine learning pipelines by exploring a large combinatorial search space of preprocessing operators, learning algorithms, and hyperparameters, typically guided by predictive performance metrics [1] [12]. Although AutoML frameworks include data reduction components, they do not explicitly model the structural assumptions, technical admissibility conditions, or intrinsic trade-offs of reduction methods, nor does it characterise their behaviour across different dataset configurations.

Overall, until now data reduction techniques are most often studied within isolated context, which prevents a unified analysis of their behaviour across comparable data conditions. While AutoML frameworks automate preprocessing within end-to-end pipelines, the selection of data reduction strategies remains implicit and driven by black-box processes. Moreover, user preferences and contextual constraints are rarely taken into account, making it impossible to make trade-offs between these criteria.

3 Decision Support System Architecture

This section describes the operational implementation of the proposed multi-objective Decision Support System (DSS). The DSS is structured around tow complementary modules, an offline module that provides an empirical knowledge base describing the behaviours of data reduction techniques under controlled conditions, and an online module that exploits this knowledge base to recommend context-consistent strategies for real datasets.

3.1 Problem Statement

We formulate the selection of a data reduction method as a multi-objective decision problem. Let d denote a dataset characterised by a technical profile $p \in \Omega = \Omega_L \times \Omega_N \times \Omega_Y \times \Omega_F \times \Omega_O$, where the dimensions describe relationship linearity, distribution normality, feature type, number of features, and number of observations. The analytical task T is treated as an external evaluation parameter and is not part of the technical space.

Let $\mathcal{R} = \{r_1, \ldots, r_k\}$ be the set of reduction methods applicable to d. Each method r produces a reduced dataset d^r_{red} and is evaluated through a vector of normalised objectives

$$g_r(d,T) = \big(g_{r,1}(d,T), g_{r,2}(d), g_{r,3}(d), g_{r,4}(r)\big) \in [0,1]^4,$$

representing performance preservation, reduction level, interpretability, and computational cost.

Given a user-defined weak preference relation $\succeq$, we define a preference-consistent aggregation operator $\Phi_{\succeq}$ and the utility

$$U_{\succeq}(r,d,T) = \Phi_{\succeq}\big(g_r(d,T)\big).$$

The recommended method is

$$r^*(d,T,\succeq) = \arg\max_{r \in \mathcal{R}} U_{\succeq}(r,d,T).$$

3.2 Objective Computation

Each objective is computed through a dedicated evaluation procedure, as summarised in Table 1. $g_{r,1}$ measures the variation in analytical performance induced by the reduction for task T using task-adapted metrics. $g_{r,2}$ quantifies the reduction in data volume based on the number of data cells. $g_{r,3}$ captures the degree to which the reduced representation preserves explicit feature semantics. $g_{r,4}$ reflects the computational cost of method r and is derived from technical documentation. Interpretability and cost are encoded as ordinal values and normalised for aggregation.

Interpretability and cost are ordinal values obtained from technical documentation. This formulation unifies heterogeneous evaluation criteria within a transparent, preference-driven framework. Unlike prior work, which focuses on single metrics or black-box benchmarks, our approach enables systematic, interpretable, and user-configurable method selection.

Table 1. Definition of objective computation for method r.

$g_{r,1}(d) = \text{Perf}(d^r_{\text{red}}, T) - \text{Perf}(d,T)$	$g_{r,3}(d) = \text{Interp}(d^r_{\text{red}})$
$g_{r,2}(d) = (F \cdot O) - (F_{\text{red}} \cdot O_{\text{red}})$	$g_{r,4}(d) = \text{Cost}(r)$

3.3 Offline Knowledge Base Construction

This module provides the empirical foundation for preference-aware recommendations. By systematically evaluating reduction techniques under controlled data conditions and for specific analytical tasks, it produces a structured repository capturing each method's applicability, performance, and behavioral characteristics. This evidence base enables a multi-dimensional analysis that jointly considers dataset properties, method behavior, and task objectives.

As presented in Algorithm 1, we construct a unified knowledge base $\mathcal{K}$ that characterizes the behaviour of data reduction techniques across controlled data conditions. We begin by selecting a comprehensive set R of 13 methods covering dimensionality reduction, feature selection, and sampling, alongside their 63 admissible ordered pairwise combinations. This ensures balanced coverage of vertical, horizontal, and hybrid reduction paradigms without prioritizing any single family.

A global technical space Ω is defined as the Cartesian product of five dataset characteristics: relationship linearity $L \in \Omega_L$, distribution normality $N \in \Omega_N$, feature-type $Y \in \Omega_Y$, number of features $F \in \Omega_F$, and number of observations $O \in \Omega_O$. Each element of this space represents a dataset profile $p \in \Omega$. For every reduction method $r \in R$, an applicability profile $P(r) \subseteq \Omega$ is constructed by specifying the admissible values for each characteristic, reflecting constraints such as numerical-only requirements, linearity assumptions, or incompatibility with mixed-type features.

The intersection of these two structures yields the set of feasible dataset profiles $\mathcal{P}_{\text{feas}}$, i.e., data conditions for which at least one method is applicable. Each feasible profile $p \in \mathcal{P}_{\text{feas}}$ is associated with its set of compatible reduction methods M_p, and these associations are stored in the compatibility map $\mathcal{M}$. This ensures methodological coherence and prevents the evaluation of techniques on unsuitable datasets.

For each feasible profile, a synthetic dataset d satisfying the specified conditions is generated, forming the synthetic universe $\mathcal{D}$. To ensure full coverage of the technical space Ω, the knowledge base could be populated using either real-world datasets spanning diverse profiles or synthetically generated datasets conditioned on feasible real-world configurations; however, to guarantee a controlled and exhaustive exploration of dataset characteristics, we adopt a synthetic data generation strategy. All compatible reduction methods $r \in M_p$ are then applied to these datasets, and a multi-objective performance vector is computed for each reduced output.

To identify recurrent behavioural patterns, all performance vectors are assembled in the set G and clustered in the multi-objective space. Each cluster is represented by a centroid summarising its characteristic trade-off structure, and method frequencies within clusters quantify the stability of techniques across similar regimes.

The Pareto-optimal centroids—those not dominated across the four objectives—form the final abstraction layer of the knowledge base P, providing a non-redundant set of efficient behavioural profiles. They retain the intrinsic trade-offs without arbitrary weighting and constitute the basis on which the online recommendation module aligns dataset characteristics with user priorities. Figure 1 depicts the resulting structure of the knowledge base.

Algorithm 1. Unified Knowledge Base Construction for Data Reduction Techniques

Inputs:
$\Omega_T, \Omega_L, \Omega_N, \Omega_C, \Omega_F, \Omega_O$ *// Dataset condition spaces*
$R = \{r_1, \ldots, r_k\}$ *// Reduction methods*
Output:
$\mathcal{K}$ *// Unified knowledge base*
Initialize $\mathcal{K} \leftarrow \emptyset$

1. Definition of the Global Technical Space
$\Omega \leftarrow \Omega_T \times \Omega_L \times \Omega_N \times \Omega_Y \times \Omega_F \times \Omega_O$
Each dataset profile is a tuple $p = (T, L, N, Y, F, O)$ with $p \in \Omega$

2. Construction of Method Profiles
for each method $r \in R$ **do**
  for each dimension $X \in \{T, L, N, Y, F, O\}$ **do**
    Determine admissible subset $P_r(X) \subseteq \Omega_X$
  end for
  $P_r \leftarrow P_r(T) \times P_r(L) \times P_r(N) \times P_r(Y) \times P_r(F) \times P_r(O)$
end for
3. Identification of Feasible Dataset Profiles
$\mathcal{P}_{\text{feas}} \leftarrow \emptyset$
$\mathcal{M} \leftarrow \emptyset$ *// method-profile compatibility map*
for each profile $p \in \Omega$ **do**
  $M_p \leftarrow \emptyset$
  for each method $r \in R$ **do**
    if $p \in P_r$ **then**
      Add r to M_p
    end if
  end for
  if $M_p \neq \emptyset$ **then**
    Add p to $\mathcal{P}_{\text{feas}}$
    Store (p, M_p) in $\mathcal{M}$
  end if
end for
4. Synthetic Universe Construction
$\mathcal{D} \leftarrow \emptyset$
for each profile $p \in \mathcal{P}_{\text{feas}}$ **do**
  Generate synthetic dataset d satisfying p
  Add (p, d) to $\mathcal{D}$
end for
5. Method Evaluation
for each $(p, d) \in \mathcal{D}$ **do**
  Retrieve compatible set M_p from $\mathcal{M}$
  for each method $r \in M_p$ **do**
    Apply r to d to obtain d_{red}
    Compute performance vector $g_r(d_{\text{red}})$
    Store (d, r, g_r) in $\mathcal{K}$
  end for
end for

Algorithm 2. Unified Knowledge Base Construction (Part 2, continued)

6. Clustering of Performance Behaviors
$\mathcal{G} \leftarrow \{g_r(d) : (d, r, g_r) \in \mathcal{K}\}$
Cluster $\mathcal{G}$ into $\mathcal{C} = \{C_1, \ldots, C_K\}$
7. Computation of Method Frequency
for each cluster C_k **do**
  for each method $r \in C_k$ **do**
    Compute frequency $f_r(C_k) = |\{g_r(d) \in C_k\}|$
  end for
end for
8. Cluster Centroids
for each cluster C_k **do**
  $c_k = \frac{1}{|C_k|} \sum_{g \in C_k} g$ // *Compute centroid*
end for
9. Trade-offs Optimization
$\mathcal{P} = \{c_k : \nexists c_j \text{ with } c_j \succ c_k\}$ // $c_j \succ c_k \iff c_j^i \geq c_k^i \ \forall i$ *and* $\exists i : c_j^i > c_k^i$
10. Final Knowledge Base Update
Store $\mathcal{C}$, $\{f_r(C_k)\}$, $\{c_k\}$, and $\mathcal{P}$ in $\mathcal{K}$

3.4 Online Recommendation Engine

The second module operationalises the multi-objective decision process by evaluating real datasets and user preferences against the empirical structures constructed offline, as presented in Fig. 2.

The online module begins by extracting the technical profile of the user's dataset through the same characterisation procedure used in the offline phase, ensuring that any incoming instance is mapped to the technical space underpinning the behavioural knowledge base. Based on this profile, the system retains only the behavioural clusters whose descriptors are compatible with the dataset and whose associated methods satisfy its structural constraints, thereby producing a focused set of admissible behavioural profiles.

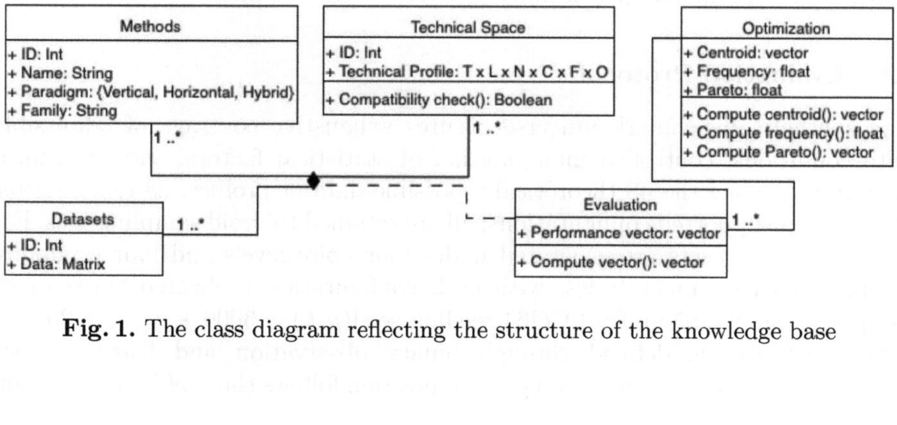

Fig. 1. The class diagram reflecting the structure of the knowledge base

The user then specifies a priority ordering over the four objectives. Although such a structure typically induces a weak preference relation, the system adopts a strict total order: since Pareto filtering in the offline phase has already resolved objective trade-offs among candidate centroids, modelling indifference would provide no additional benefit while increasing complexity. A lexicographic order is therefore applied, comparing the highest-priority objective first and using lower-priority ones to break ties.

The top-ranked centroids correspond to behavioural profiles whose empirical trade-offs best align with the user's priorities. Within each selected cluster, methods are ordered by their frequency of occurrence, reflecting the consistency with which each technique achieves the trade-off pattern represented by the centroid. This yields the final ranked list of recommended methods, providing interpretable, empirically grounded guidance aligned with both the dataset's technical characteristics and the user-defined objectives.

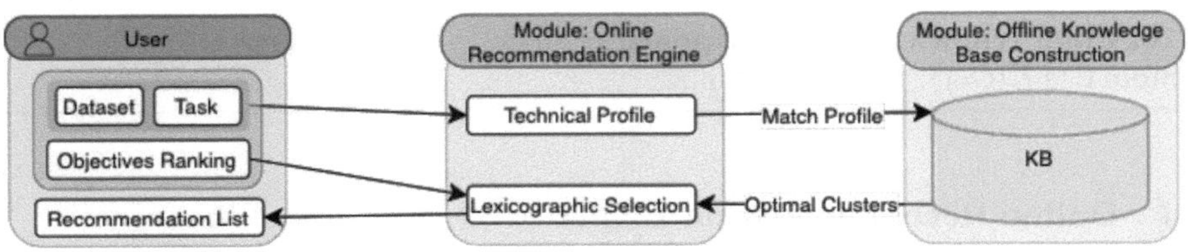

Fig. 2. The architecture of the Online Recommendation Engine

4 Experimental Evaluation

This section assesses the effectiveness of the proposed Decision Support System (DSS) on real datasets. The objective is to determine whether the system (i) produces recommendations aligned with user-defined priorities, (ii) selects methods that are technically compatible with the data, and (iii) yields reduced datasets that preserve predictive performance relative to the original.

4.1 Evaluation Protocol

The synthetic benchmark universe ensures exhaustive coverage of admissible data conditions, controlled manipulation of statistical factors, and full reproducibility. Among the 48 theoretically possible dataset profiles, 28 remain after removing unsupported configurations; all are retained to avoid sampling bias. For each profile, datasets are generated under four noise levels and four randomly sampled linearity magnitudes, with each configuration replicated three times using deterministic NumPy PCG64 seeding $\text{seed}(p,k) = 3000 + p_{\text{index}} + 20k$.

Dataset size is defined through binary observation and feature scales (O_S, O_B) and (F_S, F_B). Feature-type composition follows the profile specification (numerical, categorical, or mixed with equal proportions). Numerical features follow multivariate normal distributions when $n =$ True; otherwise, one of four distortions (log-normal, exponential, power transformation, outliers) is applied with equal probability. The linearity parameter ℓ controls covariance structure, inducing block-structured covariances for $\ell = 1$ and non-linear dependencies for $\ell = 0$. Categorical variables are generated by discretising latent Gaussian scores into $K_j \in \{2, 3, 4\}$ levels while preserving dependence structure.

The full set of 13 base methods and all 63 ordered two-step permutations is evaluated, each applied only to profiles satisfying its structural requirements. For the 28 admissible profiles, combined with the four size configurations and three noise levels, this yields 112 instantiated profile settings, each replicated three times. For each admissible (dataset, method) pair, the four-dimensional performance vector is computed. Predictive performance $g_{m,1}$ is assessed by training Logistic Regression, Decision Tree, and XGBoost classifiers on both the original and reduced datasets using stratified 5-fold cross-validation. The five required metrics (Accuracy, Precision, Recall, F1-score, ROC-AUC) are averaged and converted into an accept/reject decision using the following rule-based scoring scheme:

- **Metric-level scoring.** For each of the five metrics:
  - add 2 points if $m \geq 0.80$ *(high-quality performance).* [3]
  - add 1 point if $0.60 \leq m < 0.80$ *(acceptable performance).* [2]
- **Precision–Recall balance.** If $|\text{Precision} - \text{Recall}| < 0.10$, add 1 point *(ensures that the reduced data do not create strong class-balance distortions).* [10]
- **Overfitting penalty.** If all five metrics exceed 0.96, subtract 2.5 points *(very high scores commonly indicate overfitting).* [2]
- **Final decision.** The model is accepted if the total score is at least 3, and rejected otherwise *(minimal level ensuring that multiple aspects of performance remain preserved).*

Among the three classifiers, only the best-performing one contributes to the final decision. Interpretability $g_{r,3}$ is binary, marked "yes" when the reduced representation preserves direct feature semantics and "no" otherwise. Reduction volume $g_{r,2}$ is computed as the proportion of removed data cells, capturing both vertical and horizontal reduction within a unified measure. Computational cost $g_{r,4}$ is assigned from official library documentation using categorical levels (low/medium/high). All objectives are subsequently normalised to $[0, 1]$ through min–max scaling.

Behavioural patterns are extracted through a recursive k-means procedure with $k = 2$. A cluster is subdivided when the variance of silhouette coefficients decreases after splitting, indicating improved structural coherence. The recursion produces 62 behavioural clusters, whose centroids are then filtered by Pareto dominance; 48 non-dominated centroids remain for online recommendation.

Online Evaluation Setup. This experimental design serves two main objectives: (i) empirically assess the pertinence of the generated reduction suggestions, both in terms of alignment with technical compatibility and preservation of task-specific performance, (ii) verify whether recommended reduction strategies remain effective under diverse statistical and structural dataset profiles (linearity, normality, mixed types, etc.). To evaluate the effectiveness and robustness of our decision support system, we tested our solution on real datasets.

Table 2. The Evaluation Scenarios

ID	Dataset	Domain	L	N	C	F	O	Priority Order
H	Heart attack Prediction	Healthcare	Yes	No	Mix	S	S	Reduced size → Interpretability Performance → Comput. cost
T	Hotel Booking Prediction	Tourism	No	No	Mix	B	B	Comput. cost → Performance Reduced size → Interpretability
S	Housing	Social Studies	Yes	Yes	Num	S	B	Performance → Comput. cost Reduced size → Interpretability
F	Company Bankruptcy Prediction	Finance	No	No	Num	B	B	Performance → Comput. cost Reduced size → Interpretability
A	Mushrooms Classification	Agriculture	No	No	Cat	B	B	Comput. cost → Performance Reduced size → Interpretability
R	Boston House Pricing	Real Estate	Yes	No	Num	S	S	Performance → Interpretability Reduced size → Comput. cost
E	Students Performance	Education	Yes	Yes	Mix	S	S	Reduced size → Performance Interpretability → Comput. cost

Datasets. To evaluate the decision-support system under realistic conditions, we selected seven real-world datasets from Kaggle, each matching one of the admissible statistical profiles from the technical space. The datasets span diverse domains (healthcare, tourism, finance, agriculture, education) and are summarized in Table 2. Selection was guided by profile compatibility and the availability of publicly shared classification implementations, which served to establish baseline performance on unreduced data. This set provides a representative, domain-independent basis for assessing the system under realistic conditions.

User Priority Structures. To approximate domain-specific decision-making, each dataset was assigned a priority structure reflecting how experts in its field would typically balance data-reduction objectives. These priority schemes—reported in the last column of Table 2—encode domain-driven preferences such as favouring interpretability in healthcare, accuracy in educational analytics, or computational efficiency in operational contexts. They determine the lexicographic ordering applied by the DSS, and the evaluation assesses whether the recommended reduction techniques align with the priority structure associated with each dataset's domain.

Table 3. The Evaluation Results

D	Recommendation	Rank	L	N	$g_{r,1}$	$g_{r,2}$	$g_{r,3}$	$g_{r,4}$
H	BL		**Yes**	**No**	**9.0**	**5586**		
H	**FAMD**	**1**	**Yes**	**No**	**6.0**	**2646**	**High**	**No**
H	**FAMD + T-SVD**	**2**	**Yes**	**No**	**6.0**	**2646**	**High**	**No**
T			No	No	11.0	2980250		
T	MI	1	No	No	9.0	1549730	Med	Yes
T	MI + RANDOM SAMPLING	2	No	No	9.0	464919	Med	Yes
T	MI + STRATIFIED SAMPLING	3	No	No	9.0	464919	Med	Yes
T	RANDOM SAMPLING	4	No	No	8.0	35405370	Low	Yes
T	RFE	5	No	No	8.0	118017900	High	Yes
S			Yes	Yes	11.0	109236		
S	PCA	1	Yes	Yes	6.0	54618	Med	Yes
S	PCA + CLUSTER SAMPLING	2	Yes	Yes	9.0	26390	Med	Yes
S	PCA + RANDOM SAMPLING	3	Yes	Yes	9.0	16386	Med	Yes
S	PCA + STRATIFIED SAMPLING	4	Yes	Yes	7.0	16386	Low	Yes
S	PCA + SYSTEMATIC SAMPLING	5	Yes	Yes	6.0	18205	High	Yes
F			No	No	11.0	1034835		
F	RANDOM SAMPLING	1	No	No	9.0	310460	Low	Yes
F	STRATIFIED SAMPLING	2	No	No	9.0	310460	Med	Yes
F	SYSTEMATIC SAMPLING	3	No	No	9.0	344945	Low	Yes
A			**No**	**No**	**8.5**	**570000**		
A	**RANDOM SAMPLING**	**1**	**No**	**No**	**8.5**	**171000**	**Low**	**Yes**
A	**STRATIFIED SAMPLING**	**2**	**No**	**No**	**8.5**	**171000**	**Med**	**Yes**
A	**MCA + RFE**	**3**	**No**	**No**	**NaN**	**NaN**	**High**	**No**
A	**SYSTEMATIC SAMPLING**	**4**	**No**	**No**	**8.5**	**190000**	**Low**	**Yes**
A	**MCA + MI**	**5**	**No**	**No**	**NaN**	**NaN**	**Med**	**No**
R			Yes	No	9	6578		
R	T-SVD	1	Yes	No	7.5	4554	High	No
E			**Yes**	**Yes**	**11.0**	**7000**		
E	**LDA + CLUSTER SAMPLING**	**1**	**Yes**	**Yes**	**8.5**	**365**	**High**	**No**

4.2 Results

Table 3 presents the evaluation results for the experimental datasets, with corresponding dataset profiles summarized in Table 2. Baselines (first line per dataset) serve as dataset-specific references, while subsequent rows report the recommended reduction techniques ranked by the DSS according to user-defined priorities. Although the evaluation was conducted on seven datasets, the discussion focuses on the three datasets highlighted in bold, as they correspond to distinct priority orderings and reduction methods, and they explicitly expose different trade-off structures, methodological behaviours, and pipeline limitations observed in the knowledge base.

In **Heart Attack Prediction** (Healthcare, priority: Reduced Size → Interpretability → Performance → Computational Cost), FAMD reduces the dataset by over 50% (5,586 → 2,646 cells) while maintaining performance at 6.0. Combining FAMD with T-SVD yields the same size reduction with no further performance change, demonstrating that dimensionality-based techniques satisfy priority-driven objectives with moderate accuracy loss.

In **Mushrooms Classification** (Agriculture, priority: Computational Cost → Performance → Reduced Size → Interpretability), Random and Stratified Sampling reduce the dataset from 570,000 to 171,000 cells, preserving performance at 8.5. Systematic Sampling achieves a slightly higher size (190,000) with identical performance. Hybrid methods involving MCA (MCA+RFE, MCA+MI) occasionally produce `NaN` values due to incompatibilities between linear projections and subsequent relevance-based feature selection, highlighting intrinsic method limitations rather than DSS filtering decisions.

In **Students Performance** (Education, priority: Reduced Size → Performance → Interpretability → Computational Cost), LDA combined with Cluster Sampling reduces 7,000 cells to 365 while maintaining performance at 8.5. This demonstrates that hybrid methods can optimize multiple objectives simultaneously, achieving over 90% size reduction with limited performance loss.

The evaluation shows that the DSS consistently produces technically compatible and priority-aligned reduction recommendations across heterogeneous datasets. Reductions of 50–95% in dataset size were obtained with only moderate performance decreases (typically 2–3 points), while efficiency or interpretability were preserved when prioritised. These results indicate that the system adapts its suggestions to both dataset profiles and user-defined objective rankings, generating coherent trade-offs within the multi-criteria framework (Tables 2–3).

The DSS systematically identifies Pareto-optimal techniques satisfying the lexicographic ordering of objectives. Variations in the number of recommended methods across datasets reflect the empirical distribution of techniques in the knowledge base and their trade-off structures, rather than any procedural bias of the system. The experiments also reveal method-specific limitations, particularly within certain hybrid pipelines. MCA-based combinations (MCA+RFE, MCA+MI) occasionally produced `NaN` values in datasets with low variance or insufficient discriminative structure to satisfy the assumptions of the subsequent feature-selection stage. These methods were nevertheless retained when lying on the Pareto front, indicating that the DSS exposes inherent pipeline weaknesses rather than excluding them.

Overall, the evaluation shows that the DSS meets its objective of recommending technically compatible and priority-aligned reduction strategies, while also revealing specific method-level weaknesses—most notably the instability of certain hybrid pipelines under particular data conditions. These observations highlight that the system provides coherent, context-aware guidance across heterogeneous datasets. Limitations include sensitivity of certain hybrid pipelines to low-variance projections and occasional instability in complex configurations. Future improvements could involve adaptive preprocessing, expanded knowledge base coverage, and incorporation of additional evaluation metrics (e.g., frugality, sustainability) to enhance stability and generalizability.

5 Conclusion and Future Work

This work presents a decision-support framework for data reduction that formalizes the selection process as a multi-objective optimization problem, integrating technical compatibility, empirical performance, and user preferences. The system compares vertical, horizontal, and hybrid reduction strategies using a unified set of normalized objectives, and our experiments demonstrate its robustness and relevance across diverse statistical dataset profiles. The results show that the framework enables transparent trade-offs across predictive performance, reduction volume, interpretability, and computational cost, providing a structured alternative to intuition-driven preprocessing.

A key limitation of the current system is its static nature, which does not adapt to user feedback. Future developments will focus on incorporating an explicability module and progressively collecting user interactions and domain knowledge to refine recommendations and enrich the knowledge base with domain-specific information (e.g., ontologies). This aligns with research on informed machine learning [7]. Enhancements to the evaluation framework, including sustainability and frugality oriented metrics, will provide a more comprehensive assessment of method suitability beyond classical efficiency trade-offs. Expanding and diversifying the dataset repository will further improve generalizability and allow the system to evolve toward fully adaptive recommendations that continuously learn from user preferences and evolving data characteristics [8].

Disclosure of Interests. The authors have no competing interests to declare that are relevant to the content of this article.

References

1. Alcobaça, E., de Carvalho, A.C.P.L.F.: A literature review on automated machine learning. Artif. Intell. Rev. **59**(1), 5 (2025). https://doi.org/10.1007/s10462-025-11397-2
2. Hosmer, D.W., Lemeshow, S., Sturdivant, R.X.: Applied Logistic Regression. Wiley Series in Probability and Statistics, Wiley, Hoboken, New Jersey, 3d edition edn. (2013)
3. Kuhn, M., Johnson, K.: Applied Predictive Modeling. Springer, New York, NY (2013). https://doi.org/10.1007/978-1-4614-6849-3
4. Maaten, L., Postma, E., Herik, J.: Dimensionality reduction: a comparative review. J. Mach. Learn. Res. **10**, 66–71 (2009)

5. Martins Dias, G., Bellalta, B., Oechsner, S.: A survey about prediction-based data reduction in wireless sensor networks. ACM Comput. Surv. **49** (2016). https://doi.org/10.1145/2996356
6. Maruotto, I., Ciliberti, F.K., Gargiulo, P., Recenti, M.: Feature selection in healthcare datasets: towards a generalizable solution. Comput. Biol. Med. **196**, 110812 (2025). https://doi.org/10.1016/j.compbiomed.2025.110812
7. Micallef, L., et al.: Interactive elicitation of knowledge on feature relevance improves predictions in small data sets, January 2017. https://doi.org/10.48550/arXiv.1612.02487
8. von Rueden, L., et al.: Informed machine learning – a taxonomy and survey of integrating knowledge into learning systems. IEEE Trans. Knowl. Data Eng. 1–1 (2021). https://doi.org/10.1109/TKDE.2021.3079836
9. Saeys, Y., Inza, I., Larrañaga, P.: A review of feature selection techniques in bioinformatics. Bioinformatics **23**(19), 2507–2517 (2007). https://doi.org/10.1093/bioinformatics/btm344
10. Saito, T., Rehmsmeier, M.: The precision-recall plot is more informative than the ROC plot when evaluating binary classifiers on imbalanced datasets. PLoS ONE **10**(3), e0118432 (2015). https://doi.org/10.1371/journal.pone.0118432
11. Salekshahrezaee, Z., Leevy, J.L., Khoshgoftaar, T.M.: The effect of feature extraction and data sampling on credit card fraud detection. J. Big Data **10**(1), 6 (2023). https://doi.org/10.1186/s40537-023-00684-w
12. Tian, J., Che, C.: Automated machine learning: a survey of tools and techniques. J. Ind. Eng. Appl. Sci. **2**(6), 71–76 (2024). https://doi.org/10.70393/6a69656173.323336
13. Xue, B., Zhang, M., Browne, W.N., Yao, X.: A survey on evolutionary computation approaches to feature selection. IEEE Trans. Evol. Comput. **20**(4), 606–626 (2016). https://doi.org/10.1109/TEVC.2015.2504420
14. Yu, L., Liu, H.: Feature selection for high-dimensional data: a fast correlation-based filter solution. In: Proceedings, Twentieth International Conference on Machine Learning, vol. 2, pp. 856–863. AAAI Press, January 2003

T-SE : A Method Built on Squeeze-and-Excitation Mechanisms for Convolutional Neural Networks' Energy Efficiency

Noémie Draguet[1(✉)] and Benoît Frénay[2]

[1] NaDI, MINDIT, University of Namur, Rempart de la Vierge 8, 5000 Namur, Belgium
noemie.draguet@unamur.be

[2] NaDI, PReCISE, HuMaLearn, University of Namur, Rue Grangagnage 21, 5000 Namur, Belgium
benoit.frenay@unamur.be

Abstract. As convolutional neural networks (CNNs) are increasingly used for computer vision tasks, the environmental consequences of their high energy consumption cannot be denied. CNNs especially require a lot of computational power due to their high dimensional inputs, their high number of parameters, and their large size. Existing literature introduces techniques that do not specifically focus on energy consumption as such, but rather on improving the performance, portability, or speed of CNNs. These approaches mostly focus on the testing phase. Furthermore, the inconsistency of metrics that are used for estimating or measuring energy consumption reduces the comparability of these techniques. To address the gap in existing techniques that largely overlook energy consumption, this work proposes a novel approach, T-SE, aimed at reducing the energy usage of CNNs during both the training and inference phases. This new method leverages Squeeze-and-Excitation networks [7], introducing a threshold in this architecture to reduce the number of computations performed by the CNN, and in turn the energy consumed by it. Results show that more than 500,000 convolutions per input (equivalent to around 13% of the total convolutions in the CNN) can be avoided without decreasing significantly the performance of the CNN.

Keywords: Deep Learning · Convolutional Neural Networks · Energy efficiency · Energy consumption · Squeeze-and-Excitation Networks

1 Introduction

In a world where deep learning technologies are increasingly used by both individuals and organizations, figuring out how to reconcile these energy-greedy algorithms with growing environmental concerns has become a new challenge.

M. Baratchi et al. (Eds.): IDA 2026, LNCS 16513, pp. 45–57, 2026.
https://doi.org/10.1007/978-3-032-23833-7_4

Even though AI offers numerous opportunities to support sustainability and improve resource efficiency, its use can negatively impact the environment for three main reasons [2]. Firstly, a significant amount of electricity is required to power the GPUs needed for computations. This implies an increase in carbon emissions and in the use of non-renewable energy. Secondly, servers are cooled using large amounts of water. From 2021 to 2022, Google's and Microsoft's data centers' water consumption increased by 20% and 34%, respectively [11]. These data centers power many chatbots and AI models. Finally, the use of raw materials in AI's hardware production contributes to its environmental footprint.

Large AI models' training, in addition to inference, requires significant energy. For instance, training GPT-3 reportedly consumed as much energy as 121 USA homes for a year, while producing around 550 tons of carbon dioxide [2].

In this work, the focus is on Convolutional Neural Networks (CNNs), which are extensively used in computer vision applications. Section 2 introduces existing methods that can help reduce the energy consumption of CNNs, while Sect. 3 presents T-SE Networks, a novel approach inspired by Squeeze-and-Excitation (SE) Networks and aiming at reducing the energy consumption of CNNs. Section 4 details experiments, Sect. 5 discusses the results and Sect. 6 concludes with perspectives for future research.

2 Related Works

So far, the literature on CNNs has focused on improving the accuracy of these models, or on decreasing their size [9], inference time [3] and energy consumption [15,20] for performance or portability purposes. There is a noticeable lack of attention to CNNs' energy consumption as such and to the reduction of their environmental impact. Furthermore, the decrease in energy resulting from the proposed methods is usually not measured with the same metrics. This makes comparing the success of different techniques difficult.

There are three main ways of measuring or inferring energy consumption: direct hardware measurements [15,20], proxies, or models. More specifically, proxies include the number of parameters of the model [9], the average number of operations per input [16], the number of Multiply-and-Accumulate (MAC) operations [18], or the runtime for inference [14]. These proxies do not indicate with precision the models' energy consumption [20]. As an example, scalable-effort classifiers [17] enable a 2.79x decrease in average operations per input (OPS), but only a 2.3x decrease in energy consumption. The difference in results between these two metrics can be partly explained by implementation overheads. Correctly and precisely measuring the energy consumption of a model during its training and testing phases is a complex task, and is out of the scope of this work.

The literature introduces four main types of methods that can help reduce the energy consumption of CNNs: pruning and network sparsification techniques, reduced precision networks, conditional computation, and hardware optimizations. Most of them focus exclusively on the testing phase and do not address the training phase. They are briefly discussed in the present work, but a detailed introduction to these techniques can be found in Draguet and Frénay [5].

Pruning methods trim an initially complete CNN by removing some of its components (weights, neurons, filters, or number of layers) after training it. The first paper about pruning used the metaphor of "Optimal brain damage" [10]. Years later, Iandola et al. [9] and Wen et al. [19] used this technique to successfully reduce the size of models and speed them up. Yang et al. [20] present energy-aware pruning, which effectively reduces the energy consumption of AlexNet and GoogleNet by respectively 3.7x and 1.6x, with less than 1% top-5 accuracy loss.

Reduced precision networks are models whose components are made less precise by representing them with fewer bits. The energy reduction resulting from it remains largely theoretical as modern commercial GPUs are often restricted to arithmetic operations in 16 or 32 bits. Hubara et al. [8] and Courbariaux et al. [4] use this method to increase computational speed, which is expected to increase power efficiency with suitable hardware.

Conditional computation considers the input's difficulty to dynamically adapt the components of the CNN or the input itself. Layer-based and network-based conditional computation methods consist in learning early exit or network selection decision policies. Panda et al. [14] leverage conditional deep learning (CDL) to enable a 1.84x reduction in energy consumption. Rashid et al. [15] propose an adaptive CNN for human activity recognition inspired by CDL. Compared to the baseline, the new model increases the initial accuracy and decreases energy consumption by 1.12x and 1.11x for the Opportunity and w-HAR datasets, respectively. Network-based conditional computation methods [3] can lead to speedups by up to 2.8x, with less than 1% top-5 accuracy loss. When simultaneously addressing network selection and network design for adaptive systems of CNNs [16], the energy consumption for image classification on a mobile device decreases by up to 6x compared to adaptive systems of CNNs that consider them as black boxes. Bengio et al. [1] present spatially adaptive networks, of which only some parts are activated at runtime to enable speedups in the testing phase. Verelst and Tuytelaars [18] focus on selecting the appropriate spatial region in the input to evaluate, which improves accuracy for a given number of MACs. Finally, Gao et al. [6] present dynamic channel pruning, which involves selecting the most informative channels of the CNN at runtime after having trained it completely. Results show a reduction of 5x and 2x in MACs for ImageNet classification with ResNet-18 and VGG-16, respectively.

Hardware optimizations are complementary to the above software optimizations. Therefore, they can be integrated with other techniques, such as reduced precision networks [8] or pruning [19]. X. Liu et al. [12] propose a low-power hardware architecture that re-implements the convolution operation using column-wise processing. This enables a power reduction of up to 8.45%, 49.41% and 50.64% for LeNet, AlexNet and VGG-16, respectively. On the other hand, Z.-G. Liu et al. [13] leverage structured sparsity to efficiently accelerate CNNs.

While existing methods are numerous, most do not address energy consumption directly, considering it instead as an incidental consequence of the approach.

Furthermore, the training phase, which can be a major contributor to overall energy use, is often disregarded. Thresholded-Squeeze-and-Excitation is a novel architecture based on SE Networks. It aims at increasing energy efficiency during training and testing while preserving accuracy.

3 Thresholded-Squeeze-and-Excitation Networks

Thresholded-Squeeze-and-Excitation (T-SE) Networks focus on the environmental impact of CNNs rather than on their performance. They seek to address energy consumption holistically rather than focusing solely on the testing phase, as the training phase can likewise be a major source of energy use, particularly for large models. The method is tailored to CNNs and their particular architecture. It consists in introducing a threshold in the existing Squeeze-and-Excitation architecture [7], which enables a dynamic switch-off of some of the CNN's channels.

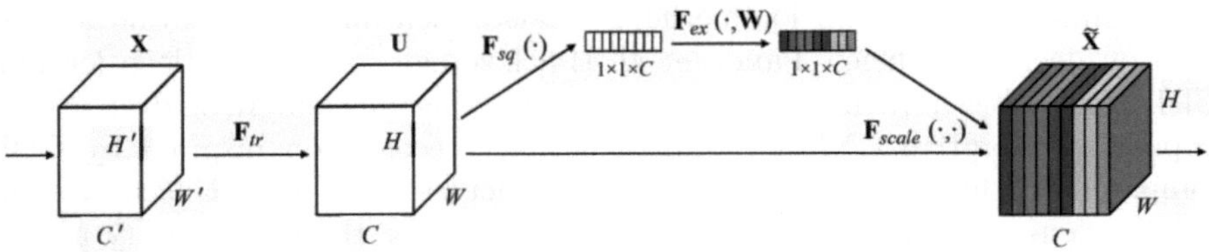

Fig. 1. Squeeze-and-Excitation Layer. Reprinted with permission from *Squeeze-and-Excitation Networks*, Hu et al. (2018), Proceedings of the IEEE Conference on Computer Vision and Pattern Recognition. © 2018 IEEE.

To describe T-SE Networks, SE Networks [7] should be introduced first. SE Networks leverage attention mechanisms, which consist in focusing on the most informative parts of the input or the network. Originally, the technique's objective was to enhance the accuracy of CNNs. This is why it had never been previously used for energy-efficiency purposes. *Squeeze-and-Excitation (SE) Layers* are added after some of the CNN's convolutional layers to put more focus on some of the channels and improve the representational power of the CNN. A SE Layer (Fig. 1) is composed of a *Squeeze* and an *Excitation* operation. These are described using the notation of Hu et al. [7]. The result of a convolutional layer can be expressed in the form

$$u_c = v_c * X = \sum_{s=1}^{C'} v_c^s * x^s, \tag{1}$$

where * is the convolution operator, u_c is one output feature map, v_c is the set of filters for the cth output feature map and X is the input, composed of C' channels. The convolutional layer result is $U = [u_1, u_2, ..., u_C]$, where C is the number of output feature maps and corresponds to the number of input channels for the next layer. u_c is a feature map of dimension $H \times W$ (see Fig. 1 for details).

The result of the *Squeeze* operation can be described as

$$z_c = F_{sq}(u_c) = \frac{1}{H \times W} \sum_{i=1}^{H} \sum_{j=1}^{W} u_c(i, j). \tag{2}$$

This corresponds to using global average pooling to end up with only one descriptor per channel [7]. After the *Squeeze* operation comes the *Excitation* phase, composed of two fully-connected layers. These layers form a bottleneck since the first reduces the dimensionality by a reduction ratio r and the second increases dimensionality back to the initial settings. The first fully connected layer has a Rectified Linear Unit (ReLU) activation function, and the second one has a sigmoid activation function. The *Excitation* operation can be expressed as

$$s = F_{ex}(z, W) = \sigma(g(z, W)) = \sigma(W_2 \delta(W_1 z)), \tag{3}$$

where δ is the ReLU function, W_1 are the weights of the first fully connected layer, W_2 are the weights of the second fully connected layer and σ is the sigmoid function. After this operation, the final output is calculated by multiplying each input channel of the SE Layer with its associated s value, which is equivalent to

$$\tilde{x}_c = F_{scale}(u_c, s_c) = s_c \cdot u_c. \tag{4}$$

As previously mentioned, T-SE adapts SE Networks to reduce the energy consumption of CNNs. As SE Networks use SE Layers to assign a weight to each output feature map of a convolutional layer, a threshold can be applied to these weights. This transforms the weights smaller than the threshold into zero values, with the intuition that feature maps associated with these zero values would not be involved in further computations. This threshold is added to each SE Layer and is applied to each element of s to produce an adapted vector:

$$s'_c = \begin{cases} s_c, & \text{if } s_c > \text{threshold} \\ 0, & \text{otherwise} \end{cases} \tag{5}$$

where threshold is user-chosen, s_c is the initial value from the vector s, and s'_c is the new value of the element after applying the threshold operation. Since the threshold operation is directly integrated into the CNN forward propagation, it is applied during both training and testing. Because every input is different, the results of the Squeeze-and-Excitation operation vary according to this input and result in various input channels being replaced by zero. The use of PyTorch's built-in Threshold function in the implementation enforces a strict inequality but enables handling the non-differentiable nature of the threshold: During backpropagation, gradients flow exclusively through channels that exceed the threshold. For the rest of the present work, "switched-off channels" refer to the feature maps (or channels) that were replaced by a zero value. Filters are never modified, since the technique only adjusts the values of the input features (or channels) for the next layer.

4 Experimental Section

This section conducts experiments to test several values for the user-chosen threshold, in order to determine which threshold values enable switching off a meaningful part of the CNN while maintaining sufficient performance.

4.1 Experimental Settings

Since a GitHub repository with the code to implement Squeeze-and-Excitation networks in PyTorch is readily available[1], the same parameters as the authors are used in the experiments. Particularly, most of the setup from the `cifar.py` file is employed, since the performance of our model is tested for a classification task on the CIFAR-10 dataset. The training and testing loops have been reimplemented due to the outdated nature of the provided code. Each experiment features 100 epochs, a batch size of 64, a reduction ratio of 16, and 2 data loading workers. The model is trained with the Stochastic Gradient Descent (SGD) optimiser, using an initial learning rate of 0.1, a momentum of 0.9, and a weight decay of 0.0001. A StepLR scheduler is set up with a step size of 25 and a γ of 0.33.

To have enough granularity in the results without exceeding computational resources, 30 linearly distributed thresholds ranging from 0 to 0.95 are tested. Chosen values are between 0 and 1 since this is the range of values that can be obtained after the *Squeeze* and *Excitation* operations. The value of 1 is excluded as it would switch off all channels in the network and result in a CNN without any training capacity. When the threshold is set to 0, the architecture is equivalent to the initial Squeeze-and-Excitation architecture.

Experiments use the SE-ResNet20 model from `cifar.py`. This CNN starts with a classic convolutional layer. Then, it is composed of three main stages. Each stage is composed of three basic blocks, and one block is composed of two convolutional layers and one SE Layer. Batch normalization and a Rectified Linear Unit (ReLU) activation function are usually applied after convolutional layers of the CNN. A residual is added at the end of each block. The first stage uses 16 output feature maps for each convolutional layer, while the second and third stages use 32 and 64 of them, respectively. Stages two and three use a stride of two in the first convolutional layer of their first block, which reduces the size of the feature maps. The kernel size used for convolutional layers is 3×3 throughout the whole network. Convolutional layers have the most parameters, since one filter is used for linking each input channel with each output feature map. This is where reducing computations can have the most impact.

After the three main stages, the output is passed through a pooling layer and a final fully connected layer. The SE-ResNet20 model counts 274,490 parameters, which illustrates that even compact CNNs can have a substantial number of parameters. The reduction ratio r in the bottleneck is 16 by default. This value is the one that gives the best trade-off between the number of added parameters and the accuracy of the model [7]. Results obtained with our T-SE approach are

[1] https://github.com/moskomule/senet.pytorch/.

compared against the baseline ResNet20 from the GitHub repository, which is equivalent to the SE-ResNet20 model without its SE Layers.

The experiments are conducted using PyTorch 2.6.0 (CUDA 12.4) and Torchvision 0.21.0, with an NVIDIA A100-SXM4-40GB GPU. A GitHub containing the implementation of T-SE and scripts for results analysis is provided at https://github.com/noemiedraguet/T-SE.git. Models are trained for 100 epochs, and results are reported for the final epoch only. The T-SE architecture is run with 30 different threshold values. For each setup, results are averaged across five iterations to increase representativeness. Graphs from Figs. 2 and 3 also show the standard deviation of obtained values. Standard deviation was chosen rather than a confidence interval, considering the low number of repetitions. To calculate the number of "switched-off channels", the percentage of channels set to zero in each SE Layer is computed and averaged over the number of training or testing images. The average number of switched-off channels over the entire CNN is computed based on the percentages obtained for each layer.

The number of switched-off channels is a biased measure, as their impact on the number of computations varies across layers. This is why the "number of avoided computations" is also calculated. An avoided computation is defined as a value that would not need to be computed in an output feature map of the convolutional layer following the SE Layer, as a result of setting some input channels to zero. In other words, it usually corresponds to a convolution output that is not calculated in subsequent convolutional layers. The term "computations" should be clearly distinguished from "arithmetic operations", which designates single multiplications and additions. The number of avoided computations in the CNN can be calculated as

$$\text{Avoided\_computations} = \sum_{l=1}^{L} s_l \cdot n_{l+1} \cdot dim_{l+1}^2, \tag{6}$$

where L is the number of SE Layers in the network, s_l is the average number of switched-off channels per input in the lth SE Layer (rounded to an integer) at the last epoch of training or testing, n_{l+1} is the number of output feature maps of the convolutional layer following the lth SE Layer and dim_{l+1} is the dimension (width and height) of each of these output feature maps. Since each output feature map value in the $l+1^{th}$ layer is obtained by summing the results of convolution operations across all channels, the total number of avoided convolutions (i.e. computations) in that layer is obtained by multiplying the number of ignored channels (i.e. the number of avoided convolutions for one value) by the total number of values across all feature maps ($n_{l+1} \cdot dim_{l+1}^2$). The last SE Layer is only followed by an average pooling layer, and not a convolutional layer. Therefore, only for this layer, one avoided pooling operation on one feature map is considered as an "avoided computation".

4.2 Results

The number of switched-off channels and avoided computations was studied for 30 thresholds. Thresholds around 0.5 are the ones that allow the best accu-

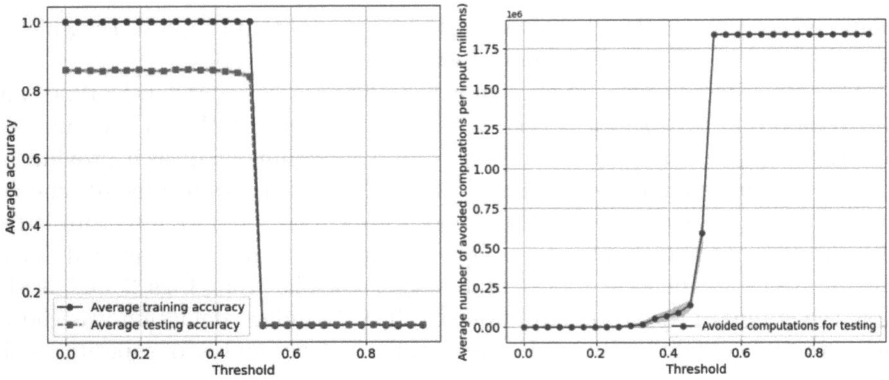

Fig. 2. Relationship between the applied threshold and (a) accuracy or (b) the average total number of avoided computations per input at testing. Markers indicate average across five iterations; Shaded areas show standard deviation.

racy/efficiency trade-offs (see Fig. 2). The maximum number of channels that can be switched off in SE layers is 336, which can lead to a maximum of 1,835,072 avoided computations. An "avoided computation" is made of a convolutional operation, which consists of 9 multiplications and 8 additions in an architecture with 3×3 filters. The approximate number of avoided arithmetic operations can thus be obtained by multiplying the number of avoided computations by 17.

Impact on Training and Testing Accuracies. Increasing the value of the threshold naturally decreases the training and testing accuracies, since the CNN performs fewer computations (Fig. 2). Thresholds equal to or above 0.5241 result in switching off all channels of the first SE Layer, which means that the inputs of all subsequent layers take a value of zero. This results in an accuracy of around 10%, equivalent to a random guess for a ten-class dataset. The maximum testing accuracy for the last epoch (displayed on Fig. 3) reaches 85.97% for a threshold of 0, which is equivalent to the baseline SE architecture (without any threshold, switched-off channels or avoided computations). Maximum training accuracy is equal to 100% and is maintained until the threshold reaches a value of 0.4586. The model overfits the data, probably because the SE-ResNet20 is an architecture that is complex for the relatively simple CIFAR-10 dataset. Given the high testing accuracy (over 80%), overfitting does not appear to affect generalisation significantly, but future experiments should validate this claim.

To limit testing accuracy loss to one percentage point, the threshold must remain equal to or below 0.4586. This threshold value switches off an average of 31 channels per input and avoids an average of around 140,089 computations per input, both at training and testing. This represents around 9% of the maximal number of channels that can be switched off in SE Layers, and 8% of the maximal number of computations that can be avoided with this architecture.

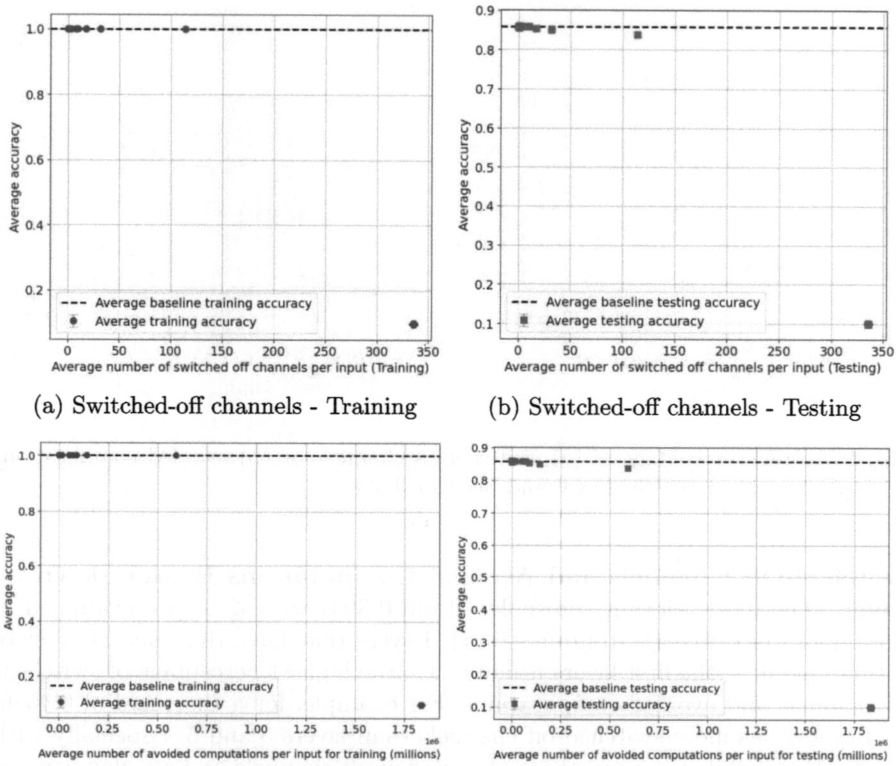

(a) Switched-off channels - Training (b) Switched-off channels - Testing

(c) Avoided computations - Training (d) Avoided computations - Testing

Fig. 3. Relationship between the average total number of switched off channels per input (a and b) or the average total number of avoided computations per input (c and d) and accuracy. Dashed line is the baseline. Markers indicate average across five iterations; Error bars show standard deviation.

If one allows a testing accuracy loss of at most 3% points, the threshold can reach a value of 0.4914, which allows switching off an average of around 114 channels per input at training and testing, while avoiding around 593,938 computations per input at training and 592,300 computations per input at testing. These numbers represent around 34% of the maximal number of channels that can be switched off in SE Layers, and 32% of the maximal number of computations that can be avoided. Results therefore show that a significant amount of computations can be avoided without significant accuracy loss. Notice that performances significantly and quickly drop for higher thresholds.

Experiments were repeated with a reduction ratio r of 8 in the bottleneck. This did not avoid more computations when testing accuracy loss is limited to one percentage point, which is why the discussion is based on $r = 16$. The previous reduction ratio of 16 introduces fewer parameters into the architecture, which helps to reduce the complexity of the CNN [7] and its energy consumption [9].

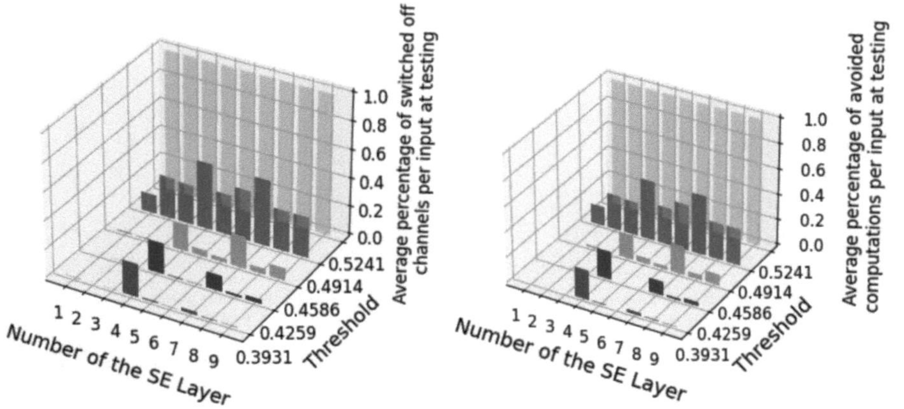

Fig. 4. Average percentage of (a) switched-off channels or (b) avoided computations per input at testing per SE Layer and per threshold.

Switched-Off Channels and Avoided Computations Broken down by Layer. The most relevant thresholds (from 0.3931 to 0.5241) are examined in more detail (see Fig. 4). In groups of SE Layers that have the same number of output channels, the first layers usually have the highest percentage of switched-off channels and avoided computations. For example, for a threshold of 0.4586, SE Layer 4 has more switched-off channels than layers 5 and 6. Especially with smaller thresholds, the percentage of switched-off channels and avoided computations tends to be greater in later layers. This is probably because low-level features from early layers have to be preserved to maintain the CNN's accuracy.

5 Discussion

Experiments showed that ignoring more computations leads to a decrease in accuracy. Once the threshold reaches a certain value, testing and training accuracies drop significantly. This was anticipated since ignoring parts of the CNN usually leads to an initial loss in accuracy [19]. The sudden increase in avoided computations and switched-off channels at threshold 0.5241 can be explained by the fact that from this value, all channels are switched off in the first SE Layer, resulting in the subsequent layers being entirely switched off as well.

Results show that a certain number of channels can be switched off while keeping the accuracy loss within one percentage point of the maximum obtained testing accuracy. In this case, it is possible to avoid around 140,000 computations (i.e., convolutions) per input at training and testing, which can be translated to around 2.4 million arithmetic operations per input during training and inference phases. For less than 3% points of testing accuracy loss, this could reach up to 590,000 avoided computations (i.e., convolutions) per input at training and testing. This accounts for around 13% of the total number of convolutions in the CNN. It translates to around 10 million avoided arithmetic operations (i.e., additions and multiplications) per input during training and testing. As previously mentioned, results are hardly comparable between different techniques. However, this represents a considerable amount of computations that could be avoided, leading to a decrease in energy with appropriate low-level software and hardware implementation.

Achieving meaningful energy savings would likely require a custom CUDA implementation. Specifically, re-implementing arithmetic operations at a lower level is necessary to effectively skip computations involving zero values. This is, however, out of the scope of this article. Custom hardware could also help translate the method's optimisations into tangible energy reduction. By default, the matrix multiplication operation in GPUs for deep learning is not adapted to detect zero values easily. Therefore, all multiplication operations might be performed, even if one of the multiplied values is equal to zero - as is the case when applying T-SE. A dedicated implementation could incorporate zero-detection mechanisms to bypass unnecessary computations. This is expected to enable tangible energy savings, as previous works [16] have shown that conditional computation techniques could lead to energy consumption reduction with appropriate hardware implementation [13].

6 Conclusion

While CNNs gain in popularity, their energy consumption continues to grow. This might interfere with current efforts to reduce carbon emissions and limit climate change. In the literature, several methods have already been explored to speed up CNNs, make them more portable, or increase their performance. However, these methods do not specifically address energy consumption as such. It results in techniques that are not adapted to this goal, which makes them less relevant when it comes to decreasing the environmental impact of CNNs.

To solve this, the present work proposes a new method, Thresholded-Squeeze-and-Excitation (T-SE), aimed at potentially reducing the energy consumption of CNNs. The T-SE approach differentiates itself from previous works, and more specifically from pruning techniques, by being dynamic, input-dependent, and enabling energy consumption reduction during both training and testing phases. This new method leverages Squeeze-and-Excitation Networks, which are CNNs augmented with SE Layers, themselves composed of a *Squeeze* and an *Excitation* operation. A threshold is inserted at the end of each SE Layer, which allows for the selection of the most relevant output channels of a layer and the replacement of the rest of the channels with a value of zero. This method has the potential to decrease computations at both training and testing times. The results show that more than 140,000 computations (i.e., convolutions) per input could be avoided during training and testing while staying within one percentage point of the maximum testing accuracy. Around 450,000 more computations (i.e., convolutions) per input could be avoided at both training and testing times if allowing a loss in testing accuracy of 3% points. This would lead to around 10 million avoided arithmetic operations (i.e., additions and multiplications) per input. The number of avoided convolutions under these conditions corresponds to around 13% of the total number of convolutions in the model. However, for these avoided computations to lead to an actual decrease in energy, an associated low-level software implementation in CUDA should be set up. Even though concrete energy savings have not been presented in this paper, it represents a solid proof of concept as a first step toward more energy-aware CNN architectures.

Based on this work, several avenues for future research can be identified. First of all, the proposed method should be re-implemented using adapted low-level software and adapted hardware to obtain an actual decrease in energy consumption. Energy consumption should preferably be measured using actual hardware measurements, provided there are no constraints for it. If any, energy can be approximated with the CodeCarbon library. The conducted experiments could also be extended to new setups, with larger models or larger datasets. Other avenues for future research include realising a benchmark of existing techniques using the same experimental setup and the same measure for energy, and developing new techniques specifically tailored to CNNs that would allow reducing energy consumption both during the training and the inference phases.

Disclosure of Interests. The authors have no competing interests to declare that are relevant to the content of this article.

References

1. Bengio, E., Bacon, P.-L., Pineau, J., Precup, D.: Conditional computation in neural networks for faster models. In: Workshop track - ICLR (2016)
2. Bolón-Canedo, V., Morán-Fernández, L., Cancela, B., Alonso-Betanzos, A.: A review of green artificial intelligence: towards a more sustainable future. Neurocomputing **599**, 128096 (2024)
3. Bolukbasi, T., Wang, J., Dekel, O., Saligrama, V.: Adaptive neural networks for efficient inference. In: Proceedings of the ICML, vol. 70, pp. 527–536 (2017)
4. Courbariaux, M., Bengio, Y., David, J.-P.: BinaryConnect: training deep neural networks with binary weights during propagations. In: Proceedings of the NIPS, vol. 28, pp. 3123–3131 (2015)
5. Draguet, N., Frénay, B.: Making convolutional neural networks energy-efficient: an introduction. In: Proceedings of the ESANN (2025)
6. Gao, X., Zhao, Y., Dudziak, Ł., Mullins, R., Xu, C.-Z.: Dynamic channel pruning: feature boosting and suppression. In: Proceedings of the ICLR (2019)
7. Hu, J., Shen, L., Sun, G.: Squeeze-and-excitation networks. In: Proceedings of the IEEE CVPR, pp. 7132–7141 (2018)
8. Hubara, I., Courbariaux, M., Soudry, D., El-Yaniv, R., Bengio, Y.: Quantized neural networks: training neural networks with low precision weights and activations. J. Mach. Learn. Res. **18**(187), 1–30 (2018)
9. Iandola, F.N., Han, S., Moskewicz, M.W., Ashraf, K., Dally, W.J., Keutzer, K.: SqueezeNet: AlexNet-level accuracy with 50x fewer parameters and $<$ 0.5 MB model size. arXiv:1602.07360 (2016)
10. LeCun, Y., Denker, J.S., Solla, S.A.: Optimal brain damage. In: Proceedings of the NIPS, vol. 2, pp. 598–605 (1989)

11. Li, P., Yang, J., Islam, M.A., Ren, S.: Making AI less "Thirsty": uncovering and addressing the secret water footprint of AI models. arXiv:2304.03271 (2023)
12. Liu, X., Cao, C., Duan, S.: A low-power hardware architecture for real-time CNN computing. Sensors **23**(4), 2045 (2023)
13. Liu, Z.-G., Whatmough, P.N., Zhu, Y., Mattina, M.: S2TA: exploiting structured sparsity for energy-efficient mobile CNN acceleration. In: Proceedings of the IEEE HPCA, pp. 573–586 (2022)
14. Panda, P., Sengupta, A., Roy, K.: Conditional deep learning for energy-efficient and enhanced pattern recognition. In Proceedings of the DATE, pp. 475–480 (2016)
15. Rashid, N., Demirel, B. U., Abdullah Al Faruque, M.: AHAR: adaptive CNN for energy-efficient human activity recognition in low-power edge devices. IEEE Internet Things J. **9**(15), 13041–13051 (2022)
16. Stamoulis, D., et al.: BinaryConnect: training deep neural networks with binary weights during propagations. In: Proceedings of the ICCAD, pp. 1–8 (2018)
17. Venkataramani, S., Raghunathan, A., Liu, J., Shoaib, M.: Scalable-effort classifiers for energy-efficient machine learning. In: Proceedings of the DAC, pp. 1–6 (2015)
18. Verelst, T., Tuytelaars, T.: Dynamic convolutions: Exploiting spatial sparsity for faster inference. In: Proceedings of the IEEE/CVF CVPR, pp. 2320–2329 (2020)
19. Wen, W., Wu, C., Wang, Y., Chen, Y., Li, H.: Learning structured sparsity in deep neural networks. In: Proceedings of the NIPS, vol. 29, pp. 2082–2090 (2016)
20. Yang, T.-J., Chen, Y.-H., Sze, V.: Designing energy-efficient convolutional neural networks using energy-aware pruning. In: Proceedings of the IEEE CVPR, pp. 6071–6079 (2017)

Detecting Propensity Score Shifts Across Groups in Positive–Unlabeled Data

Illia Tesliuk[1(✉)] and Paweł Teisseyre[1,2]

[1] Polish Academy of Sciences, Warsaw, Poland
tesliukillia@gmail.com, teisseyrep@ipipan.waw.pl
[2] Warsaw University of Technology, Warsaw, Poland

Abstract. Positive and Unlabeled (PU) data arise in many real-world settings where only a subset of positive cases are labeled, while the rest, including both unlabeled positives and true negatives, remain unannotated. This situation is common in domains such as self-reported health events, adverse drug effect reporting, and medical image annotation. A key element required for PU data analysis is modeling the labeling mechanism, which is described by the propensity score function, defined as the probability of labeling a positive observation. Estimating the propensity score is a significant challenge, especially when the labeling mechanism depends on the characteristics of individual instances. In this work, we focus on heterogeneous PU data composed of two groups that may differ in their labeling mechanisms due to external factors such as interventions or annotator differences. We propose a statistical framework to test whether the propensity score functions are equal across groups. Our approach consists of a two-step procedure: a preliminary comparison of labeling proportions, followed by a novel test for the equality of conditional distributions based on McFadden's pseudo-R^2. Through simulation experiments, we demonstrate that our method reliably detects shifts in labeling behavior, offering an effective tool for analyzing group-level heterogeneity in PU data.

Keywords: positive-unlabeled data · propensity score · distribution shift

1 Introduction

In standard binary classification, training data typically consist of both positive and negative observations (PN data), such as sick and healthy patients. In contrast, positive and unlabeled data (PU data) [2] include confirmed positive cases alongside a second group of unlabeled cases, which may in fact be either positive or negative.

PU data often arise naturally in situations where information is collected through reporting. For instance, consider the problem of predicting disease episodes in patients (e.g., migraine attacks) where episodes are self-reported via

M. Baratchi et al. (Eds.): IDA 2026, LNCS 16513, pp. 58–71, 2026.
https://doi.org/10.1007/978-3-032-23833-7_5

mobile applications [12]. Reported attacks can be treated as labeled positives, but many attacks go unreported. Thus, the set of unreported cases contains both unreported positive episodes and true negatives (no attack). It is therefore natural to treat all unreported cases as unlabeled observations. Similar scenarios occur in a variety of domains. For example, in medical imaging, some X-ray scans with pleural effusions (positives) may be overlooked by radiologists, in which case images without a confirmed diagnosis are effectively treated as unlabeled (Fig. 1).

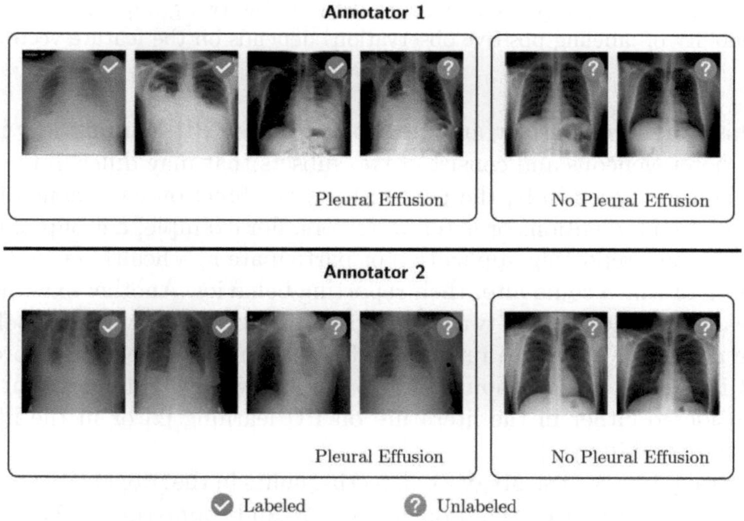

Fig. 1. Chest X-ray scans of lungs with pleural effusion (left) and healthy lungs (right). When annotators are uncertain whether the samples indicate pleural effusion rather than another disease, such scans may remain unlabeled. The annotators may rely on different criteria when labeling, which leads to differences in the labeling mechanisms.

A key challenge in PU data analysis is modeling the mechanism by which positive cases are labeled. This mechanism is described by the propensity score function, the probability that a positive observation receives a label. Many earlier works assume this probability is constant across all positive instances (the Selected Completely at Random assumption, SCAR) [2,4]. A more realistic setting allows the labeling probability to depend on instance-specific features, known as the Selected at Random assumption (SAR) [3,7]. For example, the likelihood of a migraine being reported may vary with age, gender, pain severity, or proficiency with the reporting app. When labeling depends on instance characteristics, the labeled positives are not representative of the full set of positives. Such data are referred to as biased PU data [7]. Estimating the propensity score in this setting is difficult because the complete set of positives is unknown. Nevertheless, several approaches for estimating propensity scores have been developed in recent years [3,5,6,14].

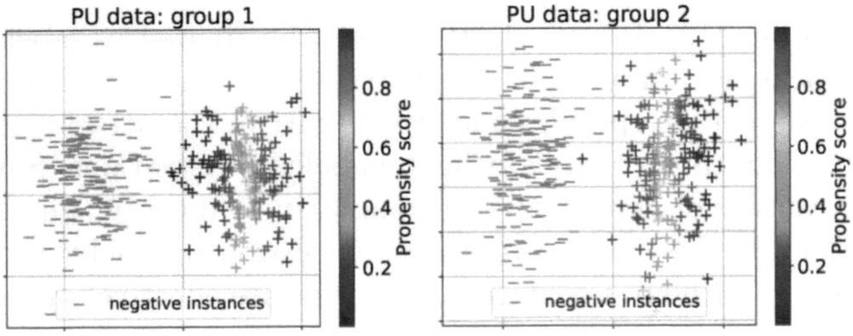

Fig. 2. Propensity score is the probability of labeling a positive instance. The visualization shows the shift in propensity score between the two groups. For both groups, the probability of labeling positive observations depends on the feature vector.

In this work, we consider an extension of biased PU learning in which the data are heterogeneous and consist of two subsets that may differ in their labeling mechanism. Specifically, the propensity score function can change between groups due to interventions or external factors. For example, a group of patients can access a new reporting application or participate in a health awareness campaign, both of which could alter their reporting behavior. Another example arises in medical annotation, where two radiologists may differ in how they label images with the disease (Fig. 1). We refer to such situations as propensity score shifts. Figure 2 illustrates this phenomenon. To our knowledge, this problem has not been considered either in the literature on PU learning [2] or in the literature on distribution shifts [11].

One straightforward strategy for detecting shifts in the propensity score could involve directly estimating the propensity score and comparing it across groups. Alternatively, one could include a group membership indicator as an additional covariate and then check whether it is significant in the model describing the propensity score function. This method requires estimating the propensity score on the pooled data from both groups. A challenge is that existing methods for estimating the propensity score are often computationally intensive or rely on strong, sometimes unrealistic, assumptions. For example, SAR-EM [3], LBE [6], and JERM [14] are based on iterative, computationally expensive EM-type algorithms, where a model is fitted in each iteration. As we will demonstrate later in this paper, direct estimation of the propensity score is not required. Instead, it is sufficient to rely on posterior probabilities of the labeling indicator.

The goal of this paper is to detect whether the labeling mechanism differs between two groups of observations. We address this problem using a statistical hypothesis testing framework, where the null hypothesis states that the propensity scores of the two groups are equal. Our contributions are as follows. 1) We formalize the problem of testing the equality of propensity scores in PU data. 2) We propose a two-step testing procedure: first, a preliminary test compares the proportions of labeled examples between groups; second, we test for equality of conditional distributions. 3) We introduce a novel test of the equality of conditional distributions based on the comparison of McFadden coefficients. 4) We validate the proposed method through simulation experiments, analyzing both statistical power and type I error compared to existing conditional distribution tests such as KL_{DR} [13], WRS [9], MMD [8] and CLF_{2S} [10].

2 Propensity Score Shift Detection

2.1 Propensity Score Shift in PU Learning

In Positive-Unlabeled (PU) learning, each observation is represented by the triple (X, S, Y), where $X \in \mathbb{R}^d$ is the feature vector, $Y \in \{0, 1\}$ is the true class label (with $Y = 1$ indicating the positive class), and $S \in \{0, 1\}$ is label indicator, which indicates whether the instance is labeled. Specifically, $S = 1$ denotes a labeled positive instance, and $S = 0$ denotes an unlabeled instance, which could either be positive or negative. In PU learning, it is assumed that negative examples cannot be labeled, i.e., $P(S = 1|Y = 0) = 0$. The fraction of positive instances that are labeled is determined by the labeling frequency $c = P(S = 1|Y = 1)$. We assume that independent and identically distributed (iid) random vectors (x_i, y_i, s_i) for $i = 1, \ldots, n$ are drawn from an unknown distribution $P(X, Y, S)$. The PU data is given by $\mathcal{D} = \{(x_i, s_i) : i = 1, \ldots, n\}$, since y_i is not directly observed. Labeling mechanism is described by the propensity score function

$$e(x) = P(S = 1|X = x, Y = 1),$$

being the instance-specific probability that a positive example's label will be observed. A high value of the propensity score indicates that a positive observation described by vector x will be assigned a label with a high probability.

Unlike in classical PU learning [2,4], in this work we consider a more general situation in which the PU data are not homogeneous and the propensity score function may change for two different subgroups. Let us consider a variable $T \in \{1, 2\}$ that indicates membership in one of two groups: 1 or 2. We further assume that the propensity score functions may be group-specific, i.e., $e_t(x) := P(S = 1|X = x, Y = 1, T = t)$, for $t = 1, 2$. A propensity score shift occurs if $e_1(x) \neq e_2(x)$. For example, two annotators select positive cases for labeling differently. Formally, we assume that the observed data is $\mathcal{D} = \{(x_i, s_i, t_i) : i = 1, \ldots, n\}$, where t_i corresponds to random variable $T \in \{1, 2\}$.

2.2 Statistical Hypothesis Framework

Our main goal is to test whether propensity score shift occurs. The task is not trivial because the variable Y is not observable, and therefore direct estimation of the propensity score function for the groups and their comparison across groups is problematic. Our goal is to verify the null hypothesis

$$H_0 : P(e_1(X) = e_2(X)) = 1 \tag{1}$$

The testing procedure will be based on the following fact. Let $s_t(x) = P(S = 1|X = x, T = t)$ be the posterior probability for label indicator.

Theorem 1. *Assume that T is conditionally independent from Y given X. Then $e_1(x) = e_2(x)$ if and only if $s_1(x) = s_2(x)$.*

Proof. The conditional independence of T from Y given X implies that the class posterior does not change $P(Y = 1|X = x, T = t) = P(Y = 1|X = x) = y(x)$. Using the law of total probability we have that

$$s_t(x) = P(S = 1|X = x, T = t) = P(S = 1|X = x, Y = 1, T = t)y(x) \\ + P(S = 1|X = x, Y = 0, T = t)(1 - y(x)) = e_t(x)y(x),$$

which follows from the fact that negative observations cannot be labeled in PU learning. The above equality gives the assertion.

The above Theorem thus shows that testing H_0 can be reduced to the problem of testing the equality of conditional distributions for an observed variable S. This is a well-recognized problem in both the statistics and machine learning literature [8–10,13].

The simplest possible propensity score shift occurs when the propensity score function is constant across groups: $e_1(x) = c_1$ and $e_2(x) = c_2$, for some constants $c_1, c_2 \in (0, 1)$. This corresponds to the so-called SCAR (Selected Completely at Random) assumption in each group. In this situation, testing H_0 can be reduced to testing for equality of proportions $P(S = 1|T = 1) = P(S = 1|T = 2)$, which follows from the following Lemma. The proof is analogous to the proof of Theorem 1; it suffices to omit conditioning on $X = x$.

Lemma 1. *Assume that T is independent from Y and that $e_1(x) = P(S = 1|Y = 1, T = 1) = c_1$, $e_2(x) = P(S = 1|Y = 1, T = 2) = c_2$. Then H_0 is true if and only if $P(S = 1|T = 1) = P(S = 1|T = 2)$.*

However, if the SCAR assumption is not satisfied in at least one of the groups, it may happen that H_0 is not true, even though the equality holds $P(S = 1|T = 1) = P(S = 1|T = 2)$. The following example illustrates such a situation. Assume that propensity score in the first group is constant $e_1(x) = P(S = 1|Y = 1, T = 1) = c$, whereas the propensity score in the second group is described by sigmoid activation function $e_2(x) = \sigma(\gamma + \theta^T x)$, for some parameters $\gamma \in R$ and $\theta \in R^p$, which is a common assumption used in many propensity score estimation methods [3,6,14]. For a fixed value of θ, one can determine such a value of γ that the labeling frequencies in both groups are the same, i.e. $P(S = 1|Y = 1, T = 2) = c$. Since $P(S = 1|Y = 1, T = 2) = E_X e_2(X)$, it is enough to determine such γ that the equality $c = E_X \sigma(\gamma + \theta^T x)$ with respect to γ holds. Assuming independence of T and Y, the equality of labeling frequencies in both groups implies the equality $P(S = 1|T = 1) = P(S = 1|T = 2)$.

Finally, note that the assumptions of Theorem 1 and Lemma 1 are satisfied in the situation where T is completely independent of (X, Y), which holds in typical randomized experiments where group assignment is random.

2.3 Proposed Two-Step Test

The above discussion leads to a novel two-step procedure called PSST (**P**ropensity **S**core **S**hift **T**esting). The main idea is to verify the hypothesis of equality of proportions of positive observations in two groups in the first step. Rejection of the hypothesis of equality of proportions is sufficient to detect a shift in the propensity score. In the case of equality of proportions, we perform a second test in which we verify the equality of conditional distributions. The motivation is that the first test is very fast and if we reject the first null hypothesis, there is no need to perform the second test which is more computationally complex. The whole procedure is described in Algorithm 1. Below we describe the details.

Algorithm 1: Propensity **S**core **S**hift **T**esting (PSST)

Input: PU training data $\mathcal{D} = \{(x_i, s_i, t_i) : i = 1\ldots, n\}$, significance level α, number of repetitions K
Significance level of the single test $\alpha_0 = 1 - \sqrt{1-\alpha}$
Compute test statistic $\widehat{T}_A$ and p-value$_A$
if p-value$_A < \alpha_0$ **then**
    $A = 1$
else
    Compute test statistic $\widehat{T}_B$ and p-value$_B$
    for $k = 1, \ldots, K$ **do**
        Generate the set $\mathcal{D}^{(k)}$ from the set $\mathcal{D}$ by randomly permuting the rows
        Split data $\mathcal{D}^{(k)} = \mathcal{D}_1^{(k)} \cup \mathcal{D}_2^{(k)}$
        Use $\mathcal{D}_1^{(k)}$ and $\mathcal{D}_2^{(k)}$ to compute $R_1^{(k)}, R_2^{(k)}$, respectively using (2)
        Compute the test statistic $\widehat{T}_B^{(k)} = |R_1^{(k)} - R_2^{(k)}|$
    end for
    Compute p-value$_B = \left(1 + |\{1 \le k \le K : \widehat{T}_B^{(k)} > \widehat{T}_B\}|\right)/(1+K)$
    if p-value$_B < \alpha_0$ **then**
        $A = 1$
    else
        $A = 0$
    end if
end if
Output: Detect propensity score shift iff $A = 1$.

Let $\mathcal{D} = \mathcal{D}_1 \cup D_2$, where $D_t = \{(x_i, s_i) : t_i = t\}$, where $t_i \in \{1, 2\}$ indicates that the i-th observation belongs to one of the groups 1 or 2. First, we perform a test to verify a null hyphotesis $H_0^A : P(S = 1 \mid T = 1) = P(S = 1 \mid T = 2)$. We can employ a well known test for equality of proportions [15] using test statistic

$$\widehat{T}_A = \frac{\widehat{P}(S = 1 \mid T = 1) - \widehat{P}(S = 1 \mid T = 2)}{\sqrt{(\frac{1}{|\mathcal{D}_1|} + \frac{1}{|D_2|})a(1-a)}},$$

where

$$a = \frac{|\mathcal{D}_1|\widehat{P}(S=1 \mid T=1) + |\mathcal{D}_2|\widehat{P}(S=1 \mid T=2)}{|\mathcal{D}_1| + |\mathcal{D}_2|}$$

and $\widehat{P}(S = 1 \mid T = t)$ is a fraction of labeled examples in group $t = 1, 2$. The corresponding p-value is denoted by p-value$_A$.

In the case where we do not reject H_0^A, it may still happen that a propensity score shift occurs, which is discussed at the end of the previous section. Therefore, if H_0^A is not rejected, we perform the additional test to verify $H_0^B : P(s_1(X) = s_2(X))$, which is more computationally demanding. We propose the new test statistic which is based on comparing McFadden coefficients of determination [1] between two groups. It is defined as follows. Consider empirical risk function for group $t = 1, 2$:

$$\widehat{\mathcal{L}}_t(\gamma, \theta) = -\frac{1}{|\mathcal{D}_t|}\sum_{i \in \mathcal{D}_t} s_i \log \sigma(\gamma + \theta^T x_i) + (1 - s_i) \log[1 - \sigma(\gamma + \theta^T x_i)]$$

and its minimizer $\widehat{\gamma}_t, \widehat{\theta}_t = \arg\min_{\gamma,\theta} \widehat{\mathcal{L}}_t(\gamma, \theta)$. Moreover, the risk function for the null model is $\widehat{\mathcal{L}}_t^{null}(\gamma) = -|\mathcal{D}_t|^{-1}\sum_{i \in \mathcal{D}_t} s_i \log \sigma(\gamma) + (1 - s_i) \log[1 - \sigma(\gamma)]$ and its minimizer is $\widehat{\gamma}_t^{null} = \arg\min_{\gamma} \widehat{\mathcal{L}}_t^{null}(\gamma)$. McFadden coefficient of determination

$$\widehat{R}_t = 1 - \frac{\widehat{\mathcal{L}}_t(\widehat{\gamma}_t, \widehat{\theta}_t)}{\widehat{\mathcal{L}}_t^{null}(\widehat{\gamma}_t^{null})} \qquad (2)$$

is based on a comparison of the empirical risk function for a model that includes feature vector x_i and a null model that ignores x_i. In case of $s_1(x) \neq s_2(x)$, we expect a discrepancy between the values of R_1 and R_2. Hence, the final test statistic is defined as $\hat{T}_B = |\hat{R}_1 - \hat{R}_2|$. To determine the distribution of the test statistic under the null hypothesis H_0^B we use the bootstrap method, as described in Algorithm 1.

In the above procedure we perform a sequence of two tests, therefore it is important to determine the significance level α of the entire procedure. Assuming that the significance level of the individual tests H_0^A and H_0^B is α_0 and assuming that the corresponding test statistics are independent, the probability of rejecting the null hypothesis is $\alpha = \alpha_0 + (1 - \alpha_0)\alpha_0$. Solving the simple quadratic equation we obtain that the significance level of the individual test should be $\alpha_0 = 1 - \sqrt{1 - \alpha}$.

In practice, the two tests are not independent. Moreover, different significance levels may be considered for each of the two tests. However, assuming independence and equal significance levels allows for a simple derivation of α_0. Importantly, our experiments show that, despite the lack of independence, the resulting α_0 enables effective control of the overall rejection rate of the procedure at the nominal level α in practice.

3 Experiments

3.1 Experimental Setup

The methods are assessed on 3 benchmark datasets (Spambase, CIFAR10 and MNIST) and two artificial datasets. We construct tabular versions of CIFAR10 and MNIST by extracting ImageNet-pretrained ResNet-50 embeddings and selecting the 50 most informative features (using simple mutual information-based filter), with vehicles and even digits serving as the positive classes, respectively. In artificial datasets, feature vectors x are generated from multivariate normal distributions. Subsequently, we generate binary target variable Y. In the artificial linear dataset, $Y \in \{0, 1\}$ is generated according to the probability $P(Y = 1|X = x) = \sigma(\theta^T x)$, where $\theta = (1, \dots, 1)$. In the artificial non-linear dataset, $P(Y = 1|X = x) = 0.2 \cdot I(x^{(1)} \in [-0.5, 0.5]) + 0.8 \cdot I(x^{(1)} \notin [-0.5, 0.5])$, where I is indicator function, and $x^{(1)}$ is the first coordinate of vector x.

From each of the original datasets, we randomly sampled two subsets, each containing 2000 observations. Next, from each of the two datasets, we generate PU data by selecting a subset of positive observations for labeling using the corresponding functions $e_1(x)$ and $e_2(x)$, described below. We consider four cases of propensity score shift between groups.

- **Scheme S1**: $e_1(x) = 0.1$ and $e_2(x) = 0.1 + g$, where g is a parameter.
- **Scheme S2**: $e_1(x) = 0.5$ and $e_2(x) = \sigma(\gamma + g \cdot \theta^T x)$, where $\theta = (1, \dots, 1)$ and γ is intercept.
- **Scheme S3**: $e_1(x) = \sigma(\gamma + \theta^T x)$ and $e_2(x) = \sigma(\gamma + (1 + g) \cdot \theta^T x)$, where $\theta = (1, \dots, 1)$ and γ is intercept.
- **Scheme S4**: $e_1(x) = F_{Cauchy}(\gamma + \theta^T x)$ and $e_2(x) = F_{Cauchy}(\gamma + (1+g) \cdot \theta^T x)$, where $F_{Cauchy}(\cdot)$ is CDF of the Cauchy distribution with location 0 and scale 1, whereas $\theta = (1, \dots, 1)$ and γ is intercept.

In Schemes S2-S4, intercept γ is chosen in such a way that $P(S = 1|Y = 1, T = t) = 0.5$.

Note that the value $g = 0$ means that there is no shift, while increasing g increases the discrepancy between the propensity scores between groups. Scheme S1 corresponds to the SCAR assumption in both groups. Schemes S2-S4 are more challenging because the labeling frequency is the same in both groups, making it impossible to detect the shift using a simple test for equality of proportions. This allows us to evaluate the effectiveness of the second test in the PSST method.

Schemes S3 and S4 are similar, however, in S4 we use a flatter activation function. Let us note that although in scheme S3 the propensity score function is a logistic activation function, the posterior probability for the labeling indicator $s_t(x)$ will not be described by a logistic function, because $s_t(x) = e_t(x)y(x)$ (see the proof of Theorem 1), and $y(x) = P(Y = 1|X = x)$ can be an arbitrary function. Thus, the logistic model that we use in the PSST procedure will be misspecified. This shows that in none of the considered labeling schemes do we favor our method.

We compare PSST against KL_{Boot}, which combines Kullback–Leibler divergence with bootstrap technique used to determine p-values, as well as four conditional distribution tests KL_{DR} [13], CDD [18], MMD [8], and CLF_{2S} [10], to evaluate their effectiveness in detecting propensity score shifts.

3.2 Power and Type I Error Analysis

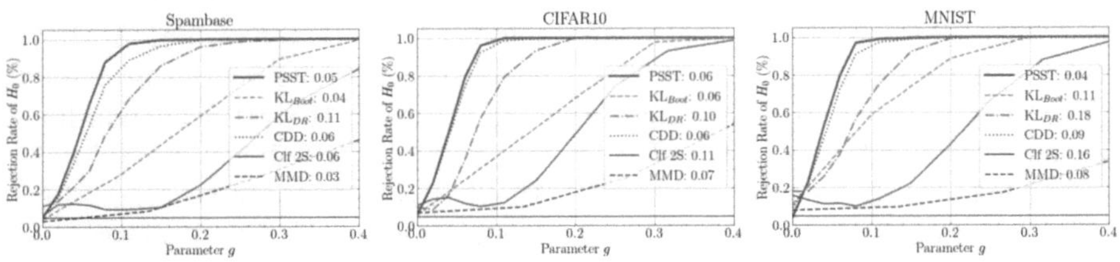

Fig. 3. Rejection rate as a function of shift magnitude g for the S1 scheme. The numbers in the legend indicate rejection rates for $g = 0$.

In the first experiment, we examine the dependence of the rejection rate on the magnitude of the shift. The significance level is set to $\alpha = 0.05$. We compare the considered methods in their ability to detect distributional changes, with the best method being the one that can identify shifts at lower magnitudes of g. Across the three propensity score shift schemes, PSST consistently demonstrates strong performance. In the first scheme, PSST and CDD outperform the other methods, showing rapid growth in rejection rates and achieving full detection at smaller shift levels (Fig. 3). PSST consistently maintains type I error below the 0.05 threshold, keeping the number of false positives low (the rejection rates for $g = 0$ are given in the legend). In the more complex scheme S2, PSST also achieves high rejection rates across datasets (Fig. 4).

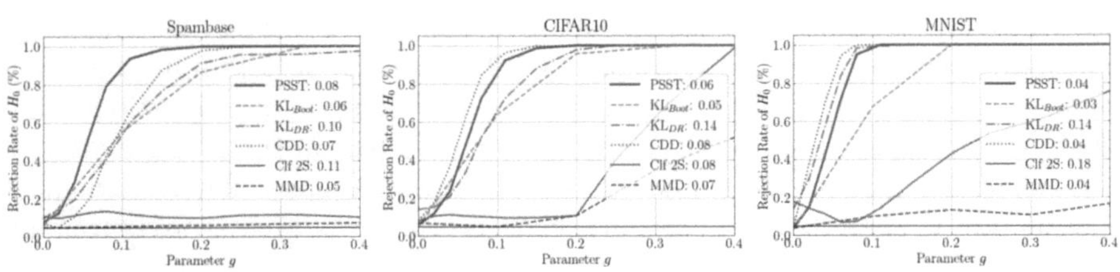

Fig. 4. Rejection rate as a function of shift magnitude g for the S2 scheme. The numbers in the legend indicate rejection rates for $g = 0$.

In the most challenging schemes, S3 and S4, PSST consistently outperforms competing methods, achieving full rejection more rapidly across all datasets

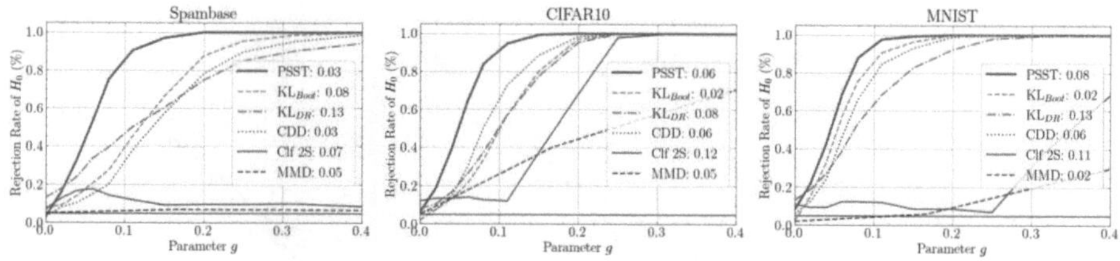

Fig. 5. Rejection rate as a function of shift magnitude g for the S3 scheme. The numbers in the legend indicate rejection rates for $g = 0$.

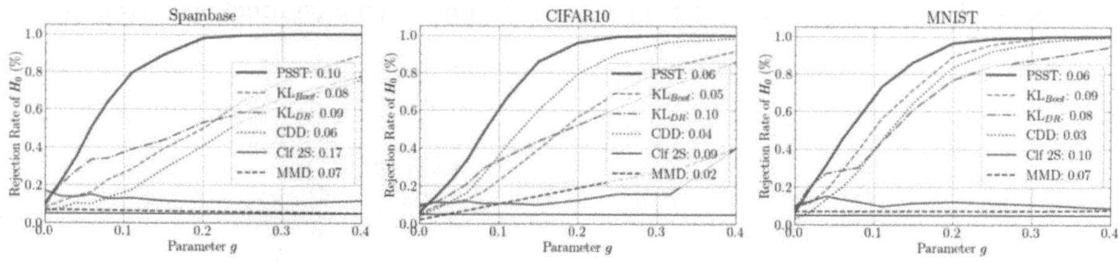

Fig. 6. Rejection rate as a function of shift magnitude g for the S4 scheme. The numbers in the legend indicate rejection rates for $g = 0$.

(Figs. 5 and 6, respectively). Overall, across all four labeling schemes, PSST consistently ranks among the top two or three methods and frequently outperforms the baselines, demonstrating both robustness and effectiveness in detecting distribution shifts.

In the second experiment, we analyzed type I error rates, that is, the probability of incorrectly rejecting the null hypothesis in the absence of a shift (i.e., for $g = 0$). We evaluated hypothesis rejection rates across a range of significance levels from 0 to 1. Ideally, each method's rejection rate should match the chosen α, though minor deviations are acceptable. Overall, with the exception of CLF_{2S}, all methods produce appropriate rejection curves across the datasets (Fig. 7). The PSST generally maintains an approximately linear relationship between rejection rate and α, with minor fluctuations similar to those observed for the other methods.

Finally, the effect of sample size on test power is analyzed (Fig. 8). Using $g = 0.1$, we generate linear and non-linear *Artificial* datasets, with sizes of 200–10000. For most methods, the power increases with the sample size. In the case of the non-linear data set, a clear advantage of the PSST method over other competitors is observed. In this case, MMD and CLF_{2S} perform ineffectively, even for large sample sizes.

3.3 Application to Medical Image Data

Finally, we evaluate our method, PSST, alongside the two most promising baselines, KL_{Boot} and CDD, in a simulated experiment using real chest X-ray scans from the ChestMNIST dataset [16,17]. Although each image is annotated with

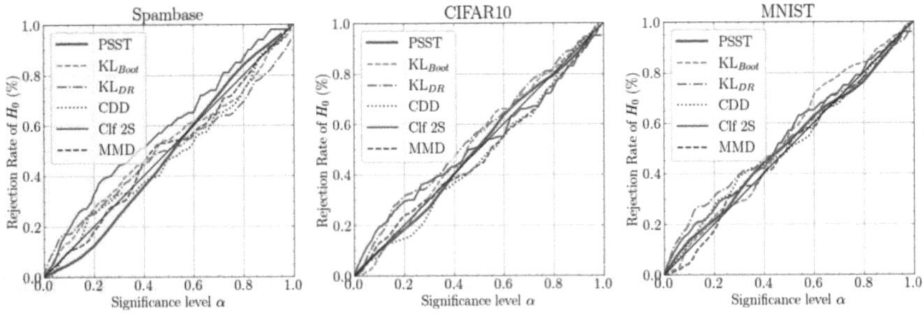

Fig. 7. Type I Error Analysis: rejection rate for varying significance levels α.

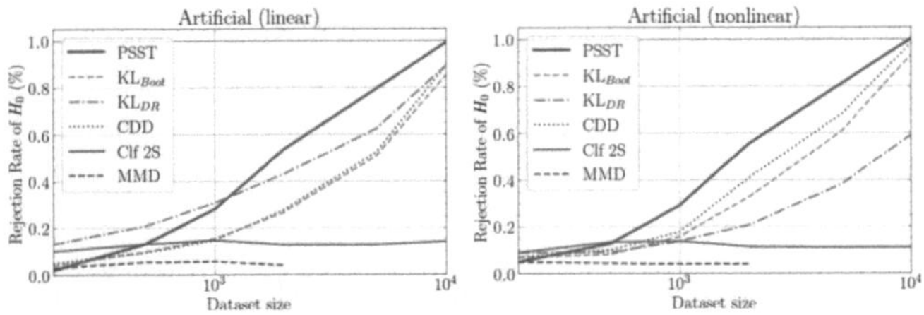

Fig. 8. Rejection rate as a function of sample size for $g = 0.1$ under the S3 scheme.

one or more of 14 thoracic diseases, we focus exclusively on a single disease, *infiltration*, thereby converting the task into a binary classification problem and labeling patients diagnosed with infiltration as the positive class ($Y = 1$).

To simulate PU data, we define a labeling mechanism for positive observations that depends on the presence of two additional diseases, *atelectasis* (Z_1) and *effusion* (Z_2). We consider two groups of images (A and B), corresponding to two radiologists, for whom the labeling behavior may potentially differ. In the case where only the target disease (infiltration) is present, and neither of the two co-occurring diseases is observed ($Z_1 = Z_2 = 0$), the images are always labeled:

$$P(S = 1 \mid Y = 1, Z_1 = 0, Z_2 = 0) = 1. \tag{3}$$

However, when at least one of the selected co-occurring diseases is present ($Z_1 + Z_2 > 0$), the probability of labeling a positive observation is controlled by a group-specific parameter p_i:

$$P(S = 1 \mid Y = 1,\ Z_1 + Z_2 > 0,\ T = i) = p_i, \quad i \in \{A, B\}. \tag{4}$$

Here, p_A and p_B are parameters whose values we vary across experiments. Naturally, whenever $p_A \neq p_B$, we observe a propensity score shift. A low value of p_A or p_B indicates that the radiologist fails to diagnose infiltration, attributing the observed pathological changes to one of the two other diseases. On the other hand, a high value of the parameter indicates that the radiologist is confident in diagnosing the target disease, despite the presence of co-occurring conditions, which may suggest greater experience of the radiologist.

Scenarios in which radiologists fail to detect a disease in an affected lung (false negatives) are not considered in the simulation. Note that the labeling mechanism depends on external factors (co-occurring diseases) that are not observed in the training data, which distinguishes the above situation from those described by the S1–S4 schemes and makes it more challenging.

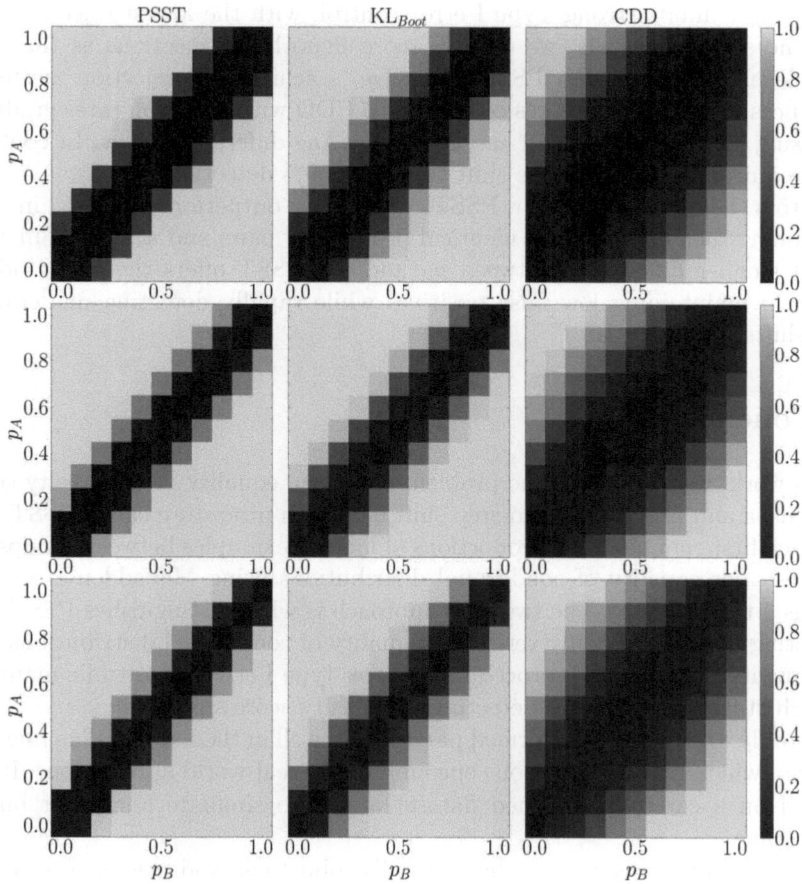

Fig. 9. Rejection rates computed on ChestMNIST subset at $\alpha = 0.005$ (row 1), 0.0025 (row 2), and 0.05 (row 3). Each cell corresponds to a single pair of parameter values, p_A and p_B, representing the probabilities of diagnosing the *infiltration* disease in groups A and B when at least one co-occurring disease is present.

The X-ray images are tabularized using the same embedding and feature selection procedure as applied to MNIST and CIFAR-10. We consider 11 values for each parameter p_A and p_B, resulting in a total of 121 possible combinations. We perform 100 simulations for each parameter pair and each method.

An ideal method would produce a dark blue diagonal, indicating zero rejections when $p_A = p_B$, and yellow off-diagonal cells, indicating a 100% rejection rate when $p_A \neq p_B$ (Fig. 9). We expect the optimal method to achieve high rejection rates even for the smallest differences between p_A and p_B.

Rejection occurs when the p-value falls below the specified significance level α. While our primary interest lies in a conventional $\alpha = 0.05$ (Fig. 9, third row), we also report results for more rigorous $\alpha = 0.005$ and $\alpha = 0.0025$ to assess consistency across different thresholds.

KL_{Boot} exhibits strong Type I error control, with the lowest rejection rates under no shift. Naturally, we observe more hypothesis rejections as the confidence level increases. Both PSST and KL_{Boot} achieve full rejection at smaller differences in propensity scores compared to CDD with rejection rates gradually increasing for small shifts and accelerating as the differences grow. In contrast, CDD requires nearly twice the shift to reach 100% detection.

Across the entire range of α, PSST consistently outperforms KL_{Boot} in rejection power when compared at identical probability pairs and achieves full rejection at smaller differences between p_A and p_B. PSST offers the best trade-off overall in maintaining low false positives while rapidly detecting moderate to large shifts.

4 Conclusions

In this work, we formalized the problem of testing equality of propensity scores in PU data and proposed a two-step shift detection procedure called PSST. Our approach first compares the proportions of labeled examples between groups and then tests for equality of conditional distributions using McFadden's R coefficient as a test statistic. The two-step approach is what distinguishes PSST from competing tests, which only verify the equality of conditional distributions. Our experiments show that the procedure controls type I error rates while maintaining high statistical power in detecting propensity score shifts.

Notably, the method performed particularly well in the most challenging shift scenario, which reflects situations encountered in real-world applications. It also excelled on a carefully designed dataset aimed to simulate annotator bias in X-ray scans.

These results demonstrate the method's robustness and practical utility for detecting subtle and complex shifts in PU data, making it a promising approach for real-world distribution shift detection.

The proposed method can be extended in future research to more than two groups, which will require addressing the multiple testing problem and introducing appropriate adjustments, such as the Bonferroni correction.

Disclosure of Interests. The authors have no competing interests to declare that are relevant to the content of this article.

References

1. Agresti, A.: Categorical Data Analysis. Wiley, 3rd edn. (2012)
2. Bekker, J., Davis, J.: Learning from positive and unlabeled data: a survey. Mach. Learn. **109**, 719–760 (2020)
3. Bekker, J., Robberechts, P., Davis, J.: Beyond the selected completely at random assumption for learning from positive and unlabeled data. In: Proceedings of the 2019 European Conference on Machine Learning and Principles and Practice of Knowledge Discovery in Databases, ECML 2019, pp. 71–85 (2019)
4. Elkan, C., Noto, K.: Learning classifiers from only positive and unlabeled data. In: Proceedings of the 14th ACM SIGKDD International Conference on Knowledge Discovery and Data Mining, KDD 2008, pp. 213–220 (2008)
5. Furmańczyk, K., Mielniczuk, J., Rejchel, W., Teisseyre, P.: Double logistic regression approach to biased positive-unlabeled data. In: Proceedings of the European Conference on Artificial Intelligence, ECAI 2023, pp. 764–771 (2023)
6. Gong, C., Wang, Q., Liu, T., Han, B., You, J., Yang, J., Tao, D.: Instance-dependent positive and unlabeled learning with labeling bias estimation. IEEE Trans. Pattern Anal. Mach. Intell. 1–16 (2021)
7. Gong, C., Zulfiqar, M.I., Zhang, C., Mahmood, S., Yang, J.: A recent survey on instance-dependent positive and unlabeled learning. Fundamental Res. (2022)
8. Gretton, A., Borgwardt, K.M., Rasch, M.J., Schölkopf, B., Smola, A.: A kernel two-sample test. J. Mach. Learn. Res. **13**(1), 723–773 (2012)
9. Hu, X., Lei, J.: A two-sample conditional distribution test using conformal prediction and weighted rank sum. J. Am. Stat. Assoc. **119**(546), 1136–1154 (2024)
10. Lopez-Paz, D., Oquab, M.: Revisiting classifier two-sample tests. In: 5th International Conference on Learning Representations. In: ICLR (2017)
11. Moreno-Torres, J.G., Raeder, T., Alaiz-Rodriguez, R., Chawla, N.V., Herrera, F.: A unifying view on dataset shift in classification. Pattern Recogn. **45**(1), 521–530 (2012)
12. Park, J.W., Chu, M.K., Kim, J.M., Park, S.G., Cho, S.J.: Analysis of trigger factors in episodic migraineurs using a smartphone headache diary applications. PLoS ONE **11**(2), 1–13 (2016)
13. Polo, F.M., Izbicki, R., Lacerda, E.G., Jr., Ibieta-Jimenez, J.P., Vicente, R.: A unified framework for dataset shift diagnostics. Inf. Sci. **649**, 119612 (2023)
14. Rejchel, W., Teisseyre, P., Mielniczuk, J.: Joint empirical risk minimization for instance-dependent positive-unlabeled data. Knowl.-Based Syst. **304**, 1–21 (2024)
15. Snedecor, G.W., Cochran, W.G.: Statistical Methods. Wiley-Blackwell (2009)
16. Yang, J., Shi, R., Ni, B.: Medmnist classification decathlon: a lightweight automl benchmark for medical image analysis. In: IEEE 18th International Symposium on Biomedical Imaging (ISBI), pp. 191–195 (2021)
17. Yang, J., et al.: Medmnist v2-a large-scale lightweight benchmark for 2d and 3d biomedical image classification. Sci. Data **10**(1), 41 (2023)
18. Yu, S., Shaker, A., Alesiani, F., Principe, J.C.: Measuring the discrepancy between conditional distributions: methods, properties and applications. In: Proceedings of the Twenty-Ninth International Joint Conference on Artificial Intelligence, IJCAI 2020 (2020)

On Sample-Wise Strict Monotonicity with a Gradient Update

O. Taylan Turan[1(✉)], Marco Loog[2], and David M. J. Tax[1]

[1] Delft University of Technology, Delft, The Netherlands
{o.t.turan,d.m.j.tax}@tudelft.nl
[2] Radboud University, Nijmegen, The Netherlands
marco.loog@ru.nl

Abstract. Learning curves describe how the performance of a model evolves with increasing training data. Although more data is generally expected to improve model performance, in practice models can exhibit non-monotonic behavior where additional data leads to performance degradation. Sample-wise double descent is one particular example. We address the question of how a learner can have a provably monotone learning curve. For isotropic Gaussian covariates under a Gaussian noise model and a linear predictor, we prove that a single step of steepest descent guarantees sample-wise monotonicity in the learning curve, if the step size does not exceed an upper bound. Furthermore, we present a practical procedure that ensures monotonicity without explicit regularization or cross-validation, using initialization from the previous training set size. Experiments on real-world datasets demonstrate that this method achieves monotone behavior and improved sample efficiency compared to ordinary least squares and optimally regularized ridge regression. We also explore extensions to binary classification, where monotonicity depends on the chosen performance metric. While our guarantees are derived under simplifying assumptions, they provide both theoretical and practical insights for constructing monotone learners and for understanding and mitigating sample-wise double descent behavior.

Keywords: Learning curve · Expected performance · Double descent · Monotonicity · Linear regression

1 Introduction

Does additional data always help? Intuitively, it should, but in practice, for some problems and some models, it does not. There are classification and regression problems both in supervised [9] and in semi-supervised learning [7] where the expected performance can get worse with additional training data. In general, it is desirable to have models get better with data, as it makes it easier to predict, for a given level of model performance, how much more data should be collected. This dependence of expected performance on training set size is studied using learning curves. Learning curves, which plot expected performance as a function

M. Baratchi et al. (Eds.): IDA 2026, LNCS 16513, pp. 72–83, 2026.
https://doi.org/10.1007/978-3-032-23833-7_6

of training set size, should not be confused with training curves commonly used in the deep learning literature, where performance is shown as a function of training epochs.

Although this observation of non-monotonic behaviour dates back to 1989 [9], its connection to deep learning [1] and to questions of model complexity renewed interest in it. In [3], the double descent phenomenon with respect to model complexity is examined in detail for linear regression, trees, and boosting. Their analysis shows that the conclusions drawn in [1] were rather flawed, because multiple axes of complexity were conflated in the original study.

We are concerned with the sample-wise double descent phenomenon for linear regression which exhibits an unconventional bias-variance trade-off: as the number of samples increases, the bias decreases while the variance increases [10]. Despite this atypical behavior, optimal ridge regularization has been shown to mitigate the effect [11].

We show that for ordinary least squares (OLS) the problem of double descent along learning curves can also be mitigated without explicitly regularizing the loss function. First, we prove that for an initial parameter vector and step size independent of the training points, if only one gradient step is taken, the expected performance decreases monotonically provided that the step size does not exceed a certain value. We also present, for Gaussian covariates, what this maximum step size is that ensures sample-wise monotonicity. Although this proof is under strict assumptions about the data-generating process, we propose a way to make monotone linear regression models. In this approach, we still take one gradient update, but with exact line search starting from the parameters of the average estimator obtained from the previous training set size.

An illustration of our method is shown in Fig. 1, which displays the learning curves of several models on isotropic Gaussian data. In this plot, the green curve corresponds to our proposal, where only one gradient update is performed using exact line search, starting from the average estimator of the previous training set size (a warm start). The orange curve depicts the optimally regularized ridge estimator proposed in [11], and the blue curve shows the standard OLS estimator. As seen in the same figure, the OLS estimator exhibits sample-wise double descent behavior, which disappears under optimal regularization. Likewise, our proposed estimator does not display this behavior.

Similar efforts on making general learners more monotone have been undertaken in [2,16]. Both of these works make the learners monotone via a smart selection strategy. Our proposal is not model-agnostic as we restrict the model to linear regression. We are only focused on making linear regression problems monotone. Moreover, we do not require a strategy other than using the previous training set size for the initial guess. Hence, the closest work to our effort is [11], where, similarly, a supervised linear regression problem is proven to be monotone with optimal regularization. In practice, the optimal regularization parameter is of course not available and one needs to resorts to techniques like cross-validation to estimate it.

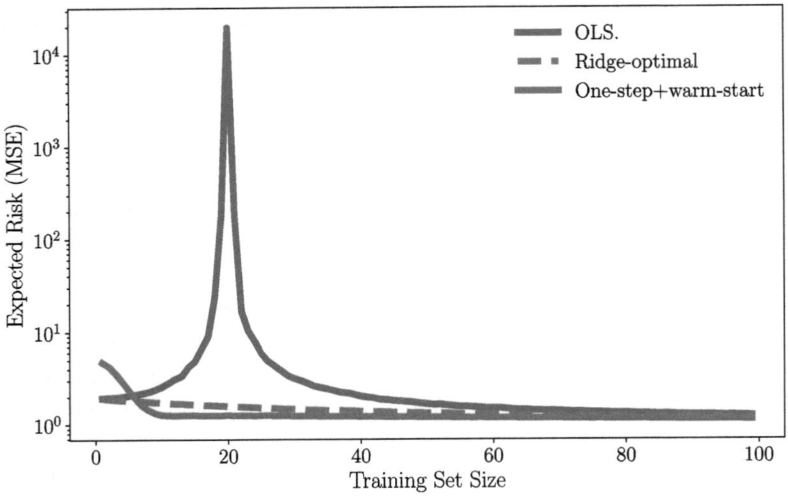

Fig. 1. Expected MSE with respect to the training set size n, where inputs come from an isotropic Gaussian $\boldsymbol{x} \sim \mathcal{N}(0, I_d)$ and outputs follow $y = \boldsymbol{\beta}^{*\top}\boldsymbol{x} + \mathcal{N}(0, \sigma^2)$, with $d = 20$, $\sigma^2 = 1$, and $\|\boldsymbol{\beta}^*\|_2 = 1$.

Our paper is organized as follows. Section 2 introduces expected risk and monotonicity along learning curves for the linear regression problem, and we show that for the optimal step size, the expected risk decreases monotonically. In Sect. 3, we make use of our theoretical findings to create monotone learners in practice. Finally, we reflect on our findings and point out open questions.

2 Strict Monotonicity for One-Step Gradient Update

We consider the linear regression problem where the input $\boldsymbol{x} \in \mathbb{R}^d$ is from a random distribution and the output (or response) is generated according to the model:

$$y = \boldsymbol{\beta}^{*\top}\boldsymbol{x} + \varepsilon, \tag{1}$$

where ε is random noise, and $\boldsymbol{\beta}^* \in \mathbb{R}^d$ is an unknown parameter vector. Let $\mathcal{D}$ denote the joint distribution of $(\boldsymbol{x}, y)$. In practice, we are given a training dataset $\mathcal{D}_n := \{(\boldsymbol{x}_i, y_i)\}_{i=1}^n$, drawn independently from $\mathcal{D}$. The goal is to learn a linear model $f_{\boldsymbol{\beta}}(\boldsymbol{x}) = \boldsymbol{\beta}^\top\boldsymbol{x}$, with $\boldsymbol{\beta} \in \mathbb{R}^d$ that minimizes the population mean-squared error:

$$\mathcal{R}(\boldsymbol{\beta}) := \mathop{\mathbb{E}}_{(\boldsymbol{x},y)\sim\mathcal{D}}\left[\left(\boldsymbol{\beta}^\top\boldsymbol{x} - y\right)^2\right]. \tag{2}$$

Let's assume $\boldsymbol{X} \in \mathbb{R}^{n\times d}$ to be the data matrix whose rows are the covariates $\mathbf{x}_i^\top$, and let $\boldsymbol{y} \in \mathbb{R}^n$ be the corresponding vector of responses. For any estimator $\hat{\boldsymbol{\beta}}_n = \boldsymbol{\beta}_n(\boldsymbol{X}, \boldsymbol{y})$ expected risk is defined as:

$$\bar{\mathcal{R}}_n := \mathop{\mathbb{E}}_{\boldsymbol{X},\boldsymbol{y}\sim\mathcal{D}}\left[\mathcal{R}(\hat{\boldsymbol{\beta}}_n)\right]. \tag{3}$$

Assume input data comes from a zero mean isotropic Gaussian distribution $\boldsymbol{x} \sim \mathcal{N}(\mathbf{0}_d, \mathbf{I}_d)$ and the noise is homoscedastic and Gaussian $\varepsilon \sim \mathcal{N}(0, \sigma^2)$. We consider ordinary least-squares estimator (OLS) with one gradient update step. Recall that the OLS estimator is given by

$$\hat{\boldsymbol{\beta}}_n := \arg\min_{\boldsymbol{\beta}} \left\{ \|\boldsymbol{X}\boldsymbol{\beta} - \boldsymbol{y}\|_2^2 \right\}. \tag{4}$$

Defining the Gram matrix as $\boldsymbol{\Sigma}_n = \boldsymbol{X}^\top \boldsymbol{X}$, the gradient for the OLS objective is given by,

$$\boldsymbol{r}_n := \boldsymbol{X}^\top \boldsymbol{y} - \boldsymbol{\Sigma}_n \bar{\boldsymbol{\beta}}. \tag{5}$$

If $\bar{\boldsymbol{\beta}}$ is the initial guess, then our estimator takes the following form:

$$\hat{\boldsymbol{\beta}}_n = \bar{\boldsymbol{\beta}} + \alpha \boldsymbol{r}_n, \tag{6}$$

where α is the step size. Finally, we assume that initial guess $\bar{\boldsymbol{\beta}}$ and step size α are independent of covariates and corresponding responses. Our starting point is to obtain the expected risk given in Eq. 3 with $\hat{\boldsymbol{\beta}} = \bar{\boldsymbol{\beta}} + \alpha(\boldsymbol{X}^\top \boldsymbol{y} - \boldsymbol{X}^\top \boldsymbol{X} \bar{\boldsymbol{\beta}})$ and we represent the difference between the initial guess and the optimal parameters as $\widetilde{\boldsymbol{\beta}} = \bar{\boldsymbol{\beta}} - \boldsymbol{\beta}^*$.

Theorem 1. *For the above setting, the expected risk of well-specified isotropic linear regression problem is sample-wise monotone, that is $\bar{\mathcal{R}}_{n+1} \leq \bar{\mathcal{R}}_n$, if $0 \leq \alpha \leq \alpha_{max}$, where*

$$\alpha_{max} := \frac{2\|\widetilde{\boldsymbol{\beta}}\|_2^2}{d\sigma^2 + (2 + d + 2n)\|\widetilde{\boldsymbol{\beta}}\|_2^2}.$$

Lemma 1. *For the above setting, the expected risk can be represented as a function of $\alpha, \bar{\boldsymbol{\beta}}, \boldsymbol{\beta}^*, d$ and n as,*

$$\bar{\mathcal{R}}_n = (\alpha^2(n^2 + n + dn) - 2\alpha n + 1)\|\widetilde{\boldsymbol{\beta}}\|_2^2 + (\alpha^2 nd + 1)\sigma^2$$

Proof of Lemma 1. For isotropic Gaussian covariates the risk of an estimator for a given n training samples is $\mathcal{R}_n := \|\hat{\boldsymbol{\beta}}_n - \boldsymbol{\beta}^*\|_2^2 + \sigma^2$. Then, the expected risk for any given training set size n can be obtained as follows [11]:

$$\mathcal{R}_n = \|\hat{\boldsymbol{\beta}}_n - \boldsymbol{\beta}^*\|_2^2 + \sigma^2$$

Using our one-step steepest descent estimator $\boldsymbol{\beta}^*$ from the initial guess $\bar{\boldsymbol{\beta}}$,

$$\begin{aligned}
\bar{\mathcal{R}}_n &= \mathop{\mathbb{E}}_{\boldsymbol{X},\boldsymbol{y}} \left[\|\alpha \boldsymbol{X}^\top \boldsymbol{y} - \alpha \boldsymbol{X}^\top \boldsymbol{X} \bar{\boldsymbol{\beta}} + \widetilde{\boldsymbol{\beta}}\|_2^2 \right] + \sigma^2 \\
&= \mathop{\mathbb{E}}_{\boldsymbol{X}, \boldsymbol{\eta} \sim \mathcal{N}(\mathbf{0}, \sigma^2 \boldsymbol{I}_n)} \left[\|\alpha \boldsymbol{X}^\top (\boldsymbol{X}\boldsymbol{\beta}^* + \boldsymbol{\eta}) - \alpha \boldsymbol{X}^\top \boldsymbol{X} \bar{\boldsymbol{\beta}} + \widetilde{\boldsymbol{\beta}}\|_2^2 \right] + \sigma^2 \\
&= \underbrace{\mathop{\mathbb{E}}_{\boldsymbol{X}} \left[\|\alpha \boldsymbol{X}^\top \boldsymbol{X} \boldsymbol{\beta}^* - \alpha \boldsymbol{X}^\top \boldsymbol{X} \bar{\boldsymbol{\beta}} + \widetilde{\boldsymbol{\beta}}\|_2^2 \right]}_{\text{A}} + \underbrace{\mathop{\mathbb{E}}_{\boldsymbol{X}, \boldsymbol{\eta}} \left[\|\alpha \boldsymbol{X}^\top \boldsymbol{\eta}\|_2^2 \right]}_{\text{B}} + \sigma^2
\end{aligned}$$

Since the elements of $\boldsymbol{X}$ are drawn from an isotropic Gaussian distribution ($\mathcal{N}(0, \mathbf{I}_d)$), the Gram matrix is a Wishart distribution $\boldsymbol{\Sigma} = \boldsymbol{X}^\top \boldsymbol{X} \sim \mathcal{W}(\mathbf{I}_d, d)$. The first and the second moments of Wishart distribution are given by

$$\mathop{\mathbb{E}}_{X}[\boldsymbol{\Sigma}] = n\mathbf{I}_d, \quad \text{and}$$
$$\mathop{\mathbb{E}}_{X}[\boldsymbol{\Sigma}\boldsymbol{\Sigma}] = \mathop{\mathbb{E}}_{X}\left[\boldsymbol{\Sigma}^2\right] = nd\mathbf{I}_d + (n^2+n)\mathbf{I}_d,$$

respectively [6]. Expanding the squared norm in A and using the expressions for moments of the Wishart distribution leads to

$$\begin{aligned} \mathrm{A} &= \alpha \mathop{\mathbb{E}}_{\boldsymbol{X}}\left[\alpha \boldsymbol{\beta}^{*\top}\boldsymbol{\Sigma}^2\boldsymbol{\beta}^* - 2\alpha\boldsymbol{\beta}^{*\top}\boldsymbol{\Sigma}^2\bar{\boldsymbol{\beta}} + 2\boldsymbol{\beta}^{*\top}\boldsymbol{\Sigma}\widetilde{\boldsymbol{\beta}} + \alpha\bar{\boldsymbol{\beta}}^\top\boldsymbol{\Sigma}^2\bar{\boldsymbol{\beta}} - 2\bar{\boldsymbol{\beta}}^\top\boldsymbol{\Sigma}\widetilde{\boldsymbol{\beta}}\right] + \widetilde{\boldsymbol{\beta}}^\top\widetilde{\boldsymbol{\beta}} \\ &= \alpha^2(n^2+n+dn)[\|\boldsymbol{\beta}^*\|_2^2 - 2\boldsymbol{\beta}^{*\top}\bar{\boldsymbol{\beta}} + \|\bar{\boldsymbol{\beta}}\|_2^2] + 2\alpha n(\boldsymbol{\beta}^{*\top}\widetilde{\boldsymbol{\beta}} - \bar{\boldsymbol{\beta}}^\top\widetilde{\boldsymbol{\beta}}) + \|\widetilde{\boldsymbol{\beta}}\|_2^2 \\ &= (\alpha^2(n^2+n+dn) - 2\alpha n + 1)\|\widetilde{\boldsymbol{\beta}}\|_2^2. \end{aligned}$$

We can obtain the B as

$$\begin{aligned} \mathrm{B} &= \mathop{\mathbb{E}}_{\boldsymbol{X},\boldsymbol{\eta}}\left[\|\alpha\boldsymbol{X}^\top\boldsymbol{\eta}\|_2^2\right] = \mathop{\mathbb{E}}_{\boldsymbol{X}}\left[\mathop{\mathbb{E}}_{\boldsymbol{\eta}}\left[\|\alpha\boldsymbol{X}^\top\boldsymbol{\eta}\|_2^2|\boldsymbol{X}\right]\right] \\ &= \alpha^2\sigma^2\mathop{\mathbb{E}}_{\boldsymbol{X}}\left[\|\boldsymbol{X}\|_{\mathrm{F}}^2\right] = \alpha^2\sigma^2\mathrm{tr}(\mathop{\mathbb{E}}_{\boldsymbol{X}}[\boldsymbol{\Sigma}]) = \alpha^2\sigma^2 nd. \end{aligned}$$

Then, the expected performance at training set size n can be expressed as,

$$\bar{\mathcal{R}}_n = (\alpha^2(n^2+n+dn) - 2\alpha n + 1)\|\widetilde{\boldsymbol{\beta}}\|_2^2 + \alpha^2\sigma^2 nd + \sigma^2$$

Now, we provide the proof for Theorem 1.

Proof of Theorem 1. Using Lemma 1, $\bar{\mathcal{R}}_{n+1} \leq \bar{\mathcal{R}}_n$ takes the form after some rearranging,

$$\bar{\mathcal{R}}_{n+1} - \bar{\mathcal{R}}_n := \underbrace{[(2n+2+d)\|\widetilde{\boldsymbol{\beta}}\|_2^2 + \sigma^2 d]}_{a}\alpha^2 - \underbrace{2\|\widetilde{\boldsymbol{\beta}}\|_2^2}_{b}\alpha \leq 0, \tag{7}$$

which is quadratic in the step size $a\alpha^2 - b\alpha \leq 0$ with respect to step size. Since we want to update the initial guess with one gradient update, step size must be bigger than 0 $\alpha > 0$. Then, the solution can be found as;

$$\left(b < 0 \,\wedge\, a < 0 \,\wedge\, \alpha \geq \frac{b}{a}\right) \vee$$
$$\left(b \geq 0 \,\wedge\, \left((a \leq 0 \,\wedge\, \alpha > 0) \,\vee\, (a > 0 \,\wedge\, 0 < \alpha \leq \frac{b}{a})\right)\right).$$

Since, $b < 0$ and $a < 0$ is not possible values for our case given in Eq. 7, the only valid interval for our case can be found as $0 < \alpha \leq \frac{b}{a}$.

To support our theoretical findings, in Fig. 2 we show for various dimensionalities and noise-levels, the sample-wise monotonicity with respect to OLS estimator and the alignment of our theoretical results with empirical observations. These plots are obtained with the exact setting described above. In the figure the green line represents the empirical expected risk for mean squared error metric and the dotted orange line shows Theorem 1. For the empirical results, we repeat the sampling of the training set 1000 times. As can be observed, our empirical findings using the maximum step size are closely aligned with our theoretical findings, when one gradient update is used for training with maximum step size $\alpha_{\max}$. These observations are consistent among different dimensionality and noise-variance.

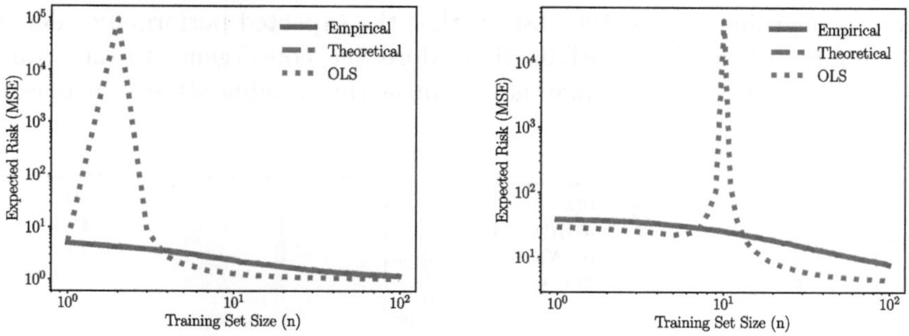

(a) $d = 2, \quad \sigma^2 = 1$ (b) $d = 10, \quad \sigma^2 = 4$

Fig. 2. Expected MSE with respect to training set size. We simulate the setting of Theorem 1 and use the expression derived in Lemma 1 with the maximum step size $\alpha_{\max}$. In this simulation, the data-generating parameters are drawn from a normal distribution $\boldsymbol{\beta}^* \sim \mathcal{N}(0, 1)$, and the initial parameter vector is also drawn independently from a normal distribution $\bar{\boldsymbol{\beta}} \sim \mathcal{N}(0, 1)$.

3 Monotonicity in Practice

Obtaining the expected risk given in Eq. 3 in a real-world setting is impossible since $\mathcal{D}$ is not known. Generally, bootstrapping or cross-validation methods are employed to get an estimate of the expected risk while creating learning curves [14]. To calculate the expected risk empirically, we first randomly sample 20% of the dataset for the test set. Then, the remaining 80% of the dataset containing N_{train} samples is used for training. In our experiments, we sample a new training set without replacement for each training set size 1000 times, in order to get a reliable estimate of the expected risk. We repeat the same procedure for every sample size in the range $n \in [1, 0.8 \times N_{\text{train}}]$ to create the learning curves. Finally, for training, we normalize both the features and the outputs[1].

Note that the maximum step size $\alpha_{\max}$ found in Theorem 1 depends on the variance of the noise σ^2 and the norm of the difference between the initial guess for the estimator and the optimal parameter vector $\|\widetilde{\boldsymbol{\beta}}\|_2$, which means

[1] Code used for the experiments is available at https://github.com/taylanot/monotonicity.git.

it depends on the data generating process. To make our proposal applicable in practice, we propose using exact line search since our objective function is convex. Then, the step size for our estimator with n training samples takes the form $\alpha_{\text{ls}} := \frac{\boldsymbol{r}_n^\top \boldsymbol{r}_n}{\boldsymbol{r}_n^\top \boldsymbol{\Sigma}_n \boldsymbol{r}_n}$.

In Fig. 3, we present results for two datasets obtained from OpenML [13]. For each dataset, we consider two models with fixed initializations: one drawn from a normal distribution and the other from a uniform distribution. Each model performs a single gradient update using exact line search, which results in monotonic learning curves. We observe that the expected performance depends on the quality of the fixed initialization. Moreover, convergence to the ordinary least squares (OLS) solution may not occur as the training set size increases.

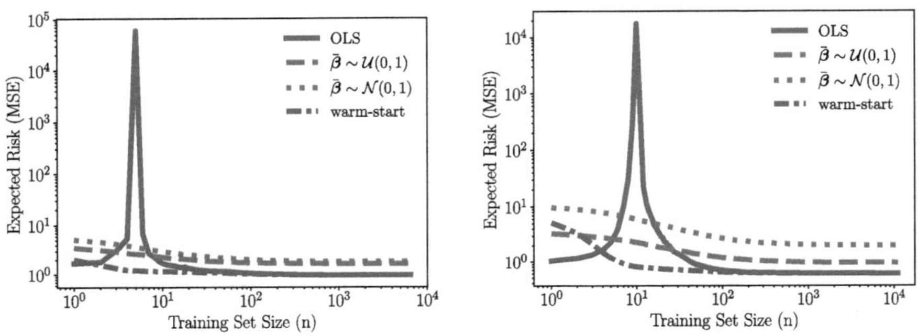

(a) Sulfur dataset (b) Bike sharing dataset

Fig. 3. Learning curves for two OpenML [13] datasets, Sulfur and Bike Sharing. We compare fixed initial guesses for the parameter vector (sampled from uniform and normal distributions) with a warm start, all using a single gradient update with exact line search.

To mitigate the effect of the starting point, we propose a warm start for each training set size n by using the average solution obtained using $n-1$ samples on top of the exact line search with one gradient update. We cannot have a warm start for the first sample, and we initialize $\bar{\boldsymbol{\beta}}_1$ with a random sample from a standard normal distribution. For the rest of the training set sizes, the average parameter vector from the previous training set size will be used; $\bar{\boldsymbol{\beta}}_n := \frac{1}{m}\sum_{i=1}^{m} \hat{\boldsymbol{\beta}}_{n-1}$, where m is the number of training sets drawn from a given training set size for obtaining $\bar{\mathcal{R}}_n$. Then, the proposed estimator for training set size n takes the form,

$$\hat{\boldsymbol{\beta}}_n = \bar{\boldsymbol{\beta}}_n + \alpha_{\text{ls}} \boldsymbol{r}_n \tag{8}$$

As can be seen from Fig. 3, the warm starting point makes the solution converge to the OLS solutions as the training set size increases. We observed the same in four more random supervised regression problems from OpenML [13] in

Fig. 4, where the green line represents the OLS solution that gives the minimum-norm solution when $d \geq n$ and the orange dotted line represents our proposed solution using one gradient update, exact line search and warm initialization.

For all the datasets considered, our estimator is monotone, where the sample-wise double descent is not present. Moreover, our proposal converges to the OLS solution that is achieved after the double descent rapidly in terms of sample-size needed. Finally, we observe that this monotonicity does not always come at the cost of low performance as well.

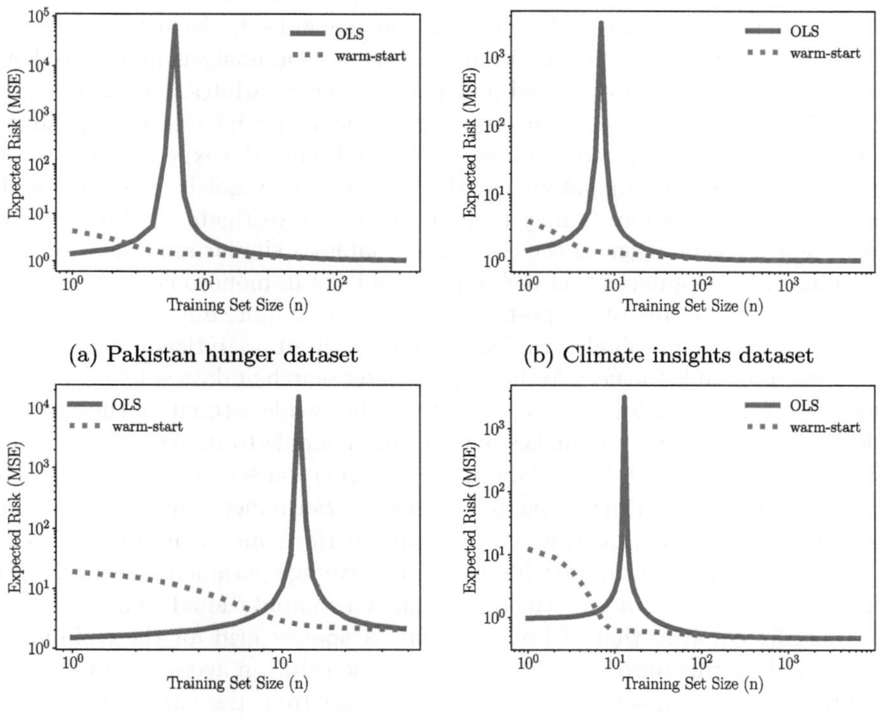

(a) Pakistan hunger dataset

(b) Climate insights dataset

(c) Sleep deprivation dataset

(d) Climate change impact dataset

Fig. 4. Learning curves for four other OpenML [13] datasets: Pakistan hunger, climate insights, sleep deprivation, and climate change impact. Across all learning curves, our proposed approach exhibits monotonic behaviour compared to its OLS counterparts.

4 Discussion and Conclusion

We tried to answer one of the open questions presented in [15]: when is a learner provably monotone? We proved that for isotropic Gaussian covariates, a linear regression with a Gaussian noise model and one step of steepest descent can result in a sample-wise monotone learning algorithm, given that the maximum step size is not exceeded. We also provide this limiting step size for which this monotonicity holds. However, in practice, we of course typically do not know the optimal step size. Hence, we propose a practical way to set this size. We showed on real-world datasets that if one step of steepest descent is used with fixed parameters, it yields sample-wise monotonic learner.

Our proof relies on three major assumptions. The initial parameter vector and step size are independent of the training points and, the training points are drawn from an isotropic Gaussian distribution with Gaussian noise model. Both the warm start and the exact line search we suggest to be used in practice violate the independence assumption. We showed empirically that both the step size dependence alone and the combined warm starting and step size dependence does not violate the monotonicity in practice. A theoretical analysis may be difficult when both parameters are allowed to depend on the covariates. We suspect that, although the exact line search decision for the step size introduces dependence on the training points, it may still be possible to bound the expected step size to guarantee sample-wise monotonicity. However, this may not be possible for the warm start initialization. It may be easier to first investigate whether relaxing the Gaussian assumption on the covariates enables a similar analysis.

Unlike other regularization techniques resulting in monotonicity, our practical proposal of one step of steepest descent does not require any cross-validation. Instead, it relies on a single gradient update with an analytically derived step size. This eliminates the need for hyperparameter search or data splitting, both of which can introduce additional sources of variance while estimating the expected risk. The absence of tuning makes our approach simple to implement, and well-suited for scenarios with limited data, where validation sets are limited. Despite its simplicity, the resulting estimator exhibits performance comparable to optimally regularized models. However, to have performance converging to OLS estimate in practice our approach relies on the average parameters from the previous training set sizes $n-1$, which introduces a computational complexity that is essentially equal to that of leave one out. A sparser grid for the training set size can reduce this burden, although it may also result in average solution from the previous training set sizes to be further apart from the current one, which can lead to non-monotonic behaviour.

We investigated only the monotonicity aspect. However, as shown in Figs. 1 and 4, our method is not only monotone but also stabilizes more quickly to a lower generalization error as the training set size increases, compared to the other methods. Similar observations, where *gradient descent* outperforms the analytical OLS solution, are also reported in [17].

All of our results pertain to the regression setting with the MSE loss. When other measures are used for obtaining the learning curve, some non-monotone behaviour can be observed (although not as severe as for the OLS). In Fig. 5 we allude to such cases for two random binary classification problems (multi-class classification problems, require additional decisions to determine the warm starting point of the parameter vector), with same experimental setting apart from the loss measure provided in Sect. 3. The non-monotonic behavior of our approach could stem from the mismatch between the MSE considered during training, while testing is done using accuracy. [8,14] have more generally indicated that metric mismatch between training and testing can introduce unexpected behavior. Expected error of Pseudo Fisher Linear Discriminant (PFLD) is presented in [12], where also analytically, the peaking behaviour [5] is observed. Our proposal addresses the non-monotonic behaviour introduced by the pseudo-inverse operation in the OLS estimator. For this reason, it may also be possible to guarantee monotonicity for linear classifiers that rely on the same pseudo-inverse, such as PFLD, for error rate or accuracy.

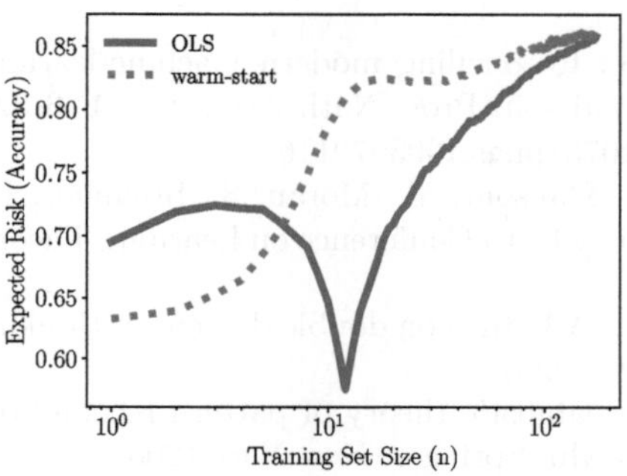

(a) **Heart Stats Metric:** Accuracy

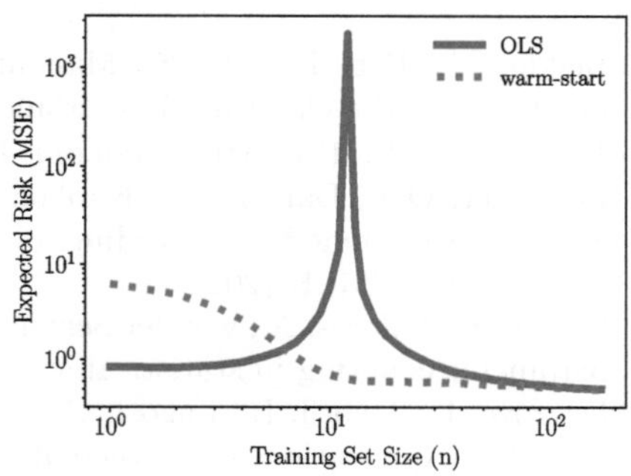

(b) **Heart Stats/Metric:** MSE

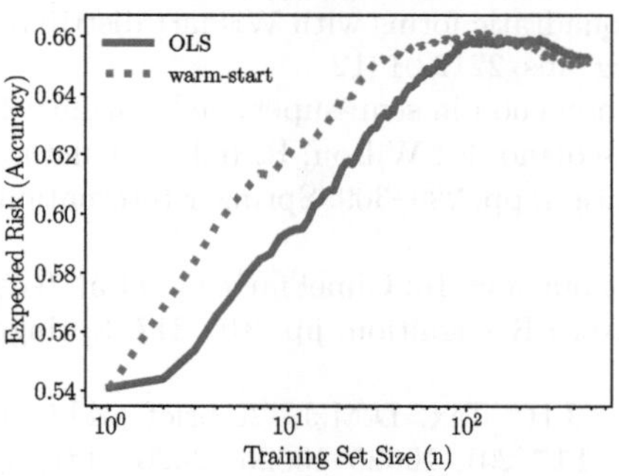

(c) **Blood Trans./Metric:** Accuracy

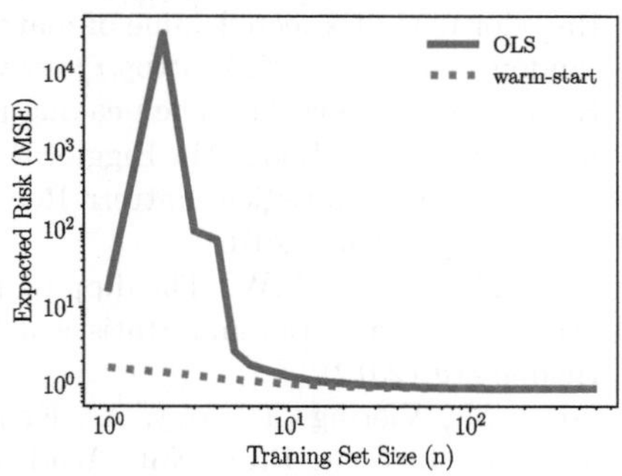

(d) **Blood Trans./Metric:** MSE

Fig. 5. The effect of metric choice (Accuracy vs. MSE) across two datasets (heart stats and blood transfusion) from OpenML [13]. Each row corresponds to a dataset, with the left column showing expected Accuracy and the right column showing expected MSE as functions of training set size.

All in all, we have given a case when a learner is provably monotone in a constrained setting. However, our observations provide no universal guarantees: there might exist a distribution $\mathcal{D}$ for which any estimator performs poorly even for arbitrarily large training sizes n [4]. Thus, outside the narrow regimes where we can guarantee monotonicity, the very existence of a universally monotone learners remains unknown. For the MSE, one may even wonder if such learner at all exists.

Acknowledgements. We thank the reviewers for their valuable feedback, which helped us significantly improve our work.

Disclosure of Interests. The authors have no competing interests to declare that are relevant to the content of this article.

References

1. Belkin, M., Hsu, D., Ma, S., Mandal, S.: Reconciling modern machine-learning practice and the classical bias-variance trade-off. Proc. Natl. Acad. Sci. **116**(32), 15849–15854 (2019). https://doi.org/10.1073/pnas.1903070116
2. Bousquet, O.J., Daniely, A., Kaplan, H., Mansour, Y., Moran, S., Stemmer, U.: Monotone learning. In: Proceedings of Thirty Fifth Conference on Learning Theory, pp. 842–866. PMLR (2022)
3. Curth, A., Jeffares, A., van der Schaar, M.: A U-turn on double descent: rethinking parameter counting in statistical learning (2023)
4. Devroye, L., Györfi, L., Lugosi, G.: A probabilistic theory of pattern recognition. No. 31 in Applications of Mathematics, 3 edn. Springer, New York (2008)
5. Duin, R.P.W.: Small sample size generalization. In: Proceedings of the 9th Scandinavian Conference on Image Analysis. Uppsala, Sweden (1995)
6. Hagedorn, M.: Expected value of matrix quadratic forms with Wishart distributed random matrices (2022). https://arxiv.org/abs/2212.01412
7. Krijthe, J.H., Loog, M.: The peaking phenomenon in semi-supervised learning. In: Robles-Kelly, A., Loog, M., Biggio, B., Escolano, F., Wilson, R. (eds.) Structural, Syntactic, and Statistical Pattern Recognition, pp. 299–309. Springer International Publishing, Cham (2016)
8. Loog, M., Duin, R.P.W.: The dipping phenomenon. In: Gimel'farb, G., et al. (eds.) Structural, Syntactic, and Statistical Pattern Recognition, pp. 310–317. Springer, Heidelberg (2012)
9. Loog, M., Viering, T., Mey, A., Krijthe, J.H., Tax, D.M.J.: A brief prehistory of double descent. Proc. Nat. Acad. Sci. **117**(20), 10625–10626 (2020). https://doi.org/10.1073/pnas.2001875117. https://www.pnas.org/doi/abs/10.1073/pnas.2001875117
10. Nakkiran, P.: More Data Can Hurt for Linear Regression: Sample-wise Double Descent (2019)
11. Nakkiran, P., Venkat, P., Kakade, S., Ma, T.: Optimal Regularization Can Mitigate Double Descent (2021)
12. Raudys, S., Duin, R.P.W.: Expected classification error of the Fisher linear classifier with pseudo-inverse covariance matrix. Pattern Recogn. Lett. **19**(5), 385–392 (1998). https://doi.org/10.1016/S0167-8655(98)00016-6
13. Vanschoren, J., van Rijn, J.N., Bischl, B., Torgo, L.: OpenML: networked science in machine learning. SIGKDD Explor. **15**(2), 49–60 (2013). https://doi.org/10.1145/2641190.2641198
14. Viering, T., Loog, M.: The shape of learning curves: a review. IEEE Trans. Pattern Anal. Mach. Intell. 45(6), 7799–7819 (2023). https://doi.org/10.1109/TPAMI.2022.3220744, https://doi.org/10.1109/TPAMI.2022.3220744

15. Viering, T., Mey, A., Loog, M.: Open problem: monotonicity of learning. In: Proceedings of the Thirty-Second Conference on Learning Theory, pp. 3198–3201. PMLR (2019)
16. Viering, T.J., Mey, A., Loog, M.: Making learners (more) monotone. In: Berthold, M.R., Feelders, A., Krempl, G. (eds.) Advances in Intelligent Data Analysis XVIII, pp. 535–547. Springer International Publishing, Cham (2020)
17. Wu, J., Bartlett, P.L., Lee, J.D., Kakade, S.M., Yu, B.: Risk comparisons in linear regression: implicit regularization dominates explicit regularization (2025). https://doi.org/10.48550/arXiv.2509.17251

PromptFusionSR: Multimodal Enhancement of Low-Resolution Images with Automatic Prompt-Guided Diffusion

Chang Qu[1], Ilhwan Kwon[1], Karthick Thiyagarajan[2], Mukesh Prasad[1], and Ali Braytee[1(✉)]

[1] School of Computer Science, University of Technology Sydney, Ultimo, NSW 2007, Australia
{Chang.Qu-1,Ilhwan.Kwon}@student.uts.edu.au,
{Mukesh.Prasad,Ali.Braytee}@uts.edu.au

[2] Smart Sensing and Robotics Laboratory (SensR Lab), Centre for Advanced Manufacturing Technology, School of Engineering, Western Sydney University, Kingswood, Sydney, NSW 2747, Australia
K.Thiyagarajan@westernsydney.edu.au

Abstract. Image super-resolution is a technique aimed at enhancing the resolution of images and is widely utilized in fields such as remote sensing and medical imaging. We view image super-resolution as a reconstruction problem for inferring missing structural information from incomplete observations. This perspective highlights the need for methods that do not merely generate pixel-level corrections, but instead analyze and infer latent geometric and semantic patterns embedded in multimodal data. In this work, we present PromptFusionSR (PF-SR), a novel framework for text-assisted image super-resolution. In contrast to conventional approaches that rely solely on visual features, our method leverages the BLIP3 vision language model to extract contextual prompts-short textual descriptions-directly from low-resolution images. A periodic semantic–structural fusion mechanism enables the model to integrate complementary information channels, enhancing its ability to interpret high-frequency patterns under severe degradation. These textual and information cues guide a pretrained stable diffusion model for 4× upscaling, yielding high-resolution outputs with improved fidelity and richer visual textures. This work further explores how complementary information channels can be systematically integrated during image reconstruction to improve robustness and perceptual fidelity. On our benchmark datasets, our method surpasses the current state-of-the-art methods across most evaluation metrics, with robust gains in perceptual quality metrics.

Keywords: Computer vision · Visual-language model · Prompt generation · Multimodal image enhancement

M. Baratchi et al. (Eds.): IDA 2026, LNCS 16513, pp. 84–96, 2026.
https://doi.org/10.1007/978-3-032-23833-7_7

1 Introduction

Reconstructing missing or uncertain information from incomplete observations is a fundamental challenge in inverse problems and image restoration [1–3]. Across scientific imaging, remote sensing, medical inspection, and real-world sensing pipelines, raw data often undergo substantial loss of structural information during acquisition or transmission. Recovering these latent structures is essential not only for visual clarity, but also for downstream interpretation, decision making, and knowledge extraction, echoing long-standing work on probabilistic inference and reasoning under uncertainty in the community [4,5].

From a data-centric perspective, image super-resolution (SR) can be viewed as a form of inverse structural inference: given degraded observations, the goal is to estimate the underlying data manifold that best explains the visible evidence. Classical SR approaches and modern deep learning methods [6] primarily optimize pixel-level fidelity, without explicitly modeling the semantic or contextual factors that govern structural variation in the data. This lack of semantic awareness limits their ability to resolve ambiguous or repetitive textures, a limitation that parallels observations in multimodal and sequential inference problems studied in recent work [7].

Diffusion-based generative models provide an alternative by modeling data distributions through iterative denoising, and have recently been adapted to the SR setting [8,9]. While these methods capture natural image statistics more effectively than conventional deep networks, they still lack mechanisms to interpret the data's semantic and structural properties: their refinement dynamics largely depend on local likelihoods and do not naturally incorporate high-level evidence or contextual cues. To address this limitation, we propose PromptFusionSR (PF-SR)—a multimodal structural inference framework that integrates semantic priors derived from BLIP style vision–language models [10,11] with latent diffusion inference [12]. PF-SR uses high-level prompts as auxiliary evidence to guide the reconstruction of ambiguous regions of data. A periodic fusion mechanism selectively injects semantic information into the diffusion trajectory, enabling the model to combine global semantic cues with local structural signals in a data-aware manner. Our contributions are listed as follows:

- We propose PF-SR, a unified semantic–structural diffusion framework that introduces automatic prompt synthesis and dual-latent periodic fusion for text-guided image super-resolution.
- We design a periodic semantic–structural fusion mechanism, which stabilizes diffusion trajectories and effectively mitigates hallucinations, enabling faithful texture synthesis under ambiguous degradations.
- We conduct extensive quantitative and qualitative evaluations, including ablations on fusion scheduling, diffusion steps, and prompt sources, demonstrating consistent perceptual gains across three datasets.

2 Related Work

2.1 CNN and Transformer-Based Super-Resolution

Early learning-based SISR methods are predominantly built upon convolutional neural networks (CNNs), such as EDSR [13] and RCAN [14], which enhance reconstruction fidelity through deep residual learning and channel attention. More recently, Transformer architectures, including SwinIR and HAT, further strengthen long-range dependency modeling. Despite strong distortion-oriented performance, these deterministic models often produce over-smoothed textures and struggle to recover fine-scale details under complex degradations.

2.2 GAN-Based and Diffusion-Based Super-Resolution

To improve perceptual realism, GAN-based methods [15] introduce adversarial learning, achieving visually plausible textures at the cost of training instability and hallucination. Diffusion-based approaches, such as StableSR [9], DiffBIR [16], and PASD [17], further advance perceptual SR by leveraging powerful generative priors. However, diffusion models conditioned solely on low-resolution inputs remain prone to semantic drift, often synthesizing plausible yet incorrect details due to stochastic sampling.

2.3 Semantic-Guided Super-Resolution

Recent work incorporates semantic cues to mitigate hallucination. SeeSR [18] injects segmentation-aware prompts into the diffusion process, while SRSR employs spatially re-focused text conditioning. Although these methods improve semantic consistency, semantics is typically treated as a passive conditioning signal that directly modulates the diffusion trajectory. As a result, semantic guidance remains tightly coupled with stochastic sampling, limiting controllability and robustness under ambiguous content.

In contrast, PF-SR introduces a semantic–structural dual-latent diffusion framework. We first synthesize textual semantics from the low-resolution input via BLIP3 and project them into a structured semantic latent space using CLIP. This semantic latent is then periodically fused with a structural latent during diffusion, forming a stable guidance mechanism that decouples semantic abstraction from generative sampling. By transforming semantics into an explicit control signal rather than a passive condition, PF-SR effectively suppresses hallucination while preserving high-frequency realism, enabling robust and controllable super-resolution.

3 Methodology

The proposed components operate as complementary information channels, enabling PF-SR to integrate semantic, structural, and latent cues within a unified multimodal reconstruction framework. We propose PF-SR, a three-branch framework illustrated in Fig. 1. It comprises a Prompt Pathway that extracts semantic cues from the low-resolution image as a text embedding, a Fusion Pathway that encodes the bicubic-upsampled input to preserve the original structure, and a Diffusion Pathway that generates high-frequency details. These branches collaborate to enhance perceptual fidelity by improving visual quality and realism, ensure semantic fidelity by maintaining consistent scene content, and preserve structural fidelity by accurately representing geometry.

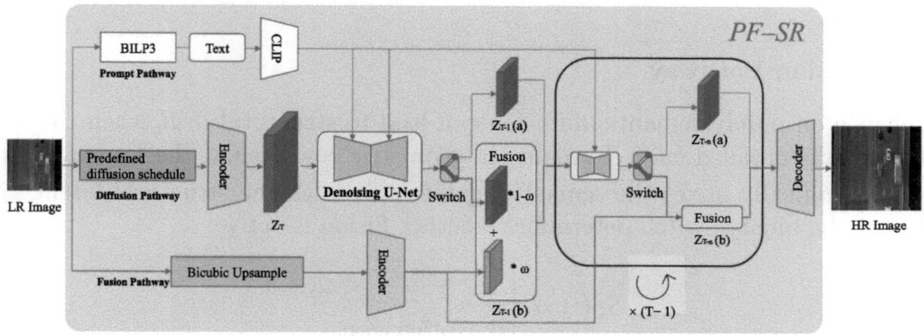

Fig. 1. Overall architecture of PromptFusionSR. The low-resolution input is first processed by a vision–language model to generate an automatic descriptive prompt. A fusion module periodically injects structural information from the LR image to preserve geometry while enhancing details. Here, z_t is the latent at step t, $z_{t-1}^{(a)}$ the prompt-conditioned update, $z_{t-1}^{(b)}$ the fused update, z_0 the initial bicubic latent, and $\omega \in [0, 1]$ the fusion weight.

3.1 Preliminaries

Given an LR input $x_{\text{LR}} \in \mathbb{R}^{C \times H \times W}$, we aim to generate a $4\times$ super-resolved image $\hat{x}_{\text{HR}} \in \mathbb{R}^{C \times (4H) \times (4W)}$. We first compute a bicubic-upsampled estimate $x_{\text{bic}} = \text{Bicubic}(x_{\text{LR}})$, which serves as a coarse geometric prior. The latent diffusion model operates in a compact latent space via the encoder $E : \mathbb{R}^{C \times (4H) \times (4W)} \to \mathbb{R}^D$, mapping x_{bic} to the structural latent $z_0 \in \mathbb{R}^D$, used as the geometry reference throughout denoising. Let z_t denote the latent at diffusion timestep $t \in \{0, \ldots, T\}$.

Simultaneously, we extract semantic descriptions from the LR input using BLIP3 and encode them with CLIP into a text embedding $c \in \mathbb{R}^d$. A scalar fusion weight $\omega \in [0, 1]$ controls the strength and timing of structural injection.

To avoid ambiguity, we denote the initial structural latent by z_0 and the final denoised latent by Z_0. Although identical in standard diffusion, PF-SR modifies the trajectory through fusion, making the distinction necessary.

3.2 Prompt Pathway

The Prompt Pathway provides semantic conditioning by converting the LR image into a descriptive prompt: $P = \text{BLIP3}(x_{\text{LR}}), c = \text{CLIP}_{\text{text}}(P)$. The text embedding enters the U-Net via cross-attention at every timestep, yielding the semantic update:

$$z_{t-1}^{(a)} = U_t(z_t, c), \tag{1}$$

where U_t denotes the U-Net block at step t. Using BLIP3 ensures that the text prompt is image-grounded and avoids the hallucinations commonly introduced by handwritten prompts.

3.3 Fusion Pathway

Reliance on purely semantic diffusion can lead to structural drift when the LR image lacks detailed cues. To maintain geometric consistency, the Fusion Pathway periodically mixes the semantic update $z_{t-1}^{(a)}$ with the structural latent z_0. A periodic binary switch determines whether fusion is active:

$$S_\omega(t) = \begin{cases} 1, & \frac{t \bmod T}{T} < \omega, \\ 0, & \text{otherwise}, \end{cases} \tag{2}$$

and the fused update becomes

$$z_{t-1}^{(b)} = S_\omega(t) \left[\omega\, z_0 + (1-\omega)\, z_{t-1}^{(a)}\right] + (1 - S_\omega(t))\, z_{t-1}^{(a)}. \tag{3}$$

This formulation unifies two common strategies—continuous latent blending and early-step fusion—under a single parameter ω. Small ω encourages sharper textures at the risk of hallucinations, while large ω enforces stronger structure but may oversmooth fine details. PF-SR achieves the best fidelity–realism balance with moderate fusion strengths.

3.4 Diffusion Pathway and Integration

PF-SR follows a standard latent diffusion process [19]. The forward noising schedule is defined as:

$$z_t = \sqrt{\bar{\alpha}_t}\, z_0 + \sqrt{1-\bar{\alpha}_t}\, \epsilon, \qquad \epsilon \sim \mathcal{N}(0, I), \tag{4}$$

where $\bar{\alpha}_t = \prod_{i=1}^{t} \alpha_i$. During reverse denoising, for each timestep $t = T, \ldots, 1$, PF-SR performs:

1) Semantic refinement

$$z_{t-1}^{(a)} = \text{UNet}(z_t, c). \tag{5}$$

2) Structure–semantic fusion

$$z_{t-1} = S_\omega(t)\, z_{t-1}^{(b)} + (1 - S_\omega(t))\, z_{t-1}^{(a)}. \tag{6}$$

After completing all timesteps, the final latent Z_0 is decoded:

$$\hat{x}_{\mathrm{HR}} = \mathrm{Dec}(Z_0). \tag{7}$$

Since PF-SR reuses the pretrained diffusion backbone, only the fusion modules are trained. The model is optimized using the standard diffusion objective:

$$\mathcal{L} = \mathbb{E}_{t,\epsilon}\left[\|\epsilon - \epsilon_\theta(z_t, t, c)\|^2\right]. \tag{8}$$

4 Experiments and Results

4.1 Datasets

We conduct our training on both synthetic and real-world SR benchmarks. Specifically, we use the DIV2K [20] dataset, which contains 900 synthetic HR–LR pairs generated via MATLAB's `imresize` bicubic downsampling. Following an 80/10/10 split, we use 720 images for training, 90 for testing, and 90 for validation. For real-world SR, we include DRealSR [21], which consists of 840 real-capture, pixel-aligned HR–LR pairs, yielding 672 training images, 84 test images, and 84 validation images. We also use the RealSR dataset [22], containing 595 HR–LR pairs of real degradations, split into 475 for training, 60 for testing, and 60 for validation. All datasets are randomly shuffled using a fixed seed to ensure reproducibility.

For evaluation, we report quantitative results on the reserved test splits of each benchmark: 90 images for DIV2K, 84 for DRealSR, and 60 for RealSR. The corresponding validation splits (also 90, 84, and 60 images) are used exclusively for hyperparameter tuning, such as selecting the fusion weight ω and the number of diffusion sampling steps T. All experiments assume a 4× upscaling setting unless otherwise noted.

4.2 Baseline Models

We compare PF-SR with five state-of-the-art baselines spanning both CNN and diffusion paradigms: EDSR [13] (deep residual CNN), StableSR [9] (structure-guided Stable Diffusion), DiffBIR [16] (blind SR with a degradation-aware encoder), PASD [17] (progressive adversarial sampling), and SeeSR [18] (semantics-aware diffusion). All methods are evaluated using the authors' official 4× pretrained weights and inference settings, so performance differences reflect algorithmic design rather than training or tuning variability.

4.3 Evaluation Metrics

We report five widely used image-quality metrics. PSNR [23] (Peak Signal-to-Noise Ratio) measures mean-squared error in decibels; higher values indicate lower distortion. SSIM [23] (Structural Similarity) compares luminance, contrast, and structure between two images; a score closer to 1 denotes higher fidelity. LPIPS and DISTS [24,25] are deep-feature perceptual metrics; lower scores mean the generated image is perceptually closer to the reference. NIQE [26] is a no-reference natural-image quality evaluator; lower is better and reflects fewer unnatural artifacts.

4.4 Results

Table 1. Results on synthetic (DIV2K-Val) and real-world (RealSR, DRealSR) benchmarks at 4× upscaling (↑ higher is better, ↓ lower is better).

Dataset	Method	PSNR↑	SSIM↑	LPIPS↓	DISTS↓	NIQE↓
DIV2K-Val	EDSR [13]	**26.39**	**0.8166**	0.4743	0.3264	5.2297
	StableSR [9]	23.37	0.5796	0.3113	0.2646	**4.6598**
	DiffBIR [16]	23.74	0.5664	0.3524	0.2594	4.7295
	PASD [17]	23.18	0.5534	0.3571	0.2667	4.8461
	SeeSR [18]	23.69	0.6088	0.3194	0.2377	4.9287
	PF-SR (Ours)	25.43	0.7350	**0.2604**	**0.1570**	4.8107
RealSR	EDSR [13]	27.78	**0.8257**	0.4596	0.2946	6.0578
	StableSR [9]	25.07	0.7190	0.3150	0.2184	5.9288
	DiffBIR [16]	25.67	0.6682	0.3681	0.2390	5.5754
	PASD [17]	24.72	0.6688	0.3468	0.2306	5.3652
	SeeSR [18]	25.25	0.7328	0.3044	0.2255	**5.1928**
	PF-SR (Ours)	**28.10**	0.7729	**0.2931**	**0.1740**	5.4080
DRealSR	EDSR [13]	**29.55**	0.8140	0.4895	0.3088	7.1599
	StableSR [9]	28.93	0.7578	0.3363	0.2304	6.5427
	DiffBIR [16]	27.14	0.6576	0.4729	0.2892	6.3008
	PASD [17]	28.27	0.7229	0.4071	0.2599	5.8629
	SeeSR [18]	28.92	0.7830	0.3200	0.2403	6.4008
	PF-SR (Ours)	29.21	**0.8373**	**0.3075**	**0.2212**	**5.3490**

Table 1 reports quantitative results across the three evaluation datasets. On the synthetic DIV2K-Val benchmark, the performance gap between convolutional and diffusion approaches is most evident. The EDSR-based CNN [13] achieves the highest pixel–level fidelity (PSNR ≈26.3 dB, SSIM ≈0.81), but its outputs look overly smooth, resulting in the poorest perceptual similarity (LPIPS >0.47). Pure diffusion baselines, such as StableSR [9], DiffBIR [16], and PASD [17], restore sharper textures and therefore reduce the perceptual error by about one-third, yet they sacrifice roughly 3 dB in PSNR. SeeSR [18] reduces this fidelity gap by injecting handcrafted semantic prompts, but still trails behind our method. PF-SR strikes a more favourable balance: it closes most of the PSNR deficit while delivering the lowest LPIPS and DISTS scores (LPIPS ≈0.26, DISTS ≈0.157) among all models, and keeps NIQE low, indicating that its high-resolution reconstructions are both structurally accurate and perceptually faithful with virtually no visible artefacts.

On the Real-Capture RealSR dataset, which is dominated by complex DSLR degradations, all methods experience shifts in metric rankings. PF-SR maintains a clear edge, producing the highest pixel fidelity (PSNR ≈28.1 dB) and the best perceptual similarity (LPIPS ≈0.29, DISTS ≈0.17). Its no-reference quality (NIQE ≈5.40) also remains competitive, confirming that the automatically generated BLIP3 prompts help the diffusion process preserve realistic details without hallucinating structures. By contrast, EDSR retains high SSIM (≈0.82) but suffers from pronounced perceptual gaps (LPIPS ≈0.45), while other diffusion models either fall short in PSNR or introduce more perceptual noise.

The DRealSR benchmark is considered the most challenging due to heavy optical distortions, further underlining the robustness of our approach. PF-SR secures the highest structural similarity (SSIM ≈0.83) and the best perceptual scores (LPIPS ≈0.30, DISTS ≈0.22) while keeping NIQE at the lowest value (≈5.34). Although its PSNR is slightly lower than EDSR's, it outperforms all diffusion-based competitors and avoids the texture loss typically associated with CNN-based upscaling. These results indicate that fusing BLIP3-derived semantic cues with recurrent low-resolution anchors enables PF-SR to recover fine details and suppress artefacts more effectively than prior art across both synthetic and real-world scenarios.

4.5 Qualitative Evaluation

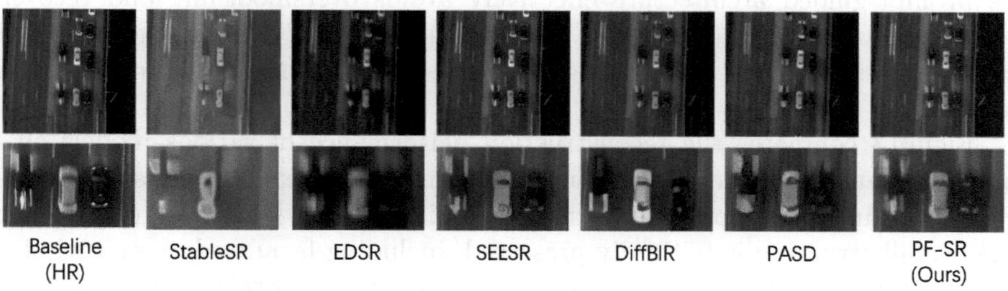

Fig. 2. Visual comparison on a highway scene (4×) with a zoomed vehicle region. PF-SR preserves lane markings and taillight structure with improved edge consistency and fewer color overflows relative to baseline methods.

As shown in Fig. 2, the super-resolution results across models differ significantly for the highway scene. EDSR, optimized using pixel-level fidelity metrics such as mean squared error, produces overly smoothed outputs, causing fine structures like lane markings to appear blurry and indistinct [13]. In contrast, diffusion-based models that incorporate prior knowledge—such as StableSR, SeeSR, DiffBIR, and PASD—reproduce richer texture details [9,16–18], thereby enhancing the visual quality of road surfaces and vehicles. However, because they are stochastic, these models may introduce slight inconsistencies or hallucinated details that deviate from the original content. For example, in the zoomed vehicle region, some diffusion models generate taillight highlights with distorted shapes or color overflow. These artifacts reflect a typical form of diffusion hallucination, in which strong generative priors and stochastic sampling introduce

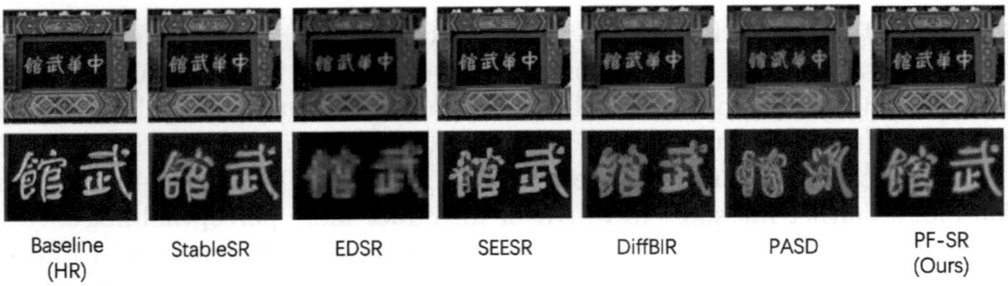

Fig. 3. Visual comparison on a building scene with embedded text (4×). PF-SR preserves facade geometry and improves text legibility. The zoomed region highlights clearer character strokes and fewer ghosting artifacts than competing methods.

visually plausible yet semantically inconsistent details. Such deviations become more evident under high magnification, particularly in regions with ambiguous structures or complex lighting.

By contrast, PF-SR delivers superior reconstruction in both global and local perspectives. The lane markings remain straight and sharp, free from blur or trailing artifacts. This demonstrates PF-SR's clear advantage in terms of color consistency, detail preservation, and structural fidelity. On one hand, its prompt-guided architecture effectively avoids oversmoothing and recovers semantically meaningful textures. On the other hand, it exhibits excellent artifact suppression—even at high magnification—without false details or edge artifacts.

Figure 3 compares the reconstruction performance of various models on a building scene containing fine textures and embedded text. Among all methods, PF-SR produces the most visually natural and consistent high-resolution output: the overall structure is faithfully preserved, including facade edges and window frames, without distortion or geometric misalignment. Notably, PF-SR demonstrates exceptional proficiency in recovering embedded textual signage within architectural scenes. In contrast, competing models exhibit notable deficiencies in this complex setting. EDSR, limited by its shallow reconstruction capacity, produces overly smoothed wall textures and heavily blurred text, making character strokes nearly unrecognizable [13]. While StableSR and DiffBIR recover moderately detailed textures, they introduce hallucinated structures or ghosting in the text regions due to the lack of semantic targeting [9,16], often resulting in incomplete or corrupted character shapes.

Overall, PF-SR demonstrates a clear advantage in both structural and semantic fidelity. It preserves consistent textures across the building facade and achieves superior text reconstruction, with minimal artifacts and no hallucinations. These strengths highlight PF-SR's ability to handle challenging scenes with both structural complexity and semantic content more effectively than all compared baselines.

4.6 Ablation Studies

We conduct ablations to analyze the influence of three key factors in PF-SR: the fusion weight ω, the contribution of BLIP3-generated prompts, and the number of diffusion sampling steps.

Table 2. Effect of fusion weight ω on DIV2K 4× validation set. $\omega = 0$ corresponds to pure diffusion (no LR guidance) while larger ω applies stronger guidance.

ω	PSNR↑	SSIM↑	LPIPS↓	DISTS↓	NIQE↓
0.0	24.713	0.667	**0.258**	**0.152**	**4.355**
0.1	24.940	0.722	0.261	0.159	4.505
0.2	25.202	0.723	0.259	0.163	4.629
0.3	25.438	0.735	0.260	0.157	4.810
0.4	25.993	0.729	0.263	0.160	4.942
0.5	26.340	0.747	0.279	0.169	5.628
0.6	26.337	**0.775**	0.304	0.176	5.939
0.7	**26.432**	0.762	0.346	0.174	6.105
0.8	26.381	0.757	0.362	0.183	6.610
0.9	26.393	0.760	0.414	0.189	7.274

Fusion Weight. Table 2 shows that very small ω reduces perceptual distortion but weakens fidelity, while very large ω forces the model to closely follow the LR input at the expense of texture richness. A mid-range value provides a better compromise, preserving structural consistency without suppressing high-frequency detail. We therefore use $\omega = 0.3$.

Prompt Semantics. To isolate the effect of semantic conditioning, we compare PF-SR with the original SDX4 upsampler, which uses no text guidance. As indicated in Table 3, adding a BLIP3-derived prompt consistently improves reconstruction fidelity and produces more semantically aligned outputs. Even with a small number of sampling steps, PF-SR benefits from richer textual context, leading to more faithful scene content.

Number of Diffusion Steps. We evaluate different sampling budgets for both SDX4 and PF-SR. As shown in Table 3, using substantially more iterations does not improve quality; instead, performance deteriorates once the sampling trajectory becomes overly long. In practice, around 20 steps strike the best balance between accuracy and efficiency, and we adopt this setting in all experiments.

Fusion Strategy. Table 4 compares two injection schedules: continuous blending at every step versus activating fusion only in the early phase. Both strategies produce nearly identical results across all ω values, confirming that PF-SR's behavior is dominated by the choice of ω rather than the specific blending schedule. This equivalence allows us to unify the design under a single ω without additional hyperparameters.

Table 3. Influence of sampling steps on reconstruction quality (DIV2K 4× validation set). Each entry reports SDX4 / PF-SR, where the better result is underlined for <u>SDX4</u> and highlighted in bold for **PF-SR.**

Steps	PSNR↑	SSIM↑	LPIPS↓
20	23.95 / **25.43**	0.661 / **0.735**	<u>0.214</u> / 0.260
50	<u>23.70</u> / 23.68	0.634 / **0.712**	0.233 / **0.232**
250	18.45 / **22.47**	0.593 / **0.683**	0.326 / **0.314**
500	15.27 / **20.65**	0.334 / **0.622**	0.340 / **0.308**

Table 4. Comparison of PSNR (dB) between two fusion strategies on the DIV2K 4× validation set (with vertical separators).

	0.0	0.1	0.2	0.3	0.4	0.5	0.6	0.7	0.8	0.9
Fusion (All Steps)	24.713	24.950	25.210	25.450	26.000	26.350	26.350	26.440	26.390	26.400
Switch (Early Steps)	24.713	24.940	25.202	25.438	25.993	26.340	26.337	26.432	26.381	26.393

5 Conclusion

We have introduced PF-SR, a framework that marries automated vision–language prompting with diffusion-based super-resolution. Our results indicate that integrating semantic evidence into the reconstruction process enhances the model's ability to analyze and infer latent structures, particularly in ambiguous or high-uncertainty regions. PF-SR demonstrates how multimodal learning can strengthen generative reconstruction pipelines by providing richer cues about the underlying data manifold. In the future, we will explore faster sampling schemes, more robust prompt generation, and adaptive structure–semantics fusion to reduce inference costs further and enhance reliability, paving the way for a unified, real-time SR solution.

Disclosure of Interests. The authors have no competing interests to declare that are relevant to the content of this article.

References

1. Freeman, W.T., Pasztor, E.C., Carmichael, O.T.: Learning low-level vision. Int. J. Comput. Vision **40**, 25–47 (2000)
2. Park, S.C., Park, M.K., Kang, M.G.: Super-resolution image reconstruction: a technical overview. IEEE Signal Process. Mag. **20**(3), 21–36 (2003)
3. Mena, F., Arenas, D., Dengel, A.: Missing data as augmentation in the earth observation domain: a multi-view learning approach. Neurocomputing **638**, 130175 (2025)
4. Vexler, J., Vieten, B., Nelke, M., Kramer, S.: Integrating inverse and forward modeling for sparse temporal data from sensor networks. In: International Symposium on Intelligent Data Analysis, pp. 318–329. Springer (2025)

5. Liu, F., Hu, X., Bu, C., Yu, K.: Fuzzy Bayesian knowledge tracing. IEEE Trans. Fuzzy Syst. **30**(7), 2412–2425 (2021)
6. Wang, Z., Chen, J., Hoi, S.C.: Deep learning for image super-resolution: a survey. IEEE Trans. Pattern Anal. Mach. Intell. **43**(10), 3365–3387 (2020)
7. Persson, D., Wahlberg, W., Vettoruzzo, A., Nowaczyk, S.: Bridging spatial and temporal contexts: sparse transfer learning. In: Krempl, G., Puolamäki, K., Miliou, I. (eds.) Advances in Intelligent Data Analysis XXIII, pp. 330–342. Springer Nature Switzerland, Cham (2025)
8. Saharia, C., Ho, J., Chan, W., Salimans, T., Fleet, D.J., Norouzi, M.: Image super-resolution via iterative refinement. IEEE Trans. Pattern Anal. Mach. Intell. **45**(4), 4713–4726 (2022)
9. Wang, J., Yue, Z., Zhou, S., Chan, K.C., Loy, C.C.: Exploiting diffusion prior for real-world image super-resolution. Int. J. Comput. Vision **132**(12), 5929–5949 (2024)
10. Xue, L., et al.: xGen-MM (BLIP-3): a family of open large multimodal models. arXiv preprint arXiv:2408.08872 (2024)
11. Zhao, Y., Prasad, M., Braytee, A.: DualPrompt-MedCap: a dual-prompt enhanced approach for medical image captioning. In: International Conference on Medical Image Computing and Computer-Assisted Intervention, pp. 205–215. Springer (2025)
12. Rombach, R., Blattmann, A., Lorenz, D., Esser, P., Ommer, B.: High-resolution image synthesis with latent diffusion models. In: Proceedings of the IEEE/CVF Conference on Computer Vision and Pattern Recognition (CVPR), pp. 10684–10695 (2022)
13. Lim, B., Son, S., Kim, H., Nah, S., Mu Lee, K.: Enhanced deep residual networks for single image super-resolution. In: Proceedings of the IEEE Conference on Computer Vision and Pattern Recognition Workshops, pp. 136–144 (2017)
14. Zhang, Y., Li, K., Li, K., Wang, L., Zhong, B., Fu, Y.: Image super-resolution using very deep residual channel attention networks. In: Proceedings of the European Conference on Computer Vision (ECCV), pp. 286–301 (2018)
15. Chen, C., et al.: Real-world blind super-resolution via feature matching with implicit high-resolution priors. In: Proceedings of the 30th ACM International Conference on Multimedia, pp. 1329–1338 (2022)
16. Lin, X., et al.: DiffBIR: toward blind image restoration with generative diffusion prior. In: European Conference on Computer Vision, pp. 430–448. Springer (2024)
17. Yang, T., Wu, R., Ren, P., Xie, X., Zhang, L.: Pixel-aware stable diffusion for realistic image super-resolution and personalized stylization. In: European Conference on Computer Vision, pp. 74–91. Springer (2024)
18. Wu, R., Yang, T., Sun, L., Zhang, Z., Li, S., Zhang, L.: SeeSR: towards semantics-aware real-world image super-resolution. In: Proceedings of the IEEE/CVF Conference on Computer Vision and Pattern Recognition, pp. 25456–25467 (2024)
19. Ho, J., Jain, A., Abbeel, P.: Denoising diffusion probabilistic models. In: Advances in Neural Information Processing Systems, vol. 33, pp. 6840–6851 (2020)
20. Agustsson, E., Timofte, R.: NTIRE 2017 challenge on single image super-resolution: dataset and study. In: Proceedings of the IEEE Conference on Computer Vision and Pattern Recognition (CVPR) Workshops, July 2017
21. Wei, P., et al.: Component divide-and-conquer for real-world image super-resolution. In: Computer Vision–ECCV 2020: 16th European Conference, Glasgow, UK, 23–28 August 2020, Proceedings, Part VIII 16, pp. 101–117. Springer (2020)

22. Cai, J., Zeng, H., Yong, H., Cao, Z., Zhang, L.: Toward real-world single image super-resolution: a new benchmark and a new model. In: Proceedings of the IEEE/CVF International Conference on Computer Vision, pp. 3086–3095 (2019)
23. Hore, A., Ziou, D.: Image quality metrics: PSNR vs. SSIM. In: 2010 20th International Conference on Pattern Recognition, pp. 2366–2369. IEEE (2010)
24. Zhang, R., Isola, P., Efros, A.A., Shechtman, E., Wang, O.: The unreasonable effectiveness of deep features as a perceptual metric. In: Proceedings of the IEEE Conference on Computer Vision and Pattern Recognition, pp. 586–595 (2018)
25. Ding, K., Ma, K., Wang, S., Simoncelli, E.P.: Image quality assessment: unifying structure and texture similarity. IEEE Trans. Pattern Anal. Mach. Intell., 1 (2020). https://doi.org/10.1109/tpami.2020.3045810
26. Mittal, A., Soundararajan, R., Bovik, A.C.: Making a "completely blind" image quality analyzer. IEEE Sig. Process. Lett. **20**(3), 209–212 (2012)

A Generic Complete Anytime Beam Search for Optimal Decision Tree

Harold Kiossou(✉) and Pierre Schaus

UCLouvain, Ottignies-Louvain-la-Neuve, Belgium
{harold.kiossou,pierre.schaus}@uclouvain.be

Abstract. Finding an optimal decision tree that minimizes classification error is known to be NP-hard. While exact algorithms based on MILP, CP, SAT, or dynamic programming guarantee optimality, they often suffer from poor anytime behavior—meaning they struggle to find high-quality decision trees quickly when the search is stopped before completion—due to unbalanced search space exploration. To address this, several anytime extensions of exact methods have been proposed, such as LDS-DL8.5, Top-*k*-DL8.5, and Blossom, but they have not been systematically compared, making it difficult to assess their relative effectiveness. In this paper, we propose CA-DL8.5, a generic, complete, and anytime beam search algorithm that extends the DL8.5 framework and unifies some existing anytime strategies. In particular, CA-DL8.5 generalizes previous approaches LDS-DL8.5 and Top-*k*-DL8.5, by allowing the integration of various heuristics and relaxation mechanisms through a modular design. The algorithm reuses DL8.5's efficient branch-and-bound pruning and trie-based caching, combined with a restart-based beam search that gradually relaxes pruning criteria to improve solution quality over time. Our contributions are twofold: (1) We introduce this new generic framework for exact and anytime decision tree learning, enabling the incorporation of diverse heuristics and search strategies; (2) We conduct a rigorous empirical comparison of several instantiations of CA-DL8.5—based on Purity, Gain, Discrepancy, and Top-*k* heuristics—using an anytime evaluation metric called the primal gap integral. Experimental results on standard classification benchmarks show that CA-DL8.5 using LDS (limited discrepancy) consistently provides the best anytime performance, outperforming both other CA-DL8.5 variants and the Blossom algorithm while maintaining completeness and optimality guarantees.

Keywords: Optimal Decision Trees · Beam Search · Anytime

1 Introduction

Decision trees are a fundamental machine learning model, widely adopted for their interpretability and solid performance in domains such as healthcare, finance, and education. Classic algorithms like CART [6] and C4.5 [17] induce

M. Baratchi et al. (Eds.): IDA 2026, LNCS 16513, pp. 97–109, 2026.
https://doi.org/10.1007/978-3-032-23833-7_8

decision trees greedily, selecting splits in a top-down manner based on local heuristics. While these greedy approaches are highly scalable, they often produce trees that are less accurate than their optimal counterparts [12].

Recent years have seen growing interest in exact decision tree learning algorithms, which aim to find globally optimal trees, typically minimizing classification error or another loss function. These algorithms leverage combinatorial optimization techniques from MILP [1,4], constraint programming [18], SAT [14], and dynamic programming [2,7]. While these methods offer strong generalization properties [4,12], they tend to suffer from poor anytime behavior: when interrupted before convergence, they often return poor-quality solutions.

The DL8.5 algorithm [2], based on dynamic programming and efficient caching, is a state-of-the-art method for optimal tree induction. DL8.5 explores the search space in a depth-first fashion, which often causes it to become stuck in unpromising regions, as illustrated in Fig. 1. As a result, it may return poor trees when stopped early. Greedy methods like C4.5 provide quick results but lack the capacity to improve or guarantee optimality over time. Neither approach offers the benefits of a true anytime algorithm, which should return a good initial solution quickly and improve it continuously as time allows.

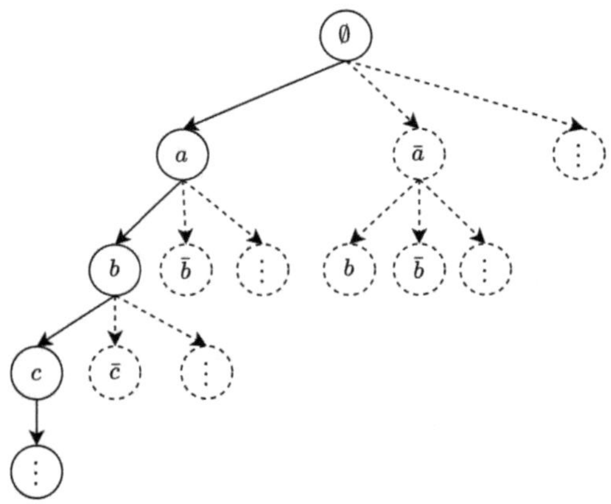

Fig. 1. DL8.5 explores the leftmost branches first, often leading to poor anytime performance when interrupted early.

To improve the anytime performance or scalability of exact methods, three notable works have been proposed. LDS-DL8.5 [11] integrates limited discrepancy search (LDS) into DL8.5, resulting in an algorithm that is both anytime and complete. Top-k-DL8.5 [5] modifies DL8.5 by restricting the candidate features at each node to the Top-k according to a ranking heuristic. This is a compromise between C4.5 and DL8.5: faster and more scalable, but unable to guarantee convergence to the optimal tree. Finally, the Blossom algorithm [8] follows a fundamentally different search strategy. It uses a depth-first approach that expands decision tree nodes level by level, offering improved anytime behavior by avoiding the possibility of highly unbalanced trees when interrupted early as in DL8.5. Blossom is guaranteed to find the optimal tree given sufficient time. Despite their promise, these three approaches have not been systematically compared in prior work, making it difficult to assess their relative strengths.

This Paper Makes Two Main Contributions. First, we introduce CA-DL8.5, a novel algorithm for decision tree learning that is: **complete** (guarantees optimality when given sufficient time), **anytime** (produces high-quality solutions early and improves them over time), and **generic** (easily instantiated with different heuristics and strategies). CA-DL8.5 builds upon the Complete Anytime Beam Search (CABS) framework [19], extending DL8.5 with an iterative weakening strategy that gradually relaxes pruning constraints across restarts. It generalizes LDS-DL8.5 and extends Top-k-DL8.5 to a complete method.

Second, we perform a rigorous empirical study of CA-DL8.5 under four heuristic strategies: Purity, Gain, Discrepancy, and Top-k. We evaluate these variants using the primal gap integral metric [3], a principled anytime evaluation measure that captures performance over time. Our results show that CA-DL8.5 instantiated with discrepancy-based search (equivalent to LDS-DL8.5) and Top-k consistently outperform other instantiations and also improves upon the Blossom algorithm.

Overall, our work provides a unified and extensible framework for designing anytime exact decision tree algorithms, offering both theoretical guarantees and strong empirical performance. The CA-DL8.5 algorithm opens up new possibilities for combining optimality with real-time responsiveness in interpretable machine learning.

2 Related Work and Background

Greedy Approaches. Traditional algorithms like CART [6] and C4.5 [17] build trees greedily using local information. While efficient, they lack optimality guarantees. Optimization-based approaches using MILP [4], SAT [14], MaxSAT [9], and CP [18] provide guarantees but struggle to scale.

Dynamic Programming Approaches and DL8.5. Dynamic Programming (DP) methods [2,7,15] improve the scalability of exact tree learning. DL8.5 [2], the foundation of our work, learns optimal decision trees on binary datasets $\mathcal{D}$ under minimum support and maximum depth constraints. A binary dataset $(\mathcal{D}, \mathcal{C})$ contains examples $\mathcal{D} \subseteq \prod_{f \in \mathcal{F}} \{f, \bar{f}\}$ over Boolean features $\mathcal{F}$, with $\mathcal{C}(x)$ the class label of example x.

For $\mathcal{S} \subseteq \mathcal{D}$ and feature f, define $\mathcal{S}(f) = \{x \in \mathcal{S} \mid f \in x\}$ and analogously $\mathcal{S}(\bar{f})$. A binary decision tree assigns features to internal nodes and labels edges with f or $\bar{f}$, each root-to-leaf path encoding a conjunction of tests. For a branch b, let $\mathcal{S}(b) = \bigcap_{f \in b} \mathcal{S}(f)$ be the subset of examples reaching that branch, and let $\mathcal{S}_c(b) = \{x \in \mathcal{S}(b) \mid \mathcal{C}(x) = c\}$ denote those examples of class c. The classification error of branch b is then $|\mathcal{S}(b)| - \max_{c \in \mathcal{C}} |\mathcal{S}_c(b)|$.

DL8.5 explores the search space via a depth-first traversal of an *OR–AND* search tree. At each OR node, a feature is selected and the two corresponding AND nodes ($\{f, \bar{f}\}$) are explored in order: first the examples satisfying the feature, then the remaining ones. This recursive filtering naturally enforces minimum-support and maximum-depth constraints, and the errors of the two subtrees are combined to compute the node's error. Efficiency comes from upper-bound pruning, which discards branches that cannot improve the current best solution, and from a trie-based cache that avoids recomputing identical subproblems, together greatly reducing the search space.

Anytime Complete Approaches. DL8.5's depth-first search prioritizes leftmost branches (Fig. 1), potentially delaying exploration of promising subtrees. With many features or large depths, the search may remain confined to early subtrees. If interrupted, DL8.5 yields unbalanced trees that may underperform greedy heuristics.

LDS-DL8.5 [11] applies iterative Limited Discrepancy Search to prioritize trees close to a greedy baseline. A discrepancy is choosing a feature differing from the greedy choice. Gradually increasing allowed discrepancies transitions from greedy to exhaustive search, identifying good trees early. Blossom [8] expands nodes layer by layer, processing the non-expanded node closest to the root. Its first solution is the greedy tree, improving anytime performance. However, it may exhibit poor diversification near the root. No empirical comparison exists between Blossom and LDS-DL8.5.

Anytime Incomplete Approaches. Top-k [5] restricts branching to the k highest-ranked features at each OR node, trading off between greedy C4.5 and full DL8.5 search.

LGDT [10] combines greedy strategies with limited-depth lookahead, improving quality without exhaustive search cost.

3 Complete Anytime Beam Search DL8.5 (CADL8.5)

Beam search is an informed search algorithm that explores a space by keeping only the most promising nodes at each level. While it improves upon breadth-first and depth-first search through heuristic pruning, overly aggressive pruning can cause it to miss solutions. To overcome this, Zhang [19] introduced Complete Anytime Beam Search (CABS), based on the principle of *iterative weakening* [16].

CABS performs successive beam search iterations with progressively relaxed pruning constraints. It starts with strict pruning, quickly identifying initial solutions, and then gradually weakens the pruning rules to explore larger portions of the search space. This process enables the algorithm to discover solutions that would have been pruned in earlier iterations.

This iterative relaxation approach gives CABS two valuable properties: (1) it provides anytime behavior by quickly finding initial solutions that improve over time, and (2) it ensures completeness by gradually reducing pruning constraints until an optimal solution is found. These properties make CABS particularly well-suited for complex optimization problems like decision tree learning, where balancing solution quality with computational efficiency is important.

In this work, we propose Complete Anytime DL8.5 (CADL8.5), which adapts the original DL8.5 algorithm by including CABS ideas. It is a generic framework that guarantees high-quality solutions even when terminated early. By integrating principles from CABS, CADL8.5 guides tree construction using adaptive pruning rules. These pruning rules are relaxed over multiple iterations, allowing the algorithm to prioritize promising regions initially while progressively exploring more diverse parts of the solution space.

Algorithm 1 presents the structure of CADL8.5. In addition to the standard inputs required by DL8.5 (dataset $\mathcal{D}$, minimum support `minsup`, and maximum depth `maxdepth`), CADL8.5 requires one additional parameter: `r`, the pruning rule or the set of pruning rules guiding the search space exploration. This allows the algorithm to balance quickly finding initial solutions and ensuring optimality through complete exploration.

Table 1. Pruning rules definitions for CADL8.5

Rule	State & Constraint	Key Functions
Purity	$\texttt{T\{purity : float\}}$ $\texttt{R\{min\_purity : float\}}$	$\texttt{update : purity} = 1 - \frac{\texttt{ctx.e}}{\lvert\texttt{ctx.S}\rvert}$ $\texttt{prune : purity} \geq \texttt{min\_purity}$ $\texttt{relax : min\_purity} \mathrel{+}= \delta$
Gain	$\texttt{T\{cum\_gap : float\}}$ $\texttt{R\{max\_gap : float\}}$	$\texttt{update : cum\_gap} \mathrel{+}= \texttt{best\_gain} - \texttt{gain}$ $\texttt{prune : cum\_gap} > \texttt{max\_gap}$ $\texttt{relax : max\_gap} \mathrel{+}= \delta$
Discrepancy	$\texttt{T\{tot\_discr : int\}}$ $\texttt{R\{max\_discr : int\}}$	$\texttt{update : tot\_discr} \mathrel{+}= \texttt{ctx.i}$ $\texttt{prune : tot\_discr} \geq \texttt{max\_discr}$ $\texttt{relax : max\_discr} \mathrel{+}= \delta$
Top-k	$\texttt{T\{feat\_idx : int\}}$ $\texttt{R\{k : int\}}$	$\texttt{update : feat\_idx} = \texttt{ctx.i}$ $\texttt{prune : feat\_idx} \geq \texttt{k}$ $\texttt{relax : k} \mathrel{+}= \delta$

CADL8.5 extends DL8.5 with an outer loop (Lines 6–11) that repeatedly invokes `BeamDL8.5` while progressively relaxing constraints. Starting with strict pruning to rapidly find a feasible solution and establish an upper bound, the `relax` function weakens pruning rules after each iteration, expanding the search space. This continues until a perfect tree is found, no further relaxation is possible, or time is exhausted.

To implement this strategy, CADL8.5 introduces two generic types: `T` and `R`. Type `T` encapsulates state information required to apply pruning decisions, while `R` stores the parameters that define the current rule constraints. These types are manipulated through five core functions:

Algorithm 1: Complete Anytime DL8.5 (CADL8.5)

```
Input  : D, rule, minsup, maxdepth
Output: Best tree satisfying minsup and maxdepth
Struct Best{error : float, ub : float, tree : Tree, state : T}
cache ← Trie < branch, Best >
ub ← +∞
c_0 ← { i: 0, e: error(∅), s: |D| }
s_0 ← initial_state(c_0)
do
    sol ← BeamDL8.5(∅, ub, c_0, s_0)
    ub ← sol.error
    rule ← relax(rule)
    s_0 ← sol.state
while sol is not optimal or rule is relaxable
return sol.tree
Procedure BeamDL8.5(b, ub, c_p, s_p)
    e ← error(b)
    if |b| = maxdepth or e = 0 then
        return Best{e, ub, leaf(b), terminal_state(s_p)}
    if time limit reached or prune(s_p, rule) then
        return Best{e, ub, leaf(b), s_p}
    node ← cache.get(b)
    if node ≠ ∅ and ub ≤ node.ub and ¬prune(node.state, rule) then
        return node
    τ ← ∅, child_ub ← ub, optimal ← true
    foreach (i, f) in F sorted by heuristic do
        if |D(b ∪ {f})| < minsup or |D(b ∪ {f̄})| < minsup then
            continue
        c_l ← { i: i, e: error(b ∪ {f̄}), s: |D(b ∪ {f̄})| }
        s_l ← update(s_p, c_l)
        left ← BeamDL8.5(b ∪ {f̄}, child_ub, c_l, s_l)
        if left.tree = ∅ then
            continue
        c_r ← { i: i, e: error(b ∪ {f}), s: |D(b ∪ {f})| }
        s_r ← update(s_p, c_r)
        right ← BeamDL8.5(b ∪ {f}, child_ub − left.error, c_r, s_r)
        if right.tree = ∅ then
            continue
        e ← left.error + right.error
        if prune(left.state, rule) or prune(right.state, rule) then
            optimal ← false
        if e < child_ub then
            child_ub ← e
            τ ← Tree(left.tree, right.tree)
            c_p.error ← e
            s_p ← update(s_p, c_p)
        if e = 0 then
            break
    if optimal then
        s_p ← terminal_state(s_p)
    node ← Best{child_ub, ub, τ, s_p}
    cache.save(b, node)
    return node
```

- update(t : T, c : Context) $\rightarrow$ T: modifies a node's state based on its context, including error metrics, dataset size, and feature selection information.

- `prune`(t : T, r : R) → `boolean`: evaluates whether a node should be pruned based on its current state and rule parameters.
- `relax`(r : R) → R: incrementally weakens rule constraints to permit wider exploration in subsequent iterations.
- `terminal_state`(t : T) → T: generates a special state marking a node as fully explored and exempt from future pruning.
- `initial_state`(c : `Context`) → T: creates the initial state for the root node based on its context.

The search begins by constructing an initial context c_0 at the root using the dataset's error and size. This context is passed to `initial_state` to generate the root's initial state t_0 (Line 5), which starts the first `BeamDL8.5` iteration with the initial upper bound.

During `BeamDL8.5`, each node's state is evaluated before expansion. The `update` function computes an updated state (s_l, s_r) based on the node's context (Lines 27 and 32), which is then evaluated using `prune` (Line 17) to check if the node violates active constraints. If so, the node is not expanded.

Nodes are also not expanded when they reach terminal conditions: either the maximum tree depth is attained or perfect classification (zero error) is achieved. In these cases, the node is explicitly marked as fully explored by applying the `terminal_state` function (Line 15). This ensures that terminal nodes are treated as optimal in future caching operations and pruning decisions, preventing unnecessary recomputation of already optimal subtrees.

CADL8.5 extends DL8.5's caching strategy to avoid redundant computations across iterations. Each branch b is associated with a `Best` object that stores the optimal subtree, its error bound, and its state. Before expanding any node, the cache is queried for previous results. If a compatible cached entry exists—one whose upper bound is consistent with the current requirements and not pruned under the current rule set—the cached result is immediately reused (Line 20). Additionally, nodes are marked as fully explored only when all their children have been optimally processed (Lines 47–49).

To control the exploration cost of nodes in CADL8.5, we implement four different pruning strategies: the *Purity* rule, the *Gain* rule, the *Discrepancy* rule, and the *Top-k* rule. Table 1 summarizes these pruning rules using the generic type structures introduced earlier.

3.1 Purity Rule

The purity rule defines a minimum purity threshold in R. Node expansion stops when purity meets or exceeds this threshold. Weakening incrementally increases the threshold by δ until 1.0. If purity remains below the threshold, search continues until maximum depth. The purity of branch b is $purity(b) = 1 - error(b)/|\mathcal{S}(b)|$.

3.2 Gain Rule

The gain rule restricts feature expansion using an information gain threshold. At a node, let τ^* be the highest gain and $\tau(f)$ the gain of feature f. The cumulative gap is $\texttt{cum\_gap} = \texttt{cum\_gap}_{\text{parent}} + (\tau^* - \tau(f))$. A feature expands only if $\texttt{cum\_gap} \leq \texttt{max\_gap}$. Setting $\texttt{max\_gap} = 0$ yields greedy C4.5-like behavior while larger values allow broader exploration. The cumulative nature of the constraint enforces a stricter selection in deeper parts of the tree, preventing getting stuck in lower parts of the search space.

3.3 Discrepancy Rule

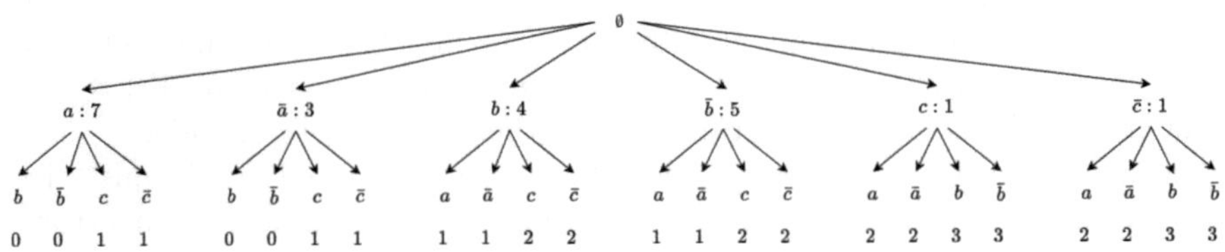

Fig. 2. Search tree for three features. Values at depth 1 show node error before expansion; at depth 2, discrepancy costs.

The Discrepancy rule employs the same principle of Limited Discrepancy Search (LDS) as in the LDS-DL8.5 algorithm [11], to control deviations from a preferred exploration order. Each node tracks the total discrepancy `tot_discr` accumulated from the root, where each feature is assigned an index i based on its position in the candidate list. The discrepancy of a path thus reflects how many times the search deviated from the leftmost option.

At each node, only features whose associated cost does not exceed the threshold `max_discr` are considered. For example, exploring only the leftmost feature at each split (with $\texttt{max\_discr} = 0$) results in a greedy tree. Increasing the discrepancy budget expands the search space and enables exploring other parts of the search space.

As illustrated in Fig. 2, if feature A is explored first, then $cost(a) = cost(\bar{a}) = 0$. Choosing B instead at the same level requires $cost(b) = cost(\bar{b}) = 1$. Similarly, deeper paths such as ba and $b\bar{a}$ have $cost = 1$ since A is the first successor of b, and $cost = 2$ for branches like bc since C is the second.

3.4 Top-k Rule

The *Top-k rule* controls the breadth of the search by limiting the number of features explored at each node. It considers only the k best candidates, where the position of a feature in the sorted list determines its index `ctx.i`. When $k = 1$, the algorithm behaves greedily, producing trees similar to CART or C4.5

depending on the heuristic used. We also propose a new variant, denoted as Top-k^* in the results, where the beam width k is halved at each level of the tree, with a minimum value of one. This allows reducing the time spent in the deeper parts of the search space in early iterations.

4 Results

Table 2. Average primal integral on depth 6

Approach	Sub	Runtime (s)					
		15	30	60	120	240	300
C4.5	–	64.3	64.3	64.3	64.3	64.3	64.3
Top-3	–	46.3	45.0	44.4	44.1	44.0	44.0
Top-5	–	37.5	35.2	34.3	33.8	33.6	33.5
DL8.5	–	46.6	40.4	34.6	31.5	29.5	28.8
Blossom	–	**27.1**	24.7	19.6	14.5	9.6	8.5
CA-Purity	–	48.8	42.5	36.8	31.6	27.7	26.7
CA-Gain	exponential	43.7	36.3	32.2	29.2	26.4	24.7
	luby	42.8	35.9	31.5	25.5	22.0	21.2
	monotonic	43.2	36.6	32.3	26.2	22.5	21.5
CA-Discrepancy	exponential	29.1	23.2	18.6	15.8	13.0	11.4
	luby	27.2	**20.7**	**16.7**	13.9	9.2	7.8
	monotonic	27.4	20.8	16.7	14.0	8.8	**7.4**
CA-Top-k	exponential	34.9	30.6	24.9	20.0	17.5	16.8
	luby	32.7	25.4	20.5	17.3	14.0	12.0
	monotonic	31.1	23.5	19.0	16.3	13.4	11.3
CA-Top-k^*	exponential	34.4	27.2	23.3	17.0	12.7	11.4
	luby	31.0	25.4	18.0	**12.2**	**8.5**	7.7
	monotonic	31.2	25.6	18.4	12.2	8.6	7.7

To evaluate the performance of CADL8.5, we conducted a series of experiments. This section presents the results. We first analyze the anytime behavior of CADL8.5 under the previously described pruning rules, and then compare its performance in finding optimal solutions.

All experiments were performed on 25 datasets from CP4IM[1] and the UCI Machine Learning Repository[2], using a minimum support threshold of 1. The algorithms were executed on a server equipped with an Intel Xeon Platinum

[1] https://dtai-static.cs.kuleuven.be/CP4IM/datasets/.

[2] https://archive.ics.uci.edu/.

8160 CPU and 320 GB of RAM, running Rocky Linux 8.4. The source code and datasets are publicly available at https://github.com/haroldks/pytrees-rs. For comparison, we include a scikit-learn implementation of C4.5[3], as well as DL8.5 and Blossom.

The runtime of C4.5 is not reported in the time-based analysis because it consistently completed in less than one second on all datasets and depths considered. This is negligible compared to the time budgets used for CADL8.5 and other exact methods. We evaluate anytime performance using the *average primal integral* [3], which measures how quickly the primal bound converges toward the optimal solution. The *primal gap* of a solution $x(t)$ at time t is defined as $\gamma(x(t)) = \frac{|x(t)-x_{\text{opt}}|}{|x(t)|}$. The primal function $p(t)$ equals 1 if no solution has been found by time t, and $\gamma(x(t))$ otherwise. The primal integral is computed as $P(T) = \int_0^T p(t)\,dt = \sum_{i=1}^{n} p(t_{i-1})(t_i - t_{i-1})$, where t_i denotes the time at which a new incumbent solution is found. The ratio $P(t_{\max})/t_{\max}$represents the average solution quality over time; smaller values indicate better anytime behavior. Here, $t_{\max}$ denotes the time limit (timeout) allocated to each run of the algorithm.

Table 3. Average primal integral on depth 7

Approach	Sub	Runtime (s)					
		15	30	60	120	240	300
C4.5	–	69.3	69.3	69.3	69.3	69.3	69.3
Top-3	–	52.9	51.7	51.1	50.8	50.7	50.6
Top-5	–	46.8	41.0	38.5	37.2	36.6	36.5
DL8.5	–	58.3	52.0	47.2	43.9	40.4	39.5
Blossom	–	37.1	30.5	24.7	21.5	18.6	17.2
CA-Purity	–	55.9	50.0	45.7	43.2	40.7	39.7
CA-Gain	exponential	50.5	45.3	39.1	35.5	33.0	32.5
	luby	54.7	51.4	45.6	41.7	38.0	36.9
	monotonic	55.1	51.6	46.2	41.9	39.3	38.5
CA-Discrepancy	exponential	33.0	28.3	25.0	22.8	19.3	17.8
	luby	**31.3**	26.9	22.8	19.0	15.1	**14.0**
	monotonic	31.5	**26.5**	**22.5**	**19.0**	**15.1**	14.0
CA-Top-k	exponential	41.8	34.5	28.5	25.1	22.4	21.1
	luby	39.4	31.2	25.3	22.1	19.1	18.0
	monotonic	37.7	29.7	24.2	20.8	17.6	16.6
CA-Top-k^*	exponential	41.8	36.7	33.9	31.3	27.7	26.6
	luby	39.6	34.7	30.7	26.7	23.4	22.3
	monotonic	41.8	35.8	31.3	26.9	23.8	22.6

[3] https://scikit-learn.org/.

Tables 2 and 3 show average primal integral evolution across time budgets (15-300s) for depths 6 and 7, excluding datasets where DL8.5 solves in under 1s. We evaluate three relaxation strategies: *Monotonic* (threshold increased by 1), *Exponential* (multiplied by 2), and *Luby* [13]. All complete anytime strategies outperform DL8.5. CA-Discrepancy (Luby/Monotonic) and CA-Top-k^* achieve best results, especially at longer timeouts. At depth 6, CA-Discrepancy with monotonic relaxation achieves 7.4 at 300 s, while CA-Top-k^* reaches 7.7.

Under short timeouts (15–30 s), Blossom produces high-quality early solutions, often outperforming CADL8.5 variants. However, CADL8.5 quickly catches up and surpasses Blossom as runtime increases. This trend becomes more pronounced at depth 7 (Table 3), where CA-Discrepancy with Monotonic and Luby relaxation achieves average primal integral values of 14.0, better than Blossom's 17.2 at 300 s. These improvements are largely due to the increased diversification in its search strategy, exploring more parts of the search space, whereas Blossom tends to remain focused on optimizing the deeper layers of a specific tree before moving elsewhere.

Among all pruning rules, *Discrepancy* consistently outperforms the others. CA-Top-k and its variant CA-Top-k^* also perform well, particularly at large timeouts. CA-Gain and CA-Purity lag behind: the Gain rule often selects larger subtrees, leading to longer subsearches; Purity may require several ineffective relaxations before contributing to meaningful diversification.

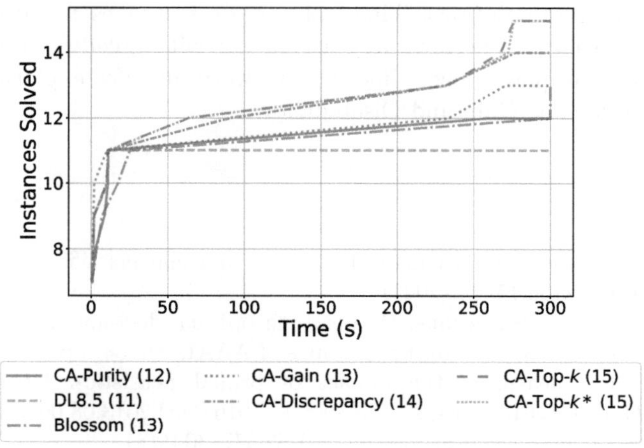

Fig. 3. Cumulative number of instances solved as a function of time with the total number of instances solved by each approach within 300 s

Greedy baselines such as C4.5, Top-3, and Top-5 deliver quick but static solutions. Among them, Top-5 performs best under tight time budgets (15–30 s), briefly outperforming DL8.5. However, none of the greedy methods improve their solutions over time. Figure 3 illustrates the cumulative termination count of each algorithm to find and prove optimality within a time budget of 300 s and a depth constraint of 5. Overall, the CADL8.5 variants demonstrate superior solving power compared to DL8.5 and Blossom. Notably, CA-Top-k^* solves the most instances (15 out of 25) and does so more quickly than the other methods across most of the timeline. CA-Discrepancy and CA-Top-k also perform strongly, solving 14 and 15 instances respectively, and surpassing other approaches beyond the 50-s mark. Blossom shows a steep initial rise, indicating strong early performance, but plateaus sooner than the CADL8.5 variants. CA-Gain achieves a similar final count as Blossom (13 instances) but shows slower progress in the early phase. DL8.5 and CA-Purity underperform both in terms of speed and total solved instances, solving only 11 and 12 datasets respectively. This shows that CADL8.5 variants especially Top-k(*) and Discrepancy do not compromise the ability to reach optimal solutions.

5 Conclusion

In this paper, we introduced CADL8.5, a complete anytime framework for decision tree learning that generalizes DL8.5, LDS-DL8.5, and Top-k. It combines DL8.5's efficient branch-and-bound pruning and trie-based caching with a restart-based search that progressively relaxes pruning criteria, guided by rule-based strategies such as node purity, Information Gain gap, Discrepancy, and Top-k. Our experiments show that CADL8.5 variants, especially CA-Top-k^* and CA-Discrepancy, deliver strong anytime performance without sacrificing the ability to reach optimal solutions. They solve more instances to optimality than DL8.5 and Blossom and perform at least on par with greedy approaches while improving solution quality over time. Future work includes exploring combined rule strategies such as Gain and Discrepancy.

References

1. Aghaei, S., Gómez, A., Vayanos, P.: Strong optimal classification trees. arXiv preprint arXiv:2103.15965 (2021)
2. Aglin, G., Nijssen, S., Schaus, P.: Learning optimal decision trees using caching branch-and-bound search. In: Proceedings of AAAI, vol. 34, pp. 3146–3153 (2020)
3. Berthold, T.: Measuring the impact of primal heuristics. Oper. Res. Lett. **41**(6), 611–614 (2013). https://doi.org/10.1016/j.orl.2013.08.007. https://www.sciencedirect.com/science/article/pii/S0167637713001181
4. Bertsimas, D., Dunn, J.: Optimal classification trees. Mach. Learn. **106**, 1039–1082 (2017)
5. Blanc, G., Lange, J., Pabbaraju, C., Sullivan, C., Tan, L., Tiwari, M.: Harnessing the power of choices in decision tree learning. In: Advances in Neural Information Processing Systems, vol. 36 (2024)
6. Breiman, L., Friedman, J., Olshen, R., Stone, C.: Classification and Regression Trees, vol. 37. Wadsworth International Group (1984)
7. Demirović, E., et al.: MurTree: optimal decision trees via dynamic programming and search. J. Mach. Learn. Res. **23**(26), 1–47 (2022)

8. Demiroviç, E., Hebrard, E., Jean, L.: Blossom: an anytime algorithm for computing optimal decision trees. In: International Conference on Machine Learning, pp. 7533–7562 (2023)
9. Hu, H., Siala, M., Hebrard, E., Huguet, M.J.: Learning optimal decision trees with MaxSAT and its integration in AdaBoost. In: IJCAI-PRICAI 2020, 29th International Joint Conference on Artificial Intelligence and the 17th Pacific Rim International Conference on Artificial Intelligence (2020)
10. Kiossou, H., Schaus, P., Nijssen, S., Aglin, G.: Efficient lookahead decision trees. In: International Symposium on Intelligent Data Analysis, pp. 133–144. Springer (2024)
11. Kiossou, H., Schaus, P., Nijssen, S., Houndji, V.R.: Time constrained DL8. 5 using limited discrepancy search. In: Joint European Conference on Machine Learning and Knowledge Discovery in Databases, pp. 443–459. Springer (2022)
12. van der Linden, J.G., Vos, D., de Weerdt, M.M., Verwer, S., Demirović, E.: Optimal or greedy decision trees? Revisiting their objectives, tuning, and performance. arXiv e-prints, arXiv–2409 (2024)
13. Lorenz, J.H.: Restart strategies in a continuous setting. Theory Comput. Syst. **65**(8), 1143–1164 (2021)
14. Narodytska, N., Ignatiev, A., Pereira, F., Marques-Silva, J.: Learning optimal decision trees with sat. In: International Joint Conference on Artificial Intelligence 2018, pp. 1362–1368. Association for the Advancement of Artificial Intelligence (AAAI) (2018)
15. Nijssen, S., Fromont, E.: Mining optimal decision trees from itemset lattices. In: KDD, pp. 530–539 (2007)
16. Provost, F.: Iterative weakening: optimal and near-optimal policies for the selection of search bias. In: Proceedings of the Eleventh National Conference on Artificial Intelligence, pp. 749–755 (1993)
17. Quinlan, J.: C4.5: Programs for Machine Learning. Elsevier (2014)
18. Verhaeghe, H., Nijssen, S., Pesant, G., Quimper, C.G., Schaus, P.: Learning optimal decision trees using constraint programming. Constraints **25**(3), 226–250 (2020)
19. Zhang, W.: Complete anytime beam search. In: AAAI/IAAI, pp. 425–430 (1998)

Section: XAI

NTS-DAGMA: A Score-Based Causal Discovery for Anomaly Detection

Navin Vincent[1,2], Abhishek Srinivasan[1,2(✉)], Anders Holst[2,3], and Sepideh Pashami[3,4]

[1] Traton AB, Södertälje, Sweden
abhishek.srinivasan@se.traton.com
[2] KTH Royal Institute of Technology, Stockholm, Sweden
[3] RISE AB, Kista, Stockholm, Sweden
[4] Halmstad University, Halmstad, Sweden

Abstract. Anomaly detection is an essential component for ensuring the safety and reliability of critical systems. Currently, most machine learning-based anomaly detection approaches rely purely on correlations among sensor signals, rather than causal relations, making them susceptible to spurious associations. This limitation can lead to poor generalization and unreliable anomaly detection in practical scenarios. To address this, we propose an anomaly detection approach which leverages score-based causal discovery, NTS-DAGMA. This causal discovery method advances over prior work by combining the network architecture from NTS-NOTEARS with the acyclicity constraint from DAGMA. Like other prediction-based anomaly detection methods, it can predict future states; however, it does this by learning and utilizing causal relations in the time series data. Through comprehensive experiments, we demonstrate that: (i) on causal discovery tasks NTS-NOTEARS and NTS-DAGMA achieve similar performances; (ii) on anomaly detection tasks NTS-NOTEARS and NTS-DAGMA also perform similar to each other, and have comparable performance to state-of-the-art ML approaches; and (iii) most importantly, our results show that NTS-DAGMA provides causally meaningful models: detects anomalies which propagate through child nodes in agreement with the inferred causal graph. Incorporating causal structure into the model enables improved interpretability and aligns anomaly detection with the physical dynamics of the system.

Keywords: NTS-DAGMA · causal discovery · anomaly detection

1 Introduction

Cyber-physical systems (CPS) such as smart power grids, autonomous cars and automated industrial systems are now ubiquitous. Actions taken by CPS have physical consequences in the real world, system failures could have drastic safety, economic and operational consequences. To avoid such failures, anomaly detection is essential for maintaining system reliability, preventing cascading faults, and ensuring safe real-world operation.

M. Baratchi et al. (Eds.): IDA 2026, LNCS 16513, pp. 113–125, 2026.
https://doi.org/10.1007/978-3-032-23833-7_9

Standard approaches for anomaly detection are predominantly knowledge-driven. These methods rely on models grounded in physics-based principles, signal processing, expert rules, and mathematical transformations such as the Fourier or wavelet transform [3]. These approaches are effective and well understood for small systems. As CPS continue to grow in scale, complexity, and dimensionality, such approach struggle to adapt, motivating a transition towards data-driven solutions [3,7].

In response to this shift, anomaly detection evolved from classical statistical models to more expressive deep learning architectures such as neural networks, recurrent neural networks (RNNs), and, more recently, graph neural networks (GNNs) [3,7]. Most of these approaches use either paradigm of forecasting-based or paradigm of reconstruction-based approaches for anomaly detection. Forecasting-based approaches learn to predict future sensor states, while reconstruction -based approaches learn to reconstruct normal behavior from a compressed state, with deviations indicating of anomalies [7]. Despite their empirical success, such models rely on correlations rather than true system dependencies, making them more sensitive to spurious associations. As a result, they often exhibit poor generalization and offer limited interpretability [4], for instance, they may detect anomalies but fail to explain why they occur or how they propagate through the system. This highlights the need for models that reflect true system dependencies.

To overcome these limitations, we propose a causality-aware anomaly detection framework that integrates causal discovery with forecasting modeling. Causal methods aim to learn the underlying generative mechanisms of data, enabling more robust reasoning under distributional shifts [2]. For anomaly detection, this is particularly valuable because anomalies often arise from disruptions in causal mechanisms rather than surface-level statistical deviations. Prior work, Invariant Causal Prediction (ICP) [6,12], has demonstrated that the models with causal knowledge generalize well in different environments and the causality could be learned directly from data. However, the existing variants of ICP currently deal only with non-time series data which limits their applicability over CPS. If the causal structure of a CPS could be estimated, similar generalization properties could be leveraged for anomaly detection. Yet CPS typically exhibit complex, high-dimensional multivariate time series, and their causal structure is unknown. This motivates the use of causal discovery.

Causal discovery methods seek to infer these relationships directly from observational data. Several algorithmic families have emerged, including constraint-based approaches (e.g., PCMCI, PCMCI$^+$) [14–16,18], and score-based methods (e.g., GES, DAGMA, NOTEARS) [1,17]. Among these, differentiable approaches such as NOTEARS and DAGMA have made causal discovery amenable to large-scale, gradient-based learning [18]. Additionally, score based approach such as NTS-NOTEARS are modeled to predict the next state of the system, which naturally mimics the paradigm of forecasting based anomaly detection.

We address the problem of generalization by developing a causality-aware anomaly detection framework that learns to predict the system's dynamics via its causal structure. Our results demonstrate that the proposed model: (i) can identify causal relations, (ii) can effectively detect anomalies, and (iii) exhibit causal interpretability on the detected anomalies: highlighting how deviations propagate across the causal graph.

2 Why Causality?

Understanding why causality matters requires moving beyond surface-level statistical relationships toward models that can reason about mechanisms and consequences. Judea Pearl's *Book of Why* [11] conceptualizes this progression into a hierarchy of three levels of causal reasoning: association, intervention, and counterfactuals [11]. These levels describe the increasing ability of a model to reason not only what is, but also what could be and what would have been.

At the first level—association, models capture statistical dependencies among variables. This involves learning to predict one variable from others based solely on observed correlations. Traditional statistical and machine learning models operate primarily at this level: they can identify that two sensors co-vary, or that certain feature combinations tend to occur during anomalies. For instance, an anomaly detector in a two-tank process might learn that when inlet pressure drops, tank level often decreases shortly after. Such association-based reasoning can detect anomalies but are susceptible to react to spurious correlations, leading it struggles to generalize under distributional shifts [4].

The second level—intervention, goes beyond prediction by reasoning about actions. It allows us to ask: *"What happens if we deliberately change a system variable?"* Formally expressed using the do-operator, do(·) [10], this level captures cause-and-effect relationships that support what-if reasoning. In the same two-tank example, an interventional model could estimate the downstream effects of fixing the pump speed or valve opening to specific values, e.g., $P(.|do(PumpSpeed = 10))$. Such reasoning enables the testing of different operational settings to evaluate system stability and spot conditions that could trigger anomalies.

The third level—counterfactual reasoning, extends this capability by enabling retrospective inference [13]. Counterfactuals allow us to imagine alternate scenarios given known exogenous factors, asking questions such as, *"Would the anomaly still have occurred if the valve pressure had been higher?"*. In the CPS context, this means that after an anomaly occurs, the model can reconstruct hypothetical system trajectories to explain why the failure happened and how it could have been prevented.

Through this hierarchy, causality enriches anomaly detection in several important ways. Association enables anomaly identification; intervention enables controlled experimentation within the model to predict outcomes of hypothetical adjustments; and counterfactual reasoning supports root-cause analysis and system diagnosis. Integrating these causal levels transforms anomaly detection from a reactive task into a proactive framework for causality-aware reliability modeling—one that predicts, explains, and ultimately can mitigate system anomalies.

In this work, as a first step towards this broader vision we focus on causal aware model for association tasks. This foundation enables more interpretable anomaly detection and lays the groundwork for future extensions toward interventional and counterfactual reasoning in cyber-physical systems.

3 Relevant Work

Recent years have seen a growing interest in causality-aware modeling for anomaly detection, particularly in cyber-physical systems (CPS). Existing approaches differ in how they incorporate causal structure into the detection pipeline, and each comes with distinct advantages and limitations.

Yang et al. (2022) [20] propose a two-stage paradigm: (i) causal discovery is applied to identify parent–child dependencies among variables, and (ii) each variable is modeled using a conditional variational autoencoder (VAE), where conditioning is performed on its causal parents. Although intuitive, this design suffers from poor scalability, as a separate model must be trained for every variable. This becomes computationally prohibitive for large sensor networks and restricts the expressiveness of interactions across variables.

Meli et al. (2024) [9] employ PCMCI, a constraint-based causal discovery method, to infer causal relations among sensor variables. The discovered graph is then used to model anomalies through learning a correlation score for each relation and deviations from these learned scores are flagged as anomalies. While this approach offers interpretability through the causal structure, its reliance on correlation score limits its ability to capture complex nonlinear dynamics that are prevalent in high-dimensional CPS data.

Han et al. (2025) [5] adopt a fundamentally different notion of causality, Granger causality, which captures predictive (rather than structural) dependencies. The method uses encoder–decoder architectures to learn latent representations associated with exogenous influences and then identifies root causes through Granger-based tests. However, this approach does not recover a structural causal model, making it unsuitable for interventions or counterfactual reasoning.

Xiao et al. (2025) [19] propose a reinforcement learning–based approach in which a Double-Q network is trained to identify causal influences in partially observed environments. Although the method aims to infer causal relations through action–reward interactions, the causal interpretation depends heavily on the RL agent's exploration dynamics and environment assumptions. This makes the recovered structure sensitive to reward shaping and does not guarantee the identification of true structural causal relations.

Overall, existing causality-informed anomaly detection methods either rely on constrained causal assumptions (e.g., Granger causality), require computationally expensive per-variable modeling, or provide only limited causal interpretability. These limitations motivate the need for a unified framework that jointly learns causal structure and predictive dynamics in a scalable manner while providing interpretable anomaly explanations suited to high-dimensional CPS.

4 Problem Statement

Consider multivariate time series data $X = \{X_t\}_{t=1}^{T}$, where each observation $X_t \in \mathbb{R}^d$ represents d variables at time t. Our goal is to learn a directed acyclic graph (DAG) structure $\mathcal{G}$ represented by an adjacency $W \in \mathbb{R}^{T\times d\times d}$, where d is the number of variable and T the number of time lags.

The predictive function f_θ models each variable's conditional distribution given its causal parents:

$$X_t^j = f_\theta(\mathrm{PA}(X_t^j), X_{t-1}, \ldots, X_{t-K}) + \epsilon_t^j, \tag{1}$$

where $\mathrm{PA}(X_t^j)$ denotes the parent set of variable X_t^j at instantaneous time-step, K is the maximum time lag, and ϵ_t^j captures residual uncertainty (including both aleatoric and epistemic).

5 Preliminaries

In this section, we provide a background on NOTEARS, NTS-NOTEARS and DAGMA, which form the basis of our approach. These methods perform score-based causal discovery, aiming to recover a weighted adjacency matrix W that represents causal relations among variables. The data fitting function $\mathcal{L}(W;X)$ is usually optimized subjected to an acyclity constraint $h(W)$. $\|W\|_1$ enforces sparsity in causal relations.

$$\min_{W\in\mathbb{R}^{d\times d}} \mathcal{L}(W;X) + \lambda\|W\|_1 \quad \text{subject to} \quad h(W) = 0.$$

The constrained optimization problem is reduced using augmented Lagrangian or penalty-based formulation:

$$\min_{W} \mathcal{L}(W;X) + \lambda\|W\|_1 + \alpha\, h(W),$$

where $\lambda > 0$ encourages sparsity, $\alpha > 0$ controls the acyclicity penalty, and $\mathcal{L}(W;X)$ is an empirical loss. NOTEARS, NTS-NOTEARS, and DAGMA target related but distinct causal discovery settings and their vital difference is in the acyclicity constraint $h(W)$. NOTEARS introduces a differentiable smooth acyclicity constraint; NTS-NOTEARS extends this formulation to time series; and DAGMA proposes an improved acyclicity characterization based on matrix analysis.

Smooth Acyclicity Constraint in NOTEARS: NOTEARS proposes a smooth acyclic constraint h_{NT} as:

$$h_{\mathrm{NT}}(W) := \mathrm{tr}\big(\exp(W \circ W)\big) - d,$$

with $\circ$ denoting the Hadamard (element-wise) product and $\exp(\cdot)$ the matrix exponential. The trace term effectively counts weighted cycles of all lengths, ensuring that $h_{\text{NT}}(W) = 0$ if and only if W is a DAG.

Extension to time series - NTS-NOTEARS: NTS-NOTEARS extends the NOTEARS formulation to time series by incorporating both inter- and intra-time dependencies into the adjacency structure. Acyclicity is imposed only on the contemporaneous adjacency matrix $B^{(0)}$, while lagged dependencies naturally avoid cycles:

$$h_{\text{NT}}\big(B^{(0)}\big) = \text{tr}\big(\exp(B^{(0)} \circ B^{(0)})\big) - d = 0.$$

DAGMA - Log-Determinant Acyclicity and Gradient-Based Optimization: While NOTEARS relies on the matrix-exponential trace constraint $h_{\text{NT}}(W)$, *DAGMA* (Directed Acyclic Graphs via Matrix Analysis) introduces an alternative, log-determinant–based acyclicity characterization, together with an optimization strategy that avoids repeatedly solving inner constrained problems.

DAGMA reparameterizes the adjacency matrix using properties of M-matrices and their inverses. Its smooth acyclicity function h_{DG} is defined as:

$$h_{\text{DG}}(W) := -\log\ \det\big(I - \gamma\, \phi(W)\big) + \text{const},$$

where $\phi(W)$ is element-wise non-negative transform of W (e.g., an element-wise square or absolute value), $\gamma > 0$ is chosen to ensure the spectral radius of $\gamma\phi(W)$ strictly less than 1, and I is the identity matrix. Under appropriate conditions, the equality $h_{\text{DG}}(W) = 0$ if and only if W is acyclic.

6 Method

We build on the time series and acyclicity formulations from Sect. 5 to introduce *NTS-DAGMA*, a time series extension of DAGMA with a CNN-based parameterization of temporal dependencies, a log-determinant acyclicity barrier on the instantaneous graph, and uncertainty-aware anomaly detection.

6.1 NTS-DAGMA: Time Series DAG with CNN Parameterization

Let $X = \{X_t\}_{t=1}^{T}$ be a multivariate time series with $X_t \in \mathbb{R}^d$. Following the auto-regressive type of model as defined in Sect. 4 and re-iterating that we model each variable X_t^j as a function of its instantaneous parents and K lagged time steps:

$$X_t^j = f_{\theta_j}\big(\text{PA}(X_t^j), X_{t-1}, \dots, X_{t-K}\big) + \epsilon_t^j, \qquad j = 1, \dots, d,\ t = K+1, \dots, T. \quad (2)$$

Similar to NTS-NOTEARS, we achieve this structure using a 1D convolutional neural network (CNN) followed by an MLP for each variable j. The CNN processes a window of past values and the current-time context:

$$\hat{X}_t^j = f_{\theta_j}\big(\{X_{t-\ell}\}_{\ell=0}^{K}\big), \quad (3)$$

where the first convolution layer has kernels whose receptive field spans all d variables across $K+1$ time points. Let $\phi_{i,j}^{(\ell)}$ denote the filter weights connecting $X_{t-\ell}^i$ to the hidden representation used to predict X_t^j. We induce a time-lag–specific adjacency matrix by

$$W_{ij}^{(\ell)} = \|\phi_{i,j}^{(\ell)}\|_2, \qquad \ell = 0, \ldots, K, \tag{4}$$

so that $W^{(0)}$ denotes instantaneous (within-slice) influences, while $\{W^{(\ell)}\}_{\ell \geq 1}$ capture lagged causal effects.

Consistent with NTS-NOTEARS (Sect. 5), we only require the *instantaneous* graph to be acyclic, while lagged connections are unconstrained because temporal order already precludes cycles. We therefore enforce acyclicity on $W^{(0)}$ using a DAGMA-style log-determinant surrogate (Sect. 5):

$$h_{\text{ldet}}\big(W^{(0)}\big) = -\log\, \det\big(sI - W^{(0)} \circ W^{(0)}\big) + d \log s, \tag{5}$$

where $s > 0$ is chosen so that $sI - W^{(0)} \circ W^{(0)}$ is positive definite and $\circ$ denotes the Hadamard product. Under these conditions, $h_{\text{ldet}}(W^{(0)}) = 0$ if and only if $W^{(0)}$ encodes a DAG, while $h_{\text{ldet}}(W^{(0)}) > 0$ penalizes directed cycles. This is the time series analogue of the DAGMA acyclicity penalty in Sect. 5.

For modeling aleatoric uncertainty, we assume a heteroscedastic Gaussian likelihood for each X_t^j, with aleatoric variance $\sigma_{\text{ale}}^{j,2}(t)$ predicted by a parallel variance head attached to f_{θ_j}. The negative log-likelihood over the sequence is

$$Q(f_\theta; X) = \frac{1}{T-K} \sum_{t=K+1}^{T} \sum_{j=1}^{d} \left(\frac{(X_t^j - \hat{X}_t^j)^2}{2\,\sigma_{\text{ale}}^{j,2}(t)} + \frac{1}{2} \log \sigma_{\text{ale}}^{j,2}(t) \right). \tag{6}$$

We promote sparsity in the induced graph and regularize network weights via

$$\mathcal{R}(W) = \sum_{\ell=0}^{K} \lambda_1^{(\ell)} \sum_{j=1}^{d} \|\phi_j^{(\ell)}\|_1, \qquad \mathcal{R}(\theta) = \frac{\lambda_2}{2} \sum_{j=1}^{d} \|\theta_j\|_2^2, \tag{7}$$

where $\phi_j^{(\ell)}$ collects all filters feeding into variable j at lag ℓ.

Following the log-det barrier approach of DAGMA, we solve a sequence of unconstrained optimization problems along a central path:

$$\theta^{(t)} = \arg\min_{\theta} \Big[Q(f_\theta; X) + \mathcal{R}(W) + \mathcal{R}(\theta) + \mu^{(t)} h_{\text{ldet}}\big(W^{(0)}\big) \Big], \tag{8}$$

with a geometrically decreasing barrier parameter $\mu^{(t+1)} = \alpha \mu^{(t)}$, $\alpha \in (0, 1)$. As $\mu^{(t)} \to 0$, the solution is driven toward an acyclic instantaneous graph, mirroring the continuous DAG learning framework in Sects. 5.

6.2 Uncertainty-Aware Anomaly Scoring

To capture epistemic uncertainty, we apply Monte Carlo dropout at test time. For each variable j and time t, we perform S stochastic forward passes and obtain

$\{\hat{X}_t^{j,(s)}\}_{s=1}^S$. The predictive mean and epistemic variance are calculated and the total predictive uncertainty combines aleatoric and epistemic components: We define an uncertainty-normalized anomaly score

$$s_t^j = \frac{|X_t^j - \hat{\mu}_t^j|}{\sigma_{\text{tot}}^j(t)}, \tag{9}$$

and apply window-based change-point detection with an auto-regressive cost function to the resulting score sequences $\{s_t^j\}_t$. In addition, the ratio $r_t^j = \frac{\sigma_{\text{epi}}^j(t)}{\sigma_{\text{ale}}^j(t)}$ serves as a sensitive indicator of break-in-dynamics events: large r_t^j highlights regimes where the learned DAG-based model is systematically uncertain about its own predictions, suggesting a violation of the causal mechanisms captured by the instantaneous graph $W^{(0)}$ and its lagged extensions $\{W^{(\ell)}\}_{\ell \geq 1}$.

7 Experiments

7.1 Datasets

We use two datasets for our experiments: synthetic data and SWaT data [8].

Synthetic Data: We generate time series data using Erdős-Rényi (ER) graphs with varying dimensions $d \in \{5, 6, 8, 10\}$ and lags $K \in \{1, 2, 3, 4\}$. Data follows additive index models of the form:

$$x = \tanh(XW_1) + \cos(XW_2) + \sin(XW_3) + z, \tag{10}$$

with sequence length $n = 5000$ samples.

SWaT Dataset: The Secure Water Treatment testbed [8], provides 11 days of continuous recordings at 1 Hz from a six-stage water treatment plant, including 51 process variables (25 analog sensors, 26 binary actuators). The dataset contains 480k normal operation samples and 450k samples with 36 distinct cyber-physical attack scenarios.

7.2 Experimental Setup

All experiments were conducted on AWS Sagemaker ml.g4dn.xlarge instances (Tesla T4 GPU, CUDA 12.4).

Causal Discovery: For NTS-DAGMA, we use MLP hidden dimensions of 10, $\lambda_1 = 0.005$, $\lambda_2 = 0.005$, $lr = 0.002$, $\mu_{\text{factor}} = 0.1$, and weight threshold 0.05. Both models use $m = 10$ kernels.

Anomaly Detection: On real-world datasets, we set $m = 20$ and $K = 100$. Post-causal training uses reconstruction fine-tuning with learning rate $lr = 10^{-3}$, batch size 1000, 25 refinement epochs, and $\lambda_2 = 0.01$. Change-point detection employs window size 5000 with severity threshold at 70% of maximum severity.

For causal discovery, we evaluate using Structural Hamming Distance (SHD), Precision (PRE), Recall (REC), and F1 score. For anomaly detection, we compute point-adjusted Precision, Recall, and F1 scores, marking entire anomaly segments as correctly classified if any point within is detected. We compare against classical methods (PCA, KNN), deep learning approaches (Autoencoder, DAGMM), and state-of-the-art models (LSTM-VAE, MAD-GAN) for anomaly detection performance.

Table 1. Causal discovery results on synthetic data (selected configurations)

Model	d	K	SHD	F1	PRE	REC
NTS-NOTEARS	5	3	0	1.00	1.00	1.00
NTS-NOTEARS	8	2	2	0.96	1.00	0.92
NTS-NOTEARS	10	4	10	0.88	1.00	0.79
NTS-DAGMA(ours)	5	3	3	0.84	0.87	0.82
NTS-DAGMA(ours)	8	2	5	0.84	0.87	0.80
NTS-DAGMA(ours)	10	4	13	0.81	0.92	0.73

8 Results

8.1 Causal Discovery Performance

Table 1 presents causal discovery results on synthetic data. NTS-NOTEARS consistently achieves higher F1-scores (0.88–1.00) and lower SHD compared to NTS-DAGMA across most configurations. However, NTS-DAGMA demonstrates approximately 38% faster epoch per step, offering significant computational advantages.

Both models effectively recover primary causal structures, with NTS-NOTEARS achieving perfect precision in most cases. The learned adjacency matrices align well with expected physical dependencies, as verified against known system schematics.

8.2 Anomaly Detection Performance

Table 2 shows the results on SWaT Dataset, the classical methods (PCA, KNN) fail to detect SWaT anomalies (F1 < 0.10). Deep generative approaches like DAGMM achieve high recall (69.52%) but suffer from low precision (27.46%). LSTM-VAE and MAD-GAN attain extremely high precision (96–99%) but limited recall (60–64%), yielding F1 scores of 0.74 and 0.77.

Our NTS-NOTEARS model delivers a balanced trade-off: 71.3% precision and 80.9% recall (F1 $= 0.75$), matching state-of-the-art performance while using dramatically fewer resources. NTS-DAGMA achieves similar performance (F1 $=$ 0.73).

Table 2. Anomaly detection performance on SWaT dataset

Method	Precision	Recall	F1
PCA	24.92	21.63	0.23
KNN	7.83	7.83	0.08
Autoencoder	72.63	52.63	0.61
DAGMM	27.46	69.52	0.39
LSTM-VAE	96.24	59.91	0.74
MAD-GAN	98.97	63.74	0.77
NTS-NOTEARS	71.30	80.90	0.75
NTS-DAGMA (ours)	68.30	78.95	0.73

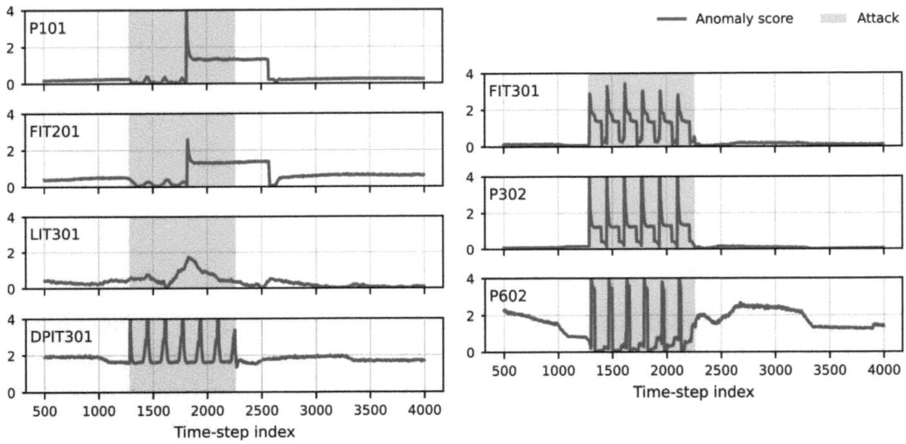

Fig. 1. Features `P101, FIT201, LIT301, DPIT301,FIT301, P302, P602` whose anomaly score (blue), attack in DPIT301 (red). Notice the delayed spikes in the nodes `P101, FIT201, LIT301` due to lack of first order connections from `DPIT301` and almost simultaneous spikes in the other nodes due to strong first order connections from `DPIT301`. (Color figure online)

8.3 Interpretability and Break-in-Dynamics Detection

DPIT301, is a differential pressure indicating transmitter in the SWAT system and during the attack, the value of DPIT301 is maliciously set above 40kpa which causes normal operation to stop and the backwash process is started repeatedly. This is reflected in the anomaly score for the feature DPIT301 in Fig. 1. The learned causal structures enable node-level anomaly isolation. For instance, during attacks on sensor DPIT301 (Fig. 1) in the SWaT dataset, nodes with strong first-order causal connections (FIT301, P302, P602) exhibit simultaneous anomaly score spikes, while weakly connected nodes (P101, FIT201, LIT301) show delayed or absent responses. This causal graph information enables precise localization of anomaly sources.

Once the attack begins, the effect of the event will trickle down to all of the other nodes as well. All the nodes having strong first order dynamic graph connection with DPIT301 will reflect the anomaly almost immediately as can be seen from Fig. 1 (FIT301, P302, P602) which can make it difficult to narrow down the exact node in which the anomaly took place if you do not have access to the causality derived information.

9 Discussion

This work explores the use of score-based causal discovery as a modeling principle for anomaly detection in multivariate time-series data from cyber-physical systems (CPS). While the proposed NTS-DAGMA framework demonstrates competitive anomaly detection performance and improved structural insight, it is important to discuss insights on assumption and limitations.

Causal Assumptions and Latent Factors. In the synthetic experiments, data are generated from a known time-expanded DAG and structural equation model (SEM), ensuring causal sufficiency by construction. In contrast, real-world CPS such as the SWaT water treatment plant are complex physical systems in which not all relevant states or disturbances may be directly measured. Environmental influences, uninstrumented physical states, or details of control logic may act as effective latent variables. In the presence of such latent confounders, NOTEARS-style structure learning methods—including NTS-NOTEARS and the proposed NTS-DAGMA—may introduce directed edges between observed variables that explain correlations induced by unobserved common causes. In these cases, the learned graph should be interpreted with care, as it may reflect predictive dependencies rather than a strictly causal structure. That said, CPS are typically engineered with extensive sensing and monitoring, particularly during normal operation, to capture the dominant physical processes governing system behavior. As a result, while latent confounding cannot be ruled out, we expect its impact to be limited in many practical settings. Domain expertise therefore plays an important role in assessing whether the inferred structure aligns with known physical relationships and operational constraints.

Hyperparameters and Design Choices. In this work an exhaustive hyperparameter sensitivity analysis has not been carried out. Instead, fixed hyperparameters are reported, and performance is examined across varying graph sizes and lag configurations in the synthetic setting. Preliminary observations indicate that performance may interact with factors such as sample size, lag length, and regularization strength. A systematic analysis of these interactions, including ablations on architectural and optimization parameters, would provide deeper insight into the conditions under which NTS-DAGMA offers clear advantages and is an important direction for future investigation.

10 Conclusion

In this work, we propose NTS-DAGMA, a score-based causal discovery approach for anomaly detection that builds upon NTS-NOTEARS and DAGMA. We demonstrate how this model can be used for anomaly detection in multivariate time-series data from cyber-physical systems. The results show that the proposed approach achieves performance comparable to classical and state-of-the-art machine learning methods, while additionally incorporating causal structure. This causal knowledge enables causal interpretability; for instance, detected anomalies propagate through the inferred graph in a manner consistent with the learned dependencies, providing insight into how faults affect connected components of the system.

Looking ahead, we identify several promising directions for advancing causal aware anomaly detection. First, extending the experimental evaluation to systematically study robustness, scalability, quantitative interpretable measures, and hyperparameter sensitivity may yield deeper insights into model behavior. Second, expanding the evaluation to additional real-world cyber-physical system datasets would further assess generalization across different system configurations and operating conditions. Finally, enhancing the framework to support explicit interventions and counterfactual reasoning would move the model higher on Pearl's ladder of causation, enabling more powerful diagnostic and prescriptive capabilities. Overall, causal models hold the potential to transform anomaly detection from reactive monitoring into proactive, explanatory, and actionable system intelligence.

Acknowledgments. This project is funded by Vinnova grant under project FAIM (Future AI-based maintenance) with grant-number 2023-01917.

References

1. Bello, K., Aragam, B., Ravikumar, P.: DAGMA: learning DAGs via m-matrices and a log-determinant acyclicity characterization. In: Advances in Neural Information Processing Systems, vol. 35, pp. 8226–8239 (2022)
2. Bühlmann, P.: Invariance, causality and robustness. Stat. Sci. **35**(3), 404–426 (2020)
3. Chandola, V., Banerjee, A., Kumar, V.: Anomaly detection: a survey. ACM Comput. Surv. (CSUR) **41**(3), 1–58 (2009)
4. Geirhos, R., et al.: Shortcut learning in deep neural networks. Nat. Mach. Intell. **2**(11), 665–673 (2020)
5. Han, X., Absar, S., Zhang, L., Yuan, S.: Root cause analysis of anomalies in multivariate time series through granger causal discovery. In: The Thirteenth International Conference on Learning Representations (2025)
6. Heinze-Deml, C., Peters, J., Meinshausen, N.: Invariant causal prediction for nonlinear models. J. Causal Infer. **6**(2), 20170016 (2018)
7. Jin, M., et al.: A survey on graph neural networks for time series: forecasting, classification, imputation, and anomaly detection. IEEE Trans. Pattern Anal. Mach. Intell. **46**(12), 10466–10485 (2024). https://doi.org/10.1109/TPAMI.2024.3443141

8. Mathur, A.P., Tippenhauer, N.O.: SWaT: a water treatment testbed for research and training on ICS security. In: 2016 International Workshop on Cyber-Physical Systems for Smart Water Networks (CySWater), pp. 31–36. IEEE (2016)
9. Meli, D.: Explainable online unsupervised anomaly detection for cyber-physical systems via causal discovery from time series. In: 2024 IEEE 20th International Conference on Automation Science and Engineering (CASE), pp. 4120–4125. IEEE (2024)
10. Pearl, J.: Causality. Cambridge University Press (2009)
11. Pearl, J., Mackenzie, D.: The Book of Why: The New Science of Cause and Effect. Basic Books (2018)
12. Peters, J., Bühlmann, P., Meinshausen, N.: Causal inference by using invariant prediction: identification and confidence intervals. J. R. Stat. Soc. Ser. B Stat Methodol. **78**(5), 947–1012 (2016)
13. Peters, J., Janzing, D., Schölkopf, B.: Elements of Causal Inference: Foundations and Learning Algorithms. The MIT Press (2017)
14. Runge, J.: Discovering contemporaneous and lagged causal relations in autocorrelated nonlinear time series datasets. In: Conference on Uncertainty in Artificial Intelligence, pp. 1388–1397. PMLR (2020)
15. Runge, J., Nowack, P., Kretschmer, M., Flaxman, S., Sejdinovic, D.: Detecting and quantifying causal associations in large nonlinear time series datasets. Sci. Adv. **5**(11), eaau4996 (2019)
16. Spirtes, P., Glymour, C.N., Scheines, R.: Causation, Prediction, and Search. MIT Press (2000)
17. Sun, X., Schulte, O., Liu, G., Poupart, P.: NTS-NOTEARS: learning nonparametric DBNs with prior knowledge. In: International Conference on Artificial Intelligence and Statistics (2021). https://api.semanticscholar.org/CorpusID:252907768
18. Vowels, M.J., Camgoz, N.C., Bowden, R.: D'ya like DAGs? A survey on structure learning and causal discovery. ACM Comput. Surv. **55**(4), 1–36 (2022)
19. Xiao, X., Shen, B., Yue, X.: Causality-informed anomaly detection in partially observable sensor networks: moving beyond correlations. arXiv preprint arXiv:2507.09742 (2025)
20. Yang, W., Zhang, K., Hoi, S.C.: Causality-based multivariate time series anomaly detection. CoRR (2022)

Analyzing Shapley Additive Explanations to Understand Anomaly Detection Algorithm Behaviors and Their Complementarity

Jordan Levy[1,2](✉), Paul Saves[1], Moncef Garouani[1], Nicolas Verstaevel[1], and Benoit Gaudou[1]

[1] Institut de Recherche en Informatique de Toulouse, Toulouse, France
{jordan.levy,paul.saves,moncef.garouani,nicolas.verstaevel, benoit.gaudou}@irit.fr

[2] TwinswHeel, Fontanes, France

Abstract. Unsupervised anomaly detection is a challenging problem due to the diversity of data distributions and the lack of labels. Ensemble methods are often adopted to mitigate these challenges by combining multiple detectors, which can reduce individual biases and increase robustness. Yet building an ensemble that is genuinely complementary remains challenging, since many detectors rely on similar decision cues and end up producing redundant anomaly scores. As a result, the potential of ensemble learning is often limited by the difficulty of identifying models that truly capture different types of irregularities. To address this, we propose a methodology for characterizing anomaly detectors through their decision mechanisms. Using SHapley Additive exPlanations, we quantify how each model attributes importance to input features, and we use these attribution profiles to measure similarity between detectors. We show that detectors with similar explanations tend to produce correlated anomaly scores and identify largely overlapping anomalies. Conversely, explanation divergence reliably indicates complementary detection behavior. Our results demonstrate that explanation-driven metrics offer a different criterion than raw outputs for selecting models in an ensemble. However, we also demonstrate that diversity alone is insufficient; high individual model performance remains a prerequisite for effective ensembles. By explicitly targeting explanation diversity while maintaining model quality, we are able to construct ensembles that are more diverse, more complementary, and ultimately more effective for unsupervised anomaly detection.

Keywords: Unsupervised Anomaly Detection · Ensemble Learning · Model Selection · Explainable AI

1 Introduction

Anomaly detection is a challenging problem, primarily because of the intrinsic nature of anomalies. In fact, anomalies are defined as deviations from what is

M. Baratchi et al. (Eds.): IDA 2026, LNCS 16513, pp. 126–140, 2026.
https://doi.org/10.1007/978-3-032-23833-7_10

considered to be normal behavior [1]. Therefore, depending on how this normality is defined, one algorithm or another may be more adapted for good detection. For instance, some methods may define normality in geometric terms and use distance-based rules to spot outliers, while some others may leverage probabilistic models that flag rare or low-probability instances as anomalies [2].

In many real-world settings, obtaining labeled anomalies is expensive or infeasible because anomalous events are rare, costly to produce, or hazardous to induce (*e.g.*, in the nuclear industry where detection must operate without waiting for faults to occur) [3]. Consequently, semi-supervised and unsupervised approaches are often preferred. In purely unsupervised anomaly detection (UAD), practitioners typically rely on machine learning methods because the "normal" behavior cannot be specified exhaustively by hand; therefore, algorithm designers must adopt implicit assumptions about normality (for instance, geometric/distance-based or probabilistic models), and no single assumption is guaranteed to be appropriate across all applications [4]. Furthermore, in a fully unsupervised setting, anomaly detection algorithms cannot be directly benchmarked due to the lack of ground truth labels and, as stated in the *no free lunch theorem*, no single detector consistently outperforms others across all datasets [5]. This theorem is also valid for anomaly detection [6]. Nevertheless, the choice of algorithm remains critical, as an inappropriate selection may lead to poor performance [4].

A common strategy to mitigate this issue is to combine multiple detectors in an ensemble framework, leveraging the strengths of each model and its diverse assumptions about what constitutes a normal behavior. Ensemble learning has been widely adopted for this reason and has demonstrated strong empirical performance across various anomaly detection scenarios [4]. The advantage of ensemble methods in anomaly detection lies in their ability to combine models that rely on different underlying assumptions about normality. By aggregating detectors with diverse inductive biases, ensembles can capture a wider variety of anomaly types. This diversity often translates into improved detection coverage [7].

A key challenge in ensemble learning is the selection of appropriate base detectors, as benchmarking with ground truth labels is typically impossible in UAD. While a robust ensemble theoretically requires models that are both diverse and performant [7], identifying such complementary behaviors remains an open issue [8].

To address this challenge, we propose a novel methodology that characterizes the behavior of anomaly detectors using SHapley Additive exPlanations (SHAP) [9]. Unlike approaches relying solely on outputs, we focus on understanding the internal decision mechanisms of the algorithms. Our study reveals that detectors with similar explanation patterns tend to produce redundant anomaly scores, whereas explanation divergence is a strong indicator of complementarity. Consequently, we demonstrate that selecting detectors based on their explanation behavior leads to improved ensemble performance. This leads to three major contributions:

- An analysis of UAD algorithms based on their SHAP explanations, which demonstrates that their behaviors are correlated to their output similarity.
- A comparison between SHAP explanations and model output similarities for selecting diverse anomaly detector models.
- An empirical study quantifying the relative importance of diversity versus individual performance, demonstrating that model quality remains a critical prerequisite.

This article is structured as follows: Sect. 2 presents the related works. The methodology of our approach is described in Sect. 3. Experimental results are introduced in Sect. 4. Finally, we conclude the article in Sect. 5.

2 Related Work

2.1 Unsupervised Anomaly Detection Algorithms

UAD is a widely studied problem [2] with a large number of algorithms, each founded on different assumptions regarding the nature of anomalies. Libraries like PyOD [10] have been introduced to standardize the usage of these models. Among these, distance-based methods (KNN, LOF, CBLOF) characterize anomalies based on proximity. Conversely, reconstruction-based algorithms (AutoEncoder, PCA) detect anomalies by learning to reconstruct normal data patterns. Other approaches, such as HBOS, ECOD, and COPOD, leverage probabilistic assumptions, while OCSVM and DeepSVDD utilize one-class classification. Finally, algorithms like Isolation Forest and LODA rely on random partitioning and space projection, respectively. Each of these algorithms has a specific hypothesis regarding the nature of anomalies; however, empirical evidence suggests that no single detector consistently outperforms all others [6]. Consequently, employing an ensemble of multiple models serves as a robust strategy, effectively leveraging diverse underlying hypotheses to improve detection capabilities [4].

2.2 Model Selection for Unsupervised Anomaly Detection

Selecting an appropriate model for an UAD task is commonly known as Unsupervised Outlier Model Selection (UOMS). UOMS is a particularly challenging problem that has gained increasing importance as the number of available models, comprising various algorithmic families and hyperparameter configurations, keeps expanding as new methods appear in the literature [2].

To that end, a common strategy is to estimate model quality directly from unlabeled data in an unsupervised way, to select the best models. These approaches fall into two main categories. *Stand-alone* methods [11,12] compute an unsupervised score for each detector independently, while *consensus-based* methods assess detectors by measuring agreement within a pool of models and selecting those that best conform to the group [13,14]. Empirical evidence indicates that stand-alone evaluation criteria often fail to deliver consistent or reliable results in practice. In contrast, consensus-based approaches, despite their higher computational cost, have emerged as a more effective and promising strategy for unsupervised outlier model selection [8].

There is limited work in the scientific literature on UOMS for ensemble learning. In [7], the authors show that diversity in algorithm hypotheses tends to give better results. To characterize diversity, the authors use a weighted Pearson correlation between the anomaly scores of the models. However, they also highlight that diversity is important, but the chosen algorithms should already have good performance on the dataset to have a performant ensemble model. From the previous assumption, the authors in [15] introduced SELECT, which, instead of trying to select models according to their diversity, selects models according to their performances from a pseudo ground-truth label created from anomaly scores. More recently, in [16], the authors perform anomaly detection in time series using multiple convolutional autoencoders as an ensemble and optimize diversity explicitly by incorporating a diversity metric based on the dissimilarity of the reconstruction outputs. They also demonstrate that this diversity-driven strategy improves detection performance.

While it has been shown that models output like anomaly scores can reflect diversity, in this work, we investigate a consensus method based on SHAP values, which characterize the behavior of different anomaly detection models across different datasets. Furthermore, in our work, we quantify the impact of diversity and propose a methodology to model the relationship between mean individual performance, diversity, and the resulting ensemble gain.

2.3 Explainability

Explainable Artificial Intelligence (XAI) seeks to make complex models transparent by producing human-interpretable accounts of how inputs, model structure, and uncertainty produce particular outputs [17]. Among local, model-agnostic explanation methods, SHapley Additive exPlanations (SHAP) are widely used because they provide axiomatic, instance-level feature attributions grounded in the Shapley value principle [9].

Beyond interpretability, SHAP representations have been exploited to compare models and to form informative embeddings: recent works use aggregated SHAP attributions to identify model families or to serve as features for downstream learning tasks [18,19]. To our knowledge, leveraging SHAP for model comparison and selection has not been systematically investigated in the UAD setting, where the lack of labels makes model interpretability and selection particularly challenging.

3 Methodology

3.1 Problem Setting

We consider the problem of UAD, where the goal is to identify anomalous samples within a dataset $\mathcal{D} = \{x_1, \ldots, x_n\}$ containing n instances in a d-dimensional feature space. We assume a set of m anomaly detection models

$\mathcal{M} = \{M_1, M_2, \ldots, M_m\}$. For a given dataset, every model M_i produces an anomaly score output vector $s_i = M_i(\mathcal{D}) \in \mathbb{R}^n$ where $s_i^{(k)}$ is the anomaly score of the kth input point according to the ith model. Our objective is to analyze the behavior of these models and to identify groups of models that share similar interpretative structures.

3.2 Model Similarity from Explanations

For each model M_i, we compute its SHAP explanation matrix $Sh_i \in \mathbb{R}^{n \times d}$, where $Sh_i^{(k)} \in \mathbb{R}^d$ is the vector representing the contribution of every feature to the anomaly score of instance x_k.

We define the behavior similarity between two models M_i and M_j as the average per-instance Pearson correlation between their SHAP vectors:

$$\rho_{ij}^{\text{PS}} = \frac{1}{n} \sum_{k=1}^{n} \text{corr}\big(Sh_i^{(k)}, Sh_j^{(k)}\big).$$

To capture ranking consistency between feature importances rather than raw magnitudes, we also compute a similarity based on the Normalized Discounted Cumulative Gain (NDCG):

$$\rho_{ij}^{\text{NDCG}} = \frac{1}{2n} \sum_{k=1}^{n} \Big(\text{NDCG}\left(|Sh_i^{(k)}|, |Sh_j^{(k)}|\right) + \text{NDCG}\left(|Sh_j^{(k)}|, |Sh_i^{(k)}|\right) \Big),$$

where NDCG evaluates the agreement in the ranked importance of features between two detectors. If both models assign high SHAP importance to the same features, their NDCG value will be close to 1, indicating consistent interpretative behavior [20].

3.3 Linking Similarities to Detection Performances

To compare the similarity of models with performances, we introduce two additional matrices: the matrix of score correlations and the matrix of Jaccard similarities.

We define the similarity between two models M_i and M_j based on the correlation of their anomaly scores. Specifically, we compute the average per-instance Pearson correlation between their score vectors as:

$$\rho_{ij}^{\text{Score}} = \frac{1}{n} \sum_{k=1}^{n} \text{corr}\big(s_i^{(k)}, s_j^{(k)}\big).$$

This results in a symmetric matrix capturing the pairwise similarity between models in terms of their scoring behavior.

To compare model outputs directly, we define A_i as the set of instances identified as anomalies by model i (where the anomaly score exceeds a threshold τ_i). The Jaccard similarity between two models i and j is defined as:

$$J_{ij} = \frac{|A_i \cap A_j|}{|A_i \cup A_j|}.$$

This metric measures the overlap between models: a value of 1 implies identical predictions, while 0 implies disjoint sets of detected anomalies.

To quantify the relationship between the different measures, we first transform each similarity matrix $P \in \{\rho^{PS}, \rho^{NDCG}, \rho^{Scores}, J\}$ into a dissimilarity matrix $D^P_{\mathcal{D}}$. This is computed as $D^P_{\mathcal{D}} = 1 - P$, representing the distance between two detectors for a given dataset $\mathcal{D}$. We then apply the Mantel test [21] to determine whether two given dissimilarity matrices are statistically correlated. The Mantel coefficient (r_M) is calculated as the Pearson correlation between the upper triangular elements of the matrices, with statistical significance established via permutation testing. In this context, a significant positive r_M implies that anomaly detectors sharing similar SHAP-based explanations also exhibit similar anomaly scores.

4 Experiments

We conducted experiments to answer the following: (1) Does explanation similarity imply prediction similarity? (2) Do SHAP-based metrics outperform raw outputs in quantifying diversity? and (3) How significantly does this diversity impact the accuracy and robustness of the resulting ensemble?

4.1 Experimental Setup

In the experiments, 14 UAD algorithms were used. Specifically, COF, KNN, LOF, IForest, PCA, CBLOF, LODA, HBOS, MCD, OCSVM, DAGMM, DeepSVDD, COPOD, and ECOD were employed. For the implementation of each model, the Python library PyOD [10] was used. The hyperparameters defined by the authors of each model were retained, and no indication of the percentage of anomaly was provided for any dataset, keeping the threshold to its default value. Regarding the explainability component, we used the model-agnostic Kernel SHAP implementation of the Python package `shap`. This choice was necessary to ensure a unified explanation framework across our diverse algorithms. Furthermore, to ensure approximation stability, the background dataset was summarized using a k-means clustering with $k = 50$ centroids.

To further strengthen the robustness of the results, each dataset was repeatedly split into a training set and a test set five times, with the training set representing 80% of the data. The random seed for each split was set equal to the iteration index to ensure reproducibility. The source code used in the experiments is available on GitHub[1].

[1] https://github.com/jordanlv/Analyzing-SHAP-UOMS.

4.2 Datasets Considered

Due to the high computational cost associated with SHAP, the dataset selection was restricted to the 50% smallest datasets available in ADBench [6]. Additionally, datasets containing more than 20 features were removed to maintain computational feasibility. This initial filtering process resulted in a set of 16 datasets from various applications: annthyroid (AN), breastw (BR), glass (GL), Hepatitis (HE), Lymphography (LY), mammography (MA), PageBlocks (PB), Pima (PI), Stamps (ST), thyroid (TH), vertebral (VE), vowels (VO), WBC (WB), Wilt (WL), wine (WN) and yeast (YE). Datasets characteristics can be found in Table 6 in Appendix 6.

4.3 Correlation Between Similarity Matrices

We computed the four similarity matrices ρ^{PS} (linear correlation between SHAP values), ρ^{NDCG} (similarity of SHAP feature importance rankings), ρ^{Scores} (linear correlation between anomaly scores), J (jaccard index between anomaly predictions) for each dataset. The mean matrices across all datasets are presented in Fig. 1, where hierarchical clustering was applied to optimize the ordering. Visually, two clusters of models emerge. First, COPOD and ECOD consistently group, which is expected as both algorithms rely on similar distributional assumptions. Second, a larger group comprising OCSVM, AutoEncoder, IForest, PCA, KNN, and CBLOF exhibits high correlation. These algorithms share underlying characteristics related to distance metrics and data encoding strategies.

The Mantel test is used to assess whether the similarity matrices introduced above are statistically correlated. Table 1 reports the mean Mantel correlations across datasets. We observe a strong correlation ($r_M = 0.83$) between ρ^{PS} and ρ^{NDCG}, confirming that SHAP-based similarity is consistent across feature importance magnitude and ranking. Similarly, ρ^{Scores} and J show a high correlation ($r_M = 0.76$), which is intuitive since models with similar score distributions tend to produce similar binary predictions. Finally, the correlation between ρ^{PS} and J ($r_M = 0.67$) suggests that detectors sharing similar reasoning patterns (as captured by SHAP) also tend to give similar anomaly predictions. However, the similarity between each model depends on the dataset. Specifically, Table 2 shows the correlations between ρ^{PS} and ρ^{Scores}, and between ρ^{PS} and J for each dataset. Usually, both pairs are correlated. Some datasets show high correlations between matrices ($r_M > 0.5$) like AN, PI, TH, or VO. Some others have moderate or low correlation between the two matrices. Finally, some interesting datasets are BR, LY or YE where $r_M(\rho^{PS}, \rho^{Scores}) << r_M(\rho^{PS}, J)$.

Overall, the similarity of models is correlated with their predictions. In the next sections, we investigate combining models that are far apart in the similarity matrices to enhance the overall results of an ensemble.

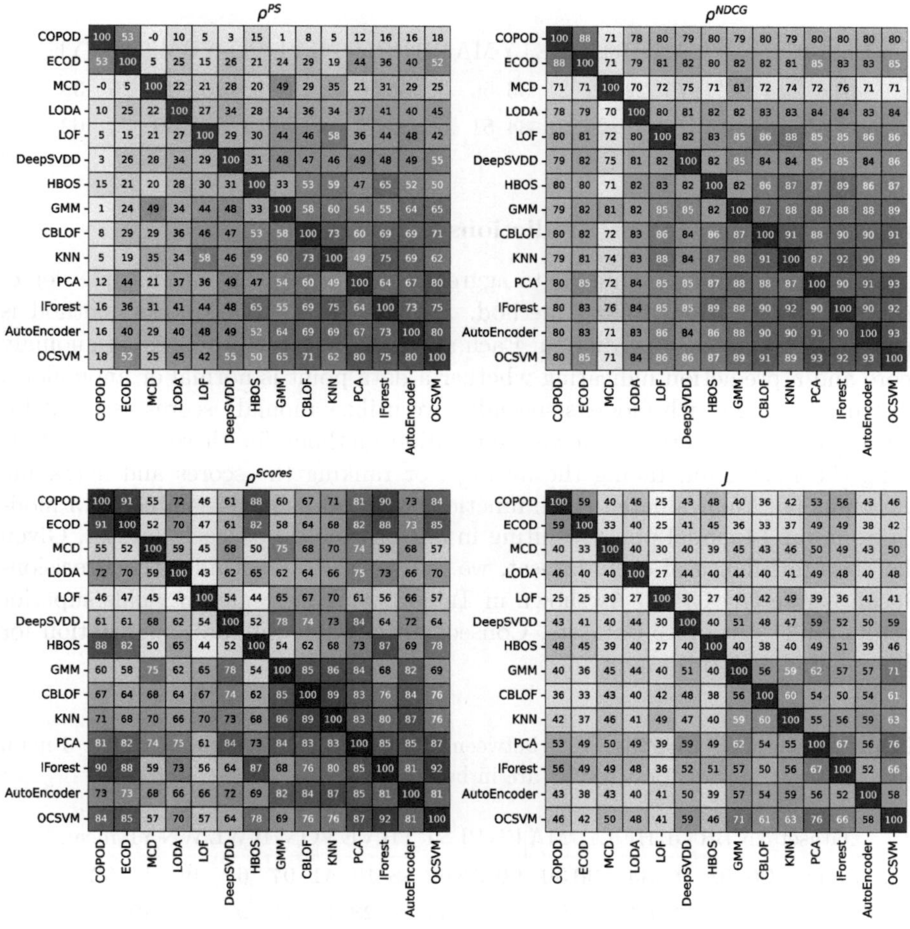

Fig. 1. Mean similarity between models across all datasets. Layout: Top-Left: correlations of SHAP values; Top-Right: NDCG of SHAP values; Bottom-Left: correlations of anomaly scores; and Bottom-Right: Jaccard similarities.

Table 1. Mean Mantel correlations between distance matrices, averaged across datasets. All correlations are significant ($p \leq 0.004$).

	ρ^{PS}	ρ^{NDCG}	ρ^{Scores}	J
ρ^{PS}	1.00	0.83	0.55	0.67
ρ^{NDCG}		1.00	0.54	0.57
ρ^{Scores}			1.00	0.76
J				1.00

Table 2. Mantel correlations ($\times 10^2$) between distance matrices for each dataset.

Dataset	AN	BR	GL	HE	LY	MA	PB	PI	ST	TH	VE	VO	WB	WL	WN	YE
$r_M(\rho^{PS}, \rho^{Scores})$	75	22	42	26	36	56	33	55	56	72	26	56	18	49	26	38
$r_M(\rho^{PS}, J)$	78	51	24	51	84	51	66	52	53	76	−4	58	15	49	30	62

4.4 Aggregate Models Predictions

The objective of an ensemble is to aggregate the predictions of multiple detectors to create a more robust method. A challenge in this kind of method is how to aggregate each prediction. Each anomaly detector produces an anomaly score and a prediction indicating whether a data point is normal or anomalous. To truly leverage each model's specialty, we utilize anomaly scores as input for our aggregation function. Known aggregation methods for these scores include using the maximum, taking the average, or ranking the scores and averaging the ranks [4]. We evaluated these functions on all possible ensembles of 3 models from our 14-model pool (resulting in 364 ensembles) on each dataset. Given the class imbalance of each dataset, we rely on the Area Under the Precision-Recall Curve (AUCPR). As shown in Table 3, rank aggregation yields superior results on 11 out of 16 datasets. Consequently, we adopt rank aggregation for the remainder of this study.

Table 3. Average AUCPR ($\times 10^2$) between all of the 364 ensembles with the 3 main aggregation strategies. Best results are in bold.

Dataset	AN	BR	GL	HE	LY	MA	PB	PI	ST	TH	VE	VO	WB	WL	WN	YE	mean
Rank	**26**	**98**	18	**84**	**100**	**30**	**60**	**55**	**62**	**58**	18	**41**	**97**	6	49	37	**52**
Max	21	59	**22**	64	68	8	35	45	35	28	**28**	11	41	**9**	**53**	**40**	35
Mean	9	49	19	47	36	4	12	42	22	5	**28**	8	25	7	31	39	24

4.5 Complementarity

To select the most dissimilar models, we computed the dissimilarity matrices $D_{\mathcal{D}}^{P}$ for each dataset $\mathcal{D}$, which gives matrices on how far two detectors are for each dataset $\mathcal{D}$. From the pool of 14 models, we constructed ensembles of size $n = 3$. Table 4 presents the correlations between the AUCPR of ensembles and the diversity distances for each similarity matrix. A positive correlation between diversity and performance implies that increasing ensemble diversity leads to superior detection results.

The table shows three key insights. Firstly, selecting diversity from SHAP values tends to give a higher diversity than model outputs (scores and predictions). On 11 out of 16 datasets, using ρ^{PS} and ρ^{NDCG} as diversity, tends to give

Table 4. Correlation ($\times 10^2$) between the distance of models and the AUCPR of the ensembles. Best results are in bold.

metric	AN	BR	GL	HE	LY	MA	PB	PI	ST	TH	VE	VO	WB	WL	WN	YE
ρ^{PS}	−20	**−0**	**5**	**−28**	15	**−9**	−18	**37**	**40**	−38	−44	−66	**51**	55	43	29
ρ^{NDCG}	**39**	−17	−2	−32	−0	−68	**21**	14	15	**19**	−30	−54	36	**62**	**48**	37
ρ^{Scores}	−1	−90	−33	−48	−29	−18	−17	30	15	−12	−12	**−0**	−11	46	7	34
J	−32	−59	−37	−31	**16**	3	−4	24	36	−48	**−7**	−3	−28	55	1	**45**

better results than using ρ^{Scores} and J. Secondly, as stated previously, diversity contributes to ensemble learning by expanding the range of detected anomalies. However, effective model selection should also account for individual model performance, a factor not explicitly optimized in this study. This limitation likely explains why the correlations between distance metrics and performance remain moderate. Furthermore, in some datasets, these correlations are consistently negative, suggesting that diversity is not always beneficial. This phenomenon occurs when a single model dominates the others; in such cases, the primary condition for a successful ensemble is the inclusion of this specific model. Additionally, negative correlations can arise in highly complex datasets where all models exhibit poor performance. In these scenarios, even high diversity cannot compensate for the lack of meaningful detection, resulting in an ineffective ensemble. Finally, it is interesting to note that SHAP values and scores do not show the same diversity as correlations vary between datasets. For example, on the dataset WB, ρ^{Scores} and J are negative, while are positive ρ^{PS} and ρ^{NDCG}. Consequently, the two metrics highlight different diversities.

4.6 Diversity and Individual Performances

While we have shown that diversity enhances ensemble performance by expanding anomaly coverage, our earlier analysis neglected individual model accuracy. Here, we refine our results to demonstrate that despite the value of diversity, individual model quality remains a critical factor. For the purpose of this section, we use ρ^{PS} to quantify diversity.

Figure 2 illustrates the importance of individual model performance for achieving an effective ensemble. Specifically, Fig. 2a presents the results for the LY dataset. A clear correlation between diversity, individual performance, and ensemble accuracy is observable on this dataset. However, this correlation is not always beneficial. In certain scenarios, enforcing diversity can be detrimental to the ensemble. For instance, Fig. 2b reveals that for the VO dataset, diversity offers negligible gain and can even lead to suboptimal results.

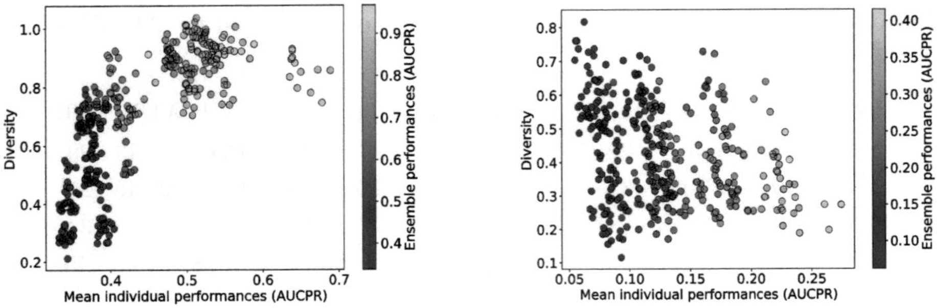

(a) Dataset Lymmography (LY).

(b) Dataset Vowels (VO).

Fig. 2. Relationship between ensemble diversity (given by ρ^{PS}) and average individual performance. Each point represents an ensemble of models. The color scale indicates the ensemble's overall performance (AUCPR).

To assess the relative impact of model quality versus diversity, we performed a linear regression to predict the ensemble performance gain using standardized mean individual performance and diversity scores. We confirmed the stability of these estimates by ruling out multicollinearity (see Appendix 7). The resulting weights, presented in Table 5, indicate that while individual performance is generally the dominant factor, diversity plays a crucial complementary role. In 12 out of 16 datasets, the diversity coefficient is positive, confirming its value as a performance booster. Notably, for the WB and LY datasets, the diversity weight rivals or even exceeds that of individual performance (Ratios of 1.2 and 0.8 respectively). However, overall, diversity plays a secondary yet valuable role in UAD ensemble design.

Table 5. Linear regression weights ($\times 10^2$) predicting ensemble performance from mean individual performance and diversity, along with their ratio (**bold**: positive contributions).

	AN	BR	GL	HE	LY	MA	PB	PI	ST	TH	VE	VO	WB	WL	WN	YE
Indiv. Perf.	2.7	6.3	1.1	6.5	10.7	3.4	5.0	1.7	5.6	5.7	0.1	7.5	2.4	0.0	8.0	0.3
Diversity	0.5	2.0	0.1	0.5	8.4	0.7	2.2	0.2	0.3	1.7	−0.0	−0.6	2.8	0.0	−0.1	−0.0
Ratio	**0.2**	**0.3**	**0.1**	**0.1**	**0.8**	**0.2**	**0.4**	**0.1**	**0.0**	**0.3**	−0.1	−0.1	**1.2**	**0.4**	−0.0	−0.0

5 Conclusion

We presented a methodology for selecting models in UAD ensembles based on the similarity of their explanations. We demonstrated that diversity can be usefuland enhance results, particularly on datasets where individual model performances are close to each other. We analyzed four different diversity metrics: two based on SHAP explanations and two based directly on model outputs. These metrics yielded different results for ensemble model selection, indicating that the diversity they highlight is different. Finally, we established that despite the benefits of diversity, individual model performance remains the deciding factor: weak but diverse models cannot outperform strong but similar models. Nevertheless, we believe that explainability should be more considered in UOMS as it provides new information on model behavior.

A limitation of our approach lies in the computational cost of SHAP, which can become prohibitive for datasets with a large number of instances or features. However, the computational cost can be alleviated by using approximation techniques or surrogate models. Additionally, our strategy is agnostic to the explanation method, allowing for the use of faster interpretability techniques if required. For instance, in [22], authors recently demonstrated how aggregating Partial Dependence Profiles (PDP) across a set of near-optimal models can provide reliable metrics for explanation uncertainty and robustness.

Furthermore, while we successfully model the diversity, our study did not address the optimization of individual models; we neither investigated the individual performances of each anomaly detector nor their best hyperparameters.

In future work, we plan to investigate the divergence between SHAP-based and raw output similarities to refine the model selection process. Furthermore, since individual performance is critical, we plan to integrate methods to characterize model performance into our methodology. Finally, we aim to extend our strategy to time series UAD, where ensemble methods can be even more powerful, as the problem is even more complex [23].

Acknowledgments. The research presented in this paper has received funding from the National Association for Research and Technology under grant number CIFRE 2023/1398 and from Soben company. The authors also thank the National Agency for Research for funding the project MIMICO under grant number ANR-24-CE23-0380.

Disclosure of Interests. The authors declare no Conflict of interest.

Appendix

6 Description of the Datasets

Table 6. Description of the datasets used.

Code	Name	# Samples	# Features	# Anomaly (%)	Category
AN	annthyroid	7200	6	534 (7.42)	Healthcare
BR	breastw	683	9	239 (34.99)	Healthcare
GL	glass	214	7	9 (4.21)	Forensic
HE	Hepatitis	80	19	13 (16.25)	Healthcare
LY	Lymphography	148	18	6 (4.05)	Healthcare
MA	mammography	11183	6	260 (2.32)	Healthcare
PB	PageBlocks	5393	10	510 (9.46)	Document
PI	Pima	768	8	268 (34.90)	Healthcare
ST	Stamps	340	9	31 (9.12)	Document
TH	thyroid	3772	6	93 (2.47)	Healthcare
VE	vertebral	240	6	30 (12.50)	Biology
VO	vowels	1456	12	50 (3.43)	Linguistics
WB	WBC	223	9	10 (4.48)	Healthcare
WL	Wilt	4819	5	257 (5.33)	Botany
WN	wine	129	13	10 (7.75)	Chemistry
YE	yeast	1484	8	507 (34.16)	Biology

7 Assessment of Multicollinearity

To ensure the reliability of the estimated coefficients in our linear regression model, we assessed the potential for multicollinearity between the predictor variables: average individual model performance and ensemble diversity. Multicollinearity can inflate the variance of coefficient estimates, making it difficult to isolate the individual influence of each predictor on the ensemble performance.

We evaluated this risk using the Variance Inflation Factor (VIF) across all examined datasets. The VIF for a prediction i is calculated as

$$VIF_i = \frac{1}{1 - R_i^2},$$

with R_i^2 the coefficient of determination of a regression of predictor i on all other predictors. Traditionally, a $VIF > 5$ or 10 indicates problematic multicollinearity. As shown in Table 7, our results consistently fall well below these thresholds. As $VIF_X = VIF_Y$ for two predictors, we reported a single value.

Table 7. VIF values for each datasets

Dataset	AN	BR	GL	HE	LY	MA	PB	PI	ST	TH	VE	VO	WB	WL	WN	YE
VIF	1.0	1.3	1.1	1.7	2.3	1.1	1.0	1.1	1.3	1.3	1.0	1.1	1.2	2.0	1.1	1.5

These results confirm the predictors are distinct, ensuring stable and interpretable estimates of their individual influences.

References

1. Chandola, V., et al.: Anomaly detection: a survey. ACM Comput. Surv. (CSUR) **41**(3), 1–58 (2009)
2. Ruff, L., et al.: A unifying review of deep and shallow anomaly detection. Proc. IEEE **109**(5), 756–795 (2021)
3. Oh, S.W., et al.: Anomalies detection by unsupervised learning using explainable artificial intelligence in nuclear power plants. Trans. Korean Nuclear Soc. Spring Meet. Jeju, Korea (2022)
4. Aggarwal, C.C.: Outlier Analysis. Springer (2016)
5. Wolpert, D.H., Macready, W.G.: No free lunch theorems for optimization. Trans. Evol. Comput. (2002)
6. Han, S., et al.: ADBench: anomaly detection benchmark. In: NeurIPS, vol. 35, pp. 32142–32159 (2022)
7. Schubert, E., et al.: On evaluation of outlier rankings and outlier scores. In: International Conference on Data Mining, pp. 1047–1058. SIAM (2012)
8. Ma, M.Q., et al.: The need for unsupervised outlier model selection: a review and evaluation of internal evaluation strategies. ACM SIGKDD Explor. Newsl. **25**(1), 19–35 (2023)
9. Lundberg, S.M., Lee, S.-I.: A unified approach to interpreting model predictions. In: NeurIPS, vol. 30 (2017)
10. Zhao, Y., et al.: PyOD: a Python toolbox for scalable outlier detection. J. Mach. Learn. Res. **20**(96), 1–7 (2019)
11. Goix, N.: How to evaluate the quality of unsupervised anomaly detection algorithms? arXiv preprint arXiv:1607.01152 (2016)
12. Marques, H.O., et al.: Internal evaluation of unsupervised outlier detection. TKDD **14**(4), 1–42 (2020)
13. Duan, S., et al.: Unsupervised model selection for variational disentangled representation learning. In: International Conference on Learning Representations (2019)
14. Lin, Z., et al.: InfoGAN-CR and ModelCentrality: self-supervised model training and selection for disentangling GANs. In: International Conference on Machine Learning, pp. 6127–6139. PMLR (2020)
15. Rayana, S., Akoglu, L.: Less is more: building selective anomaly ensembles. TKDD **10**(4), 1–33 (2016)
16. Campos, D., et al.: Unsupervised time series outlier detection with diversity-driven convolutional ensembles. Proc. VLDB Endowment **15**(3), 611–623 (2021)
17. Madsen, A., et al.: Interpretability needs a new paradigm. arXiv (2024)
18. Saves, P., et al.: Surrogate Modeling and Explainable Artificial Intelligence for Complex Systems: A Workflow for Automated Simulation Exploration. arXiv preprint (2025)

19. Garouani, M., et al.: XStacking: an effective and inherently explainable framework for stacked ensemble learning. Inf. Fusion (2025)
20. Burges, C., et al.: Learning to rank using gradient descent. In: Proceedings of the 22nd International Conference on Machine Learning, pp. 89–96 (2005)
21. Mantel, N.: The detection of disease clustering and a generalized regression approach. Can. Res. **27**, 209–220 (1967)
22. Cavus, M., et al.: Beyond the single-best model: Rashomon partial dependence profile for trustworthy explanations in AutoML. In: International Conference on Discovery Science, pp. 445–459. Springer (2025)
23. Levy, J., et al.: TimeCIEL: contextual interactive ensemble learning for time series classification. In: International Conference on Practical Applications of Agents and Multi-Agent Systems, pp. 316– 327. Springer (2025)

Counterfactual Explanations for Conformal Regression Intervals

Aicha Maalej[1,2(✉)] and Ulf Johansson[1]

[1] Jönköping University, Jönköping, Sweden
ulf.johansson@ju.se
[2] University of Skövde, Skövde, Sweden
aicha.maalej@ju.se

Abstract. Counterfactual explanations provide actionable changes to input feature values that would lead to a different model prediction. While previously studied for point prediction, their use in uncertainty-aware settings remains underexplored. In this paper, we introduce an approach for generating counterfactual explanations for conformal regression, where predictions are expressed as valid intervals. We define a set of different interval level properties that counterfactuals should satisfy and propose an optimization-based method for generating such counterfactuals. Beyond established evaluation metrics like distance and sparsity, we introduce two novel measures tailored to conformal prediction: plausibility, measured using the conformal predictive distribution, and difficulty, measured using the variability of predictions around a counterfactual. Experiments across five tabular datasets demonstrate that counterfactuals satisfying the selected property can be generated while achieving low distance, sparsity, difficulty, and high plausibility. This opens new directions for uncertainty aware explanations for conformal regression.

Keywords: Counterfactual Explanations · Conformal Prediction · Explainable AI

1 Introduction

The ability to communicate model predictions in a trustworthy way is increasingly seen as a core requirement for deploying machine learning in decision support systems [5,22]. In high-stakes applications, stakeholders demand models that are not only well-calibrated but also transparent in their reasoning. Yet, these goals are often pursued in isolation: uncertainty quantification is typically addressed by Bayesian or frequentist approaches [17,26], while explainability is usually sought through either inherently interpretable models or post-hoc explanation methods [9].

Conformal prediction (CP) provides a distribution-free framework for uncertainty quantification [26]. In regression, CP yields intervals guaranteed to include

M. Baratchi et al. (Eds.): IDA 2026, LNCS 16513, pp. 141–153, 2026.
https://doi.org/10.1007/978-3-032-23833-7_11

the true target with a probability equal to a chosen confidence level. While these intervals are statistically rigorous, they in some respects remain opaque: it is unclear why they are wide or narrow, or how to obtain a more informative one.

Normally, counterfactual explanations (CFs) [12,27] provide actionable, instance level explanations by identifying minimal changes to the input values that alter the model's output. CFs have been studied for point predictions, but they do not account for predictive uncertainty. This creates a disconnect: CFs are actionable but uncertainty-blind, while CP is uncertainty-aware but explanation-poor.

Recent works have begun to highlight this disconnect. For instance, [14] proposed calibrated explanations, which output feature importance scores accompanied by uncertainty estimates, ensuring that the explanations themselves are well-calibrated. Other studies have focused on explaining reject options, which have been defined as a way to capture the uncertainty inherent in decision-making [3]. More recently, SHAP-based approaches have been proposed to explain conformal classifiers, specifically p-values [10], while counterfactuals for conformal prediction sets have been introduced in [15]. These contributions highlight the importance of combining uncertainty quantification and explanations, but the question of how to explain interval-valued predictions, such as those from conformal regression, remains open.

This paper addresses this gap by introducing the first approach to generate uncertainty-aware CFs for conformal regression intervals. The contributions of the paper are the following:

- We introduce the task of generating counterfactual explanations for conformal regression intervals, shifting the focus from point predictions to uncertainty aware explanations.
- We formalize a set of actionable, interval-level counterfactual properties (e.g., target inclusion, non-overlapping intervals).
- We propose a new measure of plausibility where we rely on the conformal prediction distribution to favor counterfactuals that lie in high-density regions of the data.
- We introduce a novel criterion, difficulty, estimated from local target variability to capture how challenging an instance is for the model.
- We develop an algorithm to generate counterfactuals that satisfy interval properties while optimizing distance, sparsity, plausibility, or difficulty.
- We conduct a systematic study across multiple regression datasets and evaluate the quality of synthetic counterfactuals by comparison to real counterfactual from the calibration set.

To our knowledge, this is the first study to explain conformal regression intervals using CFs, integrating conformal prediction with counterfactual reasoning to make uncertainty itself explainable and actionable.

2 Preliminaries

2.1 Conformal Regression

Conformal prediction (CP) [26] produces *valid* prediction regions, i.e., given a significance level $\epsilon \in (0, 1)$, a conformal regressor produces prediction intervals that contain the true target with probability at least $1-\epsilon$. An error is committed when the target value is outside the interval.

In *split* or *inductive* conformal prediction (ICP) [26], a single model is trained on a proper training set, and a separate *calibration* set is used to compute *non-conformity* scores. *Non-conformity* functions measure the strangeness of instances ($z_i = x_i, y_i$). A basic function is the absolute error [18] in Eq. 1:

$$A(x_i, y_i) = |f(x_i) - y_i|, \tag{1}$$

where $f(x_i)$ is the prediction of x_i. The validity guarantees of ICP rely only on one assumption; that the calibration and test sets are exchangeable, which is slightly weaker than i.i.d.

Formally, an inductive conformal regression is generated as follows:

1. Split the training data $\mathcal{D}_{\text{train}}$ into a proper training set $\mathcal{D}_{\text{p-train}}$ and a calibration set $\mathcal{D}_{\text{cal}}$.
2. Train the underlying model f on $\mathcal{D}_{\text{p-train}}$.
3. For each calibration instance in $\mathcal{D}_{\text{cal}}$, calculate the non-conformity scores α_i using Eq. 1 to get the sequence $S = \{\alpha_1, \alpha_2, ..., \alpha_q\}$, where $q =\mid \mathcal{D}_{\text{cal}} \mid$ and S is sorted in descending order.
4. Given a predefined significance level ϵ and the sequence of non-conformity scores $\mathcal{S}$, compute the $(1-\epsilon)$-quantile threshold α_ϵ such that:

$$\frac{|\{z_i \in \mathcal{D}_{\text{cal}} \mid \alpha_i < \alpha_\epsilon\}| + 1}{|\mathcal{D}_{\text{cal}}| + 1} \geq 1 - \epsilon \tag{2}$$

5. Finally, for a test instance x_{n+1} with predicted output $f(x_{n+1})$ from the model f, the $(1-\epsilon)$ prediction interval is given by Eq. 3:

$$\mathcal{I}(x_{n+1}; \epsilon) = [f(x_{n+1}) - \alpha_\epsilon,\ f(x_{n+1}) + \alpha_\epsilon] \tag{3}$$

Normalized Conformal Regression. The standard inductive conformal regressor described above produces prediction intervals of identical width for all test instances. However, depending on the difficulty of the prediction (or instance), it is more reasonable to assign narrower intervals to easier instances and wider intervals to harder ones.

To achieve this, the normalized non-conformity score α_i of instance x_i is calculated as:

$$\alpha_i = \frac{|y_i - \tilde{y}|}{\sigma_i}, \tag{4}$$

where σ_i is a difficulty estimate of instance x_i. A difficulty estimate in this case could be the standard deviation of the target values among the k-nearest neighbors of an instance in the proper training set, as proposed by [19]. The prediction interval for a new input x_{n+1} is then given by:

$$\mathcal{I}(x_{n+1}; \epsilon) = [f(x_{n+1}) - \alpha_\epsilon \sigma_{n+1},\ f(x_{n+1}) + \alpha_\epsilon \sigma_{n+1}] \tag{5}$$

2.2 Counterfactual Explanations

Counterfactual explanations (CFs) provide instance-level interpretability by identifying the minimal changes in input values that result in a different model output [24,27]. Formally, given an instance x, a predictive model f, and a desired outcome y', a counterfactual x' is defined as:

$$x' = \arg\min_{z \in \mathcal{X}} \ell(f(z), y') + \Omega(x, z), \tag{6}$$

where $\ell(f(z), y')$ enforces that the counterfactual yields the desired prediction and $\Omega(x, z)$ regularizes the change from the original instance x, encoding desiderata such as proximity or sparsity.

The quality of counterfactuals can be evaluated against a set of desiderata. These include proximity and sparsity (small and few changes to feature values, respectively) [27], actionability (feasible changes) [25], diversity (multiple alternatives) [16], robustness (stability under model perturbations) [1,20], causality (respect feature dependencies) [13], and plausibility (consistent with the underlying data distribution) [11,21,23]. Plausibility is particularly important, as counterfactuals that fail to be plausible (i.e. fall outside the data manifold) risk being unreliable explanations.

To achieve this, latent space methods constrain counterfactuals by traversing a learned embedding, often obtained from generative models such as variational autoencoders [11]. Density-based approaches restrict counterfactuals to high-density regions estimated by techniques like normalizing flows or kernel density estimators [21]. Distance-based methods use distance from the counterfactual to its nearest neighbor as a measure of plausibility, under the assumption that the neighbor itself is plausible [2,9]. Finally, in [2], the size of the conformal prediction region was used as a measure of uncertainty, where smaller regions indicate higher confidence and thus higher plausibility. However, this work employs conformal prediction only to quantify uncertainty when generating counterfactuals for point predictions, not for conformal predictors.

3 Method

We consider the task of generating CFs in regression where predictions are intervals produced by conformal regression. Let $\mathcal{X} \subseteq \mathbb{R}^d$ denote the input space, and let $f : \mathcal{X} \to \mathbb{R}$ be a regression model trained on a training set $\mathcal{D}_{\text{p-train}} = \{(x_i, y_i)\}_{i=1}^n$. We employ *normalized conformal regression (NCR)* with a calibration set $\mathcal{D}_{\text{cal}} = \{(x_j, y_j)\}_{j=1}^m$. For any instance $x \in \mathcal{X}$ and significance ϵ, NCR outputs a prediction interval $\mathcal{I}(x;\epsilon) = [L(x), U(x)]$, with a validity guarantee $\Pr\{\, y \in \mathcal{I}(x;\epsilon)\,\} \geq 1-\epsilon$. Normalization in this case will adapt the interval width based on the difficulty of the instance, as introduced in Sect. 2.1, and will produce predictions (candidate counterfactuals) of different interval size.

3.1 Defining Interval-Based Counterfactuals

Given an original test instance x with interval $\mathcal{I}$, a counterfactual explanation is an alternative input $x' \in \mathcal{X}$ such that the induced interval $\mathcal{I}' = [L(x'), U(x')]$ satisfies a *property function* $\mathcal{P}$, i.e., x' is a valid counterfactual if $\mathcal{P}(x, x') = \texttt{true}$.

We define some possible classes of $\mathcal{P}$:

- Target inclusion: $\mathcal{I}'$ must include a specific value $y*$.
- Non-overlapping intervals: $\mathcal{I}$ and $\mathcal{I}'$ do not overlap.
- Shifted interval: $\mathcal{I}'$ is shifted above or below a specified threshold.

3.2 Evaluation Criteria

Counterfactuals are evaluated using four criteria. Distance and sparsity are well established in prior work. We additionally propose a plausibility measure based on conformal prediction and adapt a difficulty measure, previously used for normalization in conformal regression, as a novel evaluation criterion for interval-based counterfactuals.

Distance is measured using the Gower distance [8]:

$$\text{Proximity}(x, x') = \frac{1}{p}\sum_{j=1}^{p} d_j(x_j, x'_j), \tag{7}$$

where d_j is the standard Gower component for feature j.

Sparsity is the number of changed features:

$$\text{Sparsity}(x, x') = \sum_{j=1}^{p} \mathbf{1}\{x_j \neq x'_j\} \tag{8}$$

Plausibility. Defined by how well the counterfactual conforms to the data distribution. We introduce a new plausibility measure tailored to interval-based counterfactual explanations in conformal regression by leveraging the *conformal prediction distribution* (CPD) [26].

Let the residuals in the calibration set be $E = \{y_j - f(x_j)\}_{j=1}^m$. For a candidate x', we construct the CPD as:

$$\text{CPD}(x') = f(x') + \text{sort}(E) \tag{9}$$

The CPD is thus an empirical distribution obtained from the calibration residuals, centered at the counterfactual prediction. We approximate its empirical *cumulative distribution function* (CDF) F_{CPD} by interpolation. The plausibility of the counterfactual interval is then:

$$\text{Plaus}(x') = F_{\text{CPD}}(U(x')) - F_{\text{CPD}}(L(x')) \tag{10}$$

This corresponds to how much probability mass of the CPD is covered between $L(x')$ and $U(x')$. High plausibility indicates that the counterfactual interval lies in a high-density region of the conformal predictive distribution and is typical of the data distribution, see Fig. 1).

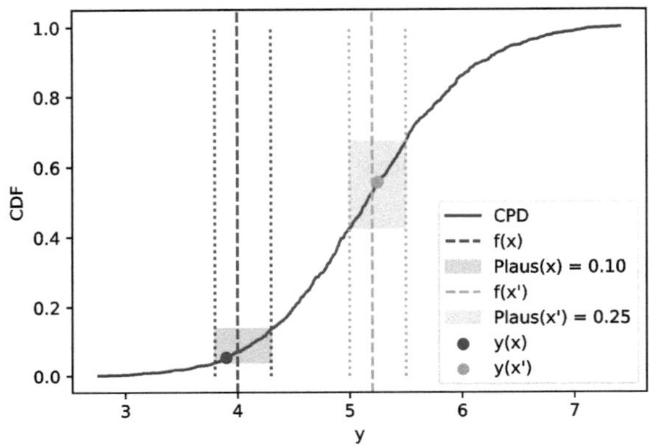

Fig. 1. Illustration of plausibility measure using the CDF. The counterfactual interval (orange), positioned in a higher-density region, achieves higher plausibility than the original (green). If the interval aligns with the densest part of the CPD, the mass is high, thus the counterfactual is plausible. If the interval falls into the tails, plausibility is low and the counterfactual does not conform to training data. for a valid comparison, both intervals need to have equal width. (Color figure online)

Difficulty. We propose *difficulty* as a novel criterion for interval-based counterfactuals that quantifies how challenging an instance is for the model to predict, measured as the expected local variability of outcomes around a counterfactual. Following [19], we estimate difficulty by the standard deviation of the target of the kth (we used the default $k = 25$) nearest neighbors of x'. If $\sigma(x')$ denotes the standard deviation of the k nearest-neighbor labels of x', then:

$$\text{Diff}(x') = \sigma(x') \tag{11}$$

Large values correspond to counterfactuals in regions of high outcome variability, implying less reliable predictions. Small values indicate easier and more predictable regions. Intervals with low difficulty yield tighter prediction intervals, and thus are more informative. In our implementation, we compute this using the `DifficultyEstimator` provided in the `crepes.extras` module [6].

3.3 Search Procedure: Finding the Counterfactual

Given an original instance x' with its conformal prediction interval $\mathcal{I}$, our objective is to identify a CF x' that (i) satisfies a predefined interval-based property $\mathcal{P}(x, x')$ and (ii) optimizes a chosen evaluation criterion $\mathcal{C}(x, x') = \{$distance, sparsity, plausibility, difficulty$\}$. We employ a two-step strategy to obtain both *real* (data-based) and *synthetic* (algorithmically generated) CFs.

1. We first search in the calibration dataset $\mathcal{D}_{\text{cal}}$ for an instance x' such that $\mathcal{P}(x, x') = \texttt{true}$. Among all such candidates, we select as the real counterfactual the one minimizing the Gower distance to the original instance:

$$x'_{\text{real}} = \arg \min_{x' \in \mathcal{D}_{\text{cal}}} \{\text{Dist}_{\text{Gower}}(x, x') \text{ s.t. } \mathcal{P}(x, x') = \texttt{true}\}$$

2. Then we construct a discrete search space between x and x'_{real}, and we use a genetic algorithm (pygad) [7] to search among all the combinations for the counterfactual such that:

$$x'_{\text{synth}} = \arg \min_{x'} \{C(x, x') \text{ s.t. } \mathcal{P}(x, x') = \texttt{true}\}$$

3.4 Experiment

We conduct experiments on six tabular regression datasets: *Abalone*, *Energy Efficiency (heating and cooling)*, *Wine Quality*, and *California Housing* [4]. Each dataset is split into training (60%), calibration (30%), and test (10%). A *RandomForest regressor* is trained on the training set and wrapped with `WrapRegressor` from the `crepes` package. To produce normalized conformal prediction intervals we use the `DifficultyEstimator` from `crepes.extras`. Prediction intervals are constructed at significance $\epsilon = 0.1$, with calibration performed using 5-fold cross-validation. The GA [7] used for searching the x'_{synth} is run with a population of 30 individuals, up to 200 generations, uniform crossover, mutation probability 0.1, elitism of two parents, and early stopping after 20 stagnant generations.

In Sect. 4, we first provide illustrative examples (plots) of the explanation method. Second, we conduct a systematic evaluation to assess the quality of the synthetic counterfactual against the real one. We sample $N = 50$ test instances per dataset and for each instance we report the ratio of each criterion of the synthetic counterfactual relative to the real one. These ratios are then averaged across the 50 test instances.

4 Results

The explanation plots (Figs. 2, 3, 4, 5) show that the proposed approach successfully generates counterfactuals consistent with the desired property $\mathcal{P}$. Importantly, the visualizations highlight how different optimization criteria affect the explanations: In Fig. 2, optimizing for distance yields CFs that remain close to the original instance, but also includes the target house value 2 in the CF interval. The CF explanation suggests that if you move to areas where MedInc was reduced from 7.11 to 5.96, and slightly reduce the value of AveRooms, AveBedrooms, and AveOccupation, it will result in a prediction interval [1.759, 3.658] that includes the target value 2.

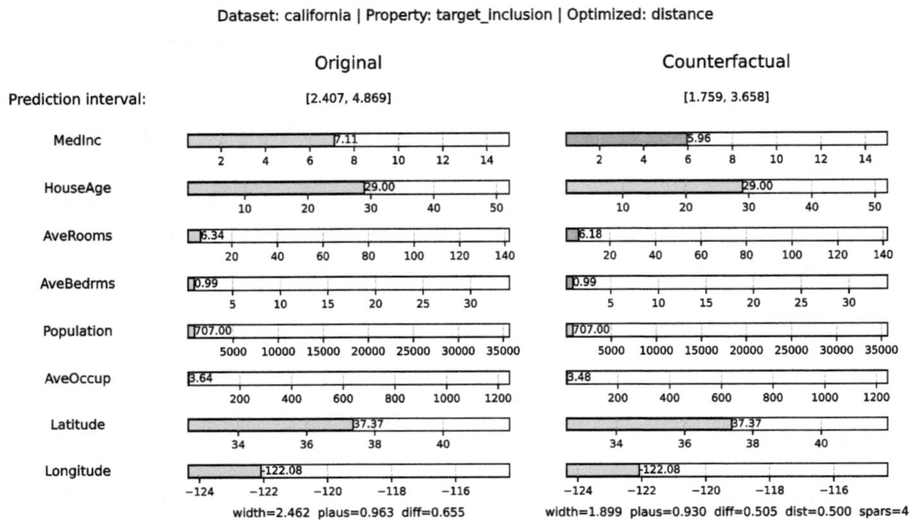

Fig. 2. Counterfactual explanation highlighting which input values to modify (in blue) to achieve the counterfactual instance that includes the target value 2. (Color figure online)

In Fig. 3, the explanation suggests that to get a non-overlapping interval with the original, at least 3 features need to be changed. In Fig. 4, plausibility score is 0.996 suggesting a CF that lies in a high density region of the CPD. It is worth mentioning that to compare plausibility of two instances, the interval width of the CF should be lower or at least equal to that of the other instance. The reason is that, if we look at Fig. 1 where the width of both examples is equal, plausibility could be higher if the interval lies on one of the edges but is wider.

In the same Fig. 4, we also see that among all candidate counterfactuals, the one with the highest plausibility required changing 6 out of 8 feature values. The same number of features were changed in Fig. 5 to result in a counterfactual with the lowest difficulty 0.206 against 0.650 in the original. This means that the variability around the counterfactual instance is low and it lies in a more predictable and easier region. This also means that its uncertainty is low, which is noticed in the decreased size of the interval.

Looking at Table 1, each optimization criterion achieves the best performance on its corresponding target metric, confirming that the optimization procedure

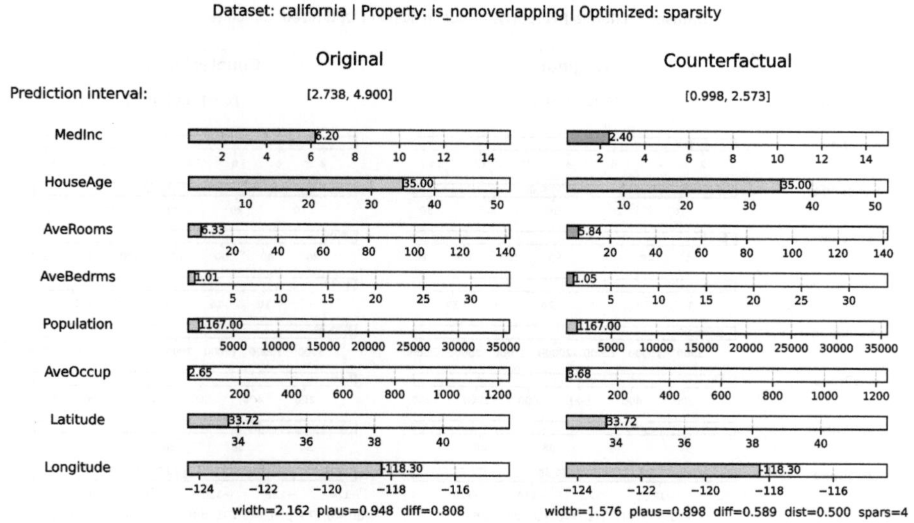

Fig. 3. Counterfactual explanation showing the minimal number of features to change (in blue) to achieve a non-overlapping interval with the original. (Color figure online)

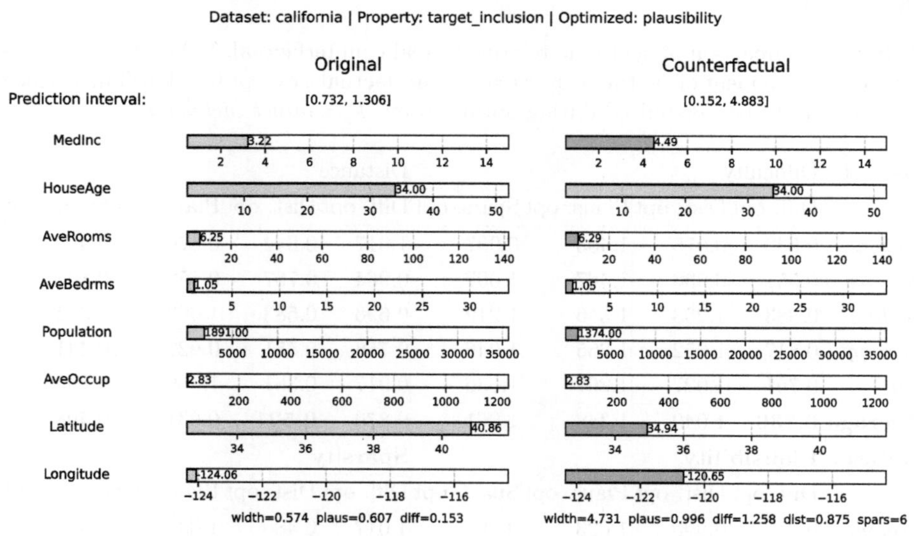

Fig. 4. Counterfactual explanation highlighting which input values to modify (in blue) to achieve a counterfactual with a highest plausibility. Target inclusion value = 3. (Color figure online)

effectively enforces the intended desiderata. Distance and sparsity optimization reliably yield lower ratios (<1) across all datasets, and this effect persists even when these criteria are not the explicit optimization target. This indicates that synthetic counterfactuals are closer and more compact than real ones.

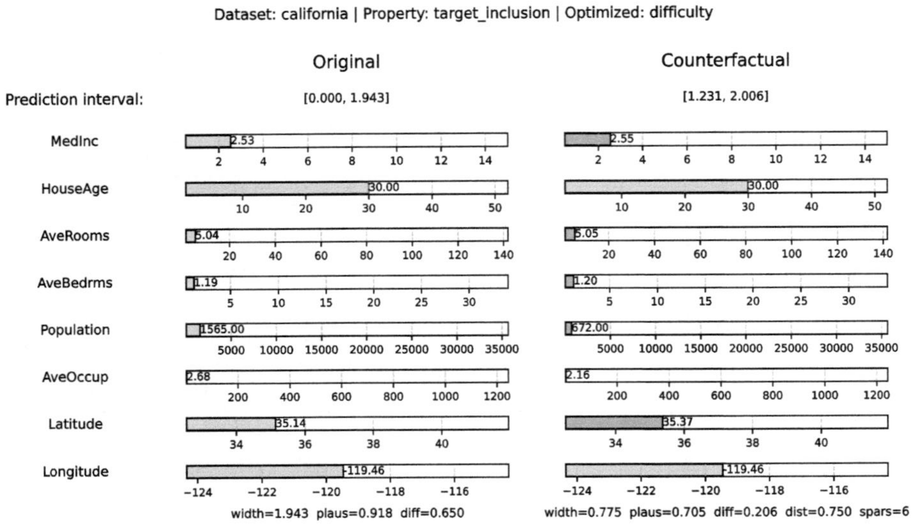

Fig. 5. Counterfactual explanation highlighting which input values to modify (in blue) to achieve a counterfactual with lowest difficulty. Target inclusion value = 2. (Color figure online)

Table 1. Comparison of synthetic relative to real counterfactual. Values below 1 indicate an improvement of synthetic over real counterfactuals, except for plausibility which is a measure to be maximized during optimization. $\mathcal{P}$ = *target inclusion*

Dataset	Difficulty				Distance			
	Diff. opt	Dist. opt	Plaus. opt	Spars. opt	Diff. opt	Dist. opt	Plaus. opt	Spars. opt
abalone	0.633	0.975	1.403	0.997	1.017	0.642	0.870	0.490
california	0.557	0.938	1.467	1.007	0.964	0.767	0.852	0.802
cooling	0.883	1.253	1.346	1.219	0.696	0.583	1.003	0.365
heating	0.817	1.052	1.265	1.046	0.756	0.451	0.922	0.441
wine	0.791	1.027	1.261	1.036	0.917	0.361	1.009	0.390
Average	**0.736**	1.049	1.348	1.061	0.870	**0.521**	0.931	0.498
Dataset	**Plausibility**				**Sparsity**			
	Diff. opt	Dist. opt	Plaus. opt	Spars. opt	Diff. opt	Dist. opt	Plaus. opt	Spars. opt
abalone	0.917	0.990	1.023	0.997	1.017	0.485	1.032	0.490
california	0.866	0.977	1.174	0.994	0.961	0.731	0.949	0.769
cooling	0.979	1.015	1.093	1.014	0.696	0.383	0.773	0.365
heating	0.946	0.984	1.127	0.984	0.756	0.451	0.722	0.441
wine	0.976	1.001	1.092	1.002	0.912	0.320	0.894	0.311
Average	0.937	0.993	**1.102**	0.998	0.868	0.514	0.874	**0.475**

When plausibility is optimized, values rise above 1 demonstrating that synthetic counterfactuals conform more strongly (than the real ones) to the data distribution, even though real counterfactuals themselves are drawn from the calibration set directly. However, difficulty values also rise. This can be explained in two ways. Either highly plausible counterfactuals tend to be in high density regions of the CPD, where variability can also be high (such counterfactuals conform to the training data but occur in regions where outcome variability is higher, making them locally harder to predict). Or, this can also be explained by the fact that wider intervals result in a higher plausibility value. In other words, if we look at Fig. 1, the wider the interval, the higher its plausibility.

It is noteworthy that plausibility improvements are modest when comparing synthetic to real counterfactuals. This is because real counterfactuals are already drawn from the calibration set and thus there is a higher probability that they inherently conform to the data distribution.

5 Conclusion

In this paper, we introduced the task of generating counterfactual explanations for conformal regression intervals, moving beyond point prediction to offer uncertainty-aware explanations. We defined interval-level counterfactual properties and proposed new evaluation criteria measures tailed to this setting, including plausibility using conformal predictive distribution and difficulty. The method proposed allows for optimizing one of the objectives (i.e. distance, sparsity, plausibility, or difficulty) while ensuring counterfactuals satisfy interval-based properties.

We run experiments across five regression tabular datasets to demonstrate our method. We first illustrate explanation examples to highlight key behaviors of each criterion. We further validated synthetic counterfactuals by comparing them to real calibration-set instances and showing that optimized synthetic counterfactuals outperform real ones on the targeted objectives.

This approach offers a promising direction for trustworthy decision support in regression tasks. Future work may explore multi-objective optimization and generating counterfactuals that are statistically valid with respect to conformal coverage guarantees.

Acknowledgments. The authors acknowledge the Swedish Knowledge Foundation, Jönköping University, and the industrial partners for financially supporting the research through the following projects: AFAIR (grant 20200223), PREMACOP (grant 20220187) and ETIAI (grant 20230036), as part of the research and education environment SPARK at Jönköping University, Sweden.

Disclosure of Interests. The authors have no competing interests to declare that are relevant to the content of this article.

References

1. Altmeyer, P., van Deursen, A., Liem, C.: Explaining black-box models through counterfactuals. arXiv preprint arXiv:2308.07198 (2023)

2. Altmeyer, P., Farmanbar, M., van Deursen, A., Liem, C.C.: Faithful model explanations through energy-constrained conformal counterfactuals. In: Proceedings of the AAAI Conference on Artificial Intelligence, vol. 38, pp. 10829–10837 (2024)
3. Artelt, A., Hammer, B.: "even if..."–diverse semifactual explanations of reject. In: 2022 IEEE Symposium Series on Computational Intelligence (SSCI), pp. 854–859. IEEE (2022)
4. Bache, K., Lichman, M.: UCI machine learning repository (2013)
5. Bhatt, U., et al.: Uncertainty as a form of transparency: measuring, communicating, and using uncertainty. In: Proceedings of the 2021 AAAI/ACM Conference on AI, Ethics, and Society, pp. 401–413 (2021)
6. Boström, H.: Conformal prediction in python with crepes. In: Proceedings of the 13th Symposium on Conformal and Probabilistic Prediction with Applications, pp. 236–249. PMLR (2024)
7. Gad, A.F.: Pygad: an intuitive genetic algorithm Python library. Multimedia Tools Appl. 1–14 (2023)
8. Gower, J.C.: A general coefficient of similarity and some of its properties. Biometrics, 857–871 (1971)
9. Guidotti, R., Monreale, A., Ruggieri, S., Turini, F., Giannotti, F., Pedreschi, D.: A survey of methods for explaining black box models. ACM Comput. Surv. (CSUR) **51**(5), 1–42 (2018)
10. Johansson, U., Maalej, A., Sönströd, C.: Explaining set-valued predictions: shap analysis for conformal classification. Proc. Mach. Learn. Res. **266**, 1–20 (2025)
11. Joshi, S., Koyejo, O., Vijitbenjaronk, W., Kim, B., Ghosh, J.: Towards realistic individual recourse and actionable explanations in black-box decision making systems. arXiv preprint arXiv:1907.09615 (2019)
12. Karimi, A.H., Barthe, G., Balle, B., Valera, I.: Model-agnostic counterfactual explanations for consequential decisions. In: International Conference on Artificial Intelligence and Statistics, pp. 895–905. PMLR (2020)
13. Karimi, A.H., Schölkopf, B., Valera, I.: Algorithmic recourse: from counterfactual explanations to interventions. In: Proceedings of the 2021 ACM Conference on Fairness, Accountability, and Transparency, pp. 353–362 (2021)
14. Löfström, T., Löfström, H., Johansson, U., Sönströd, C., Matela, R.: Calibrated explanations for regression. Mach. Learn. **114**(4), 1–34 (2025)
15. Maalej, A., Sönströd, C., Johansson, U.: Counterfactual explanations for conformal prediction sets. Proc. Mach. Learn. Res. **266**, 1–20 (2025)
16. Mothilal, R.K., Sharma, A., Tan, C.: Explaining machine learning classifiers through diverse counterfactual explanations. In: Proceedings of the 2020 Conference on Fairness, Accountability, and Transparency, pp. 607–617 (2020)
17. Neal, R.M.: Bayesian Learning for Neural Networks, vol. 118. Springer, Cham (2012)
18. Papadopoulos, H., Proedrou, K., Vovk, V., Gammerman, A.: Inductive confidence machines for regression. In: Elomaa, T., Mannila, H., Toivonen, H. (eds.) ECML 2002. LNCS, vol. 2430, pp. 345–356. Springer, Cham (2002). https://doi.org/10.1007/3-540-36755-1_29
19. Papadopoulos, H., Vovk, V., Gammerman, A.: Regression conformal prediction with nearest neighbours. J. Artif. Intell. Res. **40**, 815–840 (2011)
20. Pawelczyk, M., Datta, T., Van-den Heuvel, J., Kasneci, G., Lakkaraju, H.: Probabilistically robust recourse: navigating the trade-offs between costs and robustness in algorithmic recourse. arXiv preprint arXiv:2203.06768 (2022)

21. Poyiadzi, R., Sokol, K., Santos-Rodriguez, R., De Bie, T., Flach, P.: Face: feasible and actionable counterfactual explanations. In: Proceedings of the AAAI/ACM Conference on AI, Ethics, and Society, pp. 344–350 (2020)
22. Ribeiro, M.T., Singh, S., Guestrin, C.: "why should i trust you?" Explaining the predictions of any classifier. In: Proceedings of the 22nd ACM SIGKDD International Conference on Knowledge Discovery and Data Mining, pp. 1135–1144 (2016)
23. Schut, L., et al.: Generating interpretable counterfactual explanations by implicit minimisation of epistemic and aleatoric uncertainties. In: International Conference on Artificial Intelligence and Statistics, pp. 1756–1764. PMLR (2021)
24. Spooner, T., Dervovic, D., Long, J., Shepard, J., Chen, J., Magazzeni, D.: Counterfactual explanations for arbitrary regression models. arXiv preprint arXiv:2106.15212 (2021)
25. Ustun, B., Spangher, A., Liu, Y.: Actionable recourse in linear classification. In: Proceedings of the Conference on Fairness, Accountability, and Transparency, pp. 10–19 (2019)
26. Vovk, V., Gammerman, A., Shafer, G.: Algorithmic Learning in a Random World. Springer, Cham (2005). https://doi.org/10.1007/b106715
27. Wachter, S., Mittelstadt, B., Russell, C.: Counterfactual explanations without opening the black box: automated decisions and the GDPR. Harv. JL Tech. **31**, 841 (2017)

Extending Information Bottleneck Attribution to Video Sequences for Deepfake Detection

Veronika Solopova(✉), Lucas Schmidt, Vera Schmitt, and Dorothea Kolossa

Technische Universität Berlin, Berlin 10623, Germany
veronika.solopova@tu-berlin.de

Abstract. As the quality of deepfakes increasingly matches authentic video, reliable detection requires automated video classifiers. Yet these models often operate as black boxes, making it hard to assess their trustworthiness in high-stakes security and forensic settings.

We introduce VIBA (Video Information Bottleneck Attribution), an explainable video classification method that extends Information Bottleneck Attribution (IBA) [19] to jointly capture spatial and temporal dependencies. VIBA is post-hoc and model-agnostic, producing relevance and optical flow maps that reveal manipulated regions and motion inconsistencies across frames.

We apply VIBA to deepfake detection with three architectures: Xception for spatial features, a VGG11-based optical flow model for motion dynamics, and a CViT visual transformer for long-range temporal reasoning. Across models, VIBA yields more temporally stable and spatially precise explanations than Grad-CAM [20], and aligns more closely with expert human annotations. By making deepfake detector outputs easier to analyse and interpret, VIBA supports secure, transparent deployment of video analysis systems in digital forensics and misinformation monitoring.

1 Introduction

The proliferation of AI-generated videos, particularly deepfakes, poses a growing threat to security, trust, and public safety. EU legislative efforts, most notably the EU AI Act, now explicitly regulate manipulated content (including deepfakes), imposing transparency obligations on both providers and deployers. As high-quality synthetic media is increasingly weaponized for disinformation, identity theft, political manipulation, and financial fraud, reliable detection becomes essential, yet ever more challenging as generated content approaches the fidelity of authentic video.

Video classification models therefore play a critical role in safeguarding digital ecosystems, but their usefulness in high-stakes settings depends not only on accuracy, but also on transparency and interpretability. Existing explainable AI

M. Baratchi et al. (Eds.): IDA 2026, LNCS 16513, pp. 154–168, 2026.
https://doi.org/10.1007/978-3-032-23833-7_12

(XAI) techniques are largely designed for static images and struggle to capture the spatiotemporal dependencies that define video-based tasks, leaving a blind spot for applications such as deepfake detection, where subtle spatial manipulations and temporal inconsistencies must be understood jointly.

To address this gap, we introduce VIBA (Video Information Bottleneck Attribution), a video-specific adaptation of the Information Bottleneck for Attribution (IBA) [19]. VIBA produces post-hoc, model-agnostic visual explanations that integrate spatial and temporal information over sequences. We evaluate VIBA on deepfake detection using three complementary architectures: Xception [18] for spatial features, a VGG11-based optical flow model [21] for motion dynamics, and a visual transformer (CViT), fine-tuned for deepfake detection by Wodajo et al. [25], to capture long-range spatiotemporal relationships. On a dataset incorporating recent deepfake generation techniques, we show that VIBA effectively highlights keyframes and motion patterns critical for detecting manipulations, and that its explanations are more temporally stable and more aligned with human annotations than Grad-CAM [20]. This provides a pathway towards interpretable video models for deepfake detection and related security applications. Our code is available in a GitHub repository[1].

2 Related Work

In recent years, deep learning has achieved strong performance in tasks such as object detection and medical imaging, but its "black-box" nature complicates use in sensitive domains [17].

XAI methods specifically tailored to video are less explored than approaches for images. Chen et al. [2] use space-time attention to localize audio–visual cues, while Lee et al. [13] propose a transformer-based spatiotemporal attention network for time-series and video explanations. Vanneste et al. [24] adapt Grad-CAM to produce video heatmaps, yet these approaches are not designed to explicitly control information flow or to assess temporal explanation stability.

Deepfake detection has recently emerged as a key application area for explainable models. Prior work has adapted model-agnostic methods such as LRP and LIME [16], compared Grad-CAM variants, SHAP, LIME, and SOBOL under adversarial settings [23]. New datasets and benchmarks for explainable deepfake detection have also been proposed [7]. However, these works predominantly operate at the image level or rely on gradient-based attributions, without explicitly leveraging information bottlenecks to study temporal consistency in video explanations. To the best of our knowledge, the benefits of IBA-based explanations for video models have not yet been investigated.

[1] https://github.com/anonrep/IBA-for-Video-Sequences. The trained models will be made available upon acceptance.

3 Methods

3.1 Model Architecture and Training

We structured the analysis into two main paths (Fig. 1): a spatial stream using a pre-trained either Xception or CVit models and a temporal stream using a VGG11-based model trained on optical flow maps. This setup allows us to test VIBA on both appearance- and motion-based deepfake detection. Xception is widely used in deepfake detection and relies on depthwise separable convolutions, making it strong at capturing fine-grained spatial artifacts such as blending boundaries and texture inconsistencies. In contrast, VGG11 represents a classical, sequential convolutional architecture with dense feature maps and minimal architectural shortcuts, which is particularly suitable for analyzing optical flow representations where preserving local motion patterns is important.

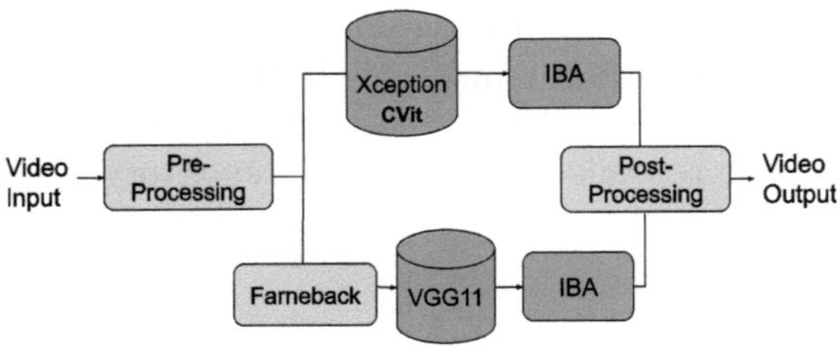

Fig. 1. Implemented VIBA pipeline.

The Xception model was initialized with publicly available weights from Liu et al. (2023) [15]. The artefact detection model was trained with a VGG11 backbone on dense optical flow maps derived from manipulated and authentic videos (Fig. 2). Over 10,000 frame pairs (keyframe and subsequent frame) were sampled from real and deepfake clips, resulting in approximately 5,000 optical flow maps. The model is a standard 2D CNN; temporal information is encoded solely via the flow representation.

Training used a fixed learning rate of 0.001, batch size 16, Adam optimizer, cross-entropy loss, and a maximum of 50 epochs with early stopping (patience 7) on validation loss. In practice, training stopped at epoch 24. The Xception and VGG11 models were trained and evaluated independently without feature- or decision-level fusion, allowing us to isolate the explanatory behaviour of each stream.

3.2 Dataset Construction

For our experiments we assembled a deepfake detection dataset[2] from established sources: FaceForensics++ [18], Celeb-DF [14], DFDC [3], DFD [4], DeeperForensics [9], FakeAVCeleb [10], AV-Deepfake1M [1], and KoDF [12]. We aimed for

[2] Due to licensing constraints, the dataset can not be made publicly available, but it can be shared with vetted academics for reproducibility purposes upon request.

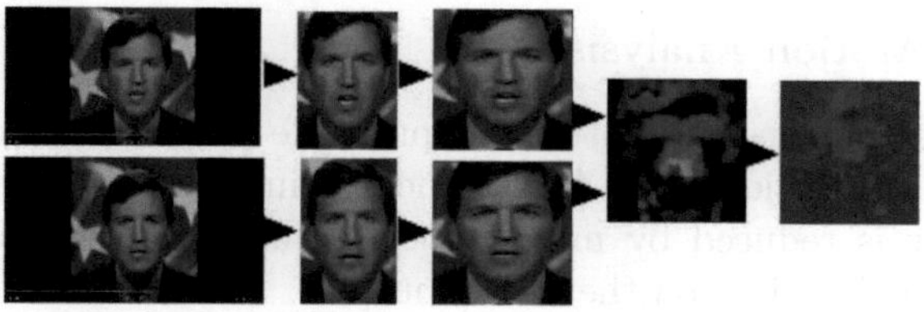

Fig. 2. Pre-processing pipeline for optical flow maps.

roughly 50 videos per dataset (see Table 1), but some entries were no longer accessible, a common issue for social media–based benchmarks. To increase diversity and include more recent manipulations, we added 30 deepfakes from attested fake YouTube videos and short clips from the TV show "Deep Fake Neighbour Wars", covering both high-quality GAN-based manipulations and lower-quality, compression-heavy content.

The dataset was randomly split into 70% train, 15% validation (for VGG11), and 15% evaluation. The *Evaluation* split was used to test both deepfake detectors and to assess VIBA explanations.

Table 1. Distribution of selected videos from various datasets with different deepfake techniques.

Dataset	# Videos	Dataset	# Videos
FaceForensics++	50	Celeb-DF	56
DFDC	50	DFD	50
FakeAVCeleb	46	AV-Deepfake1M	54
KoDF	48	YouTube	18
TV show "Deep Fake Neighbour Wars"	6		

Preprocessing was standardized across models. Videos were split into frames, and keyframes were extracted following Ibaceta (2023) [8] to reduce redundancy. Faces were detected and cropped using Dlib [11], then resized to 299×299 pixels for Xception and CViT, and 256×256 pixels for VGG11. For VGG11, dense optical flow maps between consecutive frames were computed with the Farnebäck algorithm [5], providing a dense, noise-robust motion field as temporal input representation.

4 Experimental Setup

The models were trained on two A100 GPUs with 40 GB VRAM. All inference and speed experiments were run on an NVIDIA RTX 6000 Ada (49 GB VRAM).

4.1 IBA for Motion Analysis

IBA applies the Information Bottleneck principle [22] to intermediate feature maps to obtain attributions [19]. A bottleneck is inserted into a given layer, and information flow is reduced by mixing the original representation with noise, while monitoring the effect on the model output.

Let R be a feature map of shape (N, C, H, W). We sample Gaussian noise

$$\epsilon \sim \mathcal{N}(\mu_R, \sigma_R^2) \tag{1}$$

with the same shape as R, where μ_R and σ_R are the mean and standard deviation of R. The perturbed representation Z is obtained via

$$Z = \lambda(X)R + (1 - \lambda(X))\epsilon, \tag{2}$$

where X is the input, and $\lambda(X)$ has the same dimensionality as R, controlling how much signal vs. noise is retained element-wise.

The objective is to find $\lambda(X)$ that preserves predictive performance while minimizing information flow:

$$\min_{\lambda(X)} \mathcal{L}(\theta, \lambda(X)) + \beta I(R; Z), \tag{3}$$

where $\mathcal{L}$ is the task loss (e.g. cross-entropy), $I(R; Z)$ is the mutual information between R and Z, and β trades off prediction accuracy against compression. Relevance maps are derived from learned λ, indicating which regions must remain informative for the prediction (Table 2).

Table 2. Approximate intersection over union across bottleneck injection layers for VGG11 and Xception models (lower scores mean worse spatial alignment).

Model	Placement	IoU
Xception	block 4 (used)	0.83
	bn3	0.75
	block 12	0.71
	conv3	0.68
	bn4	0.70
VGG11	layer 9 (used)	0.80
	layer 12	0.74
	layer 16	0.72

Injection Layer Identification. We identified the appropriate layer for bottleneck injection based on a balance between feature abstraction and spatial detail, guided by the methodology in Schulz et al., 2020 [19].

Injection Layer Identification. We followed Schulz et al. [19] to select injection layers, balancing feature abstraction and spatial detail. For Xception, we probed several candidate layers (block 4, bn3, conv3, block 12, bn4) on 60 frames, comparing IoU and visual clarity of the resulting maps. Block 4 consistently produced the most interpretable, fine-grained heatmaps and the highest IoU (Fig. 3), and was therefore chosen for all experiments.

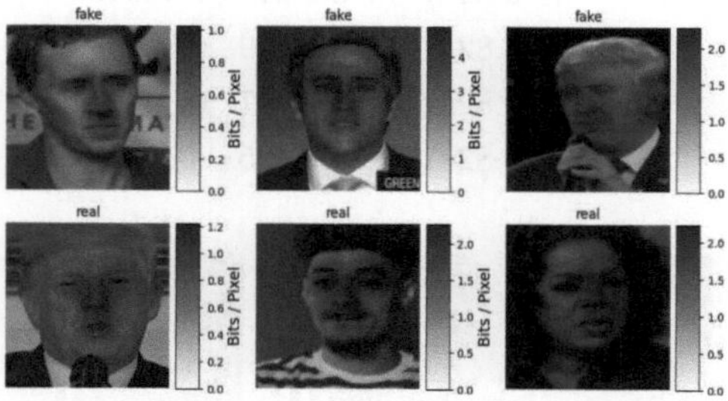

Fig. 3. Relevance maps generated with Xception and with bottleneck injection after block 4.

For VGG11, we repeated the procedure using optical flow inputs, probing layers 9, 12, and 16. Earlier layers (9, 12) retained more spatial and temporal detail and achieved higher IoU, with layer 9 performing best (Fig. 4). We thus used block 4 (Xception) and layer 9 (VGG11) as the default injection points.

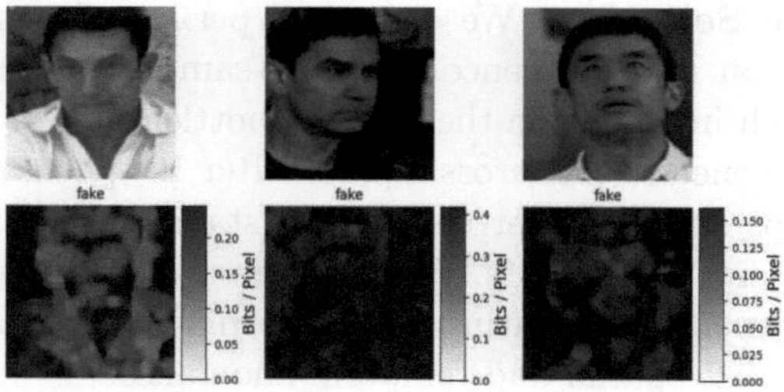

Fig. 4. Optical flow maps generated with VGG11 and bottleneck injection after layer 9, showing frames used to create optical-flow maps.

For CViT, IoU across candidate layers was very similar, so we additionally measured confidence drops after perturbation to identify the most influential layer. We selected the earliest layer with the largest drop, trading off interpretability and impact; this yielded layer 5 as the optimal placement (Table 3, Fig. 5).

Table 3. Top confidence drops across transformer layers of the CViT model with varying μ_R values (Eq. 1). A higher drop indicates more influential layer.

Layer	μ_R	Original	Perturbed	Drop
5	0.3	0.824707	0.820801	0.003906
5	0.1	0.824707	0.822754	0.001953
4	0.3	0.824707	0.823730	0.000977
4	0.5	0.824707	0.824219	0.000488
5	0.5	0.824707	0.824219	0.000488
4	0.1	0.824707	0.824707	0.000000
2	0.1	0.824707	0.825195	-0.000488

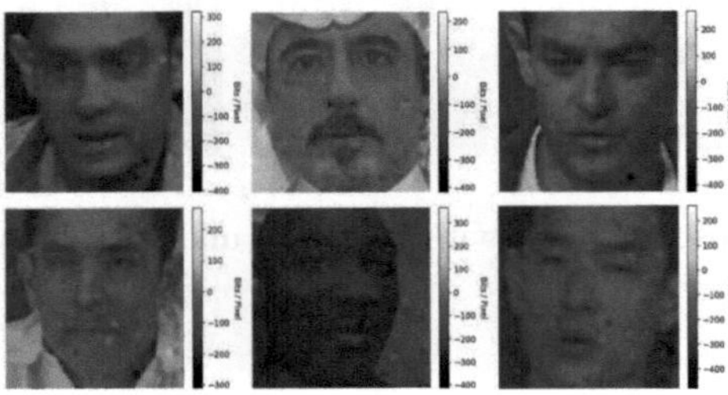

Fig. 5. Relevance maps generated with CViT and bottleneck injection after layer 5. We chose the red-and-yellow scale for CViT VIBA visualizations, as it was found to be optimal for the visibility of smaller highlighted patches. (Color figure online)

Bottleneck Type Selection. We compared per-sample and readout bottlenecks using Xception as a reference. The per-sample variant optimizes $\lambda(X)$ individually for each input, while the readout bottleneck learns a separate prediction network to generalize λ across inputs. After 10 epochs on the *Train* split, readout-based maps showed lower clarity and stability than per-sample maps, consistent with observations in [19]. We therefore use per-sample bottlenecks for all reported experiments, generating relevance maps that can be reassembled into videos to visualize spatial and temporal anomalies.

4.2 Grad-CAM Baseline

We use Grad-CAM as a baseline explanation method for the models. Grad-CAM produces class-discriminative localization maps by using gradients flowing into the final convolutional layer. For Xception, we applied Grad-CAM to block 12; for VGG11, to the last convolutional layer before the fully connected head; in CViT, it was applied to last transformer layer. For the ablation study, we generated Grad-CAM maps on the same keyframes and optical flow maps used for VIBA.

4.3 Evaluation

We evaluate explanations along four dimensions:

1. **Comparative baseline testing.** We measure model performance with and without IBA to ensure that injected noise does not significantly harm predictive accuracy. Metrics include Accuracy, Precision, Recall, and Expected Calibration Error (ECE) [6].

2. **Saliency map consistency.** To assess spatial and temporal stability, we compute Intersection over Union (IoU), Temporal Consistency Score (TCS), and Region Instability Metric (RIM) on the evaluation set.

IoU measures spatial overlap between binary masks of consecutive frames, with values closer to 1 indicating more consistent highlighted regions.

Let $\{M_i\}_{i=1}^{N}$ be binary saliency masks for a video of N frames, where $M_i : \{1,\ldots,W\} \times \{1,\ldots,H\} \to \{0,1\}$. We define the pixel-wise persistence map

$$P(x,y) = \frac{1}{N}\sum_{i=1}^{N} M_i(x,y), \tag{4}$$

and for a threshold $\tau \in [0,1]$,

$$TCS_\tau = \frac{1}{W \cdot H}\sum_{x=1}^{W}\sum_{y=1}^{H} \mathbb{1}\left[P(x,y) \geq \tau\right], \tag{5}$$

with $\tau = 0.7$ in our experiments. TCS thus captures how many pixels remain salient in at least 70% of frames.

RIM measures normalized centroid movement of salient regions:

$$RIM = \frac{\frac{1}{N-1}\sum_{i=2}^{N} \|C_i - C_{i-1}\|}{\sqrt{W^2 + H^2}} \tag{6}$$

where C_i is the centroid of the mask at frame i. Values close to 0 indicate highly stable explanations. Together, IoU, TCS, and RIM provide complementary views of spatial overlap, temporal coherence, and stability.

3. **Ablation testing with human-annotated artifacts.** To assess whether explanations align with meaningful cues, 8 annotators (two senior researchers, five PhD students in Computer Science, and one journalist) marked regions they considered indicative in 27 deepfake videos from the *Evaluation* set. Cohen's $\kappa = 0.72$ indicated substantial agreement. Using the three most frequent region choices per video, we computed F1 (macro), Precision, Recall, and the Szymkiewicz–Simpson overlap coefficient, as well as the proportion of cases where the majority-voted region was present in the explanation. Real videos were not annotated, as they do not contain manipulation artifacts.

4. **Runtime comparison.** Finally, we compare the speed of VIBA and Grad-CAM when generating per-frame and optical-flow explanations to quantify the computational overhead of VIBA relative to a gradient-based baseline.

5 Results

Table 6 shows that injecting VIBA leaves accuracy, precision, and recall essentially unchanged for all three architectures, with only minor fluctuations in calibration (ECE). Notably, CViT benefits from a clear ECE reduction when combined with VIBA, suggesting slightly more reliable confidence estimates.

Table 4. F1 score, precision, recall, overlap coefficient, and top-1 match with the majority human vote, comparing highlighted regions of deepfakes with human annotations. Results are shown for Xception and VGG model predictions using VIBA and Grad-CAM. All values in %.

Model	F1	Precision	Recall	Overlap	Top 1
Xception (VIBA)	45.42	40.76	51.28	51.92	62.96
VGG (VIBA)	52.72	50.00	55.76	57.69	40.74
CViT (VIBA)	64.82	55.79	77.33	78.00	59.26
Xception (Grad-CAM)	31.88	32.37	31.41	19.68	18.52
VGG (Grad-CAM)	44.90	42.63	47.44	32.31	22.22
CViT (Grad-CAM)	44.78	51.28	39.74	54.48	29.63

Table 5. Average per-image explanation time for VIBA and Grad-CAM.

Model	Method	Avg Time per Video (s)
Xception	VIBA	0.29
VGG11	VIBA	0.24
Xception	Grad-CAM	0.00282
VGG11	Grad-CAM	0.00269
CViT	VIBA	1.45
CViT	Grad-CAM	1.54
CViT	non	0.50

Table 7 summarizes spatial and temporal consistency. For Xception, Grad-CAM attains higher IoU, whereas VIBA yields lower RIM and slightly higher TCS on fake videos, indicating more temporally stable saliency. For the motion-based VGG11, VIBA clearly outperforms Grad-CAM across all consistency metrics (higher IoU and TCS, markedly lower RIM), while for CViT the picture is mixed: Grad-CAM sometimes achieves higher IoU, but VIBA maintains competitive TCS and low RIM, indicating smooth, stable explanations overall.

5.1 Agreement with Human Annotations and Efficiency

Human-overlap metrics in Table 4 indicate that VIBA explanations align better with human annotations than Grad-CAM across all models, with the largest gain for CViT (F1 64.82% vs. 44.78%, and higher overlap and Top-1 match). This suggests that, despite its weaker classification performance, CViT paired with VIBA produces particularly human-relevant explanations.

Runtime measurements (Table 5) show the expected trade-off: Grad-CAM is an order of magnitude faster than VIBA for CNNs, while both methods incur higher costs on CViT. In practice, VIBA adds overhead but remains feasible for offline or analyst-in-the-loop scenarios, where interpretability quality is prioritized over raw speed.

Table 6. Values in %. Performance of Xception, VGG11, and CViT on deepfake detection. A plus represents VIBA injection while a minus indicates the standard model without VIBA.

Model	Accuracy	Precision	Recall	ECE
Xception -	81.51	77.49	89.34	0.122
Xception +	81.47	77.43	89.34	0.121
VGG -	79.96	75.42	88.95	0.124
VGG +	79.87	75.31	88.95	0.125
CViT -	73.91	87.50	58.33	0.224
CViT +	73.91	87.50	58.33	0.188

5.2 Overlap with Human Annotations

As shown in Table 4, VIBA explanations for Xception and VGG11 achieve moderate agreement with human annotations, overlapping with human labels in over 50% of videos. CViT with VIBA performs even better, reaching almost 65% agreement, suggesting that transformer-based models paired with VIBA capture human-relevant regions particularly well. The region deemed most important by the majority of annotators is recovered by the models in roughly 40% of cases for VGG11 and around 60–63% for CViT and Xception. Consistently higher recall than precision indicates that models tend to cover many human-marked regions but also highlight additional areas.

Region-wise patterns (Fig. 6) show that humans most frequently select the lips and mouth as well as brows, eyes and forehead, a tendency mirrored reasonably well by VIBA for CViT and Xception. For VGG11, VIBA focuses mainly on the eyes, brows and forehead, matching these areas and visible deepfake edges but largely ignoring the chin and neck, even when humans highlight them. CViT with VIBA aligns best overall with human annotations across (i) lips and mouth, (ii) brows, eyes and forehead and (iii) nose, cheeks and ears, but is somewhat less

Table 7. Mean saliency consistency metrics for VIBA and Grad-CAM across models and video types.

Model	Video Type	IoU	TCS	RIM
Xception VIBA	Fake	0.823	0.318	0.025
Xception Grad-CAM	Fake	0.885	0.269	0.056
Xception VIBA	Real	0.832	0.307	0.025
Xception Grad-CAM	Real	0.895	0.296	0.04
VGG VIBA	Fake	0.763	0.961	0.034
VGG Grad-CAM	Fake	0.260	0.001	0.324
VGG VIBA	Real	0.800	0.893	0.027
VGG Grad-CAM	Real	0.422	0.159	0.206
CViT VIBA	Fake	0.371	0.535	0.019
CViT Grad-CAM	Fake	0.708	0.588	0.010
CViT VIBA	Real	0.800	0.893	0.027
CViT Grad-CAM	Real	0.603	0.502	0.010

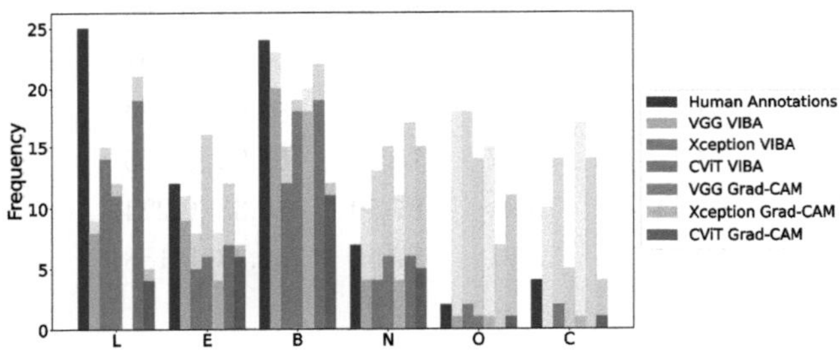

Fig. 6. Comparison of frequencies for VGG, Xception, and CViT VIBA matches to human annotation for different face regions. The transparent part of bars indicates overall frequency of such highlights produced by the method and the model. Abbreviations: L: Lips & Mouth, E: Deepfake Edges, B: Brows, Eyes & Forehead, N: Nose, Cheeks & Ears, O: Outside of Face, C: Chin & Neck.

focused on deepfake edges than VGG11. Its patch-based representation (Fig. 5) yields small, coarse but localized highlights, making interpretation somewhat more subjective while potentially useful for fact-checkers. Xception with VIBA recovers lips and mouth better than VGG11 but worse than CViT, identifies brows, eyes and forehead moderately well, and is slightly more sensitive to relevant regions outside the face. It also focuses more often on chin and neck than the other models, although this still does not systematically match human labels. Across models and methods, nose, cheeks and ears are identified with similar frequency, but less often than humans.

Compared to VIBA, Grad-CAM explanations show lower F1, precision, and Top-1 match scores and a region-frequency pattern that departs more from human annotations. For Xception, Grad-CAM concentrates on brows, eyes and forehead but almost never on lips and mouth or deepfake edges, while attributing more relevance to nose, chin and neck, and areas outside the face, which annotators rarely mark. VGG11 Grad-CAM offers slightly better alignment but still under-represents deepfake edges and spreads importance over largely non-informative regions. For CViT, Grad-CAM particularly fails to capture the lips and mouth area, whereas CViT with VIBA closely follows human annotations there. Overall, all models and methods tend to overemphasize outside-face and chin/neck regions relative to humans, but VIBA consistently provides more stable, human-aligned and interpretable explanations across architectures.

6 Discussion and Conclusion

In this work, we adapted information bottleneck attribution to video classifiers and introduced VIBA, evaluated on the task of deepfake detection across three architectures (Xception, CViT, and a VGG11 model on optical flow). Across all models, VIBA left predictive performance essentially unchanged: recall and precision remained almost identical, and ECE even improved slightly, most notably for CViT, indicating stable behaviour and improved calibration.

VIBA produces temporally and spatially consistent saliency maps, with moderately good overlap with human annotations and little reliance on off-face regions. This suggests that model-specific cues still largely reside on the face, potentially capturing artifacts that humans miss. We also observe systematic differences between real and fake videos in the highlighted regions, pointing to class-specific patterns. In contrast, Grad-CAM heatmaps fluctuate more over time and show a lower overlap with human annotations, consistent with its sensitivity to late-layer activations, gradient noise, and spatial diffusion. These effects are particularly pronounced for motion-based (VGG11 on optical flow) and transformer-based (CViT) models, where VIBA clearly yields more temporally stable and spatially coherent explanations. While optical-flow-based relevance maps effectively expose subtle temporal inconsistencies, they remain less intuitive for human inspection than static frame explanations.

Methodologically, VIBA's information bottleneck controls how much information is retained in the explanation, yielding targeted, often finer-grained attributions without being tied to any specific architecture. Its activation perturbation and focus on information flow reduce the risks associated with gradient saturation and high-confidence predictions that can bias gradient-based methods. However, these strengths come at a cost: VIBA is computationally more demanding and relies on noise injection to constrain information flow. Its attributions do not always align with human intuition, as reflected in only moderate agreement with annotators and the down-weighting of visually detailed yet prediction-irrelevant regions that humans may still consider important.

Overall, our findings indicate that VIBA is a stable, generalizable mechanism for explaining video classifiers, particularly for deepfake detection. We view the partial mismatch between VIBA and human annotations not as a flaw but as a reflection of complementary strengths: VIBA can surface subtle, temporally evolving artifacts, while humans provide contextual understanding and normative judgment. In high-stakes applications such as fact-checking, this complementarity enables human-in-the-loop verification pipelines in which systems like VIBA assist fact-checkers by flagging suspicious regions, prioritizing frames for review, and fostering trust through transparent, interpretable outputs.

7 Limitations and Ethical Considerations

XAI methods are particularly effective in human-in-the-loop (HIL) scenarios because they enable human experts to interpret model outputs, identify potential errors, and make informed decisions. This is especially relevant for deepfake detection, where nuanced judgments and context are often required to distinguish subtle manipulations. However, even in forensic settings, models trained on large-scale video data may inherit biases—such as demographic imbalances or performance gaps across ethnicities. If VIBA systematically emphasizes certain facial regions for specific groups, it risks reinforcing unfair outcomes in applications like content moderation or investigations. Future work should include fairness audits of both detection and attribution. By producing detailed attribution maps, VIBA reveals, which regions or motion cues the model relies on most. While this improves interpretability, it can make the model more vulnerable to potential adversarial attacks. Countermeasures could involve introducing randomized or masked attribution outputs in future work. VIBA-generated maps might also be misinterpreted. For instance, users might wrongly assume that low attribution scores imply authenticity or that explanations offer causal certainty. Our study shows only moderate alignment between human-marked regions and VIBA explanations, although explanations may enhance model transparency without necessarily aligning with human reasoning. Further, as VIBA currently focuses on visual features, it may miss manipulations that extend beyond pixels, such as audio tampering or narrative context, leading to over-reliance on visual cues. However, if lay users overly trust these visualizations, it could result in misplaced confidence in model predictions. Designing interfaces that support critical evaluation of AI outputs, rather than blind acceptance, is vital.

Acknowledgments. The work on this paper was performed in the scope of the "noFake" project (BMFTR, Award Identifier F16KIS1519) and "news-polygraph" project (BMFTR, Award Identifier: 03RU2U151C).

Disclosure of Interests. The authors have no competing interests to declare that are relevant to the content of this article.

References

1. Cai, Z., Ghosh, S., Adatia, A.P., Hayat, M., Dhall, A., Stefanov, K.: Av-deepfake1m: a large-scale LLM-driven audio-visual deepfake dataset (2023)

2. Chen, Y., Hummel, T., Koepke, A.S., Akata, Z.: Where and when: space-time attention for audio-visual explanations. arXiv **abs/2105.01517** (2021). https://api.semanticscholar.org/CorpusID:233714817
3. Dolhansky, B., et al.: The deepfake detection challenge (DFDC) dataset (2020)
4. Dufour, N., Gully, A.: Google research. https://ai.googleblog.com/2019/09/contributing-data-to-deepfake-detection.html/. Accessed 24 Apr 2024
5. Farnebäck, G.: Two-frame motion estimation based on polynomial expansion. In: Image Analysis, Proceedings of the Scandinavian Conference on Image Analysis (SCIA), vol. 2749, pp. 363–370 (2003)
6. Guo, C., Pleiss, G., Sun, Y., Weinberger, K.Q.: On calibration of modern neural networks. In: Proceedings of the 34th International Conference on Machine Learning (ICML), pp. 1321–1330. PMLR (2017)
7. Hondru, V., Hogea, E., Onchis, D., Ionescu, R.T.: Exddv: a new dataset for explainable deepfake detection in video (2025). https://arxiv.org/abs/2503.14421
8. Ibaceta, J.: Video-keyframe-detector (2023). https://github.com/joelibaceta/video-keyframe-detector. Accessed 11 Sept 2024
9. Jiang, L., Li, R., Wu, W., Qian, C., Loy, C.C.: Deeperforensics-1.0: a large-scale dataset for real-world face forgery detection (2020)
10. Khalid, H., Tariq, S., Kim, M., Woo, S.S.: FakeAVCeleb: a novel audio-video multimodal deepfake dataset (2022)
11. King, D.E.: Dlib-ml: a machine learning toolkit. J. Mach. Learn. Res. **10**, 1755–1758 (2009)
12. Kwon, P., You, J., Nam, G., Park, S., Chae, G.: KoDF: a large-scale Korean deepfake detection dataset (2021)
13. Lee, M.H.: Towards gradient-based time-series explanations through a spatiotemporal attention network (2024). https://arxiv.org/abs/2405.17444
14. Li, Y., Yang, X., Sun, P., Qi, H., Lyu, S.: Celeb-DF: a new dataset for deepfake forensics (2019)
15. Liu, H.: Deepfake detection (2023). https://github.com/HongguLiu/Deepfake-Detection. Accessed 11 Sept 2024
16. Malolan, B., Parekh, A., Kazi, F.: Explainable deep-fake detection using visual interpretability methods. In: 2020 3rd International Conference on Information and Computer Technologies (ICICT), pp. 289–293 (2020)
17. Montavon, G., Lapuschkin, S., Binder, A., Samek, W., Müller, K.R.: Explaining nonlinear classification decisions with deep Taylor decomposition. Pattern Recogn. **65**, 211–222 (2017). https://doi.org/10.1016/j.patcog.2016.11.008, https://www.sciencedirect.com/science/article/pii/S0031320316303582
18. Rössler, A., Cozzolino, D., Verdoliva, L., Riess, C., Thies, J., Nießner, M.: Faceforensics++: learning to detect manipulated facial images (2019)
19. Schulz, K., Sixt, L., Tombari, F., Landgraf, T.: Restricting the flow: information bottlenecks for attribution (2020). https://arxiv.org/abs/2001.00396
20. Selvaraju, R.R., Cogswell, M., Das, A., Vedantam, R., Parikh, D., Batra, D.: Grad-cam: visual explanations from deep networks via gradient-based localization. Int. J. Comput. Vision **128**(2), 336–359 (2019). https://doi.org/10.1007/s11263-019-01228-7, http://dx.doi.org/10.1007/s11263-019-01228-7
21. Simonyan, K., Zisserman, A.: Very deep convolutional networks for large-scale image recognition. CoRR abs/1409.1556 (2014). https://api.semanticscholar.org/CorpusID:14124313
22. Tishby, N., Pereira, F.C., Bialek, W.: The information bottleneck method (2000). https://arxiv.org/abs/physics/0004057

23. Tsigos, K., Apostolidis, E., Baxevanakis, S., Papadopoulos, S., Mezaris, V.: Towards quantitative evaluation of explainable AI methods for deepfake detection. Proceedings of the 3rd ACM International Workshop on Multimedia AI Against Disinformation (2024). https://api.semanticscholar.org/CorpusID:269449446
24. Vanneste, J., Verma, M., Ganguly, D.: Detecting and explaining emotions in video advertisements. In: Proceedings of the 47th International ACM SIGIR Conference on Research and Development in Information Retrieval. SIGIR '24, pp. 2734–2738. Association for Computing Machinery, New York, NY, USA (2024). https://doi.org/10.1145/3626772.3657664
25. Wodajo, D., Lambert, P., Wallendael, G.V., Atnafu, S., Mareen, H.: Improved deepfake video detection using convolutional vision transformer. In: Proceedings of the IEEE International Conference on Games, Entertainment and Media (GEM). Turin (Torino), Italy (2024)

Fast Model Selection for Interpretable Gaussian Process Models Using Laplace Approximation

Andreas Besginow[1(✉)], Thomas Pawellek[1], Jan David Hüwel[2], Christian Beecks[2], and Markus Lange-Hegermann[1]

[1] OWL University of Applied Sciences and Arts, Lemgo, Germany
{andreas.besginow,thomas.pawellek,markus.lange-hegermann}@th-owl.de
[2] Fernuniversität in Hagen, Hagen, Germany
{jandavid.huewel,christian.beecks}@fernuni-hagen.de

Abstract. Model selection aims to find the best model in terms of accuracy, interpretability or simplicity. Here, we focus on evaluating model performance of Gaussian process models, which can be used for data analysis of uncertain data. While previous work uses methods like the Marginal log likelihood, AIC or nested sampling to perform model selection, they either lack performance or have significant runtime issues, limiting their applicability. We address these challenges by introducing multiple metrics based on the Laplace approximation, where we overcome an inconsistency occurring during its naive application which causes overfitting in terms of model selection. We show how AIC can fail to recognize the more appropriate model and perform large scale experiments to conclude that our Laplace-based metrics outperform the state of the art for fast model selection. Our model selection metrics allow fast and high quality model selection of Gaussian processes.

Keywords: Gaussian Process · Model Selection · Interpretable Models

1 Introduction

Turning data into knowledge frequently requires inferring models that capture the major data characteristics. Dealing with this problem requires identifying an accurate, interpretable and simple machine learning model, we refer to this identification as *model selection*. To perform model selection automatically, it is required to have a metric that enables objective comparison of models.

In instances where data is expensive to collect and datasets are thus small, Gaussian Processes (GPs) are a good choice to model the data, owing to their interpretability and good modelling capabilities. This originates from the GP's covariance functions which can capture a wide variety of structures from diverse domains such as geometry [6], differential equations [2,4,13] and many more. These introduce prior knowledge into the GPs, resulting in reasonable models

M. Baratchi et al. (Eds.): IDA 2026, LNCS 16513, pp. 169–182, 2026.
https://doi.org/10.1007/978-3-032-23833-7_13

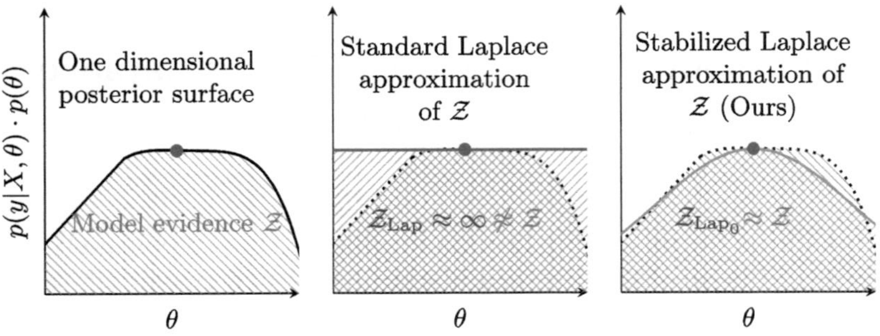

Fig. 1. A conceptual visualization of the inconsistency when naively applying the Laplace approximation and one of our suggested variants. **Left**: The posterior over parametrizations θ, with a degenerate local extremum (red dot). The model evidence $\mathcal{Z}$ is the gray shaded area. **Middle**: Naive application of the Laplace approximation around the optimum with infinitely large model evidence approximation $\mathcal{Z}_{\text{Lap}} \approx \infty$ overlaid in red. **Right**: Application of our stabilized Laplace (Lap_0) around the optimum with model evidence approximation $\mathcal{Z}_{\text{Lap}_0} \approx \mathcal{Z}$ overlaid in green.

even for scarce data, but require knowledge about the dataset and GPs. However, if such knowledge is unavailable, model selection can help select appropriate GPs.

The gold standard for GP model selection is the computation of the model evidence (Equation (2)) [14] which, due to intractability, is often approximated using expensive sampling-based strategies like nested sampling [23]. To reduce computation, approximate techniques are often employed, like the Marginal Log Likelihood (MLL), Akaike Information Criterium (AIC) [1] or Bayesian Information Criterium (BIC) [22]. Since the MLL tends to overfit in model selection, i.e. prefer models with more hyperparameters, AIC and BIC additionally penalize the MLL and try to mitigate overfitting, but often don't suffice (c.f. Sect. 4.1).

In this paper, we introduce novel model selection metrics for GPs based on the Laplace approximation, to approximately solve the model evidence integral [5,14] and allow the selection of high quality interpretable models when using small data (Sect. 3). We also discuss an inconsistency when naively applying Laplace approximation, where superfluous hyperparameters can improve the approximated model evidence, causing it to prefer those overparameterized models (c.f. Sect. 3.1). To address this inconsistency we bound the Hessian's eigenvalues, preventing the corresponding hyperparameters from contributing positively without also contributing to the model fit. We illustrate the extreme case of this inconsistency in Fig. 1, where we mitigate the infinitely large model evidence approximation of $\mathcal{Z}$ using our stabilized Laplace (Lap_0). Our experiments show that our modified Laplace approximation outperforms AIC and BIC, while being comparable in runtime, and prevents overfitting due to penalizing unstable optima through the Hessian. This enables the use of the Laplace approximation for fast and interpretable model selection by preferring simple and good models.

The core contributions are:[1] (1) introducing new model selection metrics, based on the Laplace approximation, which mitigate the original inconsistency coming from its naive application and (2) showing that our metric outperforms the state of the art in fast model selection, like AIC and BIC

2 Preliminaries

2.1 Gaussian Processes (GPs)

A GP $g = \mathcal{GP}(\mu, k)$ is a stochastic process where every finite set of realizations $g(x_i), x_i \in \mathbb{R}^d$ are jointly Gaussian [21]. Such a GP is characterized by its mean function $\mu : \mathbb{R}^d \to \mathbb{R} : x \mapsto \mathbb{E}(g(x))$ (often set to zero) and its positive semi-definite kernel $k : \mathbb{R}^d \times \mathbb{R}^d \to \mathbb{R} : (x, x') \mapsto \mathbb{E}((g(x) - \mu(x))(g(x') - \mu(x'))^T)$. Most covariance functions contain various hyperparameters θ, e.g. the SE kernel $k_{SE}(x, x') = \theta_1^2 \exp(\frac{-(x-x')^2}{2\theta_2^2})$ contains signal variance θ_1 and lengthscale θ_2, which become part of the GP, e.g. for $\mu = 0$ we have $g_{SE} = \mathcal{GP}(0, k_{SE})$.

To train the GP hyperparameters, the MLL is commonly used, defined as:

$$\log p(y|X,\theta) = -\frac{1}{2}y^T(K + \sigma_n^2 I)^{-1}y - \frac{1}{2}\log|K + \sigma_n^2 I| - \frac{n}{2}\log(2\pi) \tag{1}$$

where $X \in \mathbb{R}^{n\times d}$ and $y \in \mathbb{R}^n$ are the dataset of size n and the GPs hyperparameters θ are part of the covariance matrix K with entries $K_{ij} = k(x_i, x_j)$.

We refer to all possible parametrizations of GPs with the same kernel as a *class* of GPs, e.g. all parametrizations of g_{SE} form the class of SE GPs.

The set of kernels is closed under various operations [9,15,21]. This allows combining, for example, a periodic kernel and an SE kernel ($k = k_{PER} \cdot k_{SE}$) to enforce periodic *and* smooth behaviour, growing the set of potentially useful classes of GPs significantly and making model selection a complex endevaour.

2.2 Model Selection for Gaussian Processes

Model selection tries to find the best model for a dataset $\mathcal{D} = (X, y)$ in terms of accuracy, interpretability or simplicity. According to Occam's razor one should prefer simpler models, given comparable modelling performance. Model selection metrics, which weigh model performance against complexity, should reflect this.

Naively, one can use the optimized MLL $\widehat{\mathcal{L}} = \log p(y|X, \hat{\theta})$ to assess the model quality for a GP with optimal hyperparameters $\hat{\theta}$. This comes with the downside that increased model complexity allows to overfit the data and thus additional hyperparameters don't decrease the MLL, as exemplified in Fig. 2.

The performance of a class of GPs, for a dataset $\mathcal{D}$, can be measured objectively through the model evidence:

$$\mathcal{Z} = p(y|X) = \int p(y|X,\theta)p(\theta)d\theta \tag{2}$$

[1] https://github.com/ABesginow/LaplaceKernelSearch.

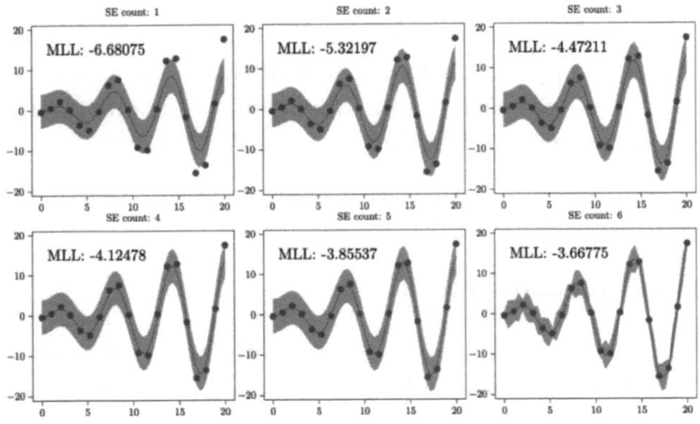

Fig. 2. GPs equipped with sums of, up to six, SE kernels as their kernel. The six summand case clearly overfits but still has the best (i.e. highest) MLL.

Since this integral is intractable one usually resorts to approximations such as nested sampling [23]. At nested sampling's core, the likelihood surface $p(y|X,\theta)$ is explored to calculate a weighted sum of likelihood evaluations. This provides a good approximation of the model evidence $\mathcal{Z}$, but at high computational cost.

Due to this, one looks into faster approximations of the likelihood. The earliest is AIC [1], which is based on the optimized MLL $\widehat{\mathcal{L}}$ and corrects the optimum for the number of hyperparameters u as follows: $\text{AIC} = 2u - 2\widehat{\mathcal{L}}$. Another common metric, closely related to AIC, is BIC, which additionally corrects for the number of datapoints as $\text{BIC} = u\log(n) - 2\widehat{\mathcal{L}}$ [22], and can be considered a simple approximation to the log model evidence (cf. §5.2.5.1 in [19]).

3 Laplace Approximation of the GP Model Evidence

In this section, we detail how to apply the Laplace approximation to approximately solve the model evidence integral in Equation (2). The Laplace approximation $\text{Lap}(f)$ aims to find a Gaussian approximation for a function f using the second order Taylor approximation in log-space of f around its optimum, i.e. a point of gradient equal to zero [5].

The Laplace approximation has been used in numerous works to train GP hyperparameters or use it to integrate out the latent GP [11,16,17], *instead of the hyperparameters* θ. Notable is its applicability to non-Gaussian likelihoods of e.g. GPs for classification, to optimize the hyperparameters and approximate the corresponding marginal likelihood (see Chaps. 3.4 and 3.5 in [21]). Outside of GPs it is also commonly used for Bayesian neural networks [8].

However, to the best of our knowledge, this work is the first to use the Laplace approximation to compute and compare the GP model evidence in Equation (2). For this, we perform the Laplace approximation of the product of likelihood and prior $f = p(y|X,\theta) \cdot p(\theta)$ to get:

$$\log \text{Lap}(p(y|X,\theta) \cdot p(\theta)) \approx \log(p(y|X,\hat{\theta}) \cdot p(\hat{\theta})) - \frac{1}{2}(\theta - \hat{\theta})^T H(\theta - \hat{\theta}) \qquad (3)$$

where $H = -\nabla\nabla \log(p(y|X,\theta) \cdot p(\theta))|_{\theta=\hat{\theta}}$ is the Hessian at a (local) optimum $\hat{\theta}$ [5, §4.4], which can be determined by standard optimizers.

To approximate the model evidence we start by replacing $p(y|X,\theta) \cdot p(\theta)$ in Equation (2) with the Laplace approximation (Equation (3)) via

$$\begin{aligned}
f(\theta) &= p(y|X,\theta) \cdot p(\theta) \\
\log(f(\theta)) &= \log(p(y|X,\theta)) \cdot p(\theta) \\
\log(f(\theta)) &\approx \log(f(\hat{\theta})) - \frac{1}{2}(\theta - \hat{\theta})H(\theta - \hat{\theta})^T \\
f(\theta) &\approx f(\hat{\theta}) \exp\left(-\frac{1}{2}(\theta - \hat{\theta})H(\theta - \hat{\theta})^T\right)
\end{aligned}$$

where $H = -\nabla\nabla \log f(\theta)$. We then insert the approximation into the model evidence integral.

$$\begin{aligned}
\mathcal{Z} &= \int_\theta f(\theta)\,\mathrm{d}\theta \\
&\approx f(\hat{\theta}) \cdot \int_\theta \exp\left(-\frac{1}{2}(\theta - \hat{\theta})H(\theta - \hat{\theta})^T\right) \mathrm{d}\theta \\
&= f(\hat{\theta}) \cdot \frac{(2\pi)^{\frac{-u}{2}} \cdot |H^{-1}|^{-\frac{1}{2}}}{(2\pi)^{\frac{-u}{2}} \cdot |H^{-1}|^{-\frac{1}{2}}} \int_\theta \exp\left(-\frac{1}{2}(\theta - \hat{\theta})H(\theta - \hat{\theta})^T\right) \mathrm{d}\theta \\
&= f(\hat{\theta}) \cdot \frac{1}{(2\pi)^{\frac{-u}{2}} \cdot |H^{-1}|^{-\frac{1}{2}}} \int_\theta (2\pi)^{\frac{-u}{2}} \cdot |H^{-1}|^{-\frac{1}{2}} \exp\left(-\frac{1}{2}(\theta - \hat{\theta})H(\theta - \hat{\theta})^T\right) \mathrm{d}\theta \\
&= f(\hat{\theta}) \cdot \frac{1}{(2\pi)^{\frac{-u}{2}} \cdot |H^{-1}|^{-\frac{1}{2}}} \cdot 1 \\
\log \mathcal{Z} &\approx \log(f(\hat{\theta})) + \frac{u}{2}\log(2\pi) + \frac{1}{2}\log(|H^{-1}|) \\
\log \mathcal{Z} &\approx \log(f(\hat{\theta})) + \frac{u}{2}\log(2\pi) - \frac{1}{2}\log(|H|)
\end{aligned}$$

where u is the number of hyperparameters.

This leaves us with:

$$\log \mathcal{Z} \approx \log \mathcal{Z}_{\text{Lap}} = \log\left(p(y|X,\hat{\theta}) \cdot p(\hat{\theta})\right) + \frac{u}{2}\log(2\pi) - \frac{1}{2}\log(|H|) \tag{4}$$

The next subsection discusses cases where this standard form of Laplace approximation fails when approximating the GP model evidence and how we mitigate this problem by introducing lower bounds on the Hessian's eigenvalues.

3.1 Overcoming Inconsistencies of the Laplace Approximation

Naively applying the standard Laplace approximation $\log \mathcal{Z}_{\text{Lap}}$ in Formula (4) can exhibit an inconsistency which causes it to favour models with superfluous hyperparameters. In extreme cases this can cause infinite model evidence approximations, as illustrated in Fig. 1 for one dimension. There, a degenerate local extremum, in the sense that one of the Hessian's eigenvalues is almost zero, causes the standard Laplace approximation to result in an infinite model evidence $\mathcal{Z}_{\text{Lap}} = \infty$. This stems from the likelihood surface to be approximated as constant at $\widehat{\theta}$. This occurs in cases of superfluous parameters directly contradicting Occam's razor and preventing its use in GP model selection.

To address this problem, we introduce various versions of the Laplace approximation, all with different interpretations. Doing so, we ensure that the mere presence of a hyperparameter does not contribute to an arbitrary improvement.

The inconsistencies are directly connected to the Hessian H and are best interpreted in one dimension ($u = 1$). For hyperparameters where the one dimensional[2] Hessian $H = [\lambda]$ is moderately small (i.e. for $\lambda < 2\pi$), we can directly observe from Formula (4): $\frac{1}{2}\log(2\pi) - \frac{1}{2}\log(\lambda)$ contributes *positively* to the model evidence just because the hyperparameter of λ *exists*, even without a contribution to the model fit. We therefore modify the Hessian matrix to mitigate this behaviour (c.f. Algorithm 1), similar to Newton methods with Hessian modification in second order optimization (cf. Sect. 3.4 in [20]).

Lemma 1. *To ensure that each hyperparameter of a GP has a minimal negative contribution r in the last two summands of Formula* (3)*, to the model evidence $\mathcal{Z}$, every eigenvalue λ of the Hessian needs to be at least:*

$$\lambda \geq \exp(-2r) \cdot 2\pi \tag{5}$$

Proof. Diagonalize the Hessian H using its eigendecomposition $H = V\Lambda V^{-1}$. Using the determinant formula $|H| = |V||\Lambda||V^{-1}| = |\Lambda|$, we reduce calculations to the one-dimensional case. In one dimension, we solve the inequality:

$$\begin{aligned} & r \geq \frac{1}{2}\log(2\pi) - \frac{1}{2}\log(\lambda) \\ \iff \quad & \lambda \geq \exp(-2r) \cdot 2\pi \end{aligned}$$

□

Algorithm 1: Eigenvalue correction in Lemma 1.

Data: Hessian H, minimal contribution r
Result: Updated Hessian $\widetilde{H}$
$\Lambda, V \leftarrow \text{eig}(H)$ `#Eigendecomposition of` H
foreach $\lambda \in \text{diag}(\Lambda)$ `#Iterate over EVs on diagonal of` Λ
do
| $\lambda \leftarrow \max(\exp(-2r) \cdot 2\pi, \lambda)$ `#Equation (5)`
end
$\widetilde{H} \leftarrow V\Lambda V^{-1}$ `#Build new Hessian`

---

[2] The result also holds for higher dimensional Hessians by disentangling the dimensions via diagonlization, as used in the proof of Lemma 1.

Based on Eq. (5) we suggest the following *stabilized* variant of the Laplace approximation which replaces all Eigenvalues lower than 2π with 2π, to compensate for the $\frac{u}{2}\log(2\pi)$ term in Eq. (4). Stabilized Laplace (Lap_0) thus prevents additional hyperparameters from improving the model evidence without also improving the model fit.

Corollary/Definition 1. *Stabilized Laplace (Lap_0): To ensure a minimal negative contribution $r = 0$ to the log model evidence $\log \mathcal{Z}_{Lap_0} \approx \log \mathcal{Z}$ every eigenvalue has to be at least $\lambda \geq 2\pi$.*

Additionally, we present two thresholds called *AIC corrected Laplace* (Lap_A) and *BIC corrected Laplace* (Lap_B), below. These thresholds are chosen due to the acceptance of AIC and BIC as model selection metrics and the historic relevance.

Corollary/Definition 2. *AIC corrected Laplace (Lap_A): To ensure a minimal negative contribution $r = -1$ to the model evidence $\log \mathcal{Z}_{Lap_A} \approx \log \mathcal{Z}$ every eigenvalue has to be at least $\lambda \geq \exp(2) \cdot 2\pi$.*

Corollary/Definition 3. *BIC corrected Laplace (Lap_B): To ensure a minimal negative contribution $r = -\log(n)$ to the model evidence $\log \mathcal{Z}_{Lap_B} \approx \log \mathcal{Z}$ every eigenvalue has to be at least $\lambda \geq \exp(2\log(n)) \cdot 2\pi = 2\pi \cdot n^2$.*

AIC corrected Laplace (Lap_A) corrects the optimal loss similarly to AIC by contributing $r = -1$ for each hyperparameter. Whereas BIC corrected Laplace (Lap_B) is correcting similarly to BIC by contributing $r = -\log(n)$ for each hyperparameter. The connection to AIC resp. BIC derives from the fact that these corrected Laplace approximations collapse into AIC resp. BIC in the case of all eigenvalues λ of the Hessian being small, i.e. all eigenvalues being replaced.

These lower bounds cause our approximations to be stricter than standard Laplace since they no longer allow a minimal contribution $r > 0$. But this now mitigates the inconsistency coming from those positive ($r > 0$) contributions, causing the standard Laplace approximation to prefer overparameterized models.

4 Evaluation

We present three experiments showing the applicability of our Laplace approximations on real and synthetic data. All GPs are written in GPyTorch [12], with standard kernels and GPyTorch's standard parameter scalings. Hyperparameter training was performed using Newton-like optimization of PyGRANSO [18], with MLL resp. Maximum A Posteriori (MAP) loss, ensuring proper optima with gradient approximately zero. Training is repeated ten times. If not stated otherwise, we employ an uninformed parameter prior $p(\theta_i) = \mathcal{N}(0, 100)$.

Table 1. The number of correctly recognized kernels, in percent, for all metrics. Higher is better. Best performance is bold, outside of $\log(\mathcal{Z})$, as it is the "ground truth", and MLL/MAP, as they are not proper model selection metrics. Lap consistently performs best and sometimes reaches the same recognition performance as the sampling-based strategy of $\log(\mathcal{Z})$. Further, Lap_0 consistently outperforms AIC and BIC, whereas Lap_A and Lap_B are about as good as them. This underlines the accuracy improvement of our stabilized Laplace (Lap_0) over AIC and BIC.

NUM DATA	($\log \mathcal{Z}$)	LAP	LAP_0(OURS)	LAP_A (OURS)	LAP_B (OURS)	AIC	BIC	(MLL)	(MAP)
10	(46.7)	**45.0**	31.7	28.3	28.3	28.3	28.3	(28.3)	(28.3)
15	(53.3)	**51.7**	38.3	30.0	30.0	31.7	31.7	(31.7)	(30.0)
25	(65.0)	**60.0**	51.7	41.7	41.7	46.7	46.7	(46.7)	(41.7)
40	(63.3)	**58.3**	56.7	51.7	51.7	51.7	51.7	(51.7)	(51.7)
50	(58.3)	**60.0**	58.3	53.3	53.3	53.3	53.3	(53.3)	(53.3)
75	(56.7)	**60.0**	**60.0**	48.3	48.3	48.3	48.3	(48.3)	(48.3)
100	(56.7)	**58.3**	**58.3**	55.0	55.0	55.0	55.0	(55.0)	(55.0)
200	(66.7)	**60.0**	**60.0**	60.0	58.3	58.3	58.3	(58.3)	(58.3)

Table 2. Percentage of instances where the standard Laplace approximation caused the inconsistency discussed in Sect. 3.1, which made model selection impossible.

	10	15	25	40	50	75	100	200
LAP [%]	2.5	0.3	1.7	0.8	0.6	0.3	1.1	0.3

Table 3. The runtime to estimate each metric, all Laplace approximations are represented by Lap_0, as their runtimes are equal. Runtime includes the metric calculation *and* respective training time using MLL and MAP, in braces. Lower is better and lowest runtime of the metrics is bold. The runtime of the fast approximations is dominated by their training time, which converges significantly faster when using MAP as the loss. This causes all Laplace approximations to be a magnitude faster than AIC and BIC and almost two magnitudes faster than nested sampling-based approximation of $\log \mathcal{Z}$.

NUM DATA	$\log \mathcal{Z}$	LAP_0(OURS)	AIC	BIC	(MLL)	(MAP)
10	62.32	**0.74**	20.09	20.09	(20.09)	(0.72)
15	59.77	**0.77**	17.72	17.72	(17.72)	(0.75)
25	54.78	**0.98**	17.82	17.82	(17.82)	(0.96)
40	57.18	**1.17**	17.28	17.28	(17.28)	(1.15)
50	58.30	**1.28**	17.17	17.17	(17.17)	(1.26)
75	60.45	**1.60**	18.31	18.31	(18.31)	(1.58)
100	60.84	**1.90**	18.77	18.77	(18.77)	(1.88)
200	72.11	**3.29**	26.22	26.22	(26.22)	(3.27)

4.1 Kernel Recognition

In this experiment we show how often the different metrics recognize the correct kernel, i.e. the kernel the data originated from, in a large scale experiment.

Our setup is as follows: We sample ten datasets each from six different GPs, each GP with a different kernel. We then train six GPs on all the sampled datasets and use the metrics to determine the most suitable GP for each dataset. We also vary the dataset size across eight different sizes, for a total of 480 different datasets $(10 \cdot 6 \cdot 8)$ and 2880 trained GPs $(480 \cdot 6)$.

Table 1 displays the number of correctly recognized kernels by each metric, clearly showing that the model evidence recognizes the most kernels for almost all dataset sizes. We also find that Lap_0 consistently outperforms AIC and BIC in terms of correctly recognized kernels and has a true advantage over MAP, which it is based on. In contrast, AIC and BIC don't perform better than MLL.

The results in Table 1 show how Lap outperforms all metrics, with stabilized Laplace (Lap_0) as a close second, while being a magnitude faster than the remaining metrics (c.f. Table 3). Still, the inconsistency of Lap remains with up to every 40th kernel causing this inconsistency (c.f. Table 2). These results underline the accuracy of the Laplace approximations as model selection metrics.

Table 4. Performance of kernels k_i according to all model selection metrics (and MLL/MAP) on Mauna Loa. Higher values indicate that k_i is evaluated as better performing, the best kernel for each metric is bold. Lap makes the same choice as nested sampling ($\log \mathcal{Z}$) while the remaining Laplace approximations choose like AIC and BIC.

k	$\log \mathcal{Z}$	LAP	LAP$_0$	LAP$_A$	LAP$_B$	$\log \mathcal{Z}_{AIC}$	$\log \mathcal{Z}_{BIC}$	(MLL)	(MAP)
k_1	–783.7	–1176.5	–1180.9	–1183.6	–1190.0	–2220.0	–2226.4	(–2217.0)	(–1180.6)
k_2	–209.2	–418.4	–426.1	–430.7	–442.8	–384.5	–397.3	(–378.5)	(–424.0)
k_3	–200.4	–196.1	**–206.9**	**–214.2**	**–232.6**	**–157.2**	**–176.4**	(–148.2)	(**–204.5**)
k_4	**–177.8**	**–193.8**	–210.3	–219.6	–242.3	–157.9	–181.3	(**–146.9**)	(–207.8)

4.2 Mauna Loa

In our second experiment we use the popular Mauna Loa dataset, which captures the CO2 levels from 1958 to 2002, of which we use the monthly average in this experiment. We take the settings used in Chap. 5.4.3 of [21], with the minor change that we add the noise (called θ_{11}) to the first kernel ($\widetilde{k}_1$) instead of the last ($\widetilde{k}_4$). For our experiment, we iteratively increase the size of the tested kernel by adding the summands from [21], shown below. We call the sums $k_n = \sum_{i=1}^{n} \widetilde{k}_i$.

$$\widetilde{k}_1 = \theta_1^2 \exp\left(-\frac{(x-x')^2}{2\theta_2^2}\right) + \theta_{11}^2\delta \quad \widetilde{k}_2 = \theta_3^2 \exp\left(-\frac{(x-x')^2}{2\theta_4^2} - \frac{2\sin^2(\pi(x-x'))}{\theta_5^2}\right)$$

$$\widetilde{k}_3 = \theta_6^2 \left(1 + \frac{(x-x')^2}{2\theta_8\theta_7^2}\right)^{-\theta_8} \qquad \widetilde{k}_4 = \theta_9^2 \exp\left(-\frac{(x-x')^2}{2\theta_{10}^2}\right)$$

where δ is the Dirac delta function that is 1 if $x = x'$ and 0 otherwise.

Just like in [21], we shift y by the mean and leave the rest untouched. For each k_i we calculate AIC, BIC, all variants of the Laplace approximation and nested sampling (and MLL/MAP for completeness) to estimate the k_i's performance and present them in Table 4. We rescale AIC and BIC by -0.5, to be on the same scale as the log model evidence, and call them $\log \mathcal{Z}_{AIC}$ and $\log \mathcal{Z}_{BIC}$.

The results of Table 4 show that our Laplace approximations select the same kernel as AIC and BIC, i.e. kernel k_3. Whereas the model evidence and Lap find that $\widetilde{k}_4$ has a significant positive contribution to the model fit and therefore choose k_4 as the most appropriate kernel for the data, similar to [21]. Unsurprisingly, this is also found by MLL, which is known to prefer more parameters.

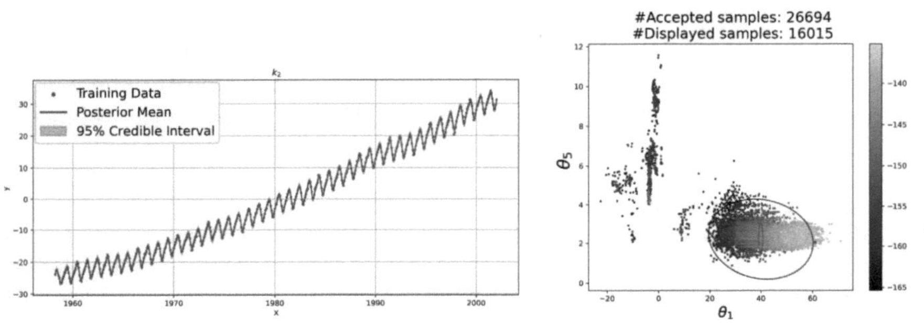

Fig. 3. (Left) The posterior GP with kernel k_2. As confirmed in Table 4, introducing the periodic kernel has a significant positive impact on the model fit. (Right) Visualization of the decreased curvature of Lap_0 (red ellipse) in contrast to Lap (black ellipse) for k_4. The ellipses can be understood as twice the standard deviation when using the normal distribution interpretation of the Laplace approximation. The colored dots are accepted samples from a nested sampling run, colored according to their likelihood. To reduce clutter only the best 60% of the accepted samples are shown. The red dot indicates the optimum found during nested sampling. The center of the ellipses indicates the optimum found during optimization. The decreased curvature causes Lap_0 to better suit the best samples with respect to θ_5 (lengthscale of the periodic kernel in $\widetilde{k}_2$), whereas it overcorrected in terms of θ_1 (signal variance of $\widetilde{k}_1$), as can be clearly seen by only covering a slim portion of the samples.

In Fig. 3 we visualize the effect of our stabilization approach, discussed in Sect. 3.1, in terms of the coverage of the likelihood surface. Figure 3 shows the accepted samples from a nested sampling run, colored according to their likelihood, for the kernel k_4. The ellipses shown are the 2σ intervals when understanding the Laplace approximation through its Gaussian distribution. The black ellipse shows Laplace (Lap) and the red ellipse shows stabilized Laplace (Lap_0). Since k_4 was misclassified by Lap_0, it is not surprising to see that Lap_0 overcorrected in terms of curvature, compared to Lap. We show this for θ_1 on the

x-axis in Fig. 3. However, we argue that for θ_5, shown on the y-axis, the correction was suitable as it now more precisely covers the relevant area of the likelihood surface, i.e. the samples that contribute to the model evidence $\log \mathcal{Z}$.

In conclusion, our Laplace variants perform the same as AIC and BIC in the selection of the kernel for the Mauna Loa dataset, preferring the simpler model. Here, the inconsistency of Lap was not an issue, allowing it to show its good performance and concur with the sampling-based model evidence approximation.

4.3 Simulated Motor Data

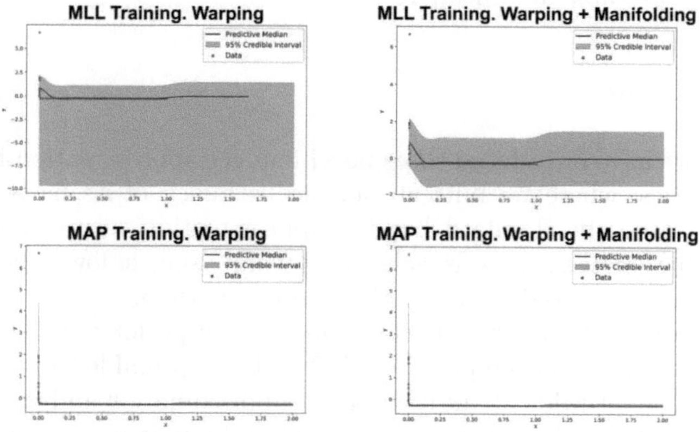

Fig. 4. Left column: The GPs with a kernel that includes warping. Right column: The GPs whose kernel includes warping and manifolding. Top row are GPs trained using MLL, bottom row are trained with MAP loss. The optima found during MLL are notably worse compared to those found during MAP, as is clear by comparing the top and bottom rows. However, despite the significantly worse posterior model of MLL (top left), AIC still prefers it over the model with tighter confidence bounds (top right), which is counterintuitive. Whereas for the MAP-trained model, there is no visible difference between the two posterior models. Hence, Lap_0's choice of preferring the model with only warping is more intuitive, as its modeling performance is as good as warping + manifolding *and* needs fewer parameters.

In the final experiment we compare AIC against stabilized Laplace (Lap_0) for GPs with warping [24] and warping + manifolding [7] on a synthetic dataset based on the real scenario of motor combustion [25]. For this experiment we adopt an informed prior crafted for the dataset to stabilize the training process.

We illustrate the results in Fig. 4 and note the significantly better optimum found with an MAP loss over the one found with MLL, with much tighter bounds that suit the data well. To emphasize: the comparison is between the models in each row, i.e. AIC compares the two top models and Lap_0 compares the two bottom models since they use the different loss functions in their approximations.

The results show that AIC makes the surprising choice of preferring the model with only warping over the model with warping + manifolding, despite its significantly worse confidence interval. We note that BIC makes the same choice, as it penalizes the model with warping + manifolding even stronger than AIC. Intuitively, given the choice between the *top* models, one would prefer the model with the tighter and more suitable confidence intervals. Despite Lap_0 also choosing the model with only warping, we emphasize that it was given the *bottom* models, which show no visible difference. It is therefore the intuitive choice to select the model with fewer parameters.

In conclusion, this shows an instance of a non-trivial dataset and established GP methods which cause a non-interpretable model choice by AIC, whereas Lap_0 makes an interpretable and intuitive choice.

5 Conclusion

In this work we have introduced three novel Laplace approximation-based model selection metrics, where we mitigate an inconsistency of its naive application which could cause the standard Laplace approximation to prefer models with superfluous parameters. This is achieved by imposing a lower bound on the Hessian's eigenvalues used in the Laplace approximation.

In our experiments we show that our metrics outperform AIC and BIC and select accurate (Sect. 4.1), simple (Sect. 4.2) and interpretable (Sect. 4.3) models. We have given an intuition about the Laplace approximation and the bounding of the eigenvalues in Fig. 3, showing how the tighter bounds cover a smaller section of likelihood surface and how this relates to the Gaussian interpretation of the Laplace approximation. This is a significant improvement over AIC and BIC which do not consider the instability of the training optima in their selection, while the Laplace approximation does so through the Hessian.

Still, standard Laplace, if it works, performs better than our metrics. It would therefore be preferable to use standard Laplace whenever possible. Therefore, an investigation into detection of the inconsistency of the standard Laplace to lower bound the eigenvalues in only those instances would be beneficial.

This method could also be beneficial for tasks like kernel search [3,10], which currently rely on MLL or AIC as model selection metrics.

Acknowledgments. Andreas Besginow is supported by the SAIL project which is funded by the Ministry of Culture and Science of the State of North Rhine-Westphalia under the grant no NW21-059C.

Disclosure of Interests. The authors have no competing interests to declare that are relevant to the content of this article.

References

1. Akaike, H.: A new look at the statistical model identification. IEEE Trans. Autom. Control (1974)
2. Alvarez, M., Luengo, D., Lawrence, N.D.: Latent force models. In: AISTATS. PMLR (2009)
3. Berns, F., Schmidt, K., Bracht, I., Beecks, C.: 3cs algorithm for efficient Gaussian process model retrieval. In: 2020 25th ICPR (2021)
4. Besginow, A., Lange-Hegermann, M.: Constraining Gaussian processes to systems of linear ordinary differential equations. In: NeurIPS (2022)
5. Bishop, C.M.: Pattern Recognition and Machine Learning, vol. 4. Springer, Heidelberg (2006)
6. Borovitskiy, V., Terenin, A., Mostowsky, P., et al.: Matérn Gaussian processes on riemannian manifolds. In: NeurIPS (2020)
7. Calandra, R., Peters, J., Rasmussen, C.E., Deisenroth, M.P.: Manifold Gaussian processes for regression. In: IJCNN. IEEE (2016)
8. Daxberger, E., Kristiadi, A., Immer, A., Eschenhagen, R., Bauer, M., Hennig, P.: Laplace redux-effortless Bayesian deep learning. In: NeurIPS (2021)
9. Duvenaud, D.: Automatic model construction with Gaussian processes. Ph.D. thesis, University of Cambridge (2014)
10. Duvenaud, D., Lloyd, J., Grosse, R., Tenenbaum, J., Zoubin, G.: Structure discovery in nonparametric regression through compositional kernel search. In: ICML (2013)
11. Flaxman, S., Wilson, A., Neill, D., Nickisch, H., Smola, A.: Fast kronecker inference in Gaussian processes with non-Gaussian likelihoods. In: ICML. PMLR (2015)
12. Gardner, J.R., Pleiss, G., Bindel, D., Weinberger, K.Q., Wilson, A.G.: Gpytorch: blackbox matrix-matrix Gaussian process inference with gpu acceleration. In: NeurIPS (2018)
13. Härkönen, M., Lange-Hegermann, M., Raiţă, B.: Gaussian process priors for systems of linear partial differential equations with constant coefficients. In: ICML (2023)
14. Jaynes, E.T.: Probability Theory: The Logic of Science. Cambridge university press, Cambridge (2003)
15. Jidling, C., Wahlström, N., Wills, A., Schön, T.B.: Linearly constrained Gaussian processes. In: NeurIPS (2017)
16. Kuss, M., Rasmussen, C.: Assessing approximations for Gaussian process classification. In: NeurIPS (2005)
17. Li, R., John, S., Solin, A.: Improving hyperparameter learning under approximate inference in Gaussian process models. In: ICML (2023)
18. Liang, B., Mitchell, T., Sun, J.: NCVX: a general-purpose optimization solver for constrained machine and deep learning (2022)
19. Murphy, K.P.: Probabilistic Machine Learning: An Introduction. MIT Press, Cambridge (2022). probml.ai
20. Nocedal, J., Wright, S.J.: Numerical Optimization. Springer, Heidelberg (1999)
21. Rasmussen, C.E., Williams, C.K.: Gaussian Processes for Machine Learning, vol. 2. MIT press, Cambridge (2006)
22. Schwarz, G.: Estimating the dimension of a model. Ann. Stat. (1978)
23. Skilling, J.: Nested sampling for general bayesian computation (2006)

24. Snelson, E., Ghahramani, Z., Rasmussen, C.: Warped Gaussian processes. In: NeurIPS (2003)
25. Thewes, S., Lange-Hegermann, M., Reuber, C., Beck, R.: Advanced Gaussian process modeling techniques. In: Design of Experiments (DoE) in Powertrain Development (2015)

Section: Representation Learning

Deep Decision Forest

Hugo Starck, Ngoc Anh Kiet David Tran, and Sławomir Nowaczyk(✉)

Center for Applied Intelligent Systems Research, Halmstad University, Halmstad, Sweden
slawomir.nowaczyk@hh.se

Abstract. Deep learning has demonstrated success in domains such as vision and speech, largely due to its ability to learn hierarchical feature representations via backpropagation. However, tree-based models, such as Random Forest and XGBoost, remain dominant for tabular data, often, but not always, outperforming deep Artificial Neural Networks (ANNs) while requiring less computational resources. It has been demonstrated that neural networks are particularly effective for data that is less heavily preprocessed and that contains important yet complex feature dependencies.

This paper introduces Deep Decision Forest (DDF), a multilayer decision tree ensemble that bridges this gap by incorporating a feedback mechanism analogous to backpropagation. We postulate that such a mechanism will enable tree-based models to capture hierarchical feature representations to a degree that was previously impossible. Each layer of trees produces an output vector that serves as the input features for subsequent layers, and final predictions are obtained through majority voting in the last layer. One key difference between tree-based models and ANNs is that, instead of making minor adjustments to all the numeric weights at once, DDF first identifies the most underperforming features across layers, and selectively retrains only the corresponding trees using specialised improvement datasets.

Experiments on seven benchmark datasets demonstrate that DDF consistently outperforms a standard Decision Tree, performs as well as or better than Random Forest, and achieves competitive accuracy compared to Deep Forest. These experiments demonstrate that integrating elements inspired by deep learning into tree ensembles is both feasible and effective, offering a new hybrid approach for tabular learning.

All the code and experiments are available open source at: https://github.com/caisr-hh/Deep-Decision-Forest.

Keywords: Decision Tree · Decision Forest · Deep Decision Forest

1 Introduction

Deep learning has achieved significant breakthroughs in solving complex problems that previously challenged the artificial intelligence community. It excels at

H. Starck and N. A. K. D. Tran—Contributed equally to the paper.

M. Baratchi et al. (Eds.): IDA 2026, LNCS 16513, pp. 185–198, 2026.
https://doi.org/10.1007/978-3-032-23833-7_14

learning hierarchical features automatically from data, where lower layers capture simple patterns (e.g., edges) and higher layers combine them into complex representations (e.g., objects or concepts). Notably, deep learning has outperformed other machine learning methods in areas ranging from image and speech recognition to drug discovery and biomedical research. The remarkable success of deep learning seems irrevocably tied to Artificial Neural Networks (ANNs), which are essentially multiple layers of learnable, nonlinear components trained through backpropagation [1]. A key element of progress in the field was understanding the representation learning inherent in ANNs and exploiting how deep, multilayer models can progressively disentangle explanatory factors across layers, thereby building hierarchies of increasingly abstract features [2].

On the other hand, deep learning is still struggling in the tabular data domain. In many cases, tree-based models such as Random Forests [3] or XGBoost [4] have been shown to outperform even quite sophisticated ANN architectures, often at a fraction of the cost [5,6]. Particularly interesting, though, is the analysis of potential reasons for these differences, as reported in [7]. Empirically, the relative performance of deep learning increases with dataset size (which is relatively well known) and with the increased complexity of feature interactions in the given task and dataset (which is much less discussed in the community). Historically, tabular datasets tended to be heavily preprocessed with carefully hand-crafted features well-designed for prediction targets, minimising the need for machine learning to discover these complex interactions. Today, however, this is changing, particularly in real-world industrial settings, and we are facing larger and larger datasets with less and less carefully engineered features becoming increasingly available due to, among others, the Internet of Things [8].

ANNs learn these hierarchical features by propagating an error signal from the output layer back throughout the whole network and applying the chain rule to compute gradients with respect to every weight and bias. Minor parameter updates, typically via stochastic gradient descent, are applied across all layers, allowing successive forward-backward passes to progressively reduce the prediction error, as measured by a loss function. This end-to-end credit assignment yields a hierarchy of internal representations: early layers capture simple, local patterns; deeper layers combine these into increasingly abstract features that are directly useful for the task.

We propose a *Deep Decision Forest* (DDF) as a layered, tree-based ensemble that brings the same notion of hierarchical representation learning to non-differentiable models. Each layer consists of multiple Decision Trees (DTs) that operate in parallel. A tree maps its input to an intermediate representation (a new feature computed from that tree's inputs), and the layer's outputs are concatenated to form the feature vector that becomes the input for the next layer. The final prediction is produced by an aggregation rule (majority vote) at the last layer. In effect, decision trees replace artificial neurons, and depth in the architecture induces progressively richer feature transformations, analogous to the feature hierarchies formed in multilayer perceptrons.

The key novelty in this paper is a feedback-driven, selective retraining mechanism that provides a cross-layer supervision signal without gradients. After training a layer stack, DDF estimates the contribution of earlier-layer features (i.e., the specific trees that generate each of them) to downstream performance. A small set of under-contributing trees is retrained from scratch with a new target (i.e., adapted to the "needs" of the subsequent trees), while the majority of the forest remains fixed. This targeted, local update serves as a "weight adjustment" in ANNs, but rather in a discrete, structural form well-suited to tree-based models.

Conceptually, DDF aims to incorporate the best aspects of several successful paradigms. Like modern artificial neural networks, it is *deep* and *end-to-end*: later layers shape earlier feature generators via feedback and iterative refinement. At the same time, DDF retains the key strengths of trees: robustness on heterogeneous tabular data, implicit feature selection, and rule-level interpretability. The result is a modular, interpretable, and data-type-friendly alternative to gradient-based deep learning while introducing hierarchical feature learning through depth and feedback.

The remainder of this paper is organised as follows: Sect. 2 discusses the existing work in this field, followed by Sect. 3, which presents our novel DDF methodology in detail, including both the inference and training procedures. The configuration and parameter settings of the experimental evaluation are covered in Sect. 4. Finally, findings and contributions are concluded in Sect. 5.

2 Related Work

This section, presenting relevant papers in the field, is divided into two parts: we first discuss the challenges deep ANNs face when learning tabular data, and then we discuss existing methods for training deep tree-based models.

2.1 Tabular Data

The primary objective of the study by Shwartz et al. [6] was to determine whether recently proposed deep learning architectures for tabular data could indeed surpass state-of-the-art models, such as XGBoost, in most real-world machine learning tasks. Their motivation stemmed from the fact that while DNNs dominate image, text, and speech domains, tabular data presents challenges for deep models that tree ensembles handle naturally. Their results show, in most cases, that deep models fail to generalise across other datasets and only perform at their best on the datasets introduced in their respective work. State-of-the-art models such as XGBoost consistently outperform deep models while also requiring significantly less hyperparameter tuning. One noteworthy finding was that ensembles combining XGBoost with deep models performed better than XGBoost alone, suggesting that hybrid models could show promise in the research area – reinforcing the motivation for DDF.

Arik et al. [8] also highlighted that standard deep learning architectures often struggle to outperform tree ensembles on tabular data. They argue, however, that deep learning can show promise and be adapted more effectively for tabular problems if its architectures are explicitly designed for this domain. To address this, they propose TabNet, which enables both representation learning and a level of model interpretability tailored to tabular inputs. In a similar spirit, DDF is a hybrid approach that incorporates elements of deep learning into a tree-based architecture while remaining well-suited for tabular data.

Grinsztain et al. [5] conducted an extensive benchmark study to better understand why tree-based models continue to outperform DNNs on tabular datasets. Across 45 datasets, they found that tree ensembles such as XGBoost and Random Forest consistently achieved higher accuracy with less computational effort, even after thorough hyperparameter tuning of deep models. The authors argue that this performance gap arises from several inherent characteristics of tabular data. Target functions in such datasets are often irregular and non-smooth, which makes them difficult for neural networks to approximate effectively. Many tabular datasets also contain a mix of informative and redundant features, to which deep networks are particularly sensitive. Furthermore, the feature-wise independence exploited by decision trees aligns naturally with the axis-aligned structure of tabular data, whereas the rotational invariance of neural networks can obscure such relationships. These findings clarify why conventional neural architectures struggle in this domain and highlight the value of exploring alternative designs such as hybrid models that better align with the properties of tabular data.

2.2 Deep Tree-Based Approaches

A topic closely related to DDF, and one that partially inspired our work, is Deep Forest, introduced in [9] by Zhi-Hua Zhou and Ji Feng. The Deep Forest model is a multi-layered tree ensemble model that uses Random Forests and Completely Random Forests to process the data. Zhou and Feng created their model based on deep learning architectures with the goal of developing a solution that would work similarly to deep learning but require less computation while also being capable of handling smaller data sizes. Each layer of Deep Forest contains an equal amount of random and completely random forests, and the resulting outputs from each forest are a single vector that becomes a feature available to the next layer. However, the Deep Forest model concatenates these output features from the previous layers with the features of the original dataset using multi-grained scanning. The final predictions are obtained with majority voting. In the experiments, the Deep Forest model has shown promising results across multiple datasets, but it particularly struggles when the dataset contains a large number of features. It is particularly relevant that the Deep Forest, while similar in architecture to deep learning models, lacks any feedback mechanism guiding early layers to create features useful for the later ones. Zhou and Feng mentioned that this is due to them not having yet figured out how to implement one that would work with their model structure. The key contribution of our paper is to bridge this gap.

Several years earlier, Miller et al. introduced the concept of Forward Thinking Deep Random Forests (FTDRF) [10], proposing an alternative approach to constructing deep models that does not rely on backpropagation. FTDRF generalises the idea of deep learning by training one layer at a time, where each layer transforms the input data into a new feature space before passing it forward. Each layer consists of random and extremely randomised trees, whose output class probability vectors serve as the input for the next layer. Compared to Zhou and Feng's Deep Forest, FTDRF passes the outputs of individual trees rather than full forests to subsequent layers. Both works demonstrate that layered, non-gradient-based tree architectures can achieve deep representations, reinforcing the potential of tree ensembles as a foundation for deep learning alternatives such as DDF.

3 Methodology

We propose DDF as a multilayer model similar to those of deep ANNs, with the core difference that DDF utilises Decision Trees (DTs) rather than artificial neurons in its nodes. Let us assume that DDF consists of L layers, each with n DTs.

We first generate $L * n$ random DTs by uniformly sampling features, split thresholds, and leaf values, akin to random initialisation in ANNs. Each DT yields a value for each sample, depending on the branches traversed by that sample and the value associated with the leaf node it ends up in. Several DTs forming the first DDF layer will therefore take as input the original features of a data instance and produce a corresponding vector of (new) features for it. These new features will serve as input to the DTs in the second layer, and so on. The final decision of DDF is made by majority voting of the DTs in the last layer. This output is then compared with the target prediction to calculate the overall error of the DDF, and also used to identify the weakest tree(s) in each layer. Only these trees are subsequently retrained: by first calculating the (locally) optimal intermediate outputs, and then approximating these optima as closely as possible.

3.1 Preliminaries

In order to make the following explanations precise, we introduce here the mathematical notation used throughout the paper. A DDF is a layered ensemble of multiple decision trees, where T_j^k represents the j-th decision tree in the k-th layer. Correspondingly, $\mathbf{X}^k$ denotes the input dataset of the k-th layer, with the natural special case of $\mathbf{X} = \mathbf{X}^1$ representing the original dataset. Let x_{ij}^k be the j-th feature of instance i in layer k. For simplicity, below we assume that $\mathbf{X}^k \in \mathbb{R}^{m \times n}$, i.e., that all of them are two-dimensional tabular datasets with m samples and n features. In practice, as demonstrated in the experiments in

Sect. 4, there is no requirement for these dimensions to be consistent; just as with ANNs, different layers are allowed to differ in size.

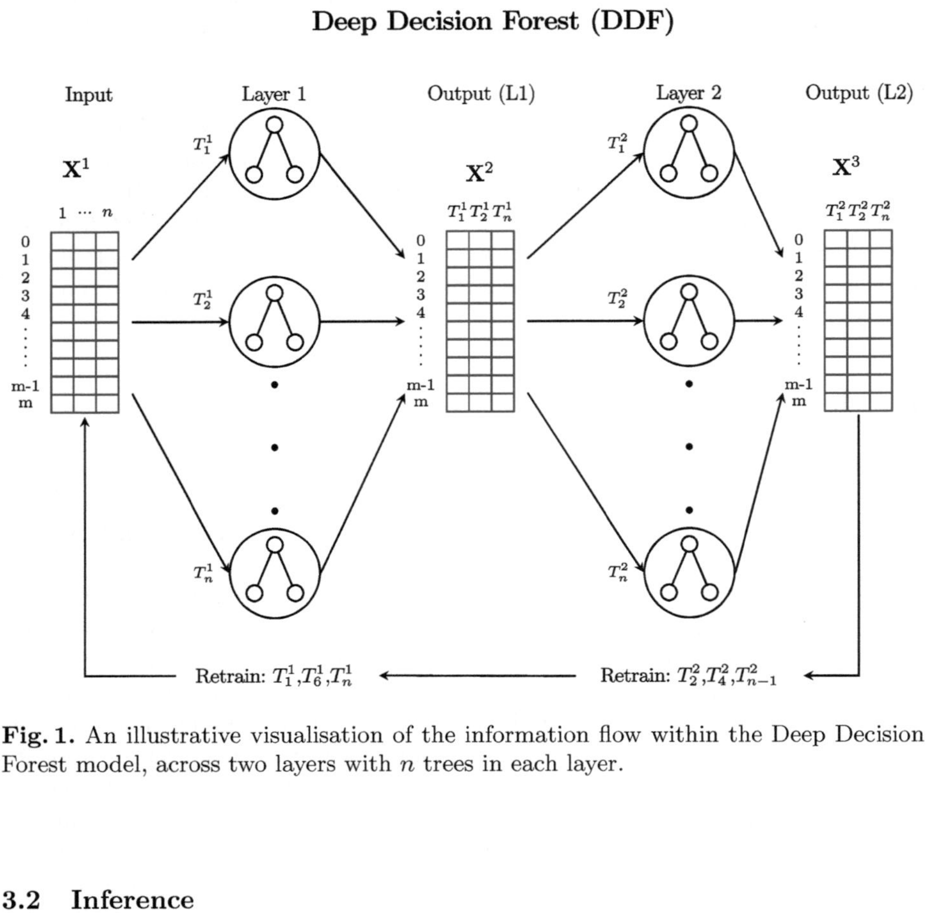

Fig. 1. An illustrative visualisation of the information flow within the Deep Decision Forest model, across two layers with n trees in each layer.

3.2 Inference

During the inference process, the original dataset $\mathbf{X} = (\boldsymbol{x}_1, \boldsymbol{x}_2, \ldots, \boldsymbol{x}_m) = \mathbf{X}^1$ of instances $\boldsymbol{x}_i$ is presented to the trees in the initial layer of the DDF, i.e., to each of $T_1^1 \ldots T_n^1$. Outputs of these trees get concatenated into the input for the second layer, $\mathbf{X}^2$. Specifically, any given tree T_j^1 will, for each instance $\boldsymbol{x}_i$, produce a single output value y_{ij}^1, which becomes the j-th feature of instance i in layer 2, thus $y_{ij}^1 = x_{ij}^2$. The same process is repeated for any layer k, with trees T_j^k processing $\mathbf{X}^k$ and producing inputs $\mathbf{X}^{k+1}$ for the trees T_j^{k+1} in layer $k+1$.

This process continues until the final layer of the DDF is reached. The predictions for individual trees are then aggregated using majority voting and compared to the ground truth. A visual representation of the model is shown in Fig. 1, indicating the inference process left-to-right and the training process right-to-left.

3.3 Theoretical Foundations

In this subsection, we present a theoretical justification for DDF, using binary functions for clarity. It is well known that decision trees can be inefficient, i.e., some binary functions require exponentially large trees, because tree structure necessitates repetitions. A layered architecture, where trees in early layers compute features useful to subsequent layers, allows DDF to avoid such a size explosion. A compact showcase is presented based on the *parity (XOR) function*, a classic example where a single decision tree must branch on every input bit, yielding an exponential number of leaves, whereas a two-layer construction can reuse intermediate features and remain small.

For (n) binary inputs $\mathbf{x} = (x_1, \ldots, x_n) \in \{0,1\}^n$, $f_n(\mathbf{x}) = x_1 \oplus x_2 \oplus \cdots \oplus x_n$. For the concrete diagram below, we use $n = 4$: $f(x_1, x_2, x_3, x_4) = x_1 \oplus x_2 \oplus x_3 \oplus x_4$.

Any decision tree for 4-bit parity must examine all four variables; every root-to-leaf path fixes (x_1, x_2, x_3, x_4), and the leaf label is the parity of that assignment.

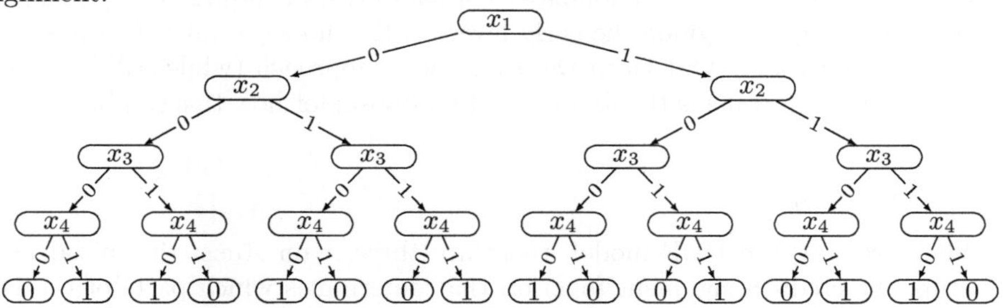

This tree has 16 leaves. For n inputs, the minimal decision tree for parity has 2^n leaves, reflecting unavoidable repetition. However, a two-layer DDF allows for feature reuse and results in three small trees total. First, two trees in Layer 1 compute pairwise XOR features $y_1 = x_1 \oplus x_2$ and $y_2 = x_3 \oplus x_4$; each 2-bit XOR fits in a tiny tree of 4 leaves:

y_1:

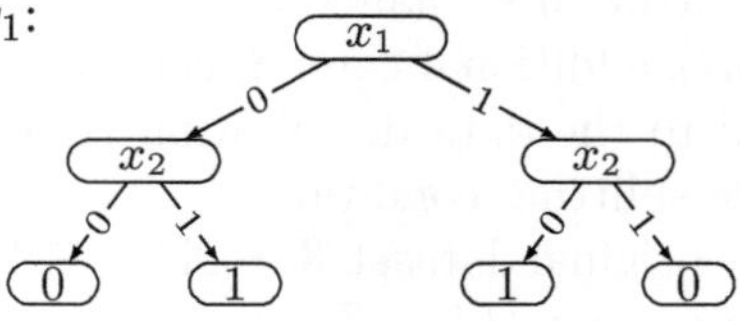

y_2:

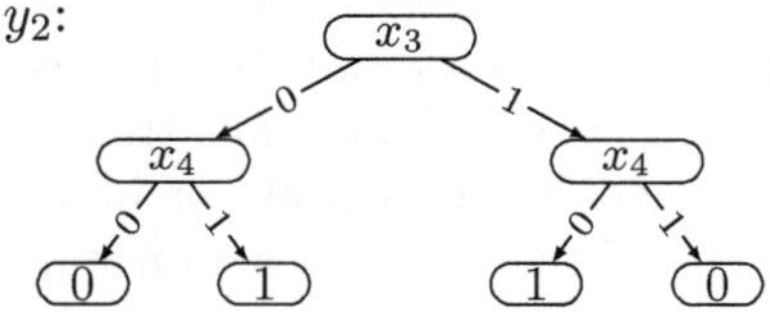

Finally, a single tree in Layer 2 combines the learned features $z = y_1 \oplus y_2 = f(x_1, x_2, x_3, x_4)$.

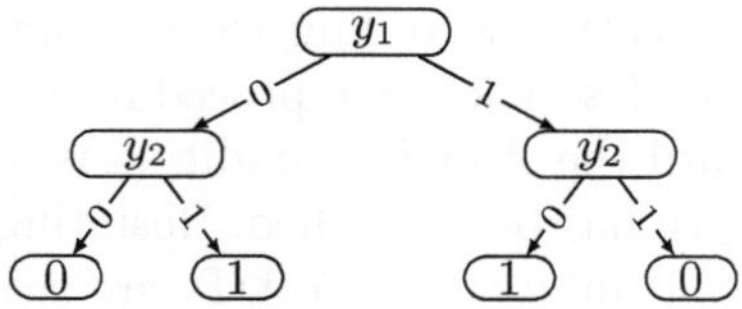

This demonstrates the key advantage of DDF: while a single tree requires $2^4 = 16$ leaves with unavoidable duplication of sub-computations, two-layer DDF only needs three tiny trees (each with 4 leaves) that reuse pairwise XOR features: a total of $3 \times 4 = 12$ leaves and substantially fewer internal tests. Considering how this scales: for $n = 8$, a single tree has $2^8 = 256$ leaves; a two-layer DDF with first-layer pairwise XORs $(y_1, \ldots, y_4)$ and a second-layer tree computing $y_1 \oplus y_2 \oplus y_3 \oplus y_4$ uses four tiny trees (first layer, 4 leaves each) plus one modest second-layer tree (16 leaves) for 32 leaves total, versus (256) leaves monolithically. And a three-layer DDF would, again, end up only using small XOR trees: 4 in Layer 1, 2 in Layer 2, and 1 in Layer 3; each two-bit XOR tree with only 4 leaves, for a total of 28.

This intuitively captures the benefit of a layered (deep) tree architecture: intermediate features computed once and reused, eliminating the exponential blow-up that arises from repeated local sub-computations in a single decision tree. However, it is unreasonable to expect such "clever" selection of features to emerge through pure randomness. Earlier works, including both [9] and [10], lack a mechanism to guide the early layers of the model towards features useful for the latter layers. Our Deep Decision Forest approach (while still in an early stage of development) is the first concrete proposal for how this can be achieved.

3.4 Training

When training the DDF model, there are three main steps: the initial model setup, the credit assignment function that determines which parts of the DDF need improvement, and the update procedure that propagates feedback and applies changes to selected trees. The following subsections provide a detailed explanation of each step.

Initialisation. At the beginning, the model is initialised with random decision trees, following the same principle as the random initialisation of ANNs. However, trees are inherently more complex structures than vectors of neuron weights, and as a result, this procedure requires additional care. Specifically, we believe it is important that trees be adapted to the data distribution (in each layer), avoiding redundant or impossible node splitting conditions.

Therefore, the initialisation starts from the original dataset $\mathbf{X} = \mathbf{X}^1 \in \mathbb{R}^{m \times n}$, which is used to generate the first layer decision trees $T_1^1 \ldots T_n^1$. Any given tree T_j^1 selects, for the root node, a random feature from $\mathbf{X}^1$ (uniformly from all n features), followed by a threshold to split on (uniformly among the midpoints between all data instances, i.e., the values in the corresponding column of $\mathbf{X}^1$). The dataset $\mathbf{X}^1$ is then divided, according to the split condition just created, into left- and right-child nodes. The same procedure is then applied recursively to create the sub-trees until the stopping condition is reached: the node is pure or the maximum tree depth has been reached. Resulting leaf nodes are assigned a random value in the $[0, 1]$ interval (since DTs are scale-invariant, there is no need to extend the range).

After the first layer of decision trees $T_1^1 \dots T_n^1$ is generated, their resulting outputs are combined into a dataset $\mathbf{X}^2$ to be used as input for the second layer, as explained in the previous section. This dataset is then used to generate random trees $T_1^2 \dots T_n^2$, using the same procedure. In other words, the key principle is that the trees in layer $k+1$ are created according to the distribution of the data that is produced by trees in the k-th layer.

Credit Assignment. The DDF feedback propagation mechanism starts by identifying the least informative features in any given layer. There are multiple strategies to find the least informative features. We have experimented with exhaustive search to find the feature that provides the greatest global performance increase, but this is clearly intractable even for relatively small DDFs. Therefore, we present two heuristic approaches that have been empirically shown to be effective. The first one is average depth, in which the model computes the average depth for each feature in a layer using Eq. 1. The features with the highest AD value are considered the most promising improvement candidates, as they participate in less important splits:

$$AD(F) = \frac{\sum_{i=0}^{k} D_i}{k} \tag{1}$$

The second strategy is the ASU (Average Symmetrical Uncertainty) approach, introduced by Park et al. [11]. It is based on IG (information gain) to calculate the quality of a node split within decision trees. Since IG inherently lacks normalisation, a concept of SU (Symmetrical Uncertainty) is introduced by normalising mutual information. $ASU(F)$ is SU aggregated over all the nodes v which are using the given feature F in all the trees of a given layer (see Eq. 2). Please note that, unlike AD, the features with the smallest ASU value are considered the least informative.

$$\begin{aligned} H(Y) &= -\sum_{y} p(y)\log p(y), \\ IG(X;Y) &= H(Y) + H(X) - H(X|Y), \\ SU(X;Y) &= 2.0 * \left[\frac{IG(X;Y)}{H(X)+H(Y)}\right], \\ ASU(F) &= \frac{1}{|N(F)|}\sum_{v \in N(F)} SU\big(X_F^{(v)};Y^{(v)}\big). \end{aligned} \tag{2}$$

Retraining. Once the weakest feature(s) in each layer are identified, the second step is to generate new *improvement datasets* to retrain the trees responsible for each of these uninformative features. For any given feature F to be retrained, originating from a tree T_j^k, we identify all nodes v in T^{k+1} trees where F is used as the split condition. Intuitively, these are the places that provide key insights into possible improvements: we want to manipulate feature F (by rebuilding tree

T_j^k) so that nodes v route data instances towards outcomes closer to the ground truth.

In particular, we collect all data instances that pass through node v, along with their F values, and their final classification by the DDF, identifying all misclassified instances. Our goal is to modify feature F in a way that maximises the overall classification accuracy. With the "current status" of all instances collected, the DDF algorithm performs a *path switch* for them: each instance originally routed to the left subtree of node v is now sent to the right subtree, and vice versa. This way, we evaluate whether an instance would fare any better had it been on the opposite side of the threshold for feature F at node v. The new results are compared with the original, and any uninformative instances are removed, i.e., those for which the prediction did not change (whether correctly or incorrectly predicted, they do not provide any useful information for retraining T_j^k). The remaining instances are stored, along with the information indicating which side of the threshold is preferred for them, forming the new *improvement datasets* to guide retraining of T_j^k. This process continues for all nodes and the selected features.

The retraining of the decision trees in DDF differs depending on whether a given tree is located in the final layer or in an intermediate layer. Trees in the final layer are trained directly using a regular CART algorithm[1], except we subsample instances and features (as is done in Random Forest) to ensure tree diversity and reduce the risk of overfitting.

The intermediate trees are retrained based on their individual *improvement datasets*, created as above. Each such dataset contains a subset of the original data instances (excluding any instances that are irrelevant to this specific tree) and information on whether the values of feature F should be higher or lower than a given threshold. This information is therefore used to create the ground truth dataset for training a new tree. Specifically, target values for each instance in the *improvement dataset* are generated randomly, subject to the sole constraint being that they adhere to the desired threshold relations. Optimising the training target values, possibly across multiple trees to be retrained simultaneously, is a promising direction for future work. We begin in the first layer of the DDF, generating new decision trees T_j^1, similarly using the standard greedy training approach. This results in a new $\mathbf{X}^2$ dataset that replaces the earlier version, now with several features better aligned to what the second layer expects. The DDF then continues forward, updating the trees layer by layer. It is essential to note that trees in layer $k+1$ are optimised with respect to already improved trees in layer k, which is crucial for building hierarchical feature representations.

Finally, the model summarises the results using majority voting to determine whether the changes are worth keeping before starting the next epoch.

[1] As implemented in the `scikit-learn` Python package https://scikit-learn.org/.

4 Experiments

This section provides a detailed description of the experimental setup and results, including dataset preprocessing, model configurations, and evaluation.

Datasets and Preprocessing. We have utilised seven publicly available datasets, all of which underwent the same preprocessing pipeline to ensure consistency across experiments. Categorical features were label-encoded into integer representations, while numerical features were scaled using Min-Max normalisation to the range $[0, 1]$. Rows containing missing or invalid entries were removed, and all data were shuffled prior to partitioning. Each dataset was split into 70% training and 30% testing. From the training portion, 20% was further separated as a validation set (used exclusively for model selection and retraining acceptance). The primary performance metric is classification accuracy, computed on the unseen test set. The mean and standard deviation reported below were calculated across five runs with different splits.

Model Configurations. The following parameters were used for the models, obtained through grid search hyperparameter tuning:
Decision Tree was implemented using `DecisionTreeClassifier` from scikit-learn, with maximum depth of 30, minimum samples per split 10, and minimum impurity decrease 5×10^{-3}
Random Forest was implemented using `RandomForestClassifier` from scikit-learn, with 60 trees and each tree parametrised as above.
Deep Forest, as proposed in [9], uses the `CascadeForestClassifier` from the `deepforest` library, with 4 forests per layer, and 100 trees per forest.
Deep Forest (restricted) was added in order to keep the size of DF comparable to DDF, by restricting the number of layers to 3 and tree depth to 10 (DF's original adaptive strategy results in much larger models).
Deep Decision Forest uses 3 layers with 60 trees, maximum tree depth 10, and training for 70 epochs.

4.1 Results

The results of these experiments can be seen in Table 1. One can clearly see that both DDF and DF outperform DT and RF on all datasets. Another interesting insight is that the performance of RF and DF (restricted) is nearly identical. This makes sense considering that DF follows analogous randomness principles to RF, and restricting both models to a similar size makes them virtually indistinguishable. On the other hand, DDF is able to maintain ensemble diversity much better than RF, and if allowed to grow 10–20 times larger, it achieves superior performance. On the other hand, DDF of size comparable with RF and DDF (restricted), manages to outperform them in 6 out of 7 datasets (with a small difference in Mushroom, Banknote, Raisin, Diabetes, and Iris, losing by 0.5% in Heart Disease but winning by 7.2% in Maternal Health dataset).

The full DF model performs slightly better than the restricted version on most datasets, outperforming DDF on 4 datasets. Nevertheless, these results demonstrate that DDF is on par with a significantly larger DF model. While training a DDF model is significantly slower than training DF (even full-sized), if inference is important, the "smarter" features extracted through DDF give it an edge over the brute force of DF.

Table 1. Table containing the final accuracy for each model when used for five different datasets. Bold values indicate the highest accuracy per dataset.

Datasets \ Models	DT	RF	DF (restricted)	DF	DDF (Ours)
Mushroom	0.990 ± 0.005	0.998 ± 0.001	0.998 ± 0.001	$\mathbf{0.999 \pm 0.002}$	0.998 ± 0.001
Banknote	0.961 ± 0.011	0.996 ± 0.003	0.996 ± 0.003	$\mathbf{0.999 \pm 0.001}$	0.990 ± 0.006
Raisin	0.820 ± 0.012	0.826 ± 0.009	0.827 ± 0.009	0.839 ± 0.002	$\mathbf{0.841 \pm 0.012}$
Diabetes	0.699 ± 0.01	0.732 ± 0.012	0.732 ± 0.012	$\mathbf{0.739 \pm 0.021}$	0.730 ± 0.034
Heart disease	0.744 ± 0.044	0.839 ± 0.038	0.839 ± 0.038	0.834 ± 0.027	$\mathbf{0.845 \pm 0.033}$
Iris	0.956 ± 0	0.950 ± 0.011	0.950 ± 0.011	0.956 ± 0.018	$\mathbf{0.966 \pm 0.013}$
Maternal health	0.685 ± 0.033	0.745 ± 0.013	0.745 ± 0.013	$\mathbf{0.817 \pm 0.035}$	0.765 ± 0.011

5 Conclusions

This paper introduces the Deep Decision Forest (DDF), a multilayer decision tree ensemble that incorporates key ideas from deep learning (hierarchical representation learning and feedback-driven refinement) into a non-differentiable setting. In DDF, each layer of decision trees produces an intermediate feature representation that is fed forward to the next layer, while a feedback mechanism selectively identifies and rebuilds underperforming trees based on their estimated contribution to downstream accuracy. In contrast to conventional tree ensembles, which rely on one-shot, greedy splitting, DDF iteratively reshapes earlier layers in response to errors observed at the output, providing a form of cross-layer credit assignment without the need for gradients.

A central contribution of this work is the design of a feedback procedure tailored to tree ensembles. We proposed two practical heuristics for assessing feature utility within a layer, allowing DDF to focus retraining effort precisely on the trees that are least informative for the global prediction task. We also provided a theoretical example (the XOR function) to illustrate how a layered tree architecture can compute certain functions exponentially more compactly than a single monolithic tree. By computing and reusing intermediate features across layers, DDF avoids the combinatorial explosion that arises from repeated local sub-computations in standard decision trees. While this example is deliberately simple, and our work is still in early stages, it highlights the potential of deep tree architectures. Empirically, DDF was evaluated on seven benchmark tabular datasets, yielding promising results and demonstrating that deep, feedback-driven tree ensembles are not only feasible but also practically effective on typical tabular learning tasks.

At the same time, our experiments revealed a clear limitation: training DDF is computationally more demanding than training standard ensembles, even when the final model size is similar. The need to iteratively identify weak features, construct improvement datasets, and retrain selected trees across multiple epochs introduces a non-trivial overhead. In applications where training time is tightly constrained, this may limit the immediate practicality of the current DDF instantiation, despite its competitive inference-time cost.

A key direction for future work is to improve the optimisation procedure underlying DDF. The current training strategy relies on largely greedy, heuristic choices, both in identifying underperforming trees and in how they are rebuilt, which can lead to overfitting, failure to identify key features, unnecessary retraining, sensitivity to hyperparameters, and suboptimal cross-layer credit assignment. We will therefore investigate more principled feedback and selection mechanisms (e.g., better estimators of marginal utility, variance-aware or budgeted retraining policies, and adaptive multi-step search with backtracking), as well as alternative search strategies that better balance exploration and exploitation across layers. The aim is to reduce the computational overhead while improving convergence stability and, ultimately, to make deep, feedback-driven tree ensembles competitive under realistic training-time constraints.

Acknowledgments. This study was partially funded by Vinnova and the Knowledge Foundation.

Disclosure of Interests. The authors have no competing interests to declare that are relevant to the content of this article.

References

1. Goodfellow, I., Bengio, Y., Courville, A.: Deep Learning. MIT press, Cambridge (2016)
2. Bengio, Y., Courville, A., Vincent, P.: Representation learning: a review and new perspectives. IEEE TPAMI (2013)
3. Breiman, L.: Random forests. Mach. Learn. **45**, 5–32 (2001)
4. Chen, T., Guestrin, C.: XGBoost: a scalable tree boosting system. In: Proceedings of the 22nd ACM SIGKDD, pp. 785–794 (2016)
5. Grinsztajn, L., Oyallon, E., Varoquaux, G.: Why do tree-based models still outperform deep learning on typical tabular data? In: NeurIPS (2022)
6. Shwartz-Ziv, R., Armon, A.: Tabular data: deep learning is not all you need. Inf. Fusion **81**, 84–90 (2022)
7. Karlsson, A., Wang, T., Nowaczyk, S., Pashami, S., Asadi, S.: Mind the data, measuring the performance gap between tree ensembles and deep learning on tabular data. In: International Symposium on Intelligent Data Analysis, pp 65–76. Springer, Heidelberg (2024)

8. Arik, S.O., Pfister, T.: TabNet: attentive interpretable tabular learning. In: AAAI, vol. 35, no. 8, pp. 6679–6687 (2021)
9. Zhou, Z.-H., Feng, J.: Deep forest. Natl. Sci. Rev. **6**(1), 74–86 (2019)
10. Miller, K., Hettinger, C., Humpherys, J., Jarvis, T., Kartchner, D.: Forward thinking: building deep random forests. arXiv arxiv:1705.07366 (2017)
11. Park, H.W., Li, D., Piao, Y., Ryu, K.H.: A hybrid feature selection method to classification and its application in hypertension diagnosis. In: International Conference on Information Technology in Bio- and Medical Informatics, pp. 11–19. Springer, Heidelberg (2017)

Graph Neural Networks for Graph-Level Regression on Heterogeneous Network Data: Use Case in Early-Stage Optimization of Software Mapping on Multicore Platforms

Oscar Roussel[1,2], Zainab Ghrayeb[2], Sébastien Le Nours[2], and Christine Sinoquet[1(✉)]

[1] LS2N/UMR CNRS 6004, Nantes University, Nantes, France
christine.sinoquet@univ-nantes.fr
[2] IETR/UMR CNRS 6164, Nantes University, Nantes, France

Abstract. Accurately predicting the power consumption and latency of software applications deployed on multicore platforms is a critical bottleneck for early-stage performance optimization, as it often relies on slow and costly simulations.

To address this challenge, we introduce a graph-based methodology that models the mappings of software applications onto multicore architectures as heterogeneous graphs. We then investigate Graph Neural Networks (GNNs) to predict power consumption and latency for these mappings. Four classes of state-of-the-art GNNs were trained under various conditions on 11 datasets, each containing multiple mappings of a given Neural-Network application. The two best-performing GNN classes achieve mean absolute percentage errors of around 2% for power-consumption prediction and 15% for latency. Inference requires only a few tens of milliseconds per prediction. This work demonstrates that GNNs offer a fast and promising approach to performance prediction, opening the door to AI-assisted optimization of software mapping on multicore platforms.

The relevance of our work extends beyond the use case: we introduce an operational framework for GNNs to support heterogeneous graph processing and multi-output regression at the graph level.

Keywords: Graph Neural Network · Heterogeneous graph · Graph-level regression · Multi-output regression · On-chip software application mapping · Multicore architecture · Early-stage performance prediction

1 Introduction

Heterogeneous graphs, with multiple types of nodes and edges, are ubiquitous and play a critical role in many applications. This work is motivated by a use case requiring integration of several heterogeneous graphs into a single unified

M. Baratchi et al. (Eds.): IDA 2026, LNCS 16513, pp. 199–212, 2026.
https://doi.org/10.1007/978-3-032-23833-7_15

graph to train a model capable of predicting two-dimensional outputs from graph instances with known target values: specifically, we aim to develop a predictive model that takes as input a heterogeneous graph representing a software application mapped onto a multiprocessor architecture, and outputs performance metrics.

Graph Neural Networks (GNNs) are designed to handle graph-structured data, capturing complex dependencies between nodes and edges. They have been applied to tasks such as graph embedding, graph classification, node classification, node clustering, link prediction, and, less commonly, scalar or multi-dimensional graph-level regression.

However, traditional GNNs are primarily tailored for homogeneous graphs. Existing approaches either split heterogeneous graphs into multiple homogeneous subgraphs or model heterogeneity via various complicated modules. Relevant works enable GNNs to handle heterogeneity by embedding relation types [11], exploiting so-called meta-paths to capture semantic relationships between nodes [3], jointly embedding nodes and edges as "edge-nodes" [14], or refining relation structures derived from multiple meta-paths [7]. Nonetheless, none of these works focus on multi-dimensional regression, making our use case underexplored.

Beyond contributions to GNN-based heterogeneous graph representation for graph-level two-dimensional regression, this work also addresses the automated and optimized deployment of software on electronic chips, emphasizing early-stage optimization. With the rise of embedded applications, including smartphones, connected devices, and smart sensors, accurately estimating performance metrics for task-to-processor assignment of parallelized software on multiprocessor systems is a major challenge. Efficient early-stage optimization can significantly reduce time- and cost-intensive testing.

Performance optimization can be carried out within either a single-objective or a multi-objective framework. The performance metrics to be optimized include, for example, power consumption, execution time (latency), memory bandwidth, and communication overhead. Importantly, these metrics depend not only on the software application and the multiprocessor architecture, but also on the mapping. A poor mapping can negate the benefits of a well-designed architecture by critically degrading overall system performance.

In this paper, we focus exclusively on static mapping. Conventional, pre-AI methods for optimizing on-chip application deployment suffer from inherent drawbacks: simulations are costly and slow, whereas analytical models are fast but less accurate. Static mapping methods include exact techniques, such as mathematical programming, which are effective only for small mapping problems. Heuristics have long remained the leading method for efficiently navigating large solution spaces. However, regardless of whether heuristics, local search, or genetic algorithms are employed, each generated mapping must be evaluated efficiently and accurately. An analytical solution to accurately evaluate a given performance metric is not always available, depending on the metric considered.

Previous work has explored Artificial Intelligence for software deployment on chips; here, we investigate a novel approach based on Graph Neural Networks

(GNNs). Since software applications, hardware architectures, and their mappings can be naturally represented as graphs, a GNN-based model is a suitable choice for predicting mapping performance. This work investigates the potential of GNNs to predict performance metrics from a heterogeneous graph that encodes a candidate mapping of a software application onto a multiprocessor architecture.

Traditional machine learning methods are now being applied to Electronic Design Automation (EDA). Within deep learning, GNNs have emerged as a natural solution for EDA challenges [8]. So far, GNN-based approaches to application-to-platform mapping optimization have been limited to a single method employing a dedicated architecture [2]. We present a unified framework featuring a generic graph-based representation of mappings and benchmark four recent state-of-the-art GNNs within this framework for predicting both latency and power consumption.

The remainder of this article is organized as follows. Section 2 introduces key definitions and provides an overview of the core principles of GNNs. Section 3 presents our methodology. Section 4 presents the experiments and discusses the corresponding results. In the last section, we conclude and provide perspectives for future work.

2 Preliminaries

2.1 Definitions

The term *platform graph* denotes the high-level abstraction of a multicore architecture whose components include processing tiles (each consisting of a processing unit and local memory), a memory shared among several tiles, communication buses, a clock generator, and a sensor. In this work, we restrict our focus to processing tiles, a unique shared memory, and a single communication bus. Figure 1(a) shows such a platform graph.

An *application graph* abstracts the key structure of a software application. A software application consists of a set of operations that sequentially process program data and produce an output. Applications can be decomposed into multiple tasks, and when data processing can be parallelized, this decomposition reflects the application's degree of parallelism. Figure 2 illustrates three application graphs derived from the same software application.

Deploying a sofware application requires mapping the tasks of a selected application graph onto tiles in a given multicore architecture. A task-to-tile mapping is valid if each task is assigned to exactly one tile.

2.2 Graph Neural Networks

We wish to encode the information of an **A**pplication graph, a **P**latform graph, and a **M**apping between them into a unified graph-based representation, enabling effective processing by Graph Neural Networks. This representation is hereafter referred to as the APM graph. Figure 1 shows an example of an APM graph.

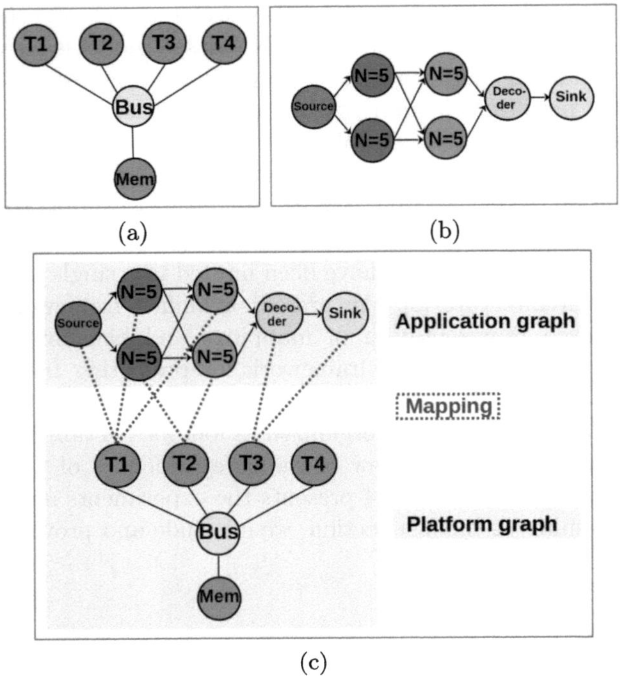

Fig. 1. Example of an APM graph. (a) Graph representation of a multicore architecture. The six nodes of this illustration consist of four processing tiles, a shared memory and a communication bus. (b) Application graph of a software application. The software application considered implements Multi-Layer Perceptron-based prediction. The seven nodes represent tasks. (c) APM graph of the application graph in (b) mapped onto the platform graph in (a). The resulting mapping is represented as a heterogeneous graph containing multiple node and edge types.

GNNs are a class of neural architectures designed to operate on data represented as graphs, where entities are modeled as nodes and their relationships as edges [6]. Unlike traditional Neural Networks that assume fixed-size, grid-like inputs, GNNs exploit the irregular structure of graphs by iteratively aggregating and combining information from each node's neighborhood. By learning rich representations directly from graph data, GNNs automatically capture both local and global graph properties. Their design incorporates key principles such as permutation invariance (order-independent processing of nodes) and local aggregation (neighbourhood-based updates), enabling efficient and scalable computation. GNNs have achieved state-of-the-art performance in various application domains, including social network analysis, molecular property prediction, recommendation systems, and traffic prediction.

We briefly review key GNN components: node-embedding-based representation learning, Message Passing mechanisms, and layer-composition strategies

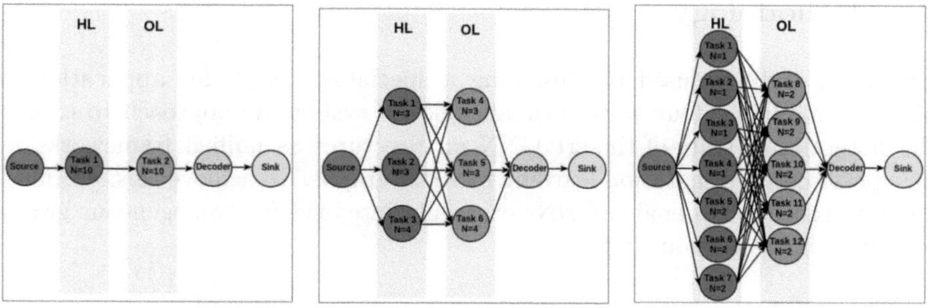

(a) (b) (c)

Fig. 2. Application graphs of a software application. The application performs prediction using a Multi-Layer Perceptron with one hidden layer (HL) and an output layer (OL), each containing 10 neurons. Three possible application graphs are shown, where each node represents a task managing operations for a set of N neurons. In each (hidden or output) layer, such tasks are performed in parallel. Application graphs differ in how 10-neuron layers are split into tasks.

Node Embedding. Standard GNNs are node-focused and primarily learn node embeddings. In these models, edges are used to propagate and aggregate information between nodes but are typically not assigned embeddings unless explicit edge features are provided. A GNN acts as a function that maps a graph and its nodes to a set of embeddings.

Message Passing. Early foundational work introduced the concept of Message Passing (MP) over graphs [4,9]. The goal of MP is to iteratively update node embeddings to capture increasingly complex information about node and edge features as well as the local topology. A GNN layer is based on this principle: at iteration k of the MP process, each node receives messages from its neighbors, which are combined by a permutation-invariant aggregation function (e.g., sum, mean, or max) and then processed by an update function to compute the new embedding. After k iterations, a node's embedding captures information from its k-hop neighborhood. The message computation and update functions are parameters of the MP framework, typically implemented as trainable Message Passing Neural Networks.

Layer Stacking for Network Design. GNN layers are stacked sequentially, where the output of one layer serves as the input to the next. Specifically, the node embeddings generated by one layer are used as node features in the MP operation of the subsequent layer. The main objective of stacking layers is to capture increasingly complex relational patterns by iteratively refining node embeddings through successive layers.

3 Methodology

The work in [2] focused on designing a specialized GNN for application-to-platform mapping. Our work, instead, takes a systematic approach to explore the potential of state-of-the-art GNN architectures as unified frameworks for unsupervised representation learning of APM graphs. To achieve this, we developed a procedure to enable GNNs originally intended for homogeneous graphs to handle heterogeneous data.

3.1 Handling APM Graphs with Graph Neural Networks

Traditional GNNs, primarily designed for homogeneous graphs with a single node and edge type, are not directly applicable to heterogeneous graphs.

To use and manipulate APM graphs within the PyTorch framework, we relied on the NetworkX Python library. This widely-used library provides flexible tools for creating, parsing, and modifying graph structures. Although NetworkX does not natively support heterogeneous graphs, the `graphml` format preserves node and edge metadata during parsing, resulting in attributing types to nodes and edges. These `type` attributes enabled us to implement type-aware processing for APM graphs. The `to_hetero` function in PyTorch Geometric library enables Message Passing between different node and edge types.

3.2 Graph Neural Network Classes Investigated

Message Passing approaches in GNNs have progressively evolved, resulting in a variety of MP layers based on different MPNN designs. In this work, we focused on four types of MP layers.

SAmple and AggreGatE (GraphSAGE). The GraphSAGE approach [5] popularized scalable neighborhood aggregation in GNN Message Passing. Its core idea is neighborhood sampling: for each node, a fixed number of neighbors are sampled to reduce computational cost while maintaining comparable performance.

Graph Attention Network (GAT). GATs extend GNN Message Passing by introducing an attention mechanism that allows each node to weight its neighbors differently when aggregating features [10], following the principles of attention in standard Neural Networks. For each edge, a learnable attention coefficient is computed, reflecting the relative importance of the neighbor's features. Node embeddings are updated as a weighted sum of neighbor features, optionally followed by a nonlinear transformation.

Graph Isomorphism Network (GIN). GINs employ a sum-based aggregation function over neighboring node features, followed by a multi-layer perceptron to update node embeddings [12]. This design allows the network to capture subtle structural differences that may be missed by mean- or max-based aggregation schemes.

Meta General Layer (GeneralConv). By defining a meta-layer abstraction capable of representing a wide range of message, aggregation, and update functions, GeneralConv enables the construction of customizable GNN architectures that subsume many existing models [13]. This approach enhances the expressiveness, generalizability, and experimental flexibility of GNNs.

3.3 Model Learning

The most commonly used loss function for training Neural Networks in regression tasks is the Mean Squared Error (MSE), defined as $MSE = \frac{1}{n}\sum_{i=1}^{n}(y_i - \hat{y}_i)^2$, where n is the number of predictions, and y_i and $\hat{y}_i$ denote the observed and predicted values.

In our case, the loss function is designed as the weighted aggregation of two MSEs. Let $\lambda \in [0, 1]$ the weight associated to the prediction of latency. The loss function writes $\mathcal{L} = \lambda\ MSE_{latency} + (1 - \lambda)\ MSE_{power}$.

4 Experiments

4.1 Data

Our benchmark consists of eleven datasets introduced in earlier research [1], in which latency and power-consumption metrics were derived through analytical modeling and simulation techniques. Each dataset contains several APM graph instances of the same application mapped onto a fixed multicore architecture. As a reminder, these instances differ in task decomposition and tile assignment. All datasets involve neural-network applications used for prediction. Figure 3 shows the five applications considered: three based on Multi-Layer Perceptrons and two based on Convolutional Neural Networks.

The main statistics for all eleven datasets are listed in Table 1. The `mlp1_expe` dataset was obtained experimentally, while the other ten were produced via analytical models and simulations. The suffixes CG (Clock Gating) and P (Polling) indicate distinct data-transfer modes: Clock Gating lets cores remain in a low-power state until data are available, whereas P continuously monitors incoming data to trigger core activation.

4.2 Implementation Details

The different GNN models were trained on an AMD Radeon RX 7900 XTX GPU using PyTorch and ROCm (CUDA library for AMD GPUs). The full computer setup included an 8-core AMD Ryzen 7 5800X3D processor (3.4 GHz) and 32GB of DDR4 RAM.

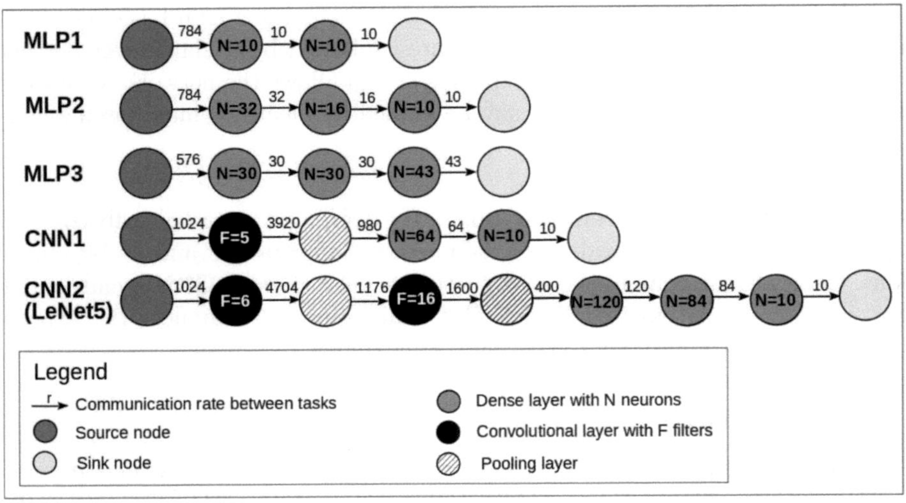

Fig. 3. The five software neural-network applications considered. These applications implement prediction with various neural networks. The five application graphs provided correspond to the coarsest level of task decomposition, in which each layer is handled by a single task. The applications are presented here before task parallelization. Each dense layer with N neurons can be parallelized into at most N tasks. This parallelization is also possible for the convolutional layers containing F filters. Each neural-network software application will lead to various application graphs by distributing the neurons or filters of each network layer differently across tasks: the different network layers can each be managed by at most N tasks (or F tasks).

Table 1. Statistics of the 11 datasets. n_i: number of instances (application-to-platform mappings).

		Nodes		Edges				Nodes		Edges	
Dataset	n_i	maximum	mean	maximum	mean	Dataset	n_i	maximum	mean	maximum	mean
MLP1_expe	18	26	20.1	114	67.7	MLP3_CG	253	31	24.1	152	90.0
MLP1_CG	85	22	28.9	78	56.7	MLP3_P	247	31	23.4	152	83.6
MLP1_P	96	22	18.9	78	56.7	CNN1_CG	224	30	22.6	116	66.8
MLP2_CG	174	30	22.1	143	76.3	CNN1_P	276	30	23.0	116	69.2
MLP2_P	288	30	23.8	143	88.5	CNN2_CG	308	45	31.6	209	108.9
						CNN2_P	192	45	28.3	209	90.3

4.3 Experimental Settings

Fixed Hyperparameters. There are no pre-processing layers. Each of the four classes of GNNs implemented is composed of 8 identical layers (GraphSAGE, GAT, GIN, and GeneralConv). Each layer incorporates a PReLU (Parametric Rectified Linear Unit) activation function. Neither Batch Normalization nor Dropout are considered in the layers. A post-processing module is added, consisting of a linear layer, a ReLU activation function, a Dropout layer with a rate of 0.5, and a final linear layer that aggregates the information into a two-dimensional output.

The MP aggregation function used was `mean`. Based on former experiments (not shown), we set the number of epochs as 400, and the learning rate as 0.01. The optimizer used was Adam. Training was performed with a batch size of 8.

Questions Addressed. The experiments described in this section were designed to answer the following key questions:

- $\mathbf{Q_1}$: what prediction accuracy do the four GNN classes achieve for power consumption and latency?
- $\mathbf{Q_2}$: how sensitive are the power consumption and latency predictions to the weighting parameter λ?
- $\mathbf{Q_3}$: how does reducing the amount of training data affect model performance?

The evaluation method used was Repeated Random Sub-sampling (RSS) which randomly splits data into training and test sets multiple times (10 times in our case). The standard MAPE (Mean Absolute Percentage Error) metric was computed on each test fold. This metric is defined as: $MAPE = \frac{100}{n}\sum_{i=1}^{n}\left|\frac{y_i-\hat{y}_i}{y_i}\right|$, where n is the number of predictions, and y_i and $\hat{y}_i$ denote the observed and predicted values. The model's performance was estimated as the average across the 10 splits. By ensuring multiple independent repetitions, RRS provides a more robust estimate of model performance, particularly on small datasets, than 10-fold cross-validation. At each repetition, we used an 80%/20% train/test split for the `MLP1_expe` dataset, and a 90%/10% split for the other ten datasets.

Due to its small size, the `MLP1_expe` was only trained on the entire training set ($\alpha = 1$). Overall, our experiments involved training 7,440 models.

To answer $\mathbf{Q_2}$, we varied the λ hyperparameter in $\{0, 0.5, 1\}$. To address $\mathbf{Q_3}$, we defined α as the proportion of the training set actually used during training. We varied α in $\{0.3, 0.6, 1\}$. For instance, with a 90%/10% train/test split, and $\alpha = 0.6$, models were trained on 60% of the 90% training portion.

4.4 Results and Discussion

Inference Time. Observed times were on the order of a few tens of milliseconds per prediction, demonstrating the framework's suitability for fast performance estimation.

Prediction Performance. Table 2 shows the MAPE values obtained for each GNN model and dataset, across all (λ, α) configurations. This table shows that all GNN models predict power consumption more easily than latency.

Table 2 also reports the prediction performance for one metric when the model was trained using exclusively the other metric: latency predictions for

Table 2. Latency and power-consumption MAPE values by GNN class, for different datasets and varying (λ, α) configurations. For a given λ value, the best GNN class is highlighted in bold and indicated by color for each of the three α values: red for $\alpha = 0.3$, green for $\alpha = 0.6$, and blue for $\alpha = 1$.

$\lambda = 0.5$

GNN type	Latency MAPE												Power consumption MAPE											
	GraphSAGE			GAT			GIN			GeneralConv			GraphSAGE			GAT			GIN			GeneralConv		
α	0.3	0.6	1.0	0.3	0.6	1.0	0.3	0.6	1.0	0.3	0.6	1.0	0.3	0.6	1.0	0.3	0.6	1.0	0.3	0.6	1.0	0.3	0.6	1.0
MLP1_expe	–	–	26.8	–	–	21.3	–	–	23.5	–	–	37.4	–	–	3.95	–	–	3.71	–	–	5.18	–	–	5.47
MLP1_CG	20.8	18.2	18.1	20.5	17.8	18.8	21.4	17.3	16.2	**18.5**	**16.1**	**14.3**	3.09	2.75	2.66	2.90	2.24	2.07	**2.35**	**1.67**	**1.66**	3.13	2.31	2.40
MLP1_P	20.4	20.4	20.7	19.7	25.1	22.0	18.5	**15.8**	**15.9**	**17.6**	18.1	16.6	1.71	1.37	1.36	**1.38**	**1.32**	**1.21**	1.43	1.37	1.26	1.81	1.61	1.48
MLP2_CG	20.7	16.7	16.1	21.3	16.4	15.3	**19.3**	**15.7**	**14.5**	24.8	21.8	22.6	**2.71**	**2.38**	2.32	2.85	2.80	2.33	2.86	2.59	**2.25**	3.27	3.07	2.99
MLP2_P	20.8	21.4	18.8	24.0	22.6	18.5	**17.6**	**16.1**	**12.9**	22.5	23.8	19.8	1.45	1.42	1.43	1.39	1.29	1.19	**1.18**	**1.15**	**1.03**	1.47	1.43	1.37
MLP3_CG	12.9	16.7	**9.5**	14.8	14.2	10.6	**12.5**	**13.4**	10.5	16.5	18.1	14.3	2.50	2.38	2.18	2.13	2.05	1.91	**2.10**	**1.95**	**1.75**	2.39	2.35	2.25
MLP3_P	18.8	17.0	14.9	16.8	**13.9**	**13.4**	**16.0**	14.8	14.1	19.3	18.3	20.8	1.68	1.38	1.52	**1.28**	**1.10**	**1.10**	1.31	1.17	1.17	1.65	1.39	1.55
CNN1_CG	38.7	29.7	26.4	24.9	22.6	19.8	**20.2**	**21.9**	**14.0**	47.7	39.3	24.1	2.79	2.42	2.02	2.26	**2.03**	1.94	**2.20**	2.18	**1.90**	3.17	2.91	2.56
CNN1_P	32.0	25.3	23.1	22.9	19.9	18.3	**21.3**	**16.3**	**14.3**	49.1	30.8	26.4	1.60	1.44	1.45	1.48	1.37	1.28	**1.36**	**1.21**	**1.11**	1.92	1.47	1.47
CNN2_CG	15.1	13.3	10.6	**13.4**	**10.4**	**9.3**	15.4	11.9	9.8	16.8	14.4	13.2	**1.69**	1.46	1.41	**1.69**	**1.41**	**1.29**	1.95	1.76	1.60	2.23	2.12	2.07
CNN2_P	19.7	16.8	17.3	17.1	**14.3**	**15.1**	**15.6**	15.2	17.1	25.8	20.1	27.1	1.32	1.17	1.14	**1.28**	**1.14**	**1.07**	1.38	1.28	1.34	1.67	1.62	1.69
Mean	22.0	19.5	17.5	19.5	17.7	16.1	**17.8**	**15.8**	**13.9**	25.8	22.1	19.9	2.05	1.82	1.75	1.86	1.67	1.54	**1.81**	**1.63**	**1.51**	2.27	2.03	1.98

$\lambda = 1$ (training controled with $MSE_{latency}$)

GNN type	Latency MAPE												Power consumption MAPE											
	GraphSAGE			GAT			GIN			GeneralConv			GraphSAGE			GAT			GIN			GeneralConv		
α	0.3	0.6	1.0	0.3	0.6	1.0	0.3	0.6	1.0	0.3	0.6	1.0	0.3	0.6	1.0	0.3	0.6	1.0	0.3	0.6	1.0	0.3	0.6	1.0
MLP1_expe	–	–	24.8	–	–	21.2	–	–	24.0	–	–	46.2	–	–	4.74	–	–	4.74	–	–	4.74	–	–	4.74
MLP1_CG	18.3	17.2	17.9	17.7	18.1	17.6	**16.3**	15.7	13.9	18.0	**14.9**	**11.9**	4.29	4.14	4.22	4.29	4.14	4.22	4.29	4.14	4.22	4.29	4.14	4.22
MLP1_P	21.9	22.3	18.4	20.8	24.2	18.6	19.5	**14.7**	**12.9**	**18.3**	18.4	16.3	4.61	4.59	4.62	4.61	4.59	4.62	4.61	4.59	4.62	4.61	4.59	4.62
MLP2_CG	19.0	15.3	15.1	20.3	16.2	15.0	**18.3**	**13.8**	**13.7**	24.4	22.2	23.1	4.71	4.69	4.68	4.71	4.69	4.68	4.71	4.69	4.68	4.71	4.69	4.68
MLP2_P	21.6	21.2	19.6	24.4	21.8	19.5	**14.9**	**15.8**	**13.6**	22.4	22.5	22.4	3.16	3.18	3.20	3.16	3.18	3.20	3.16	3.18	3.20	3.16	3.18	3.20
MLP3_CG	12.9	15.8	10.9	15.5	**13.5**	12.5	**12.1**	15.0	**9.7**	15.8	16.6	13.9	5.08	5.05	5.03	5.08	5.05	5.03	5.08	5.05	5.03	5.08	5.05	5.03
MLP3_P	20.2	16.8	17.5	**16.9**	**15.4**	15.8	20.0	16.2	**14.0**	27.2	21.4	21.4	4.05	4.07	4.10	4.05	4.07	4.10	4.05	4.07	4.10	4.05	4.07	4.10
CNN1_CG	39.7	34.1	30.8	20.3	20.9	19.8	**18.3**	**15.0**	**15.6**	40.0	31.3	24.2	5.13	5.14	5.14	5.13	5.14	5.14	5.13	5.14	5.14	5.13	5.14	5.14
CNN1_P	30.2	22.5	22.2	**19.9**	19.5	16.8	21.3	**16.7**	**14.3**	45.2	37.2	23.4	3.19	3.19	3.20	3.19	3.19	3.20	3.19	3.19	3.20	3.19	3.19	3.20
CNN2_CG	13.2	11.0	10.1	13.1	9.9	9.0	**11.8**	**9.7**	**8.7**	15.3	13.8	12.6	3.07	3.04	3.04	3.07	3.04	3.04	3.07	3.04	3.04	3.07	3.04	3.04
CNN2_P	20.4	18.9	20.8	**16.6**	**15.2**	**14.2**	17.9	16.3	16.7	29.1	30.8	40.9	2.55	2.51	2.50	2.55	2.51	2.50	2.55	2.51	2.50	2.55	2.51	2.50
Mean	21.7	19.5	18.3	18.5	17.5	15.9	**17.0**	**14.9**	**13.3**	25.6	22.9	21.0	3.98	3.96	3.97	3.98	3.96	3.97	3.98	3.96	3.97	3.98	3.96	3.97

$\lambda = 0$ (training controled with MSE_{power})

GNN type	Latency MAPE												Power consumption MAPE											
	GraphSAGE			GAT			GIN			GeneralConv			GraphSAGE			GAT			GIN			GeneralConv		
α	0.3	0.6	1.0	0.3	0.6	1.0	0.3	0.6	1.0	0.3	0.6	1.0	0.3	0.6	1.0	0.3	0.6	1.0	0.3	0.6	1.0	0.3	0.6	1.0
MLP1_expe	–	–	56.1	–	–	56.1	–	–	56.1	–	–	56.1	–	–	4.46	–	–	4.68	–	–	6.22	–	–	5.47
MLP1_CG	58.5	62.1	60.3	58.5	62.1	60.3	58.5	62.1	60.3	58.5	62.1	60.3	3.19	2.54	2.14	2.77	2.11	2.02	**2.51**	**1.76**	**1.56**	3.32	2.04	2.02
MLP1_P	57.8	57.1	56.7	57.8	57.1	56.7	57.8	57.1	56.7	57.8	57.1	56.7	1.44	1.24	1.27	1.31	**1.18**	1.19	**1.29**	**1.18**	**1.07**	1.54	1.38	1.46
MLP2_CG	59.2	59.3	60.1	59.2	59.3	60.1	59.2	59.3	60.1	59.2	59.3	60.1	**2.72**	**2.29**	**2.06**	3.14	2.53	2.31	2.77	2.51	2.07	3.73	3.21	2.89
MLP2_P	46.9	46.3	44.8	46.9	46.3	44.8	46.9	46.3	44.8	46.9	46.3	44.8	1.42	1.53	1.48	1.31	1.19	1.20	**1.15**	**1.04**	**0.91**	1.45	1.41	1.25
MLP3_CG	55.8	56.9	55.9	55.8	56.9	55.9	55.8	56.9	55.9	55.8	56.9	55.9	2.38	2.21	2.17	2.17	2.10	1.73	**2.07**	**1.94**	**1.56**	2.33	2.53	2.20
MLP3_P	48.5	48.6	47.9	48.5	48.6	47.9	48.5	48.6	47.9	48.5	48.6	47.9	1.65	1.26	1.43	1.19	1.11	1.06	**1.15**	**1.06**	**1.03**	1.68	1.25	1.49
CNN1_CG	58.9	58.5	58.5	58.9	58.5	58.5	58.9	58.5	58.5	58.9	58.5	58.5	2.32	1.92	1.92	**2.01**	1.98	1.71	2.06	**1.80**	**1.60**	3.40	2.52	2.62
CNN1_P	61.0	57.8	57.5	61.0	57.8	57.5	61.0	57.8	57.5	61.0	57.8	57.5	1.60	1.44	1.47	1.56	1.34	1.25	**1.23**	**1.19**	**1.05**	1.63	1.46	1.35
CNN2_CG	36.1	34.4	34.9	36.1	34.4	34.9	36.1	34.4	34.9	36.1	34.4	34.9	1.96	2.38	2.19	**1.55**	**1.52**	**1.23**	2.00	1.68	1.56	2.23	2.16	2.02
CNN2_P	51.1	50.7	51.4	51.1	50.7	51.4	51.1	50.7	51.4	51.1	50.7	51.4	1.33	1.24	1.09	1.32	1.12	1.03	**1.19**	**1.00**	**0.90**	1.47	1.35	1.54
Mean	53.4	53.2	52.8	53.4	53.2	52.8	53.4	53.2	52.8	53.4	53.2	52.8	2.00	1.81	1.72	1.83	1.62	1.47	**1.74**	**1.52**	**1.33**	2.28	1.93	1.89

$\lambda = 0$ and power-consumption predictions for $\lambda = 1$. As expected, these correspond to the worst predictions. However, a nontrivial observation is that the four GNN classes exhibit identical worst predictions on the alternate metric at a fixed α.

Impact of λ. Figure 4 shows the impact of moving from training optimized for a single metric ($\lambda = 1$ for latency, $\lambda = 0$ for power consumption) to bi-objective

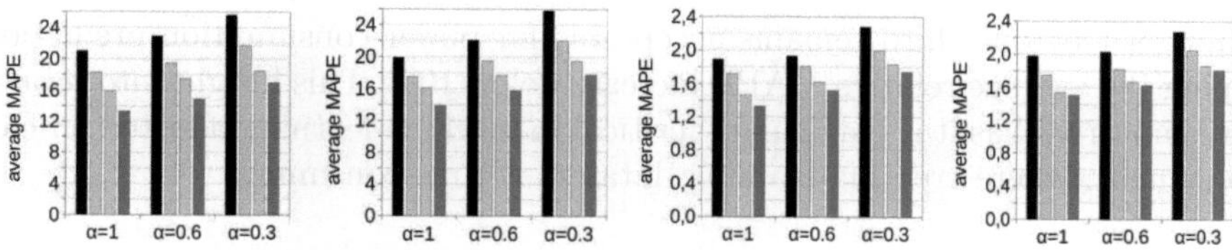

(a) latency, $\lambda = 1$ (b) latency, $\lambda = 0.5$ (c) power, $\lambda = 0$ (d) power, $\lambda = 0.5$

Fig. 4. Prediction performance of the four GNN classes across different λ and α configurations. For each GNN class, the MAPE values are averaged over the 11 datasets for $\alpha = 1$, and over the 10 simulated datasets for the two other values of α. From left to right: GeneralConv, GraphSage, GAT, GIN.

training ($\lambda = 0.5$). It presents trends in dataset-averaged MAPE values for each GNN class, λ, and α.

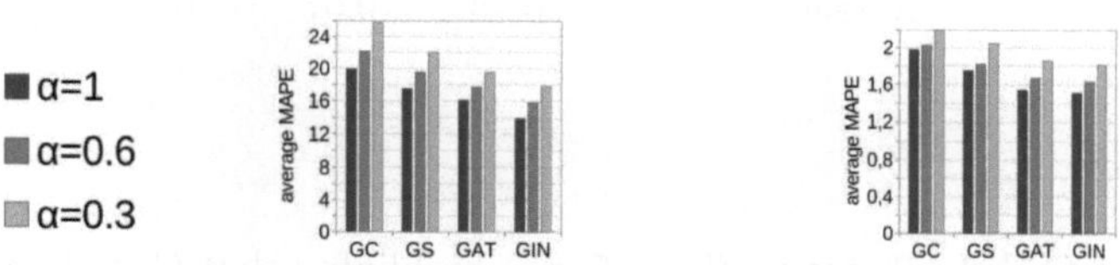

(a) latency, $\lambda = 0.5$ (b) power consumption, $\lambda = 0.5$

Fig. 5. Impact of α on prediction performance under the $\lambda = 0.5$ setting. GC: GeneralConv, GS: GraphSAGE. From left to right: α values are 1, 0.6 and 0.3.

When λ is reduced from 1 to 0.5, the latency MAPE rises for 83% of the datasets across the four GNN classes and the three α values (details not shown), with a relative increase between 0 and 6.0% (average 3.4%). Increasing λ from 0 to 0.5 results in an increase in power-consumption MAPE in 92% of the datasets, with a relative increase ranging in [0.05%,13.6%] (average 4.0%). Thus, for latency, the already low prediction accuracy at $\lambda = 1$ degrades only marginally when a weighted loss is applied. When considering relative variation, power-consumption accuracy is affected to a similar degree, which remains acceptable given the high performance under the optimal setting ($\lambda = 0$).

Impact of α. Figure 5 examines the $\lambda = 0.5$ case by comparing MAPE values averaged over datasets across 120 different configurations (4 GNN classes × 10 simulated datasets × 3 α values). The trends highlighted in this figure reflect how reducing the size of the training datasets affects prediction performance. These trends align with the behaviors observed in the detailed per-dataset results. Table 3 shows statistics about the per-dataset results.

For both metrics, reducing α from 1 to 0.6 causes a relative performance degradation comparable to that from 0.6 to 0.3. The latency MAPE relative increase is roughly twice that of power consumption. Finally, cutting the initial training set by one third remains acceptable for power-consumption prediction, keeping the average relative MAPE increase below 10%. This insight into robustness is nontrivial as it is particularly critical in machine-learning-assisted on-chip deployment, where creating training datasets is time-consuming.

Table 3. Impact of training-set reduction on model accuracy. p_{config} denotes the proportion of configurations (4 GNN classes × 10 simulated datasets) that exhibit an increase in MAPE when α is reduced. Summary statistics computed over the relative MAPE increase for this proportion are reported.

		Latency		Power consumption	
	α transition	$1 \rightarrow 0.6$	$0.6 \rightarrow 0.3$	$1 \rightarrow 0.6$	$0.6 \rightarrow 0.3$
	p_{config}	77.5 %	80 %	82.5 %	80 %
relative	minimum	0.01	0	0	0.01
increase	maximum	0.76	0.59	0.20	0.31
in MAPE (%)	average	0.16	0.18	0.07	0.10

Comparison of GNN Classes. Figure 5 also highlights that GIN delivers the best prediction accuracy for both latency and power consumption, followed by GAT. GraphSAGE and GeneralConv trail behind, with GeneralConv being the least accurate.

Comparison with Other Works. We could not compare our approach with the GNN-based method reported in [2], which used source-code graphs that are not publicly available and does not provide code to run on the benchmark used in our study. In [1], a hybrid method combining analytical models and simulations is evaluated on the same datasets as in our study. The author reports MAPE values below 3% for both latency and power-consumption predictions, albeit with inference times on the order of several tenths of a second. In contrast, our proposed GNN-based framework attains comparable power-consumption accuracy with prediction times of only a few tens of milliseconds, representing an approximate tenfold speedup.

5 Conclusion and Future Work

A GNN-based framework was developed to predict latency and power in task-to-tile mappings. Evaluation on datasets ranging from tens to a few hundred instances shows that GNN models can achieve low MAPE for power consumption, even with limited training data, while providing predictions in real time. Although latency prediction errors remain comparatively high, performance could be enhanced by enlarging the training dataset or incorporating additional information beyond structural features. Reliable latency estimation would allow fast, on-the-fly bi-criteria mapping evaluation, supporting metaheuristic optimization integration. Future work will explore the potential of our framework for multi-output regression, depending on the nature of the target performance criteria. In addition, our operational framework is not domain-specific; it is designed to support standard GNNs in processing heterogeneous graphs and performing multi-output regression at the graph level. We plan to evaluate it on additional use cases and assess whether GIN and GAT remain top performers.

Acknowledgments. The first two authors received funding from the FAISTOS research cluster of the NExT (Nantes Excellence Trajectory) I-SITE.

Disclosure of Interests. The authors have no competing interests to declare that are relevant to the content of this article.

References

1. Dariol, Q.: Early timing and energy prediction and optimization of artificial neural networks on multi-core platforms. Ph.D. thesis, Nantes University (2023)
2. Ferretti, L., Cini, A., Zacharopoulos, G., Alippi, C., Pozzi, L.: Graph neural networks for high-level synthesis design space exploration. ACM Trans. Des. Autom. Electron. Syst. **28**(2), 1–20 (2022)
3. Fu, X., Zhang, J., Meng, Z., King, I.: MAGNN: metapath aggregated Graph Neural Network for heterogeneous graph embedding. In: The Web Conference (WWW), pp. 2331–2341 (2020)
4. Gori, M., Monfardini, G., Scarselli, F.: A new model for learning in graph domains. In: IEEE International Joint Conference on Neural Networks (IJCNN), vol. 2, pp. 729–734 (2005)
5. Hamilton, W., Ying, R., Leskovec, J.: Inductive representation learning on large graphs. In: Advances in Neural Information Processing Systems (NeurIPS), pp. 1024–1034 (2017)
6. Khemani, B., Patil, S., Kotecha, K., Tanwar, S.: A review of Graph Neural Networks: concepts, architectures, techniques, challenges, datasets, applications, and future directions. J. Big Data **11**(1), 18 (2024)
7. Lin, W., Chen, Z., Chen, Y., Wang, S.: Heterogeneous Graph Neural Network with adaptive relation reconstruction. Neural Netw. **187**, 107313 (2025)
8. Sánchez, D., Servadei, L., Kiprit, G.N., Wille, R., Ecker, W.: A comprehensive survey on electronic design automation and graph neural networks: theory and applications. ACM Trans. Design Autom. Electron. Syst. **28**(2), 1–27 (2023)
9. Scarselli, F., Gori, M., Tsoi, A.C., Hagenbuchner, M., Monfardini, G.: The graph neural network model. IEEE Trans. Neural Netw. **20**(1), 61–80 (2009)
10. Velicković, P., Cucurull, G., Casanova, A., Romero, A., Liò, et al.: Graph attention networks. In: Sixth International Conference on Learning Representations (ICLR), pp. 1–12 (2018)
11. Wang, J., Guo, Y., Yang, L., Wang, Y.: Enabling homogeneous GNNs to handle heterogeneous graphs via relation embedding. IEEE Trans. Big Data **9**(6), 1697–1710 (2023)

12. Xu, K., Hu, W., Leskovec, J., Jegelka, S.: How powerful are Graph Neural Networks? In: Seventh International Conference on Learning Representations (ICLR), pp. 1–17 (2019)
13. You, J., Ying, Z., Leskovec, J.: Design space for Graph Neural Networks. In: Advances in Neural Information Processing Systems (NeurIPS), pp. 10727–10738 (2020)
14. Zhu, S., Zhou, C., Pan, S., Zhu, X., Wang, B.: Relation structure-aware heterogeneous Graph Neural Network. In: 19th IEEE International Conference on Data Mining (ICDM), pp. 1534–1539 (2019)

Guided Feature Distillation for YOLO11: Efficient Detection of Pseudomonas Aeruginosa in High-Resolution Microscopy

Mouhcine Ouaaziz[1(✉)], Dimitri Klockenbring[1], Joseph Lam-Weil[1], Cédric Wemmert[1], Benoît Naegel[1], Sarah Chouchene[1], Morgan Madec[1], and Winfried Römer[2]

[1] ICube Laboratory, CNRS/University of Strasbourg, Illkirch-Graffenstaden, France
{ouaaziz,d.klockenbring,jlam,wemmert,b.naegel,chouchene, morgan.madec}@unistra.fr

[2] Faculty of Biology, University of Freiburg, Freiburg, Germany
winfried.roemer@bioss.uni-freiburg.de

Abstract. Object detection in high-resolution microscopy is essential for automated bacterial analysis but demands computationally intensive models that challenge embedded deployment. This work addresses the problem of compressing large detectors for detecting *Pseudomonas aeruginosa* in crowded microscopic scenes, where very small and densely packed targets exacerbate the accuracy loss typically incurred by model compression. We explore guided feature knowledge distillation, focusing in particular on Masked Generative Distillation and Channel-Wise Distillation applied to the YOLO11 family, evaluating both direct teacher-student transfer and an assistant-teacher setup that bridges large capacity gaps. To better handle tiny objects, we introduce an overlapping 640×640 pixel patching strategy that expands the effective training set. Experiments demonstrate that our lightweight YOLO11-N student improves from a baseline AP_{50-95} of 71.9 to 74.0 using a single assistant teacher, outperforming direct distillation results (73.3–73.5) while approaching the teacher performance of 75.5. These results establish that feature distillation combined with patch learning effectively compresses large microscopy detectors without sacrificing accuracy, yielding efficient models suitable for resource-constrained deployment.

Keywords: Knowledge Distillation · Object Detection · Microscopic Imaging · Bacteria Detection

1 Introduction

Antimicrobial resistance and hospital-acquired infections are increasing global health threats [1,2]. Conventional diagnostics (culture, immunoassay, polymerase chain reaction) remain accurate but are often too slow or resource-intensive for

M. Baratchi et al. (Eds.): IDA 2026, LNCS 16513, pp. 213–224, 2026.
https://doi.org/10.1007/978-3-032-23833-7_16

real-time clinical use. Recent work combines optical microscopy with AI to provide rapid, portable pathogen detection (Fig. 1), for example label-free continuous monitoring [1] and video-microscopy with trajectory analysis to separate living bacteria from artifacts [2]. These image-based, machine-learning approaches are promising but can be computationally demanding for embedded deployments.

Advances in deep learning and embedded computing have enabled new approaches for microscopic pathogen detection. Object detection frameworks like YOLO [3] have been successfully adapted for microbe identification [4]. Recent anchor-free models, such as YOLO8, predict object centers and dimensions directly, simplifying architecture and improving scale generalization, and successive versions like YOLO11 build on these innovations to enhance efficiency and accuracy for real-time detection [5].

However, such models remain computationally expensive and are typically too large for resource-constrained embedded devices. Knowledge distillation (KD) has emerged as a powerful solution, transferring knowledge from a large, high-performing "teacher" network to a lightweight "student" network [6], thereby enabling deployment on edge platforms without significant performance loss. Building upon these advances in anchor-free detection and KD [5,6], and leveraging prior work in label-free pathogen detection [1,2], we propose a distilled anchor-free YOLO model for efficient pathogen detection on resource-constrained devices.

This paper makes the following contributions.

- We present, to the best of our knowledge, the first study applying Masked Generative Distillation (MGD) and Channel-Wise Distillation (CWD) to the anchor-free YOLO11 family for object detection.
- We evaluate the enhancement obtained by introducing a single assistant model between the teacher and the student, demonstrating its effectiveness in transferring feature representations.
- We introduce an overlapping 640 × 640 patch-based training scheme that boosts performance, producing compact yet accurate models suitable for future embedded systems.

2 Related Works

Early work on model compression aimed to shrink networks with minimal accuracy loss. Buciluǎ *et al.* [7] compressed ensembles into single models, and Hinton *et al.* [8] formalized KD, where a student learns a teacher's soft predictions.

Shu *et al.* [9] introduced Channel-Wise Distillation (CWD), which is illustrated in Fig. 2band is specifically tailored for dense prediction tasks [10]. Instead of aligning features in the spatial domain, as in Focal and Global Distillation (FGD) [11], CWD treats each channel as a probability distribution, normalizing channel activations via spatial softmax, and then performs channel-wise Kullback-Leibler (KL) divergence comparisons between the teacher and the student. The CWD process can be formally described in following steps:

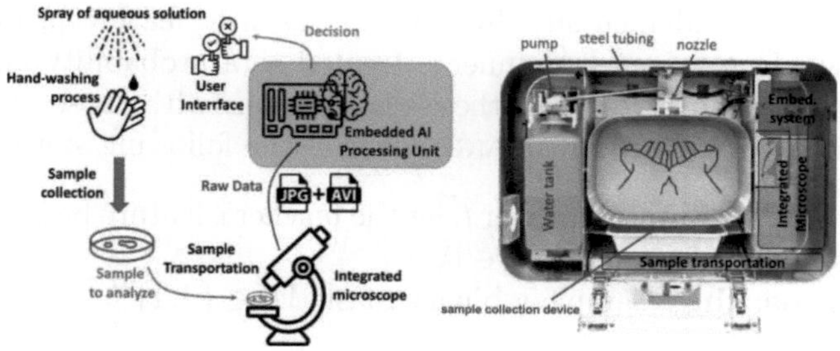

Fig. 1. Description of the complete detection system (on the left) and its integration into a commercial hand sanitizer (on the right). Note: These results are based on [1].

1. **Feature Maps:** Let $y^T \in \mathbb{R}^{C\times H\times W}$ and $y^S \in \mathbb{R}^{C\times H\times W}$ be the intermediate feature-map tensors of the teacher and student, respectively, at some layer with C channels of spatial size $H \times W$.
2. **Softmax Normalization:** Each channel's activation map is normalized using a softmax function to convert the raw activations into a probability distribution as defined in (1). This is typically done across the spatial dimensions of the feature map:

$$\phi(y_c) = \frac{\exp\left(\frac{y_{c,i}}{T}\right)}{\sum_{i=1}^{W.H} \exp\left(\frac{y_{c,i}}{T}\right)}, \quad i = 1, \ldots, HW \tag{1}$$

   where $y_{c,i}$ denotes the activation at spatial location i in channel c, and T is a temperature that smooths the distribution.
3. **Distillation Loss:** The average KL divergence is then minimized between the teacher's and student's per-channel distributions (2):

$$\mathcal{L}_{cwd} = \frac{T^2}{C} \sum_{c=1}^{C} \sum_{i=1}^{W.H} \phi(y_{c,i}^T) \cdot \log \left[\frac{\phi(y_{c,i}^T)}{\phi(y_{c,i}^S)} \right] \tag{2}$$

4. **Combined Objective:** In practice, (2) is multiplied by a hyperparameter α and added to the task loss. In our case (detection loss by the student), the overall loss can be formulated as in (3):

$$\mathcal{L}_{total} = \mathcal{L}_{task} + \alpha \cdot \mathcal{L}_{cwd} \tag{3}$$

Yang *et al.* [12] introduced Masked Generative distillation (MGD), illustrated in Fig. 2a, a method in which portions of the student model's feature map are randomly masked during training. The student is then tasked with reconstructing the missing regions to align with the teacher model's complete output. This strategy encourages the student to develop a more holistic understanding of the input by learning to recover global contextual information from partial observations.

MGD adopts a fundamentally different approach to feature processing compared to CWD. In CWD, each channel is treated as a probability map, and the student model is trained to align these distributions with those of the teacher using KL divergence. In contrast, MGD involves the following steps:

1. **Feature extraction:** From layer l, let the teacher's feature be $T^l \in \mathbb{R}^{C\times H\times W}$ and the student's feature be $S^l \in \mathbb{R}^{C\times H\times W}$
2. **Random masking:** Sample a binary mask $M^l \in \{0,1\}^{H\times W}$ by:

$$M^l_{i,j} = \begin{cases} 0, & \text{if } R^l_{i,j} < \lambda, \\ 1, & \text{otherwise,} \end{cases} \quad R^l_{i,j} \sim U(0,1) \tag{4}$$

   where λ is the masking ratio between 0 and 1.
3. **Masked input & generation:** Apply the mask as in (4) to the student's feature (after any alignment layer f_{align}) then pass it through a small projector (5) $G(\cdot)$ to predict the full teacher feature as defined in (6):

$$\hat{T}^l = G(f_{\text{align}}(S^l) \odot M^l) \tag{5}$$

   where $\odot$ denotes the element wise multiplication between feature maps and mask from layer l

$$G(F) = W_2^l \left(\text{ReLU}\left(W_1^l(F)\right)\right) \tag{6}$$

   W_1^l, W_2^l denote two 3×3 convolution layers.
4. **Reconstruction loss:** Measure how well $\hat{T}^l$ matches T^l with an L2 loss over all the layers l, channel k, and spatial positions (i,j):

$$\mathcal{C}_{\text{mgd}} = \sum_{l=1}^{L}\sum_{k=1}^{C}\sum_{i=1}^{H}\sum_{j=1}^{W}(T^l - \hat{T}^l)^2 \tag{7}$$

5. **Combined objective:** Finally, we add $\mathcal{C}_{\text{mgd}}$ to the model's original loss $\mathcal{C}_{\text{task}}$ (detection in our case) to (7) with weight α (see (8)):

$$\mathcal{C}_{\text{total}} = \mathcal{C}_{\text{task}} + \alpha \cdot \mathcal{C}_{\text{mgd}} \tag{8}$$

While CWD [9] and MGD [12] have shown effectiveness for feature distillation, their evaluations on anchor-free models do not fully address modern architectures such as YOLO. In this work, we extend and study CWD, and MGD on the anchor-free YOLO11 family, and we investigate the performance gain obtained by inserting a single intermediate assistant model between the teacher and the student.

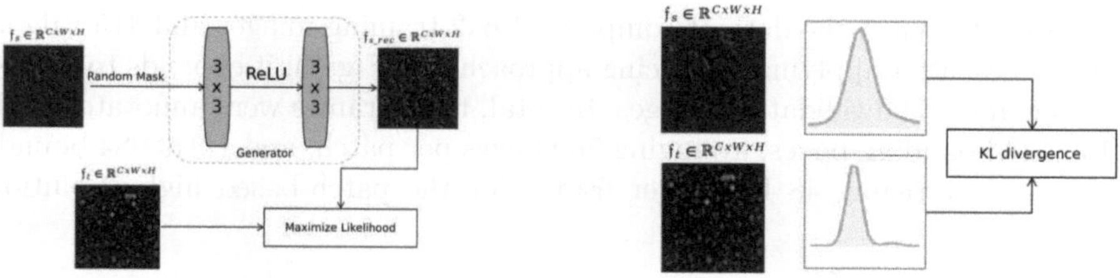

(a) Masked Generative distillation (b) Channel-wise distillation

Fig. 2. Knowledge distillation methods.

3 Material and Methods

3.1 Data

In this study, we focused on *Pseudomonas aeruginosa*, a rod-shaped, Gram-negative bacterium [13,14] known for its multidrug resistance [15] and its involvement in hospital-acquired infections, including ventilator-associated pneumonia [16] and sepsis [17]. All images were extracted from a single high-resolution video (7.26 pixels/µm) captured under controlled lighting conditions, yielding frames of size 2464 × 2056 pixels (px).

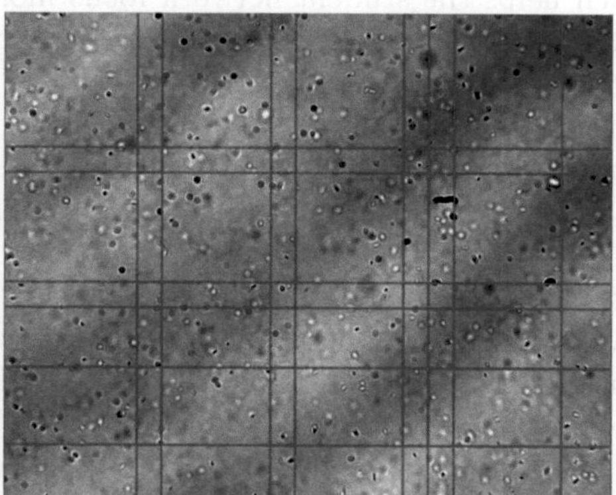

Fig. 3. Visualization of overlapping patch extraction (640 × 640 px) on an original image of size 2464 × 2056 px.

Raw data were annotated using the Cellpose 2.0 [18] human-in-the-loop workflow. Starting from a pretrained Cellpose model, we first generated automatic segmentations and then manually corrected them (fixing merges, splits, false positives, and missed cells). The corrected masks, which served as silver-truth annotations [19], were subsequently used to fine-tune the model. This annotation-retraining loop was repeated until the segmentation quality was deemed satisfactory.

Without slicing, the dataset comprises 1,642 training images and 415 validation images, after applying the slicing approach (see Fig. 3), it expands to 32,840 training and 8,300 validation images. In total, these frames were annotated with 1,150,923 bounding boxes, averaging 560 boxes per patch, and 1,940,084 bounding boxes, averaging 48 boxes per frame, for the patch-based high-resolution dataset.

3.2 Network Architectures

In this study, we employed the YOLO family of architectures (Nano, Medium, Large, and XLarge) provided by the Ultralytics framework [20]. These variants span a wide range of computational complexities and inference speeds, enabling a comprehensive investigation of knowledge transfer across models with varying capacities.

3.3 Distillation Implementation Details

We integrated KD into the Ultralytics YOLO [20] training pipeline. Two distillation methods were implemented: CWD [9] and MGD [12].

For feature alignment, we use learnable 1×1 convolutions to project the student features to the teacher's feature dimensionality, as Chen et *al.* [21] demonstrated. This approach helps the student network focus more effectively on the main feature extraction task while still utilizing the teacher's features as guidance through the projector, followed by batch normalization to standardize activations. To collect intermediate representations, we registered forward hooks on both teacher and student models and extracted feature maps at multiple depths (early, mid and late layers) to capture multiscale information.

The distillation losses were computed between aligned feature maps and added to the standard detection losses, allowing joint optimization of the student for the target task while receiving supervision from the teacher.

One of the limitations of the distillation approaches used in this work lies in the gap between the teacher and student architectures. Building on this observation and inspired by the work of Mirzadeh *et al.* [22], which demonstrated that introducing intermediate teacher-assistant networks can enhance distillation performance when a substantial capacity gap exists between teacher and student models, we examined two distinct KD paradigms: direct distillation, and single-assistant distillation, in which the assistant network simultaneously serves as a student with respect to the teacher and as a teacher with respect to the student (See Fig. 4).

4 Setup of the Experiments

We first conducted a systematic exploration of the main hyperparameters influencing the distillation process as shown in Fig. 5. For MGD, we varied the masking

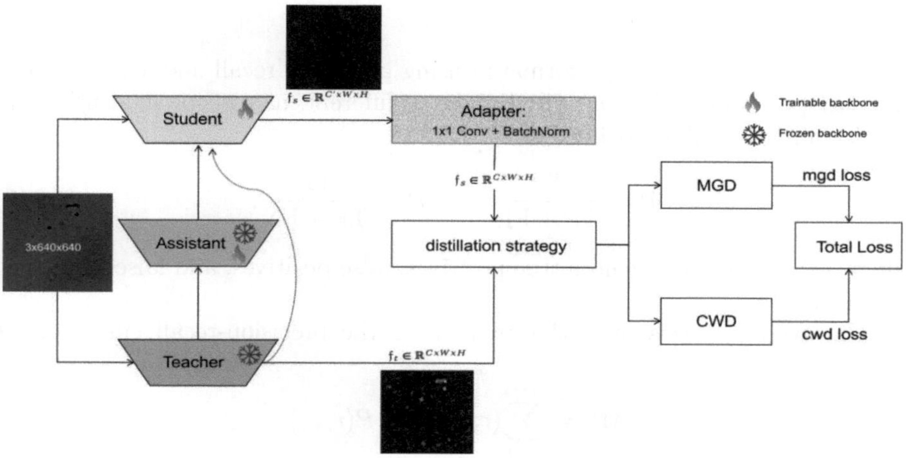

Fig. 4. Knowledge Distillation Framework with Teacher-Assistant-Student Architecture.

ratio λ and the distillation weight α to assess their impact on the reconstruction ability of student features. For CWD, we explored the temperature T and α through a grid search to identify the configuration that yields the best balance between stability and accuracy. This exploration allowed us to determine the optimal values of $T = 4$ for CWD and $\lambda = 0.9$ for MGD, with both methods showing weak sensitivity to α=1×10^{-5}.

In parallel, we evaluated two training schemes:

1. **Full-frame training**, where entire high-resolution images (2464×2056px) were used.
2. **Patch-based (sliced) training**, where the overlapped 640×640px patches were extracted from the same frames.

Although full-frame training simplifies preprocessing, it suffers from reduced recall in dense microscopic scenes containing many small targets because full frames are squished to fit the 640×640px YOLO input size. The sliced approach, by contrast, increases the effective dataset and improves object localization by providing higher spatial resolution per region (see Table 2). To eliminate duplicate detections from overlapping tiles at inference time, we map all tile detections back to full-image coordinates and apply Non-Maximum Suppression (NMS) [23] to retain the most confident prediction.

All models were trained three times independently on NVIDIA H100 4 GPUs for 100 epochs with a batch size of 16 using SGD optimizer with an initial learning rate of 0.001, a value of 0.0001 for weight decay and momentum 0.9 following Yang *et al.* [12] setup training with an early stopping patience set to 10 epochs.

Evaluation

We evaluate our detector performance using precision, recall and average precision (AP) on unseen data acquired under a different microscope lighting setup. Precision and recall are defined as

$$P = \frac{\mathrm{TP}}{\mathrm{TP} + \mathrm{FP}}, \qquad R = \frac{\mathrm{TP}}{\mathrm{TP} + \mathrm{FN}}, \tag{9}$$

where TP, FP and FN denote true positives, false positives, and false negatives, respectively.

The average precision is the area under the precision-recall curve and is computed as

$$\mathrm{AP} = \sum_{i=1}^{n-1} (r_{i+1} - r_i)\, P(r_{i+1}), \tag{10}$$

5 Results and Discussion

Table 1 presents our feature distillation results. Without distillation, the student model YOLO11-N attains a baseline $\mathrm{AP}_{50\text{–}95} = 71.9 \pm 0.1$. The teacher model YOLO11-X achieves 75.5 ± 0.3, establishing a teacher-student gap of 3.6 percentage points (pp) in $\mathrm{AP}_{50\text{–}95}$. The assistant model YOLO11-L reaches 75.2 ± 0.1, positioning it as an effective intermediate capacity model.

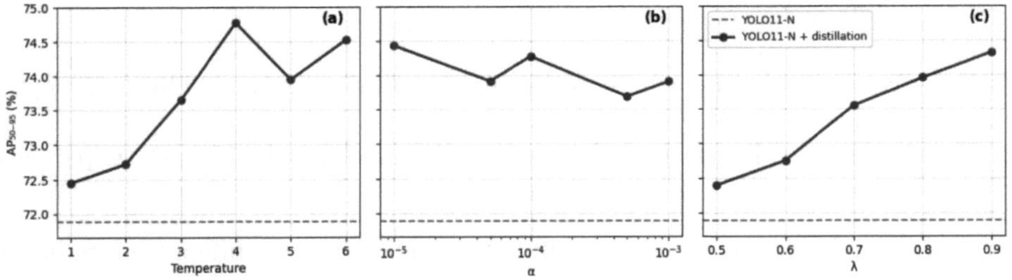

Fig. 5. Sensitivity analysis of CWD and MGD hyperparameters. The three panels (a)-(b)-(c) show the effects of temperature T (CWD), loss weight α (CWD, MGD), and masking ratio λ (MGD).

Direct distillation methods demonstrate measurable improvements over the student baseline. Direct MGD increases $\mathrm{AP}_{50\text{–}95}$ to 73.5 ± 0.1 (+1.6 pp), while direct CWD achieves 73.3 ± 0.1 (+1.4 pp).

Our proposed single assistant distillation strategy further enhances performance. Our approach improves upon direct MGD by +0.5 pp, achieving a total gain of +2.1 pp over the baseline student model (74.0 ± 0.2). Similarly, the CWD-based approach shows an improvement of +0.6 pp over direct CWD, resulting in an overall gain of +2.0 pp (73.9 ± 0.3).

Table 1. Performance comparison of KD strategies on YOLO11-N student model. Results show AP_{50-95} as mean ± standard deviation over three independent runs.

YOLO11-X	YOLO11-L	**YOLO**11-N	Gain (%)
75.5 ± 0.3	75.2 ± 0.1	Without distillation (71.9 ± 0.1)	-
		Direct MGD (73.5 ± 0.1)	1.6
		Ours (**74.0 ± 0.2**)	**2.1**
		Direct CWD (73.3 ± 0.1)	1.4
		Ours (**73.9 ± 0.3**)	**2.0**

Note: "Ours" refers to single-assistant distillation.

Table 2. Performance comparison: Full-Frame vs. Patch-Based Slicing (without distillation). All results represent mean ± standard deviation over three independent runs.

Model	Precision	Recall	AP_{50}	AP_{50-95}
Default Full-Frame Training				
YOLO11-X	96.7 ± 0.2	51.8 ± 0.1	74.7 ± 0.0	55.0 ± 0.1
YOLO11-N	95.0 ± 0.1	51.1 ± 0.0	73.7 ± 0.1	51.2 ± 0.1
Patch-Based High-Resolution Training (640×640 px)				
YOLO11-X	94.3 ± 0.0	**91.7 ± 0.2**	**97.3 ± 0.1**	**75.5 ± 0.3**
YOLO11-N	93.3 ± 0.1	**91.2 ± 0.0**	**96.1 ± 0.0**	**73.3 ± 0.1**

Overall, the introduction of an assistant teacher substantially reduces the teacher student gap. Our approach decreases the $AP_{50–95}$ gap from 3.6 to 1.5 pp (75.5 vs. 74.0), while our CWD approach reduces it to 1.6 pp (75.5 vs. 73.9). These results demonstrate the effectiveness of progressive knowledge transfer through an intermediate assistant model. This trend is also reflected in the parameter-performance landscape shown in Fig. 6a.

We selected YOLO11-L as the assistant teacher due to its optimal efficiency-accuracy trade-off (Fig. 6b). It achieved the highest accuracy (74.0% in AP_{50-95}) with the shortest training time (4.08h), outperforming YOLO11-M as an assistant by 1.9 pp while being 3× faster, and surpassing the two-assistant by 2.8 pp with 31% reduced training cost.

We conducted a grid search to identify optimal hyperparameters for the distillation process. In Fig. 5a, performance with CWD attains a maximum at $T = 4$, indicating that this value provides an optimal degree of softening of the teacher's predictive distribution for effective knowledge transfer. Values outside the $2-5$ range lead to a measurable degradation in performance. Figure 5b, the distillation procedure is largely insensitive to the choice of α in the two distillations approached studied. Figure 5c, higher masking ratios in MGD (approximately (0.8–0.9)), which occlude a larger portion of the background, produce the best performance.

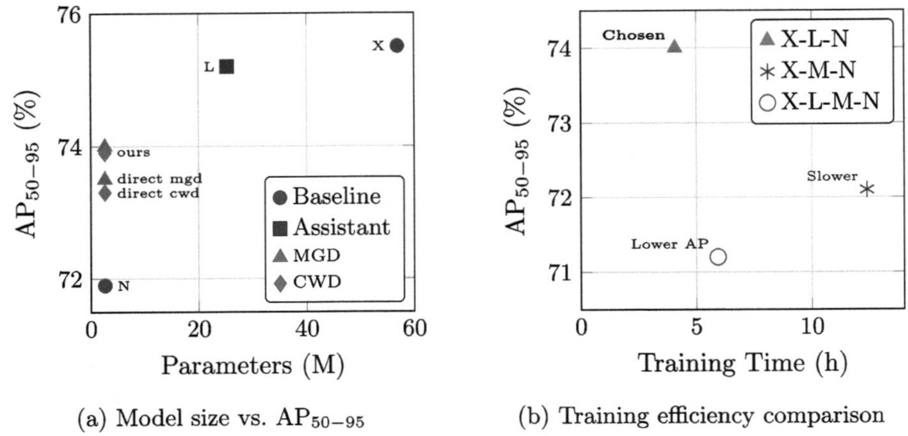

Fig. 6. Performance and efficiency analysis across model configurations. (a) shows model size vs. accuracy trade-offs, (b) shows training time vs. accuracy for assistant selection.

In contrast, patch-based training on overlapping 640×640 px slices preserves the original resolution, nearly doubling recall (from ∼51% to ∼92%) and improving AP_{50-95} by over 20 pp for both models. This substantial gain comes at minimal cost: precision decreases by only 2–3 pp. These results confirm that maintaining high spatial resolution is critical for detecting small, densely packed objects in microscopy images.

6 Conclusion

This work demonstrates that feature distillation transfers rich representations from large YOLO11 models to lightweight students for real-time microscopy detection. Two key findings emerge: (1) CWD and MGD effectively bridge the teacher-student gap in anchor-free detectors, with the YOLO11-N student improving from 71.9% to 73.5% AP_{50-95} via direct distillation and to 74.0% with an assistant teacher, approaching the teacher's 75.5% performance. (2) Patch-based training at 640 × 640px resolution significantly improves recall and AP for small objects by increasing input resolution. Combined, these strategies produce compact detectors that preserve most teacher accuracy while substantially reducing parameters and computation, establishing guided distillation as an efficient compression technique for high-resolution microscopy. Evaluation is limited to a single bacteria class. Generalization across bacterial species such as *Staphylococcus aureus* requires further validation. Additionally, runtime and energy performance on embedded hardware have not been measured and will be addressed in future work through multi-source datasets and hardware-level profiling.

Acknowledgments and Disclosure of Funding. This work was funded by the French National Research Agency (ANR) and the German Federal Ministry of Education and Research (BMBF) under the French-German joint call for proposals on Artificial Intelligence (Project LAIveBactDetect, ANR/BMBF (01IS23031A), 2022). The authors would like to acknowledge the High Performance Computing Center of the University of Strasbourg for providing scientific support and access to computing resources. Part of these computing resources was funded by the Equipex Equip@Meso project (Programme d'Investissements d'Avenir) and the CPER Alsacalcul/Big Data.

References

1. Claudinon, J., et al.: A label-free optical detection of pathogens in isopropanol as a first step towards real-time infection prevention. Biosensors **11**(1), 2 (2021)
2. Perronno, P., et al.: Innovative fast and low-cost method for the detection of living bacteria based on trajectory. Sci. Rep. **15**(1), 16535 (2025). https://doi.org/10.1038/s41598-025-95069-9
3. Jiang, P., Ergu, D., Liu, F., Cai, Y., Ma, B.: A review of Yolo algorithm developments. Procedia Comput. Sci. **199**, 1066–1073 (2022)
4. Tiwari, A., Nikolić, N., Anagnostidis, V., Gielen, F.: Label-free analysis of bacterial growth and lysis at the single-cell level using droplet microfluidics and object detection-oriented deep learning. Front. Lab-on-a-Chip Technol. **2**, 9 (2023). https://doi.org/10.3389/frlct.2023.1258155
5. Sapkota, R., Flores-Calero, M., Qureshi, R., et al.: Yolo advances to its genesis: a decadal and comprehensive review of the You Only Look Once (Yolo) series. Artif. Intell. Rev. **58**, 274 (2025)
6. Chen, G., Wang, W., Xu, Y., Deng, Z., Shi, J.: Learning efficient object detection models with knowledge distillation. In: NeurIPS Workshop (2017)
7. Bucilă, C., Caruana, R., Niculescu-Mizil, A.: Model compression. In: Proceedings of the ACM SIGKDD International Conference on Knowledge Discovery and Data Mining (KDD), pp. 535–541 (2006)
8. Hinton, G., Vinyals, O., Dean, J.: Distilling the knowledge in a neural network. In: Proceedings of the NIPS Deep Learning Workshop (2015)
9. Shu, C., Liu, Y., Gao, J., Yan, Z., Shen, C.: Channel-wise knowledge distillation for dense prediction. In: Proceedings of the IEEE/CVF International Conference on Computer Vision, pp. 5311–5320 (2021)
10. Long, J., Shelhamer, E., Darrell, T.: Fully convolutional networks for semantic segmentation. In: Proceedings of the IEEE Conference on Computer Vision and Pattern Recognition (CVPR), pp. 3431–3440 (2015)
11. Yang, Z., et al.: Focal and global knowledge distillation for detectors. In: Proceedings of the IEEE/CVF Conference on Computer Vision and Pattern Recognition (CVPR), pp. 4643–4652 (2022)
12. Yang, Z., Li, Z., Shao, M., Shi, D., Yuan, Z., Yuan, C.: Masked generative distillation. In: European Conference on Computer Vision, pp. 53–69. Springer, Heidelberg (2022)
13. Ryan, K., Ray, C.: Pseudomonas. In: Medical Microbiology. Bookshelf, National Center for Biotechnology Information (NCBI) (1995)
14. Klockgether, J., Tümmler, B.: Pseudomonas aeruginosa genomics and phenotyping analysis. Nat. Rev. Microbiol. **15**, 559–576 (2017)

15. Centers for Disease Control and Prevention (CDC). Multidrug-Resistant Pseudomonas Aeruginosa. CDC Healthcare-Associated Infections (2024)
16. Kollef, E., Kollef, M.: Ventilator-associated pneumonia due to Pseudomonas aeruginosa. Am. Rev. Respir. Dis. **153**(2), 587–593 (1996)
17. Pier, D.: Pseudomonas aeruginosa: pathogenesis and virulence factors. Clin. Infect. Dis. **45**, S26–S32 (2007)
18. Stringer, C., Wang, T., Michaelos, M., Pachitariu, M.: Cellpose: a generalist algorithm for cellular segmentation. Nat. Methods **18**, 100–106 (2021). https://doi.org/10.1038/s41592-020-01018-x
19. Maška, M., Ulman, V., Delgado-Rodriguez, P., et al.: The Cell Tracking Challenge: 10 years of objective benchmarking. Nat. Methods **20**, 1010–1020 (2023). https://doi.org/10.1038/s41592-023-01879-y
20. Jocher, G., Qiu, J.: Ultralytics Yolo," 2025, version 8.3.203. https://github.com/ultralytics/ultralytics. Accessed 30 Oct 2025
21. Chen, Y., Wang, S., Liu, J., Xu, X., de Hoog, F., Huang, Z.: Improved feature distillation via projector ensemble. In: Advances in Neural Information Processing Systems, pp. 12084–12095 (2022)
22. Mirzadeh, S.I., Farajtabar, M., Li, A., Levine, N., Matsukawa, A., Ghasemzadeh, H.: Improved knowledge distillation via teacher assistant. In: Proceedings of the AAAI Conference on Artificial Intelligence, vol. 34, no. 04, pp. 5191–5198 (2020). https://doi.org/10.1609/aaai.v34i04.5963
23. Hosang, J., Benenson, R., Schiele, B.: Learning non-maximum suppression. In: Proceedings of the IEEE Conference on Computer Vision and Pattern Recognition (CVPR), pp. 4507–4515 (2017)
24. Romero, A., Ballas, N., Kahou, S.E., Chassang, A., Gatta, C., Bengio, Y.: FitNets: Hints for thin deep nets. In: International Conference on Learning Representations (ICLR) (2015)
25. Zhang, X., Wan, F., Liu, C., Ji, R., Ye, Q.: FreeAnchor: learning to match anchors for visual object detection. In: Advances in Neural Information Processing Systems (NeurIPS) (2020)
26. Duan, K., Bai, S., Xie, L., Qi, H., Huang, Q., Tian, Q.: CenterNet: keypoint triplets for object detection. In: Proceedings of the IEEE/CVF Conference on Computer Vision and Pattern Recognition (CVPR) (2019)

Learning Molecular Structures from Infrared Spectra Through Latent Evidence Prediction

Sergio José Peresson(✉), Danilo Croce, and Roberto Basili

Department of Enterprise Engineering, University of Rome Tor Vergata, Via del Politecnico 1, 00133 Rome, Italy
sergiojose.peresson@gmail.com, {croce,basili}@info.uniroma2.it

Abstract. Infrared (IR) spectroscopy provides fast and non-destructive molecular analysis, but inferring full structures directly from spectra remains highly underdetermined, as distinct molecules often share similar vibrational patterns. Recent Transformer models show promise for IR-to-SMILES prediction, yet still face this intrinsic ambiguity. We address this issue by introducing *latent molecular evidence*: intermediate symbolic representations that encode mid-level structural cues. Auxiliary predictors infer molecular fingerprints and extended functional group descriptors from spectra and molecular formulas, and a SMILES generator conditions its decoding on these predicted signals. This two-stage multitask design encourages interpretable, data-driven reasoning between raw spectral features and molecular structure. Experiments show that integrating such complementary cues and training the generator under realistically noisy evidence, substantially improves accuracy, robustness, and chemical validity over a strong Transformer baseline.

Keywords: Infrared spectroscopy · Transformer · Molecular generation · Molecular descriptors · Representation learning · Multitask learning

1 Introduction

Spectroscopic techniques such as infrared (IR), Raman, and nuclear magnetic resonance (NMR) spectroscopy are fundamental tools for molecular identification and structural elucidation [11]. Among them, IR spectroscopy is particularly valued for its efficiency, low cost, and non-destructive nature [5]. Yet, interpreting IR spectra remains challenging: much of the structural information is encoded in broad and overlapping vibrational bands, which obscure molecular connectivity and hinder the extraction of substructural features beyond basic functional groups. Recent advances in deep learning, especially Transformer architectures [14], have enabled models that directly translate spectra into symbolic molecular representations such as Simplified Molecular-Input Line-Entry System (SMILES) [1]. While promising, these systems treat IR-to-structure prediction as a single-step sequence-generation problem. Without any explicit decomposition of the reasoning process, the model must directly infer a full molecular graph

M. Baratchi et al. (Eds.): IDA 2026, LNCS 16513, pp. 225–238, 2026.
https://doi.org/10.1007/978-3-032-23833-7_17

from a one-dimensional spectrum, an inherently demanding task, as many distinct molecules can produce remarkably similar IR signatures. This suggests that spectral data alone provide insufficient inductive bias to reliably guide structure reconstruction.

To mitigate this limitation, we introduce *latent molecular evidence*: intermediate symbolic representations that encode mid-level structural cues between spectra and molecular graphs. We augment the standard IR-to-SMILES pipeline with two complementary descriptors, molecular fingerprints (FPs) and extended functional group (EFG) labels, which are easier to infer from vibrational patterns while remaining highly predictive of the overall molecular structure. Importantly, these descriptors are automatically derived from SMILES and require no additional supervision. Because such representations are unavailable at inference time, we design a two-stage multitask framework.

First, *support predictors* learn to infer FP and EFGs from the IR spectrum and chemical formula. Then, a *SMILES generator* conditions its decoding on these predicted (and potentially noisy) cues. By training under controlled noise and partial evidence, the generator learns to integrate uncertain but chemically correlated information, improving robustness and helping to resolve ambiguities intrinsic to IR spectra. Experiments show that incorporating latent molecular evidence yields substantial improvements in accuracy, structural validity, and molecular similarity compared to a baseline Transformer trained solely on spectra and formulas. More broadly, the introduction of interpretable, data-driven intermediate reasoning bridges the gap between low-level spectroscopic patterns and high-level molecular representations.

The remainder of the paper reviews related work (Sect. 2), details the proposed framework (Sect. 3), presents experimental results (Sect. 4), and concludes with final remarks (Sect. 5).

2 Related Work

Learning From Spectroscopic Data. Spectroscopic analysis plays a central role across many areas of chemistry, from synthetic chemistry to analytical chemistry, where it is routinely used for tasks such as molecular identification. Infrared (IR) spectra, in particular, serve as valuable molecular fingerprints. However, translating these spectra into complete molecular structures remains challenging due to peak overlap and the inherently many-to-one mapping between spectra and molecular graphs.

Transformers in Chemistry. Transformer architectures [14] have become a dominant paradigm in computational chemistry, supporting reaction prediction [12], molecular property estimation [8], and *de novo* generation from SMILES or Self-Referencing Embedded Strings (SELFIES) [4]. Sequence models trained on large reaction datasets [3] capture complex structure-property relationships and have demonstrated strong performance in symbolic chemical translation tasks, making them natural candidates for spectrum-to-structure generation.

IR-to-Structure Prediction. The first end-to-end IR-to-SMILES model was introduced by Alberts *et al.* [1], who trained a Transformer on the large synthetic IR dataset they released, mapping IR spectra and chemical formulas to molecular structures. On this dataset, their baseline achieved a Top-1 accuracy of 26%, which increased to 45% after applying data augmentation, spectral-region selection, and model ensembling. The study also showed that pretraining on simulated IR spectra provides an effective initialization: after fine-tuning on experimental IR data, such as those from the NIST gas-phase library[1], the model attains meaningful structure-prediction performance, confirming simulation as a scalable training signal. More recent work has proposed architectural advances such as patch-based self-attention to improve spectral encoding by partitioning high-resolution spectra into local patches, thereby reducing the effective token length and attention cost while preserving fine-grained peak information, particularly in dense fingerprint regions [15]. Related studies have extended this paradigm to NMR and multimodal spectroscopy [2,13]. Despite these advances, the core challenge remains: the model must infer a full molecular graph directly from a one-dimensional spectrum, without intermediate reasoning steps.

Dataset Generation. To enable pretraining at scale, Alberts *et al.* generated a large and valuable synthetic dataset by sampling molecules from PubChem and simulating their IR spectra via molecular dynamics with the class II Polymer Consistent Force Field (PCFF) [1]. However, this simulation strategy also imposes clear limitations. Force-field spectra are only as accurate as the underlying parameterization and may deviate from real measurements due to missing electronic structure effects, imperfect treatment of anharmonicities across diverse chemistries, and the exclusion of entire regions of chemical space (e.g., charged species, stereochemistry, elements beyond the chosen set, larger molecules). Most importantly, computed spectra do not reflect instrument- and acquisition-specific effects, creating a domain gap that explains why zero-shot transfer from simulation to experiment performs poorly and why fine-tuning on experimental spectra is necessary [1].

Molecular Fingerprints and Functional Group Descriptors. Molecular descriptors such as extended connectivity fingerprints (ECFPs) [10] and functional group taxonomies [6,9] are widely used in QSAR/QSPR, similarity search, and virtual screening. These mid-level representations summarize local chemical environments and functional motifs, and are easier to learn than full molecular graphs. Prior work, however, has used them primarily as input instances or labels, but not as latent intermediate evidence to support spectroscopic structure prediction.

Positioning of this Work. Our approach differs from previous IR-to-structure systems by introducing *latent, learnable* intermediate representations for FPs and EFGs. To our knowledge, this is the first framework that predicts such descriptors directly from spectra and conditions a molecular generator on their

[1] https://www.nist.gov/srd/nist-standard-reference-database-35.

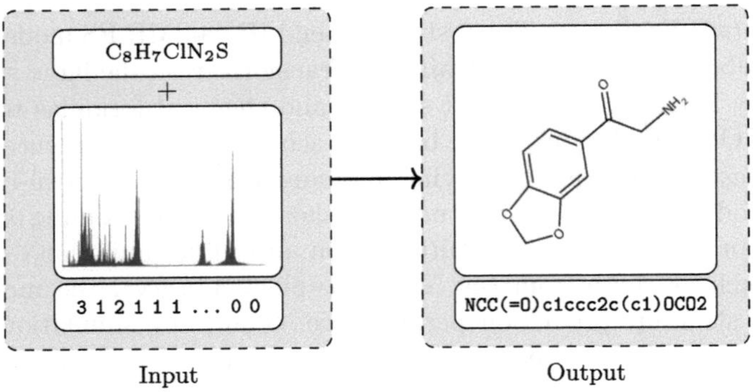

Fig. 1. Baseline IR-to-SMILES translation: the model takes as input a chemical formula along with a quantized IR spectral sequence and outputs the corresponding molecular structure in SMILES format.

(possibly noisy) predictions, effectively introducing an interpretable reasoning layer between the spectrum and the final SMILES output.

3 Learning with Latent Evidence

IR spectra convey rich information about the presence of specific functional groups, but exhibit only limited sensitivity to the overall molecular topology. Consequently, molecules with distinct connectivities or substitution patterns may exhibit highly similar IR signatures, making direct spectrum-to-SMILES prediction a fundamentally underdetermined problem. The baseline approach proposed by Alberts *et al.* [1] tackles this task using a single encoder-decoder Transformer that receives the chemical formula together with the IR spectrum and directly generates a SMILES string. As illustrated in Fig. 1, the model takes two serialized inputs: (i) the chemical formula, expressed as a character string, and (ii) the IR spectrum, represented as a sequence of tokens encoding the intensity at each wavenumber. From this minimal, purely sequential representation, lacking any explicit structural cues, the model is expected to reconstruct the full molecular graph. This requires it to implicitly infer functional groups, assemble local fragments, and resolve global connectivity without any intermediate reasoning step, placing a substantial burden on the underlying Transformer. Consequently, model performance generally declines with increasing molecular size and structural diversity, as overlapping vibrational bands and non-specific spectral motifs shared across multiple molecules make the IR signal less discriminative for identifying a unique structure. To ease this difficulty, we enrich the prediction process with an explicit intermediate reasoning layer. Instead of forcing the model to jump directly from the IR signal to the full molecular structure, we introduce *latent molecular evidence*: symbolic descriptors such as FP fragments and EFG labels that capture mid-level structural hints. These representations are easier to infer from the spectra, yet remain strongly informative

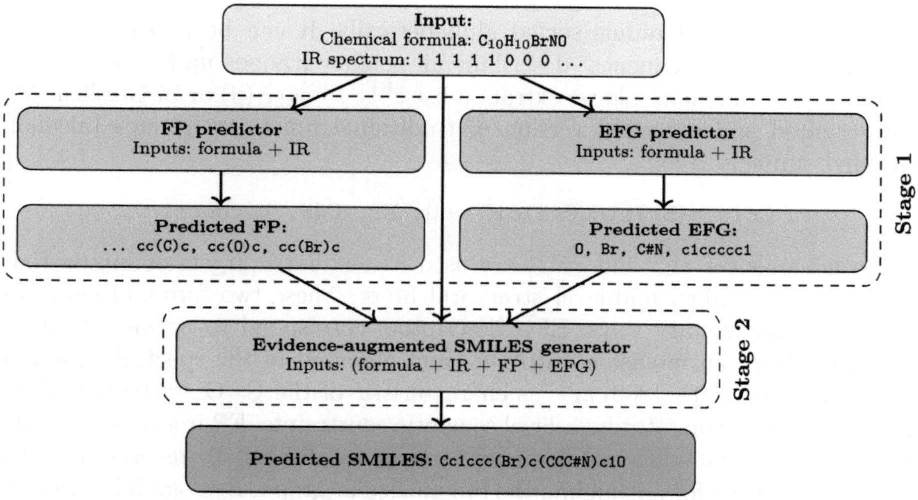

Fig. 2. Latent-evidence framework: the IR spectrum and chemical formula are fed into two support predictors that infer molecular Fingerprints (FPs) and Extended Functional Group (EFG) descriptors. The final generator receives the original inputs and the predicted evidence to produce the SMILES structure.

about the molecule's topology. Figure 2 shows the resulting two-stage framework. In Stage 1, two support predictors estimate FPs and EFGs from the input spectrum and formula. In Stage 2, an evidence-augmented SMILES generator conditions its decoding process on these predicted cues, allowing it to integrate uncertain but chemically meaningful signals when reconstructing the molecule. Each training example describes a molecule and its infrared (IR) spectrum. A sample is $(x_i, y_i) = ((s_i, c_i), y_i)$, where $s_i \in \mathbb{R}^L$ is the IR spectrum (intensity sequence at fixed wavenumbers), c_i is the chemical formula (e.g., $\mathtt{C_7H_6O_2}$), and y_i is the molecular structure as a SMILES token sequence $y_i = (y_{i,1}, \ldots, y_{i,T})$.

The baseline model, following Alberts *et al.* [1], is a standard Transformer encoder-decoder that predicts the SMILES sequence $f_\phi : (s, c) \mapsto p_\phi(y \mid s, c)$ and is trained with the usual token-level cross-entropy objective for autoregressive generation. In Stage 1, to provide intermediate, chemically meaningful cues, we derive from each reference molecule two additional symbolic representations: a FP sequence and a list of EFG descriptors. Both are deterministic functions of the target SMILES and summarize structural information that is easier to infer from spectra than the full molecular graph. For a molecule y with molecular graph $G = (V, E)$, each atom $v \in V$ defines a local environment $\mathcal{N}_r(v) = \{u \in V : \text{dist}(u, v) \leq r\}$ for radius r. The subgraph induced by $\mathcal{N}_r(v)$ is encoded into a canonical token $\tau(v, r)$ (e.g., via SMILES subfragments). In ethanol (`CCO`): $r{=}0 \rightarrow$ [`C`,`O`], $r{=}1 \rightarrow$ [`CC`,`CO`,`CCO`]. Concatenating all fragments up to $r_{\max}$ yields the FP sequence:

$$z^{\text{FP}} = g_{\text{FP}}(y) = [\tau(v_1, 0), \ldots, \tau(v_{|V|}, r_{\max})],$$

with tokens at equal radius sorted alphabetically. It can be interpreted as a "molecular subword sequence" describing all neighbourhoods up to radius $r_{\max}$.

From the same molecule we extract a list $z^{\mathrm{EFG}} = g_{\mathrm{EFG}}(y)$ of EFG tokens [9]: a generalized and extended version of traditional functional groups (alcohol, carbonyl, amine, ...) [6]:

$$\texttt{Cc1ccc(Br)c(CCC\#N)c1O} \rightarrow \texttt{O, Br, C\#N, c1ccccc1}.$$

Both sequences z^{FP} and z^{EFG} are used as auxiliary targets or inputs that provide interpretable, mid-level structural hints. These two forms of evidence help in complementary ways. EFG descriptors correspond to well-localized IR signatures (e.g., an intense absorption band observed in the spectral region of approximately 1650 - 1800 cm^{-1} is characteristic of the C=O stretching vibration), giving the generator high-level semantic constraints. FP fragments instead encode short-range connectivity patterns that are difficult to recover from IR alone but essential for assembling the full molecular graph. By combining coarse, spectrally anchored cues with fine-grained structural hints, the model operates in a substantially reduced space of plausible structures, making the final SMILES generation far more stable and accurate.

Since $(z^{\mathrm{FP}}, z^{\mathrm{EFG}})$ are available only when the original (gold-standard) SMILES is known, we introduce two *support predictors* that learn to estimate these representations directly from the observable inputs: $a_{\theta_{\mathrm{FP}}} : (s, c) \rightarrow \hat{z}^{\mathrm{FP}}$, $a_{\theta_{\mathrm{EFG}}} : (s, c) \rightarrow \hat{z}^{\mathrm{EFG}}$. Each predictor is a Transformer encoder-decoder trained on gold pairs $((s, c), z)$, where the targets are derived from y. After training, we use them to create *augmented* examples

$$(s, c, \hat{z}^{\mathrm{FP}}, \hat{z}^{\mathrm{EFG}}, y), \quad \hat{z}^{\mathrm{FP}} = a_{\theta_{\mathrm{FP}}}(s, c), \ \hat{z}^{\mathrm{EFG}} = a_{\theta_{\mathrm{EFG}}}(s, c),$$

which replicate the realistic test-time condition in which the auxiliary information must be predicted rather than provided. To obtain realistic pseudo-labels for Stage 2, we must avoid a trivial regime in which the support predictors perfectly memorize the training set. If Stage 1 is trained until convergence on the training split, its predictions on those molecules become almost indistinguishable from the gold descriptors, while remaining noticeably noisier on unseen spectra. Using such "almost-gold" evidence only on the generation training set would therefore create a mismatch between training conditions and inference-time behaviour. To mitigate this distribution shift, two strategies were adopted.

Mixed Evidence. We select the Stage 1 checkpoint according to its generalization behaviour. During training, we monitor the model's performance on both the training and validation splits. We first identify the checkpoint that achieves the best validation performance, which represents the desired noise level at inference time. We then examine earlier checkpoints and select one whose performance on the training split best matches the validation performance of the best checkpoint. This corresponds to the point where training and validation improve at a similar relative rate, no overfitting has started and no clear train-validation gap emerges. This "matched-performance" checkpoint is then frozen and applied to all training examples to generate their FP and EFG sequences, obtaining the dataset $\mathcal{D}_{\mathrm{pred}}$.

Best Alternatives. To make the generator robust to this mismatch, we form augmented training triples with predicted evidence and two "best alternative" views, corresponding to the two highest-ranked non-gold predictions produced by the support models using beam search. These samples expose the model to the same type of imperfections expected at inference time, improving generalization to predicted evidence, obtaining the dataset $\mathcal{D}_{\text{alt}}$.

Subsequently, the final generator f_ϕ produces the molecular structure conditioned on the spectrum, the formula, and the auxiliary symbolic evidence:

$$f_\phi : (s, c, z^\star) \longmapsto p_\phi(y \mid s, c, z^\star), \qquad z^\star = [\, z^{\text{FP}} \| z^{\text{EFG}} \,].$$

All modalities are serialized as token sequences and separated by sentinel markers: $s \mid c \mid z^{\text{FP}} \mid z^{\text{EFG}}$. For example:

```
C 10H 10 Br N O  |  1 1 1 ... |  ...cc(O)c, cc(Br)c  |  O, Br, C#N, ...
```

The generator is trained as a standard autoregressive Transformer, maximising the likelihood of the target SMILES under this augmented input. When $z^\star$ is obtained from the gold SMILES y, the generator receives perfect auxiliary information. This setting is not feasible at inference, since the true structure is precisely what we aim to predict, but it provides an upper bound on the performance the model could achieve if the evidence were always correct. Conversely, when $z^\star$ comes from the predictors $\hat{z}$, the generator must operate with the same imperfect, noisy evidence available at test time, thus reflecting the realistic inference scenario.

We compare four settings:

1. **Clean evidence:** gold $(z^{\text{FP}}, z^{\text{EFG}})$ derived from y, representing an upper bound;
2. **Predicted evidence:** $(\hat{z}^{\text{FP}}, \hat{z}^{\text{EFG}})$ obtained from the support predictors;
3. **Mixed training:** uses the union $\mathcal{D}_{\text{mix-pred}} = \mathcal{D}_{\text{gold}} \cup \mathcal{D}_{\text{pred}}$, so the generator learns from both perfect and noisy examples.
4. **Best alternatives:** uses the union $\mathcal{D}_{\text{mix-alt}} = \mathcal{D}_{\text{gold}} \cup \mathcal{D}_{\text{alt}}$, so the generator learns from both perfect and highest-ranked non-gold predictions.

As discussed previously, training proceeds in two stages:

- *Stage 1 (support):* train $a_{\theta_{\text{FP}}}$ and $a_{\theta_{\text{EFG}}}$ on gold data and generate cross-fitted predictions $\hat{z}$ for all training molecules;
- *Stage 2 (generator):* train f_ϕ using one of the regimes above on tuples $(s, c, z^\star, y)$ with the same autoregressive cross-entropy objective used for the baseline.

At inference time, given a new (s, c), the system first predicts $\hat{z}^{\text{FP}} = a_{\theta_{\text{FP}}}(s, c)$ and $\hat{z}^{\text{EFG}} = a_{\theta_{\text{EFG}}}(s, c)$, then generates the SMILES string $\hat{y} \sim f_\phi(y \mid s, c, \hat{z})$ via beam search.

4 Experimental Evaluation

We now assess whether latent molecular evidence effectively improves IR-to-SMILES prediction in practice. Our evaluation considers four key aspects: (i) the potential gain achievable under perfect auxiliary evidence (*upper bound*); (ii) the accuracy with which FPs and EFGs can be predicted from IR spectra (*quality of support predictors*); (iii) whether training with predicted or highest-ranked non-gold predictions improves performance under realistic conditions compared to the baseline Transformer; and (iv) the molecular regimes (in terms of size and structural complexity) for which latent evidence is most beneficial.

Experimental Setup. We evaluate the proposed approach on the dataset[2] generated by Alberts *et al.* as in [1]. Each example provides an IR spectrum s, the chemical formula c, and the corresponding molecular structure y as SMILES. The dataset contains approximately 457,000 training samples, 50,800 validation samples, and 127,000 test samples. All models are based on a Transformer encoder-decoder architecture implemented with OpenNMT [7]. We use 4 encoder layers and 4 decoder layers with hidden size 512, and 8 attention heads. Support predictors share the same architecture but have distinct vocabularies (FP vs. EFGs). Models are trained with Adam optimizer ($\beta_1 = 0.9$, $\beta_2 = 0.998$), learning rate 0.0001, batch size 8, and early stopping on validation loss. Spectra are normalized to integer values in the range from 0 to 99.

We evaluate all models in terms of both exact recovery and chemical plausibility. At inference time, the Transformer decoder generates a ranked list of candidate SMILES by beam search; Top-k accuracy is the fraction of test molecules whose gold SMILES appears among the k most probable sequences. Top-1 is the strictest case (the correct structure must be the single best prediction), while Top-5 and Top-10 reflect a more realistic scenario in which a chemist can inspect a small set of high-scoring candidates. To quantify how close a prediction is to the target even when it is not an exact match, we report the mean Tanimoto similarity (T.S.) between reference and predicted molecules, computed on ECFP4, with values ranging from 0 (no shared fragments) to 1 (identical FP). Finally, Invalid (%) denotes the fraction of generated SMILES that RDKit[3] cannot parse into a chemically valid molecule; lower values indicate more reliable generative behavior.

Results. Table 1 reports oracle results in which the generator receives gold auxiliary evidence derived from the target SMILES, i.e., perfect EFG labels, FP, or both. This setting is not attainable at test time, but quantifies the maximum gain that could be obtained if the support predictors were error-free. The *Baseline* row corresponds to applying the public implementation of Alberts *et al.* [1] (formula + IR Transformer) to our setting, without any auxiliary inputs[4].

---

[2] https://zenodo.org/records/7928396.

[3] https://www.rdkit.org/.

[4] Alberts *et al.* [1] report a Top-1 accuracy of 26% under this same setup. When rerunning their configuration, we reproduced the same score; however, switching from

Table 1. Oracle performance when the generator receives *gold* auxiliary evidence.

Auxiliary evidence (oracle)	Top-1	Top-5	Top-10	Inv (%)	T.S.
Baseline (no evidence)	30.54	54.05	60.27	2.69	0.554
Gold auxiliary descriptors					
Gold EFG	37.62	62.80	68.87	1.46	0.611
Gold FP	58.15	82.00	85.89	0.82	0.774
Gold FP + Gold EFG	59.86	81.06	83.64	1.00	0.784

Gold auxiliary evidence yields large gains over the baseline: Top-1 accuracy nearly doubles when perfect FP information is available, and further improves when combined with EFG labels, while the fraction of invalid SMILES substantially decreases and the Tanimoto similarity increases. These oracle results show that the generator can effectively exploit structured chemical cues, and that there is considerable headroom that could in principle be recovered by improving the quality of the predicted EFG and FP representations.

Table 2. Performance of support predictors on the test set.

Model	Top-1(%)	Top-5(%)	Top-10(%)
FP predictor	39.32	62.39	68.95
EFG predictor	75.40	86.47	88.62

Table 2 evaluates the two support predictors that infer FPs and EFGs directly from the IR spectrum and the chemical formula. Both models extract meaningful structural cues from the spectral signal. The EFG predictor achieves high accuracy (Top-1 75.40%), indicating that such representations are strongly anchored to characteristic IR bands. FPs, being more granular and less directly tied to isolated vibrational features, are harder to recover, yet the FP predictor still reaches 39.32% Top-1 and over 68% Top-10. Overall, both predictors provide informative and chemically relevant evidence for the generator, confirming that IR spectra contain sufficient signal to support mid-level structural inference.

In realistic usage, the generator must rely on *predicted* auxiliary evidence rather than gold descriptors. To reflect this setting, we examine three training regimes, summarised in Table 3, while ensuring that the validation and test sets always use only the outputs of the support predictors. The first regime (A) trains the generator exclusively on gold evidence and then evaluates it with predicted evidence. This configuration isolates the effect of noise at test time: since the model has only ever seen perfect descriptors, even moderate prediction errors

a dynamic to a fixed batch size (while keeping all other hyperparameters unchanged) yielded a consistent improvement to 30%.

Table 3. Performance under realistic conditions using predicted or mixed auxiliary evidence. Each block corresponds to one of the training regimes discussed in the text: (A) training on gold evidence but testing on predicted; (B) mixed gold-predicted training; (C) gold plus "best alternative" predictions. Validation and test always use predicted evidence.

Training regime	Top-1	Top-5	Top-10	Inv (%)	T.S.
Baseline - no auxiliary evidence					
Formula/IR only	30.54	54.05	60.27	2.69	0.554
(A) Train on gold evidence, test on predicted					
+ EFG	30.76	52.29	57.65	1.98	0.549
+ FP	26.76	40.23	42.79	3.41	0.534
+ FP + EFG	26.97	39.13	40.79	4.08	0.537
(B) Mixed training: 50% gold + 50% predicted					
+ EFG	32.76	55.45	60.99	1.99	0.567
+ FP	34.97	55.32	59.30	2.16	0.586
+ FP + EFG	35.01	55.56	59.90	1.82	0.586
(C) Gold + "best alternatives" predicted evidence					
+ FP	36.63	56.99	60.62	1.88	0.596
+ FP + EFG	38.61	59.25	63.01	1.86	0.613

lead to substantial degradation (e.g., Top-1 dropping to 26.76% with FP-only). This confirms that a model trained purely on ideal inputs becomes brittle and fails to generalise to realistic, noisy evidence. The second regime (B) mixes, for each molecule, one gold evidence view with its corresponding predicted one, producing a balanced 50/50 blend of clean and noisy examples. This exposes the generator to the characteristic error patterns of the support predictors while still anchoring learning on correct structural cues. Even this simple adjustment yields clear improvements over the baseline Transformer (30.54% Top-1). Fingerprints prove especially beneficial: predicted FPs raise accuracy to 34.97%, increase mean Tanimoto similarity to 0.586, and reduce invalid SMILES. Predicted EFGs also help, though their coarser granularity results in more modest gains. Finally, the third regime (C) further enhances robustness by supplementing gold descriptors with a small number of "best alternative" views, derived from the highest-ranked non-gold predictions produced by the support models. These alternatives are chemically plausible but imperfect, and including them in a controlled 3:2 ratio (gold: alternatives) acts as a regulariser: the generator learns to integrate uncertain evidence rather than overfitting to exact gold fragments. This strategy yields the best overall results, reaching Top-1 38.61%, mean Tanimoto 0.613, and the second lowest invalid-SMILES rate, with FPs again contributing the largest gains. Overall, these experiments show that, even when noisy, latent molecular evidence provides valuable structural constraints. FPs are particularly impactful: although harder to predict, their distributed, set-like nature makes them surprisingly resilient to partial errors, allowing the generator to benefit from them even when they are imperfect. The resulting system is not only more accurate but also markedly more robust and tolerant to the variability inherent in real IR spectra.

Impact of Fingerprint Quality on Generation Performance. We now analyze why, beyond the aggregate results of Table 3, FP evidence yields the largest performance gains, and under which conditions it can instead be detrimental. Unlike EFG labels, a FP is a set-like sequence of local fragments, and its quality can be directly quantified: given a predicted fingerprint $\hat{z}^{\mathrm{FP}}$ and the corresponding gold fingerprint z^{FP}, we measure their agreement through the Tanimoto similarity between the associated binary Morgan Fingerprints. This provides a scalar estimate of how much structural information the predictor has recovered for a given molecule, ranging from completely uninformative (near 0) to essentially perfect (near 1).

Table 4. Effect of fingerprint accuracy on generation quality. For each similarity range between the predicted and gold Fingerprints we report how many molecules improve or degrade w.r.t. the baseline, together with their percentages and net difference.

	Fingerprint-target Tanimoto similarity									
	0.1	0.2	0.3	0.4	0.5	0.6	0.7	0.8	0.9	1.0
Improved	8	20	27	53	119	499	1 253	2 277	3 559	12 682
Degraded	12	72	119	187	338	711	1 255	1 966	2 320	5 785
Impr. (%)	0.02	0.06	0.08	0.16	0.36	1.50	3.77	6.85	10.70	38.13
Degr. (%)	0.04	0.22	0.36	0.56	1.02	2.14	3.77	5.91	6.97	17.39
Diff.(%)	−0.02	−0.16	−0.28	−0.40	−0.66	−0.64	0.00	0.94	3.73	20.74

Table 4 groups all test molecules according to this predicted-vs.-gold Tanimoto score and reports, for each similarity bin, how often adding fingerprint evidence improves or degrades performance relative to the baseline Transformer. A clear pattern emerges. When the predicted FPs are reasonably accurate (Tanimoto ≥ 0.7), the generator benefits substantially: at similarity ≥ 0.9, improvements outnumber degradations by more than 20% points. In this regime, FPs provide reliable topological anchors, short-range connectivity fragments that constrain the decoder and help it recover the correct structure. Conversely, when the predicted FPs are noisy (Tanimoto ≤ 0.5), the auxiliary evidence tends to mislead the generator, producing more degradations than improvements. This confirms that FPs, while highly informative, are also sensitive to low-quality predictions: incorrect local fragments can push the decoder toward incompatible connectivity patterns. Taken together, these results reconcile the behaviour observed in Table 3: FPs yield larger gains than EFGs because they encode richer and more discriminative structural information, but their effectiveness depends critically on predictor quality. When fingerprint similarity is high, the generator

Table 5. Accuracy of the baseline and enhanced models as a function of molecular size (number of heavy atoms), together with absolute and relative gains.

	Number of heavy atoms							
	6	7	8	9	10	11	12	13
Baseline	59.90	54.63	46.51	40.47	36.04	33.38	29.73	22.84
Best alternatives	68.02	62.83	56.03	50.14	44.43	41.81	37.45	30.41
Absolute impr. (%)	8.12	8.20	9.52	9.67	8.39	8.43	7.72	7.57
Relative impr. (%)	13.56	15.01	20.47	23.89	23.28	25.25	25.97	33.14

receives strong mid-level constraints; when it is low, the same mechanism can become counterproductive.

Moreover, we study performance as a function of molecular size (Table 5). Both baseline and the model enhanced using both perfect and highest-ranked non-gold predictions lose accuracy as the number of heavy atoms increases, reflecting the combinatorial growth of the structural search space. Yet the relative improvement introduced by latent evidence increases with molecular size, reaching over 30% for molecules with 13 heavy atoms. This indicates that fingerprint evidence is particularly valuable for complex molecules: local substructure hints help the generator navigate a much larger and more ambiguous space of possible graphs.

5 Conclusion

We introduced a two-stage framework for molecular structure prediction from infrared (IR) spectra that incorporates *latent molecular evidence* in the form of Fingerprints (FPs) and Extended Functional Group descriptors (EFGs). By predicting these mid-level representations from the spectrum and conditioning a SMILES generator on them, the model gains access to interpretable and chemically meaningful cues that substantially ease the spectrum-to-structure translation task.

Experimental results show that latent evidence significantly improves accuracy, molecular similarity, and structural validity over a strong Transformer baseline. Even when noisy, predicted FP and EFG descriptors act as valuable constraints, reducing the search space of plausible molecular graphs and yielding more reliable generation. Training the generator under realistic noise conditions further improves robustness, leading to consistent gains across top-k metrics and lower invalid-SMILES rates. Benefits are especially pronounced for larger and more complex molecules, where the combinatorial ambiguity is greatest. Beyond performance, the framework enhances interpretability by exposing intermediate descriptors that can be inspected and analyzed, offering insight into how spectral cues influence structure reconstruction. The approach is general and extensible, and could integrate additional spectroscopic modalities (e.g., Raman, NMR, MS) or improved noise models. Overall, our findings show that incorporating chemically grounded intermediate reasoning provides a principled pathway toward more accurate, robust, and transparent spectrum-to-structure prediction.

Disclosure of Interests. The authors have no conflicts of interest to declare that are relevant to the content of this article.

References

1. Alberts, M., Laino, T., Vaucher, A.C.: Leveraging infrared spectroscopy for automated structure elucidation. Commun. Chem. **7**(1), 268 (2024). https://doi.org/10.1038/s42004-024-01341-w
2. Alberts, M., Zipoli, F., Laino, T.: Setting new benchmarks in AI-driven infrared structure elucidation. Digital Discov. **4**(7), 1936–1943 (2025). https://doi.org/10.1039/D5DD00131E
3. Andronov, M., Voinarovska, V., Andronova, N., Wand, M., Clevert, D.A., Schmidhuber, J.: Reagent prediction with a molecular transformer improves reaction data quality. Chem. Sci. **14**(12), 3235–3246 (2023). https://doi.org/10.1039/D2SC06798F
4. Bagal, V., Aggarwal, R., Vinod, P.K., Priyakumar, U.D.: MolGPT: molecular generation using a transformer-decoder model. J. Chem. Inf. Model. **62**(9), 2064–2076 (2022). https://doi.org/10.1021/acs.jcim.1c00600
5. Dutta, A.: Chapter 4 - fourier transform infrared spectroscopy. In: Thomas, S., Thomas, R., Zachariah, A.K., Mishra, R.K. (eds.) Spectroscopic Methods for Nanomaterials Characterization, pp. 73–93. Micro and Nano Technologies, Elsevier (2017). https://doi.org/10.1016/B978-0-323-46140-5.00004-2
6. Ertl, P.: An algorithm to identify functional groups in organic molecules. J. Cheminform. **9**(1), 36 (2017). https://doi.org/10.1186/s13321-017-0225-z
7. Klein, G., Kim, Y., Deng, Y., Senellart, J., Rush, A.M.: OpenNMT: open-source toolkit for neural machine translation (2017). https://doi.org/10.48550/arXiv.1701.02810
8. Liu, K., Cui, H., Yu, X., Li, W., Han, W.: Predicting cardiotoxicity in drug development: a deep learning approach. J. Pharm. Anal. **15**(8), 101263 (2025). https://doi.org/10.1016/j.jpha.2025.101263
9. Lu, J., Xia, S., Lu, J., Zhang, Y.: Dataset construction to explore chemical space with 3D geometry and deep learning. J. Chem. Inf. Model. **61**(3), 1095–1104 (2021). https://doi.org/10.1021/acs.jcim.1c00007
10. Rogers, D., Hahn, M.: Extended-connectivity fingerprints. J. Chem. Inf. Model. **50**(5), 742–754 (2010). https://doi.org/10.1021/ci100050t
11. Schrader, B.: Vibrational spectroscopy of different classes and states of compounds. In: Infrared and Raman Spectroscopy, pp. 189–410. John Wiley & Sons, Ltd (1995). https://doi.org/10.1002/9783527615438.ch04
12. Schwaller, P., et al.: Molecular transformer: a model for uncertainty-calibrated chemical reaction prediction. ACS Cent. Sci. **5**(9), 1572–1583 (2019). https://doi.org/10.1021/acscentsci.9b00576
13. Su, Y., Chen, J., Jiang, Z., Zhong, Z., Wang, L., Liu, Q.: Language models can understand spectra: a multimodal model for molecular structure elucidation (2025). https://doi.org/10.48550/arXiv.2508.08441

14. Vaswani, A., et al.: Attention is all you need. In: Advances in Neural Information Processing Systems 30: Annual Conference on Neural Information Processing Systems 2017, December 4-9, 2017, Long Beach, CA, USA. pp. 5998–6008 (2017)
15. Wu, W., Leonardis, A., Jiao, J., Jiang, J., Chen, L.: Transformer-based models for predicting molecular structures from infrared spectra using patch-based self-attention. J. Phys. Chem. A **129**(8), 2077–2085 (2025). https://doi.org/10.1021/acs.jpca.4c05665

E-PAGEC: A Differentiable Joint Attributed-Graph Embedding and Clustering Model

Imane Akdim[1(✉)], Loubna Mekouar[1], Youssef Iraqi[1], and Mohamed Nadif[2]

[1] College of Computing, University Mohammed VI Polytechnic, Ben Guerir, Morocco
{imane.akdim,loubna.mekouar,youssef.iraqi}@um6p.ma
[2] Centre Borelli UMR9010, Université Paris Cité, Paris, France
mohamed.nadif@u-paris.fr

Abstract. Attributed graphs are essential for modeling complex relational data across numerous domains. Existing clustering methods typically address embedding and clustering separately, while joint approaches often rely on fixed, non-learnable update steps. Both strategies limit the adaptability and integration of these methods into broader learning pipelines. In this work, we introduce E-PAGEC, Enhanced Power Attributed Graph Embedding and Clustering, a differentiable variant of PAGEC that jointly refines node embeddings and cluster assignments through gradient-based optimization. In contrast to prior non-adaptive designs, E-PAGEC incorporates topology, similarity of attributes, and iterative updates driven by gradients, enabling the model to dynamically adjust both embedding space and cluster boundaries during training. Experiments on real-world datasets, against state-of-the-art graph clustering baselines, demonstrate that E-PAGEC achieves superior clustering performance.

Keywords: Clustering · Embeddings · Attributed Networks

1 Introduction

Attributed graphs, where each node is associated with a feature vector, provide a natural and powerful representation for many real-world systems such as social networks [16], recommender systems [28], and biological networks [11]. The dual availability of structural and attribute information makes attributed graphs especially suitable for downstream tasks such as node classification [17], link prediction [15], anomaly detection [39], and community discovery [30].

Two core analytic objectives on attributed graphs are learning compact informative node embeddings that capture both topology and attribute similarity, and partitioning nodes into coherent clusters that reflect meaningful communities or functional groups. In practice, these objectives are deeply related: improved embeddings can lead to better clustering, and cluster-aware representations can

M. Baratchi et al. (Eds.): IDA 2026, LNCS 16513, pp. 239–252, 2026.
https://doi.org/10.1007/978-3-032-23833-7_18

enhance downstream prediction tasks [35]. Consequently, designing methods that exploit the interplay between embedding and clustering has attracted considerable attention.

Many methods treat embedding and clustering as separate stages. A typical workflow computes node representations using dimensionality reduction or graph embedding techniques and subsequently applies a clustering algorithm on the learned embeddings. This modular approach is simple and often effective, but it may miss opportunities for mutual refinement between the two tasks.

To capture this mutuality, a line of joint embedding-clustering methods has been proposed [1,14,26]. Such strategies can produce representations that are inherently more cluster-aware, improving both the quality of embeddings and the coherence of clusters. However, many joint methods ignore the graph's topology, or implement clustering updates using closed-form steps that are not embedded in a differentiable learning graph.

In this paper, we present E-PAGEC, Enhanced Power Attributed Graph Embedding and Clustering, an enhanced, differentiable variant of PAGEC [14] that embeds its closed-form updates into the computation graph, enabling end-to-end refinement of node embeddings and cluster assignments via gradients. PAGEC is a state-of-the-art algorithm in attributed graph clustering that jointly derives embeddings and cluster assignments via a powered proximity matrix. However, PAGEC is fundamentally unsupervised and non-trainable. E-PAGEC retains the strengths of the coupling between embedding and clustering while making the framework differentiable. This gradient-driven formulation not only preserves the interpretability of the original PAGEC updates but also provides the flexibility to adapt to diverse graph structures and feature distributions. Clusters and embeddings become part of a computational graph that can be optimized end-to-end via gradient descent.

Thus, we summarize our main contributions below:

- We introduce E-PAGEC, a novel differentiable reformulation of PAGEC that embeds the algorithm's closed-form refinement steps into the computation graph and enables joint optimization of node embeddings and cluster assignments via gradient-based learning.
- We evaluate E-PAGEC on multiple benchmark datasets against a broad set of competitive baselines. We report clustering metrics and embedding quality measures to provide a balanced assessment of clustering accuracy and representation learning quality.

The remainder of this paper is organized as follows: Sect. 2 reviews related work. Section 3 presents our approach. Section 4 presents our experimental results, and Sect. 5 concludes the work.

2 Related Work

Clustering seeks to partition unlabeled data into groups of similar points. Classical approaches include k-means [9], which iteratively assigns points to the nearest of k centroids and updates those centroids to minimize within-cluster variance, and spectral clustering [31], which computes the leading eigenvectors of its Laplacian to form a low-dimensional embedding, and then applies a standard clustering method. More recently, the rise of deep learning revealed deep clustering frameworks that jointly learn feature representations and cluster assignments. Deep Embedding Clustering (DEC) [35] refines autoencoder latent spaces with a clustering-oriented KL-divergence loss, while Improved DEC [8] incorporates a reconstruction term to preserve local structure. DeepCluster [4] and ClusterGAN [20] leverage self-supervision and adversarial objectives to discover cluster-friendly representations without labels. On attributed graphs, clustering reconciles node features and graph topology. One line of work employs graph autoencoders; variational graph auto-encoders [13] and marginalized graph autoencoders [33], to learn joint embeddings, then applies k-means or Gaussian mixture models (GMMs) in latent space. In [14], a unified approach, PAGEC, that simultaneously learns node embeddings and clusters via a novel powered proximity matrix is presented. This approach blends structural and content affinities to yield both latent representations and cluster assignments.

In this work, we present an enhanced variant of PAGEC and align it to downstream learning objectives.

3 Proposed Method

3.1 Preliminaries

Let $G = (\mathcal{V}, A, X)$ be an attributed, undirected graph with vertex set $\mathcal{V} = \{v_1, ..., v_N\}$, where $N = |\mathcal{V}|$. The adjacency matrix $A \in \mathbb{R}^{N \times N}$ is symmetric and a_{ij} denotes the edge between v_i and v_j. The attribute matrix is $X = [x_1, ..., x_N] \in \mathbb{R}^{N \times d}$, where $x_i \in \mathbb{R}^d$ is the d-dimensional feature vector of node v_i. Tr(.) denotes the trace of a matrix. K is the number of clusters and h is the embedding dimension. I_N is the $N \times N$ identity matrix.

To ensure that each node is connected to itself, we add self-loops to form $A_{self} = A + I_N$.

The corresponding degree matrix $D \in \mathbb{R}^{N \times N}$ is a diagonal matrix defined by:

$$D_{ii} = \Sigma_{j=1}^{N} [A_{self}]_{ij}. \quad (1)$$

The normalized structural transition matrix is computed as:

$$W = \frac{1}{2}(D^{-1} A_{self} + I_N), \quad (2)$$

3.2 Construction of S and M

To incorporate content-based similarity, we compute the cosine similarity matrix $W_r \in \mathbb{R}^{N\times N}$, where:

$$[W_r]_{ij} = \frac{\langle x_i, x_j\rangle}{\|x_i\|.\|x_j\|}, \quad \forall i,j \in \mathcal{V} \tag{3}$$

where $\langle .,. \rangle$ is the scalar product and $\|.\|$ is the Euclidean norm. We then apply a mutual k-nearest neighbor filter to retain only symmetric high-similarity edges: for each node i, let $\mathcal{N}_k(i)$ denote the set of its top-k most similar neighbors under cosine similarity. We define a binary mask $\mathcal{M}_k \in \{0,1\}^{N\times N}$ such that:

$$[\mathcal{M}_k]_{ij} = \begin{cases} 1 & \text{if } j \in \mathcal{N}_k(i) \text{ and } i \in \mathcal{N}_k(j) \\ 0 & \text{otherwise} \end{cases} \tag{4}$$

This mask preserves bidirectional high-similarity relationships. The filtered similarity matrix is then

$$W_X = W_r \odot \mathcal{M}_k \tag{5}$$

where $\odot$ is the element-wise multiplication. We construct similarity matrix S as the first matrix to combine structure and content:

$$S = W + W_X \tag{6}$$

To capture longer-range structure, we raise W to power p as in the original PAGEC [14]. We define the matrix

$$M = W^p X \tag{7}$$

as the second matrix that encodes both topology and features.

3.3 Joint Node Clustering and Embedding Objective

The objective function to be minimized is the same as in [14], and it is given by:

$$\mathcal{L} = \|M - BQ^\top\|_F^2 + \lambda\|S - GZB^\top\|_F^2 \tag{8}$$

with:

- $B \in \mathbb{R}^{N\times h}$: node embeddings and $B^\top B = I_h$.
- $Q \in \mathbb{R}^{d\times h}$: learnable projection matrix.
- $Z \in \mathbb{R}^{K\times h}$: cluster prototypes and $Z^\top Z = I_h$.
- $G \in [0,1]^{N\times K}$: cluster assignment matrix.
- $\lambda > 0$: regularization parameter.

The first term aligns embeddings with the propagation-informed matrix M, while the second term enforces that cluster assignments reconstruct the content and structure similarity S. The hyperparameter λ balances representation quality and cluster coherence.

We make use of the following proposition from [14].

Proposition 1. Given $S \in \mathbb{R}^{N\times N}$, $G \in [0,1]^{N\times K}$, $Z \in \mathbb{R}^{K\times h}$, and $B \in \mathbb{R}^{N\times h}$, we have:

$$\|S - GZB^\top\|_F^2 = \|S - SBB^\top\|_F^2 + \|SB - GZ\|_F^2 \quad (9)$$

Initialization: We initialize B as the top-h left singular vectors S, and initialize Z using the top-ranked Personalized PageRank [21] nodes for each cluster. The SVD-based initialization provides a numerically stable orthonormal starting point, while Personalized PageRank seeds are used to locate high-centrality representatives per cluster and accelerate convergence.

Cluster Assignment: Given B, Q, and Z, the objective reduces to $\min_G \|S - GZB^\top\|_F^2$. By using Proposition 1, this reduces to $\min_G \|SB - GZ\|_F^2$. Since $B^\top B = I_h$ and $Z^\top Z = I_h$, we compute the cluster logits as follows:

$$L = SBZ^\top \quad (10)$$

To obtain assignments, we use a straight-through estimator [2] that allows hard clustering in the forward pass and soft gradients in the backward pass. This allows us to reconcile the need for discrete cluster assignments during inference with the requirement for gradient-based optimization, which allows for end-to-end differentiability. Specifically:

- We compute soft assignments via softmax:

$$G_{soft} = \text{softmax}(L) \quad (11)$$

  This produces a differentiable probability distribution over clusters for each node, which is used in the backward pass for gradient computation.
- We derive hard labels through a matrix of one-hot assignments G_{hard} where for each node i, the entry corresponding to the highest scoring cluster is set to 1:

$$G_{hard,ik} = \begin{cases} 1 & \text{if } k = \text{argmax}_j L_{ij}, \\ 0 & \text{otherwise} \end{cases} \quad (12)$$

  This enforces discrete, unambiguous cluster membership during the forward computation.
- We define the final assignment matrix G as:

$$G = G_{hard} + (G_{soft} - \text{stop\_gradient}(\text{G}_{\text{soft}})) \quad (13)$$

  where stop_gradient(.) denotes an operator that prevents gradients from propagating through its argument.

Algorithm 1. E-PAGEC: Differentiable joint embedding & clustering

Require: adjacency $A \in \mathbb{R}^{N \times N}$, features $X \in \mathbb{R}^{N \times d}$, clusters K, embedding dimension h, hyperparameters λ, p, k
Ensure: embeddings $B \in \mathbb{R}^{N \times h}$, cluster assignments $G \in \mathbb{R}^{N \times K}$
$(S, M) \leftarrow \text{ConstructSAndM}(A, X; p, k)$ (Eq. (6) and (7))
Initialize learnable params $Q \in \mathbb{R}^{d \times h}$; initialize $B \in \mathbb{R}^{N \times h}, Z \in \mathbb{R}^{K \times h}$
for epoch $= 1, \ldots, T$ **do**
  Inner loop (embedding–clustering refinement):
  for iter $= 1, \ldots, T_{\text{inner}}$ **do**
    scores $\leftarrow S\,B\,Z^\top$ (Eq. (10))
    $G_{\text{soft}} \leftarrow \text{softmax}(\text{scores})$ (Eq. (11))
    $G_{\text{hard}} \leftarrow \text{one\_hot}(\arg\max \text{scores})$ (Eq. (12))
    $G \leftarrow G_{\text{hard}} + (G_{\text{soft}} - \text{stop\_grad}(G_{\text{soft}}))$ (Eq. (13))
    $U, \_, V^\top \leftarrow \text{SVD}(G^\top SB)$ (Eq. (14))
    $Z \leftarrow UV^\top$
    $\hat{U}, \_, \hat{V}^\top \leftarrow \text{SVD}(M^\top Q + \lambda\, SGZ)$ (Eq. (15))
    $B \leftarrow \hat{U}\hat{V}^\top$
  end for
  Loss & parameter update:
  $\mathcal{L} \leftarrow \|M - BQ^\top\|_F^2 + \lambda\|S - GZB^\top\|_F^2$ (Eq. (8))
  Update learnable parameters by backpropagation and optimizer step
end for
return B, G

Z **Update:** By fixing G and B, the objective function becomes $\min_Z \|S - GZB^\top\|_F^2$. From Proposition 1, this reduces to $\min_Z \|SB - GZ\|_F^2$, which under the orthonormality constraint $Z^\top Z = I_h$, is equivalent to $\max_Z \text{Tr}(G^\top SBZ)$.

The optimal solution is obtained by projecting $G^\top SB$ onto the Stiefel manifold via the SVD-based update as in [14]:

$$\text{SVD}(G^\top SB) = U\Sigma V^\top \quad \text{and} \quad Z = UV^\top \tag{14}$$

B **Update:** Given G, Q, and Z, the objective reduces to $\text{Tr}((M^\top Q + \lambda SGZ)B^\top)$ such that $B^\top B = I_h$. As in [14], and similarly to Z's update, we update B as follows:

$$\text{SVD}(M^\top Q + \lambda SGZ) = \hat{U}\hat{\Sigma}\hat{V}^\top \quad \text{and} \quad B = \hat{U}\hat{V}^\top \tag{15}$$

This completes one refinement iteration. We unroll this process for a fixed number of steps during training, allowing the model to converge to high-quality embeddings and cluster structures in a fully differentiable manner.

4 Experiments

4.1 Datasets

We run our experiments on six real-world attributed graph datasets: Cora [19], CiteSeer [7], Wiki [36], Pubmed [27], DBLP [23], and Flickr [10]. Table 1 shows their statistics.

Table 1. Statistics of datasets

Datasets	Nodes	Features per node	Edges	Classes
Cora	2708	1433	5294	7
Citeseer	3312	3703	4732	6
Wiki	2405	4973	17981	17
Pubmed	19,717	500	88,648	3
DBLP	17,716	1,639	105,734	4
Flickr	7,575	12,047	479,476	9

4.2 Baseline Methods

We compare our clustering performance against state-of-the-art clustering methods: k-means [9], Spectral Clustering [31], DeepWalk [24], Deep Neural Networks for Graph Representation (DNGR) [3], Graph Autoencoders (GAE) and Variational Graph Autoencoders (VGAE) [13], Adversarially Regularized Graph Autoencoder (ARGA) and Adversarially Regularized Variational Graph Autoencoder (ARVGA) [22], Marginalized Graph Autoencoder (MGAE) [33], Deep Attention Embedding Graph Clustering (DAEGC) [32], Adaptive Graph Convolution (AGC) [38], PAGEC [14], Dynamically Fusing Self-Supervised Learning (DyFSS) [40], and HACD [37].

4.3 Evaluation Metrics

As evaluation metrics, we report clustering accuracy (ACC) [35], normalized mutual information (NMI) [29], F1-macro [18], and Adjusted Rand Index (ARI) [29] as external measures of clustering quality. We also report clustering Purity [6], which measures, for each predicted cluster, the fraction of samples correctly assigned. In addition, we evaluate representation quality using internal, label-free metrics, including Trustworthiness (T) [12] to assess local structure preservation, the Silhouette Width Criterion (SWC) [25], which quantifies cluster compactness, and the Davies-Bouldin Index (DB) [5], which evaluates inter-cluster similarity.

4.4 Implementation Details

We adopt the AdamW optimizer for model training. We use the Optuna Python library for hyperparameter tuning, and we rely on SWC to estimate the hyperparameters. The experiments were repeated for 50 runs, 400 epochs each. We conduct the experiments on NVIDIA A100 GPU with PyTorch v2.6.0. The code is available in a public repository[1].

[1] https://github.com/ImaneCT/E-PAGEC.

Table 2. Clustering performance comparison (ACC%, NMI%, F1-macro%, ARI%). Bold indicates best results, underline indicates second best. Statistical significance between best and second best is marked with * ($p < 0.05$). – indicates unavailable results.

Approach	Input	Cora				Citeseer				Wiki			
		ACC ↑	NMI ↑	F1 ↑	ARI ↑	ACC ↑	NMI ↑	F1 ↑	ARI ↑	ACC ↑	NMI ↑	F1 ↑	ARI ↑
k-means	X	35.92	16.83	31.58	9.69	42.92	20.22	39.92	15.79	30.17	26.66	17.31	3.66
Spectral	W	44.02	32.59	41.33	10.72	35.08	14.94	28.32	5.35	37.54	33.60	27.31	10.61
DeepWalk	W	28.59	11.58	29.57	6.85	22.65	2.61	22.37	1.40	35.98	32.11	32.36	16.79
DNGR	W	53.04	42.26	51.56	28.93	27.97	8.70	23.72	3.56	33.51	29.51	29.86	12.88
GAE	X, W	58.50	42.56	59.57	31.94	41.24	22.31	36.22	15.76	29.19	26.50	20.80	4.53
VGAE	X, W	65.75	45.07	35.54	40.19	36.78	18.62	32.92	11.56	–	–	–	–
ARGA	X, W	63.73	45.06	55.39	38.24	49.14	24.38	47.56	20.83	24.74	21.87	18.50	6.19
ARVGA	X, W	53.17	44.69	49.13	35.20	56.50	31.25	53.58	27.91	–	–	–	–
MGAE	X, W	68.49	51.19	53.84	44.50	65.79	41.09	51.56	41.10	51.14	**48.12**	**39.96**	24.50
DAEGC	X, W	68.28	53.14	63.21	46.81	67.21	40.74	**63.37**	41.52	36.88	24.96	25.90	15.16
AGC	X, W	66.87	53.08	61.35	43.39	65.77	41.22	60.91	41.85	47.28	45.37	39.59	15.06
PAGEC	X, S	72.38	55.52	57.84	48.74	64.38	42.23	52.51	43.04	40.46	38.97	30.68	20.49
DyFSS	X, W	59.24	44.59	54.76	35.05	66.37	41.74	63.29	40.95	24.01	12.59	16.09	1.80
HACD	X, W	27.07	31.19	18.89	21.67	42.89	25.46	33.20	17.27	18.92	16.38	12.03	4.70
E-PAGEC	X, S	**73.22***	**55.77***	**64.43***	**53.03***	**71.16***	**43.95***	61.56	**46.45***	**55.42***	47.33	32.95	**27.11***

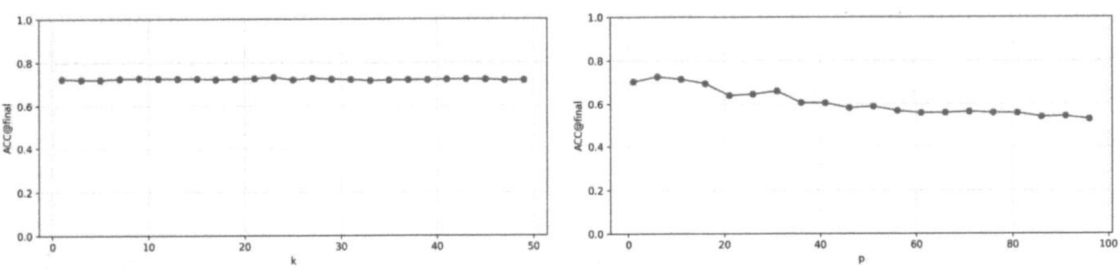

(a) Impact of k on accuracy (b) Impact of p on accuracy

Fig. 1. Sensitivity analysis on Cora.

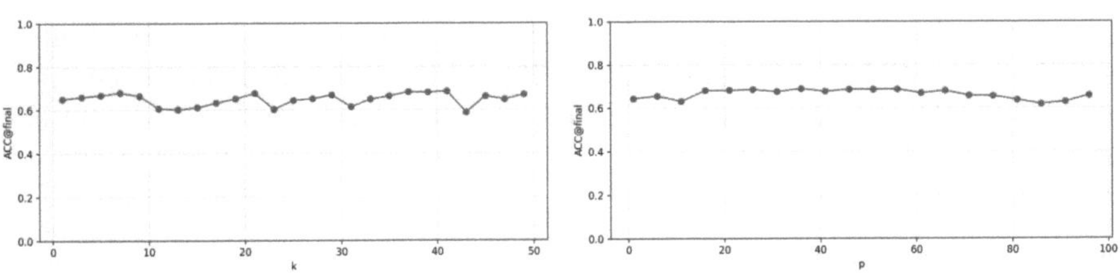

(a) Impact of k on accuracy (b) Impact of p on accuracy

Fig. 2. Sensitivity analysis on Citeseer.

Table 3. Clustering performance comparison (Purity%, T%, SWC , DB). For each dataset, bold numbers indicate leading results, while underlining highlights second-best scores. Statistical significance between the leading and second-best scores is indicated by * for a $p < 0.05$. – indicates unavailable results.

Approach	Input	Cora				Citeseer				Wiki			
		Purity ↑	T ↑	SWC ↑	DB ↓	Purity ↑	T ↑	SWC ↑	DB ↓	Purity ↑	T ↑	SWC ↑	DB ↓
k-means	X	40.21	63.4	0.005	6.14	46.24	60.69	−0.005	8.41	34.31	59.15	−0.15	1.88
Spectral	W	51.9	60.09	−0.18	3.78	39.82	57.93	−0.09	4.98	41.29	57.12	−0.3	3.33
DeepWalk	W	39.46	65.52	0.007	4.30	25.64	64.95	−0.01	4.23	36.95	61.04	−0.03	3.48
DNGR	W	48.99	67.74	0.04	3.98	31.83	63.93	−0.03	3.68	45.96	67.53	0.02	2.30
GAE	X, W	42.18	72.67	0.21	2.05	39.07	94.62	0.41	1.42	39.89	86.32	**0.58**	0.79
VGAE	X, W	57.95	69.52	0.15	2.23	42.98	72.73	0.21	1.89	-	-	-	-
ARGA	X, W	51.25	63.63	0.04	3.08	36.21	62.09	0.006	3.52	30.76	77.88	−0.19	0.94
ARVGA	X, W	66.35	71.85	0.07	2.32	47.79	70.68	0.02	2.58	-	-	-	-
MGAE	X, W	69.99	84.69	0.26	1.03	69.16	86.19	0.22	0.90	**62.20**	81.98	0.06	1.26
DAEGC	X, W	64.48	**97.56**	0.14	**0.25**	65.46	**97.96**	0.16	**0.17**	34.97	97.77	−0.18	**0.56**
AGC	X, W	68.51	77.99	0.16	1.02	**70.50**	77.68	0.13	0.98	52.78	70.32	−0.08	1.01
PAGEC	X, S	71.64	83.11	0.26	1.21	69.76	76.71	0.19	1.12	41.41	82.50	−0.08	0.91
DyFSS	X, W	62.75	84.29	0.11	0.83	68.28	92.12	0.13	0.91	30.84	**98.14**	−0.13	0.86
HACD	X, W	52.98	95.55	0.18	1.21	34.54	91.54	−0.02	1.94	23.16	85.32	−0.15	2.82
E-PAGEC	X, S	**72.20***	91.11	**0.38***	0.89	69.84	96.36	**0.43***	0.85	58.47	84.70	0.22	1.24

Table 4. Clustering performance comparison (ACC%, NMI%, F1-macro%, ARI%). Bold indicates best results, underline indicates second best. Statistical significance between best and second best is marked with * ($p < 0.05$). – indicates unavailable results. OOM denotes an Out-Of-Memory issue.

Approach	Input	Pubmed				DBLP				Flickr			
		ACC ↑	NMI ↑	F1 ↑	ARI ↑	ACC ↑	NMI ↑	F1 ↑	ARI ↑	ACC ↑	NMI ↑	F1 ↑	ARI ↑
k-means	X	59.5	31.23	58.15	28.09	41.23	6.57	26.81	1.36	12.78	2.29	4.48	0.03
Spectral	W	56.67	13.28	47.62	13.33	46.16	20.04	37.97	10.97	15.59	3.96	10.73	0.34
DeepWalk	W	35.71	0.09	34.35	0.13	26.95	0.18	24.81	0.15	33.56	19.73	32.49	11.76
DNGR	W	OOM	OOM	OOM	OOM	OOM	OOM	OOM	OOM	29.01	15.48	26.01	8.91
GAE	X, W	62.32	20.6	60.72	20.58	64.55	33.88	59.3	36.39	–	–	–	–
VGAE	X, W	61.56	22.27	61.8	20.36	58.37	34.43	50.87	35.8	–	–	–	–
ARGA	X, W	65.34	25.04	65.83	24.56	64.18	35.85	62.16	30.62	17.3	4.02	13.67	1.65
ARVGA	X, W	38.94	0.34	31.4	0.53	43.84	27.56	36.92	23.74	17.96	7.08	14.05	5.05
MGAE	X, W	43.13	7.94	42.04	3.54	OOM	OOM	OOM	OOM	OOM	OOM	OOM	OOM
DAEGC	X, W	66.45	25.32	65.54	26.73	OOM	OOM	OOM	OOM	25.24	11.83	18.6	7.33
AGC	X, W	68.74	31.81	67.98	30.69	64.41	34.67	64.92	28.35	26.61	17.05	18.02	7.08
PAGEC	X, S	OOM	OOM	OOM	OOM	57.31	36.58	47.53	33.83	37.62	30.49	27.62	13.02
DyFSS	X, W	63.14	23.31	63.75	22.03	56.08	37.07	51.86	32.12	29.78	17.74	27.52	8.77
HACD	X, W	**83.83**	**56.29**	**83.98**	**55.89**	OOM	OOM	OOM	OOM	11.6	0.47	2.84	0.08
E-PAGEC	X, S	73.08	32.78	71.99	36.91	**77.46***	**46.54***	**74.16***	**52.59***	**46.78***	**33.32***	**44.13***	**39.29***

Table 5. Clustering performance comparison (Purity%, T%, SWC , DB). For each dataset, bold numbers indicate leading results, while underlining highlights second-best scores. Statistical significance between the leading and second-best scores is indicated by * for a $p < 0.05$. – indicates unavailable results. OOM denotes an Out-Of-Memory issue.

Approach	Input	Pubmed				DBLP				Flickr			
		Purity ↑	T ↑	SWC ↑	DB ↓	Purity ↑	T ↑	SWC ↑	DB ↓	Purity ↑	T ↑	SWC ↑	DB ↓
k-means	X	59.5	61.27	−0.001	5.16	47.49	65.04	−0.001	4.41	12.72	50.53	−0.09	2.45
Spectral	W	49.32	61.11	−0.02	7.03	47.42	62.55	−0.11	3.04	16.77	56.6	−0.25	3.22
DeepWalk	W	41.06	60.58	−0.005	5.29	44.71	56.51	−0.06	6.53	35.57	63.42	−0.04	3.33
DNGR	W	OOM	OOM	OOM	OOM	OOM	OOM	OOM	OOM	31.57	66.92	−0.01	5.43
GAE	X, W	68.36	74.36	0.11	2.66	64.94	68.7	0.07	**0.29**	–	–	–	–
VGAE	X, W	55.42	80.29	0.14	2.21	62.28	69.05	0.08	2.9	–	–	–	–
ARGA	X, W	66.95	79.53	0.08	1.87	63.84	69.14	0.02	3.11	30.12	90.12	−0.05	1.099
ARVGA	X, W	53.67	98.2	−0.04	0.61	69.28	84.04	0.08	1.61	29.86	91.55	−0.05	1.08
MGAE	X, W	OOM	OOM	OOM	OOM	OOM	OOM	OOM	OOM	OOM	OOM	OOM	OOM
DAEGC	X, W	60.67	**99.51**	0.07	**0.16**	OOM	OOM	OOM	OOM	21.35	**99.66**	−0.38	1.29
AGC	X, W	66.85	98.49	0.13	0.78	70.16	85.72	0.18	0.91	26.7	73.65	−0.06	**0.09**
PAGEC	X, S	OOM	OOM	OOM	OOM	63.53	82.5	0.21	0.91	37.54	79.65	−0.23	1.16
DyFSS	X, W	63.14	98.33	0.09	0.64	70.24	**99**	0.1	0.85	30.42	94.7	−0.06	0.99
HACD	X, W	54.25	96.01	**0.61**	–	OOM	OOM	OOM	OOM	11.75	99.15	−0.19	–
E-PAGEC	X, S	**75.2***	98.24	0.15	2.05	**78.28***	93.77	**0.41***	1.21	**47.43***	85.96	**0.22***	1.49

4.5 Results

Performance Evaluation. We present the results of the comparison of E-PAGEC with the baseline methods in Tables 2, 3, 4, and 5. Across Cora, Citeseer, and Wiki, E-PAGEC consistently achieves top or second-best results on almost all external metrics. On Wiki, it remains competitive and achieves the best ARI. On internal metrics, E-PAGEC also delivers robust performance. It leads on SWC on Cora and Citeseer and shows high Trustworthiness across all three datasets. These results confirm that E-PAGEC improves both the external clustering quality and the internal consistency of the learned representations. On the larger datasets, E-PAGEC demonstrates strong scalability. It ranks second on Pubmed and significantly outperforms all baselines on DBLP. On Flickr, E-PAGEC shows clear performance gains over the baselines. Its internal metrics on these datasets further highlight stable structures.

To qualitatively examine the learned latent space, we project the embeddings into two dimensions using UMAP with 15 neighbors in Fig. 3. This projection preserves meaningful local structure while revealing cluster separation.

Statistical Significance. All results are averaged over 50 independent runs with different random seeds. To test whether the improvements of E-PAGEC over the second-best baseline are due to chance, we perform two-sided Welch's t-tests

[34] on the per-run evaluation scores for each metric. We consider differences significant at $p < 0.05$. All reported improvements are statistically significant under this criterion.

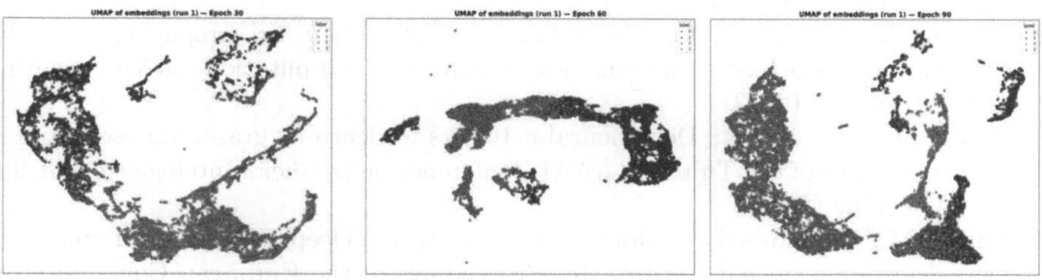

(a) Epoch 30 (ACC=64.69%) (b) Epoch 60 (ACC=65.42%) (c) Epoch 90 (ACC=69.15%)

Fig. 3. UMAP visualization on DBLP at different epochs.

Sensitivity Analysis. We perform sensitivity analyses by sweeping the power p in {1,...,100} and k, the number of most similar neighbors, in {1,...,50} on Cora and Citeseer (see Figs. 1 and 2). For Cora, ACC is largely stable across the examined range of k. On the other hand, ACC varies substantially with p as it keeps degrading for larger p values. For Citeseer, we notice that the ACC changes with different k values but stays mostly stable across the examined range of p.

5 Conclusion

In this work, we introduced E-PAGEC, an enhanced learnable clustering framework for attributed graphs that integrates features, topology, and end-to-end optimization. Through experiments on multiple benchmark datasets, we demonstrated that E-PAGEC consistently produces high-quality embeddings and well-structured clusters. Our findings highlight the potential of combining graph structure, learned representations, and clustering objectives within a unified framework.

Acknowledgments. This international collaboration is supported by the privileged partnership between Université Paris Cité and Mohammed VI Polytechnic University, funded by the ANR-18-IDEX-0001 grant under the France 2030 Investment Plan.

Disclosure of Interests. The authors have no competing interests to declare that are relevant to the content of this article.

References

1. Allab, K., Labiod, L., Nadif, M.: Power simultaneous spectral data embedding and clustering. In: Proceedings of the SIAM International Conference on Data Mining, pp. 270–278. SIAM (2016)
2. Bengio, Y., Léonard, N., Courville, A.: Estimating or propagating gradients through stochastic neurons for conditional computation. arXiv preprint arXiv:1308.3432 (2013)
3. Cao, S., Lu, W., Xu, Q.: Deep neural networks for learning graph representations. In: Proceedings of the Thirtieth AAAI Conference on Artificial Intelligence, vol. 30, pp. 1145–1152 (2016)
4. Caron, M., Bojanowski, P., Joulin, A., Douze, M.: Deep clustering for unsupervised learning of visual features. In: Proceedings of the European Conference on Computer Vision (ECCV), pp. 132–149 (2018)
5. Davies, D.L., Bouldin, D.W.: A cluster separation measure. IEEE Trans. Pattern Anal. Mach. Intell. **2**, 224–227 (2009)
6. Forestier, G., Wemmert, C., Gançarski, P.: Background knowledge integration in clustering using purity indexes. In: International Conference on Knowledge Science, Engineering and Management, pp. 28–38. Springer, Cham (2010)
7. Giles, C.L., Bollacker, K.D., Lawrence, S.: Citeseer: an automatic citation indexing system. In: Proceedings of the third ACM Conference on Digital Libraries, pp. 89–98 (1998)
8. Guo, X., Gao, L., Liu, X., Yin, J.: Improved deep embedded clustering with local structure preservation. In: Proceedings of the 26th International Joint Conference on Artificial Intelligence, pp. 1753–1759. IJCAI (2017)
9. Hartigan, J.A., Wong, M.A.: Algorithm as 136: a k-means clustering algorithm. J. R. Stat. Soc. Ser. c (Appl. Stat.) **28**(1), 100–108 (1979)
10. Huang, X., Li, J., Hu, X.: Label informed attributed network embedding. In: Proceedings of the tenth ACM International Conference on Web Search and Data Mining, pp. 731–739 (2017)
11. Karim, M.R., et al.: Deep learning-based clustering approaches for bioinformatics. Brief. Bioinform. **22**(1), 393–415 (2021)
12. Kaski, S., Nikkilä, J., Oja, M., Venna, J., Törönen, P., Castrén, E.: Trustworthiness and metrics in visualizing similarity of gene expression. BMC Bioinform. **4**(1), 48 (2003)
13. Kipf, T.N., Welling, M.: Variational graph auto-encoders. arXiv preprint arXiv:1611.07308 (2016)
14. Labiod, L., Nadif, M.: Power attributed graph embedding and clustering. IEEE Trans. Neural Netw. Learn. Syst. **35**(1), 1439–1444 (2022)
15. Li, J., Cheng, K., Wu, L., Liu, H.: Streaming link prediction on dynamic attributed networks. In: Proceedings of the Eleventh ACM International Conference on Web Search and Data Mining, pp. 369–377 (2018)
16. Liao, L., He, X., Zhang, H., Chua, T.S.: Attributed social network embedding. IEEE Trans. Knowl. Data Eng. **30**(12), 2257–2270 (2018)
17. Liu, Y., Li, M., Li, X., Giunchiglia, F., Feng, X., Guan, R.: Few-shot node classification on attributed networks with graph meta-learning. In: Proceedings of the 45th International ACM SIGIR Conference on Research and Development in Information Retrieval, pp. 471–481 (2022)
18. Manning, C.D.: Introduction to Information Retrieval. Syngress Publishing, Rockland (2008)

19. McCallum, A.K., Nigam, K., Rennie, J., Seymore, K.: Automating the construction of internet portals with machine learning. Inf. Retrieval **3**(2), 127–163 (2000)
20. Mukherjee, S., Asnani, H., Lin, E., Kannan, S.: ClusterGAN: latent space clustering in generative adversarial networks. In: Proceedings of the AAAI Conference on Artificial Intelligence, vol. 33, pp. 4610–4617 (2019)
21. Page, L., Brin, S., Motwani, R., Winograd, T.: The pagerank citation ranking: Bringing order to the web. Technical report, Stanford infolab (1999)
22. Pan, S., Hu, R., Long, G., Jiang, J., Yao, L., Zhang, C.: Adversarially regularized graph autoencoder for graph embedding. arXiv preprint arXiv:1802.04407 (2018)
23. Pan, S., Wu, J., Zhu, X., Zhang, C., Wang, Y.: Tri-party deep network representation. In: 25th International Joint Conference on Artificial Intelligence (IJCAI-16). AAAI Press/International Joint Conferences on Artificial Intelligence (2016)
24. Perozzi, B., Al-Rfou, R., Skiena, S.: Deepwalk: online learning of social representations. In: Proceedings of the 20th ACM SIGKDD International Conference on Knowledge Discovery and Data Mining, pp. 701–710 (2014)
25. Rousseeuw, P.J.: Silhouettes: a graphical aid to the interpretation and validation of cluster analysis. J. Comput. Appl. Math. **20**, 53–65 (1987)
26. Rozemberczki, B., Davies, R., Sarkar, R., Sutton, C.: GEMSEC: graph embedding with self clustering. In: Proceedings of the IEEE/ACM International Conference on Advances in Social Networks Analysis and Mining, pp. 65–72 (2019)
27. Sen, P., Namata, G., Bilgic, M., Getoor, L., Galligher, B., Eliassi-Rad, T.: Collective classification in network data. AI Mag. **29**(3), 93–93 (2008)
28. Sharma, K., et al.: A survey of graph neural networks for social recommender systems. ACM Comput. Surv. **56**(10), 1–34 (2024)
29. Strehl, A., Ghosh, J.: Cluster ensembles—a knowledge reuse framework for combining multiple partitions. J. Mach. Learn. Res. **3**(Dec), 583–617 (2002)
30. Sun, H., et al.: Network embedding for community detection in attributed networks. ACM Trans. Knowl. Discov. Data (TKDD) **14**(3), 1–25 (2020)
31. Von Luxburg, U.: A tutorial on spectral clustering. Stat. Comput. **17**, 395–416 (2007)
32. Wang, C., Pan, S., Hu, R., Long, G., Jiang, J., Zhang, C.: Attributed graph clustering: a deep attentional embedding approach. In: Proceedings of the 28th International Joint Conference on Artificial Intelligence, pp. 3670–3676 (2019)
33. Wang, C., Pan, S., Long, G., Zhu, X., Jiang, J.: MGAE: marginalized graph autoencoder for graph clustering. In: Proceedings of the ACM Conference on Information and Knowledge Management, pp. 889–898 (2017)
34. Welch, B.L.: The generalization of 'student's'problem when several different population varlances are involved. Biometrika **34**(1–2), 28–35 (1947)
35. Xie, J., Girshick, R., Farhadi, A.: Unsupervised deep embedding for clustering analysis. In: International Conference on Machine Learning, pp. 478–487 (2016)
36. Yang, C., Liu, Z., Zhao, D., Sun, M., Chang, E.Y.: Network representation learning with rich text information. In: IJCAI, pp. 2111–2117 (2015)
37. Zhang, A., Wang, X., Zhao, Y.: HACD: harnessing attribute semantics and mesoscopic structure for community detection. In: Proceedings of the Eighteenth ACM International Conference on Web Search and Data Mining, pp. 616–624 (2025)
38. Zhang, X., Liu, H., Li, Q., Wu, X.M.: Attributed graph clustering via adaptive graph convolution. arXiv preprint arXiv:1906.01210 (2019)

39. Zhou, S., Tan, Q., Xu, Z., Huang, X., Chung, F.L.: Subtractive aggregation for attributed network anomaly detection. In: Proceedings of the 30th ACM International Conference on Information & Knowledge Management, pp. 3672–3676 (2021)
40. Zhu, P., Wang, Q., Wang, Y., Li, J., Hu, Q.: Every node is different: dynamically fusing self-supervised tasks for attributed graph clustering. In: Proceedings of the AAAI Conference on Artificial Intelligence, vol. 38, pp. 17184–17192 (2024)

Section: Data mining

Exploiting Treatment Similarities for Enhanced Multi-treatment Uplift Prediction

Nathan Le Boudec[1,2](✉), Nicolas Voisine[1], and Bruno Crémilleux[2]

[1] Orange Labs, 22300 Lannion, France
{nathan.leboudec,nicolas.voisine}@orange.com
[2] UNICAEN, ENSICAEN, CNRS, GREYC, Normandie Univ, 14000 Caen, France
bruno.cremilleux@unicaen.fr

Abstract. Uplift modeling estimates the effect of a treatment on an individual's outcome. However, accurately estimating this effect is challenging in multi-treatment scenarios due to data sparsity and class imbalance. While existing methods handle treatments independently, we introduce a novel paradigm that explicitly models and exploits local treatment similarities. To this end, we propose a non-parametric Bayesian criterion that automatically identifies and groups treatments with similar causal effects on specific subpopulations. We leverage this grouping strategy in two distinct ways: as an information pooling technique to improve state-of-the-art methods, and as a standalone uplift estimator that directly predicts treatment effects. Experimental results on diverse synthetic datasets demonstrate that our approach significantly improves the accuracy of uplift estimation compared to state-of-the-art methods.

Keywords: Uplift Modeling · Multi-Treatment · Causal Inference · Non-Parametric Bayesian · Treatment Grouping · Machine Learning

1 Introduction

Uplift modeling is a technique for predicting how a specific action, or treatment, will change an individual's behavior. This type of modeling is valuable in fields like personalized medicine and advertising. When used in recommendation systems, measuring uplift helps determine which marketing campaign will produce the most positive result for a given customer. The focus is not just on predicting whether someone will respond to a campaign, but on estimating the incremental effect of that campaign compared to others. Therefore, uplift models improve the allocation of marketing resources by focusing on individuals for whom a campaign's impact will be the greatest. Prior research in uplift modeling has predominantly focused on the single-treatment setting, where the effect of one treatment is measured against a control group using methods like decision trees [14] and Random Forests [7]. In contrast, this paper tackles the more complex multi-treatment scenario: the objective is to rank and quantify the impact of several treatments relative to a control.

M. Baratchi et al. (Eds.): IDA 2026, LNCS 16513, pp. 255–268, 2026.
https://doi.org/10.1007/978-3-032-23833-7_19

In this context, a major challenge lies in data imbalances. Some campaigns are observed more frequently than others, certain subpopulations receive few treatments, and some profile-treatment combinations may be entirely absent. These imbalances lead to greater uncertainty in the estimation of the individual treatment effect, limiting the performance of the models.

To address this objective, we explore the possibility of leveraging the similarity between treatments. The idea is to group treatments that have comparable effects on certain individuals in order to pool the available information and improve the uplift estimation. This grouping helps to compensate for the lack of instances for certain treatments or in certain areas of the data space. In other words, the more abundant and varied the available data, the better the uplift estimation will be. When information is shared between similar treatments, the model indirectly benefits from a larger volume of relevant data, leading to a more reliable estimation of the treatment effect.

To the best of our knowledge, this work is the first to propose a method that automatically discovers and exploits local treatment similarities for uplift modeling. Our main contribution to implementing this grouping strategy in the univariate setting is a novel, parameter-free Bayesian criterion derived from the MODL (Minimum Optimized Description Length) [4] methodology. This criterion, named MUBU (Multi-treatment Uplift Bayesian Univariate), is designed to automatically identify the optimal local group of treatments without any parameter. From MUBU, we propose MUBU-G, a model-agnostic strategy, enabling any existing uplift method to leverage the identified treatment groups. Based on MUBU, we define MUBU-E, an uplift model that inherently utilizes the resulting discretization and groupings. Experimental results show that this approach improves the quality of the uplift estimation, particularly where data is limited. MUBU-G enhances the performance of existing uplift learners when used as a pre-processing strategy. Furthermore, MUBU-E also shows highly competitive performance in uplift estimation.

This paper is organized as follows. Section 2 provides the necessary background and Sect. 3 reviews the state of the art. Section 4 introduces our proposed criterion and methods. Finally, Sect. 5 presents the experimental results.

2 Background and Preliminaries

2.1 Definitions and Notations

Uplift and the Conditional Average Treatment Effect. Uplift modeling aims to estimate the causal effect of a treatment on an individual's outcome [14,16]. Let T denote the treatment variable. Let $Y_i(t)$ be the outcome for an individual i with features $\mathbf{x_i}$ if they receive treatment $T = t$. The true effect for this individual is the Individual Treatment Effect (ITE), defined as $Y_i(1) - Y_i(0)$ in the single-treatment scenario ($T \in \{0, 1\}$). Unfortunately, the ITE is unobservable since we cannot observe both $Y_i(1)$ and $Y_i(0)$; this is known as the fundamental problem of causal inference [17]. The goal of uplift modeling is to estimate the Conditional Average Treatment Effect (CATE) for a subpopulation of individuals who share the same features $\mathbf{x_i}$: $\tau(\mathbf{x_i}) = \mathbb{E}[Y_i(1) - Y_i(0)|X = \mathbf{x_i}]$. Under standard assumptions[1], the CATE corresponds to the uplift [19], usually defined for the single-treatment case as: $\tau(\mathbf{x_i}) = \mathbb{E}[Y|T = 1, X = \mathbf{x_i}] - \mathbb{E}[Y|T = 0, X = \mathbf{x_i}]$. A positive CATE value suggests that the treatment has a beneficial effect on average for individuals with features $\mathbf{x_i}$, while a negative value indicates a detrimental effect.

While the literature has predominantly focused on this single-treatment setting, where T is binary, this framework is insufficient when there are several competing treatments ($T \in \{0, 1, ..., K\}$). In the multi-treatment context, there is no longer a single uplift to consider, but K distinct uplifts—one for each treatment relative to the control ($T = 0$). If we denote the treatments as values from 1 to K, the uplift of treatment k for an individual with features $\mathbf{x_i}$ is defined as: $\tau_k(\mathbf{x_i}) = \mathbb{E}[Y|T = k, X = \mathbf{x_i}] - \mathbb{E}[Y|T = 0, X = \mathbf{x_i}]$.

2.2 Multi-treatment Uplift Modeling

In our analysis, we compare the top-performing methods for multi-treatment uplift modeling [11,12]. These methods, which are meta-learners, use standard supervised learning algorithms as base estimators to infer causal effects. This section provides a brief overview of the learners implemented.

S Learners. The S Learner [2,10] uses a single model $\hat{\mu}(x, t)$, treating the treatment T as a feature. The uplift of treatment k for individual x is the difference in the predictions: $\hat{\tau_k}(x) = \hat{\mu}(x, k) - \hat{\mu}(x, 0)$. While computationally efficient, this approach may bias the treatment effect estimate towards zero if the treatment signal is weak relative to other features.

T Learners. The T Learner [8] trains a separate model $\hat{\mu}_k(x)$ for each treatment k and the control group. The uplift of the treatment k is the difference between the treatment k model and the control model: $\hat{\tau_k}(x) = \hat{\mu}_k(x) - \hat{\mu}_0(x)$. This method avoids the regularization bias of the S-Learner but can suffer from high variance if sample sizes for specific treatments are small.

X Learners. The X Learner [10] extends the T Learner by weighting the effect estimates using propensity scores [16]. It involves a two-stage estimation process designed to perform well even when treatment groups are imbalanced or when the causal effect is structurally different from the baseline outcome. Since it builds upon the T-Learner strategy, it remains dependent on the quality of the initial separate models. Consequently, it may still face stability issues if the sample size for a specific treatment is insufficient to train a reliable base estimator.

[1] These assumptions include *Unconfoundedness* (no unobserved confounders influence the treatment assignment), *Positivity* (every individual has a non-zero probability of receiving any treatment), and the *Stable Unit Treatment Value Assumption* (one individual's outcome is not affected by others individuals' treatment assignments) [19].

2.3 MODL

The MODL (Minimum Optimized Description Length) approach is a non-parametric Bayesian method designed for discretization and conditional probability estimation [4]. Unlike heuristic methods that require manual parameter tuning (e.g., the number of bins), MODL treats discretization as a formal model selection problem. It relies on the Minimum Description Length (MDL) principle [6] to identify the optimal partition of the data by maximizing the posterior probability of the model. In the MODL framework, an evaluation criterion is established to find the best trade-off between accuracy and complexity. Then a search algorithm is employed to evaluate all possible discretization models, allowing for the selection of the one that achieves the best score.

3 State of the Art

The state of the art for uplift estimation itself is well-established, primarily through meta-learners (like the S, T, and X Learners described in Sect. 2) [2,10]. These methods focus on estimating the effect τ_k for each treatment k independently. However, in a multi-treatment scenario, this independence assumption can be inefficient, particularly when data is sparse for certain treatments. These models do not exploit the fact that some treatments might have similar or identical effects on the population. Our work addresses this specific gap by proposing a method to automatically identify and leverage these similarities.

Alternatively, tree-based methods directly partition the data space to isolate subgroups with heterogeneous treatment effects. Originally designed for single-treatment settings [7,14], this framework was generalized to the multi-treatment setting in [18] by introducing splitting criteria based on divergence measures to maximize the difference between multiple treatment distributions. While effective at splitting the population based on features, these methods typically keep all treatments separate throughout the tree growth, rather than grouping similar ones together to pool information.

Recent deep learning approaches move beyond independent estimation by using shared representations to tackle data sparsity. For structured treatments such as graphs or images, Kaddour et al. [9] proposed leveraging representation learning to generalize estimation across interventions. Similarly, to capture synergies between concurrent treatments in the biomedical domain, Aoki et al. [1] introduced *M3E2*, an architecture based on multi-task learning. Furthermore, Zhao et al. [20] proposed a Multiple Instance Learning framework to regularize noisy ITE estimation. However, these methods rely on implicit parameter sharing in latent spaces rather than explicitly grouping treatments. Moreover, they are specifically designed for high-dimensional or structured inputs (e.g., images, graphs) and are thus outside the scope of our univariate benchmark focused on tabular data partitioning.

Data augmentation approaches, such as SMOTE [5], are generally not suitable for estimating causal effects. Their philosophy relies on generating new synthetic samples, often through interpolation or more recent generative models (e.g., GATE [13]). This process can fail to respect the treatment assignment mechanism or the causal structure linking X, T, and Y, which is a necessary condition for uplift modeling. The approach we propose in this paper follows a different philosophy. Instead of fabricating new data, our aim is to better leverage the existing data. Our method focuses on revealing the true causal structure, already hidden in the real data, by pooling information from treatments that are identified as having similar effects.

Closer to our methodological framework, Rafla et al. [15] applied non-parametric Bayesian discretization to uplift modeling. While this method effectively addresses feature selection and noise reduction, it is strictly limited to single-treatment scenarios contrary to our approach which enables the multi-treatment setting.

4 Our Approach: MUBU, MUBU-E, MUBU-G

In this section, we present our criterion MUBU, which forms the foundation of our approach. Then we present MUBU-E and MUBU-G.

4.1 MUBU: Multi-treatment Uplift Bayesian Univariate

We propose a new criterion called MUBU designed to discretize the continuous feature X while locally grouping the treatments T. The objective is to select the model M that is most probable given the data D. The optimal model M is the one that minimizes $MUBU(M, D)$:

$$MUBU(M, D) = \log N + \log \binom{N + I - 1}{I - 1} + I \log |\mathcal{T}| + \sum_{i=1}^{I} \log \mathcal{B}(|\mathcal{T}|, G_i) + \sum_{i=1}^{I} \sum_{g=1}^{G_i} \left[\log \binom{N_{ig} + J - 1}{J - 1} + \log(N_{ig}!) - \sum_{j=1}^{J} \log(N_{igj}!) \right] \quad (1)$$

where:

- N is the total number of instances,
- I is the number of intervals,
- $|\mathcal{T}|$ is the cardinality of all treatments $\mathcal{T}$,
- G_i is the number of treatment groups in interval i,
- $\mathcal{T}_{ig}$ is the set of treatments included in group g of interval i,
- $\mathcal{B}(|\mathcal{T}|, G_i)$ is the Stirling number of the second kind,
- N_{ig} is the number of instances in group g of interval i,
- N_{igj} is the number of instances belonging to class j in group g of interval i.

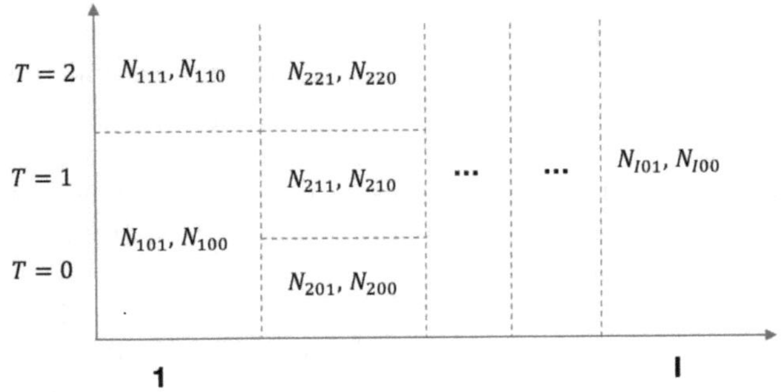

Fig. 1. Parameters of the MUBU criterion. The plot illustrates the decomposition of the data space: the continuous feature X is discretized into I intervals (horizontal axis), while the vertical axis (treatments T) contains the observed counts N_{igj}, where N_{igj} is the number of instances in interval i, belonging to treatment group g, and having outcome class j.

We define an uplift discretization model M by the number of intervals, the bounds of the intervals, the number of groups and treatment groups, class frequencies per interval, or per treatment group within each interval. In other words, a model M is defined by the hierarchy of parameters (cf. Fig. 1):

$$\{I, \{N_i\}, \{G_i\}, \{T_{ig}\}, \{N_{igj}\}\}$$

Derivation. We provide here the proof for the MUBU criterion defined in Eq. 1. According to Bayes' theorem, maximizing the posterior $P(M \mid D)$ is equivalent to maximizing the joint probability $P(M, D) = P(M)P(D \mid M)$. This is equivalent to minimizing: $MUBU(M, D) = -\log P(M) - \log P(D \mid M)$. We express $P(M)$ and $P(D|M)$ according to the parameters of our multi-treatment uplift discretization model. We introduce a prior distribution by exploiting the hierarchy of the model parameters. Assuming the independence of the local distributions across the intervals, we obtain the following decomposition:

$$P(M) = P(I) \times P(\{N_i\}|I) \times \prod_{i=1}^{I} \left[P(G_i|I) \times P(\{T_{ig}\}|G_i) \times \prod_{g=1}^{G_i} P(\{N_{igj}\}|N_{ig}) \right] \tag{2}$$

We express each of the terms of Eq. 2 according to the parameters of M assuming a uniform distribution for each parameter. Assuming that the number of intervals I is uniformly distributed between 1 and N, the first term becomes:

$$P(I) = \frac{1}{N} \tag{3}$$

Given a number of intervals I, all the discretizations into I intervals are equiprobable. Computing the probability of an interval set leads to a combinatorial calculation of the number of ways of distributing the N instances in the I intervals. The second term is then:

$$P(\{N_i\}|I) = \frac{1}{\binom{N+I-1}{I-1}} \tag{4}$$

For a given interval i, we assume that the number of treatment groups G_i is chosen uniformly between 1 and $|\mathcal{T}|$. We obtain:

$$\prod_{i=1}^{I} P(G_i|I) = \left(\frac{1}{|\mathcal{T}|}\right)^{I} \tag{5}$$

Given a number of groups G_i, the treatments must be partitioned into these groups. The number of ways to partition the set of treatments $|\mathcal{T}|$ into G_i non-empty subsets is given by the Stirling numbers of the second kind $\mathcal{B}(|\mathcal{T}|, G_i)$. Assuming uniformity, we get:

$$P(\{\mathcal{T}_{ig}\}|G_i) = \frac{1}{\mathcal{B}(|\mathcal{T}|, G_i)} \tag{6}$$

Finally, for each formed group g, the model describes the distribution of the outcome variable. Given the number of examples N_{ig} in the group, and assuming that each of the class distributions is equiprobable, we end up with a combinatorial problem using multiset coefficients:

$$P(\{N_{igj}\}|N_{ig}) = \frac{1}{\binom{N_{ig}+J-1}{J-1}} \tag{7}$$

After defining the model prior, we define the likelihood $P(D|M)$ of the data given the model. For each group g in interval i, we assume that all possible observed data sequences consistent with the multinomial counts $\{N_{igj}\}$ are equiprobable. Using multinomial terms, we obtain the following likelihood term:

$$P(D|M) = \prod_{i=1}^{I} \prod_{g=1}^{G_i} P(D_{ig}|M) = \prod_{i=1}^{I} \prod_{g=1}^{G_i} \frac{1}{N_{ig}!/\prod_{j=1}^{J} N_{igj}!} \tag{8}$$

Combining the prior $P(M)$ with the likelihood $P(D|M)$, we obtain the joint probability $P(M, D)$. Taking the negative logarithm yields the criterion presented in Eq. 1. □

To find the optimal model M, we adapt the optimization algorithms used by MODL [4]. Since finding the global optimum is NP-hard, we employ a greedy bottom-up heuristic that optimizes the MUBU criterion in three steps:

1. *Initialization (Interval Search)*: We first determine the structure of X. To ensure speed, we treat all treatments as distinct and perform a standard discretization where the target classes are the pairs $(Treatment, Outcome)$. We employ a multi-start greedy algorithm that starts from the most granular level and iteratively merges adjacent values to maximize the model probability [4]. This bottom-up merging phase is followed by a post-optimization splitting phase to refine the partitioning and escape local optima, ensuring a robust initial structure.

2. *Local Optimization (Treatment Grouping)*: Within each fixed interval, we minimize the criterion by grouping treatments that show similar outcome statistics [3]. This refines the model by exploiting local similarities.
3. *Post-Optimization*: We perform a final pass to merge adjacent intervals if the grouping in Step 2 allows for a better overall score.

The overall complexity is $\mathcal{O}(N \log N)$, as the computational cost is dominated by the interval search phase.

4.2 From Criterion to Uplift Estimation: MUBU-G and MUBU-E

The MUBU criterion (Eq. 1) provides the theoretical foundation for finding an optimal model M. Once this optimal model is identified, we propose MUBU-G and MUBU-E, two methods addressing the uplift estimation.

MUBU-G (Grouping Strategy). MUBU-G utilizes the optimal model M as a model-agnostic pre-processing strategy designed to enhance existing uplift learners. MUBU-G has two steps: (1) *Model Optimization:* We first minimize the criterion MUBU on the training data to identify the optimal intervals and the optimal grouping of treatments within each interval. (2) *Treatment Grouping:* We construct an augmented training set where, within each interval, instances are duplicated such that a unit observed with treatment $T = t$ is replicated for every other treatment belonging to the same group defined in that interval.

MUBU-E (Standalone Estimator). Our second method, MUBU-E, is a standalone, non-parametric uplift model that directly uses the structure of M and the observed counts to predict uplift. We estimate the conditional probabilities for a new instance x into interval i using the observed frequencies within the treatment group g_k (containing treatment k) and the control group g_0 (containing treatment 0):

$$\hat{P}(Y = 1 \mid x \in i, T = k) = \frac{N_{ig_k 1}}{N_{ig_k}} \quad \text{and} \quad \hat{P}(Y = 1 \mid x \in i, T = 0) = \frac{N_{ig_0 1}}{N_{ig_0}} \tag{9}$$

where N_{ig} is the total number of instances in interval i and group g, and N_{ig1} is the number of positive outcomes among them. The uplift is then simply computed as:

$$\hat{\tau}_k(x) = \hat{P}(Y = 1 \mid x \in i, T = k) - \hat{P}(Y = 1 \mid x \in i, T = 0) \tag{10}$$

5 Experiments

This section experimentally evaluates our approach by answering the following research questions:

- **RQ1:** How does MUBU-E perform as a direct uplift estimator compared to other methods?
- **RQ2:** Does using MUBU-G as a model-agnostic preprocessing strategy enhance the performance of existing meta-learners (e.g., S Learner, T Learner)?

The source code to generate datasets and execute the experiments is available at: https://anonymous.4open.science/r/Treatment-Similarities-Uplift/

5.1 Experimental Setup

In order to answer our research questions, we benchmark the performance of MUBU-G and MUBU-E w.r.t state-of-the-art (SOTA) methods. We evaluate three categories of models:

- *SOTA uplift estimation methods:* S Learner, T Learner, and X Learner, trained on the original, unmodified data. For all meta-learners, we use *XGBoost* as the base estimator to ensure a fair comparison.
- *SOTA uplift estimation methods enhanced by MUBU-G:* S Learner, T Learner, and X Learner, but trained on the augmented dataset by MUBU-G.
- *MUBU-E:* Our non-parametric standalone estimator (defined in Sect. 4.2).

5.2 Uplift Evaluation Protocol

As previously stated, due to the fundamental problem of causal inference [17], the true individual uplift is unknown, rendering standard error metrics inapplicable. To overcome this limitation and enable precise quantitative evaluation, we resort to synthetic data. This is the only setting that allows providing the ground-truth uplift $\tau_k(\mathbf{x_i})$ for every individual, whereas in real-world scenarios, the true effect is inherently unobservable. Using synthetic datasets for uplift estimation is advantageous as they provide a controlled environment where the ground truth is explicitly known, enabling the precise validation and benchmarking of uplift methods. In contrast, real-world data often lack this certainty, as the unobservable nature of the true uplift is compounded by various biases and confounders.

Access to this ground truth allows us to employ the Root Mean Square Error (RMSE). It quantifies the deviation between predicted and true uplift values across the K treatments: $\text{RMSE}(\mathbf{x_i}) = \sqrt{\frac{1}{K}\sum_{k=1}^{K}\left(\hat{\tau}_k(\mathbf{x_i}) - \tau_k(\mathbf{x_i})\right)^2}$ where $\hat{\tau}_k(\mathbf{x_i})$ is the predicted uplift and $\tau_k(\mathbf{x_i})$ is the true uplift for treatment k.

5.3 Datasets

We generated 10 synthetic datasets to evaluate robustness across diverse data characteristics and difficulties. These include both interval-based (e.g. Figs. 2a, 2b, 2d) and continuous functions, linear and sinusoidal (e.g. Fig. 2c) to ensure

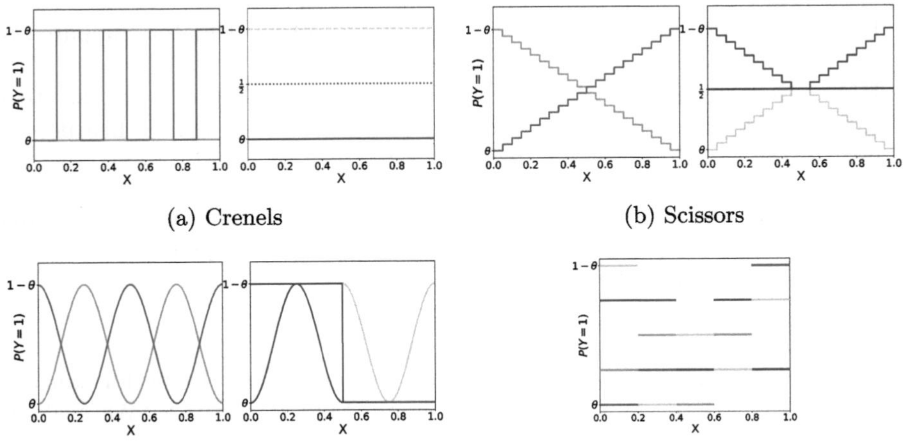

(a) Crenels (b) Scissors

(c) Sinus (d) Random

Fig. 2. Ground truth probabilities for three synthetic datasets. The figure displays the conditional probability $P(Y = 1 \mid X = x, T = t)$ as a function of the continuous feature X across five treatments. For clarity, the datasets *Crenels*, *Scissors* and *Sinus* are presented in two side-by-side plots separating treatments $T \in \{0, 1\}$ from $T \in \{2, 3, 4\}$. Each color corresponds to a treatment.

a fair evaluation as our methods are based on a discretization. Furthermore, these scenarios cover cases where treatment grouping is beneficial (i.e., effects are similar) as well as cases where there are no groupings to be made (i.e. no groupings is optimal). Each dataset defines a unique underlying causal structure and uplift pattern. The resulting datasets are: *Crenels*, *Crenels no group*, *Scissors discrete*, *Scissors continuous*, *Sinus*, *Sinus no group*, *Random sinus*, *Random sinus no group*, *Random* and *Random no group*. The datasets *Random* involve dividing the input space (X) into multiple intervals and assigning a different data generator (crenels or sinus) to each interval, chosen randomly. It allows for generating as many different treatments as desired, while ensuring that within each interval, treatments can have the same behavior or not.

To rigorously assess the limits of the estimators, we vary two key parameters. First, we introduce a parameter $\theta \in \{0, 0.1, 0.2, 0.3, 0.4\}$ that controls the uplift amplitude of the conditional probabilities. A higher θ results in a smaller divergence between treatment outcomes (i.e., lower uplift amplitude), making the uplift significantly harder to estimate. Second, we vary the number of treatments such that $T \in \{4, 8, 16, 32\}$ for the *Random* and *Random Sinus* families, allowing us to evaluate scalability and performance in high-dimensional treatment spaces, as illustrated in Fig. 2d.

For each dataset, we generate 10 training samples for each of the 8 sample sizes (ranging from 400 to 51,200 individuals per treatment), resulting in 80 training files per dataset. A large, corresponding test set of 10^5 instances per treatment is also generated to ensure a stable evaluation. The data generation

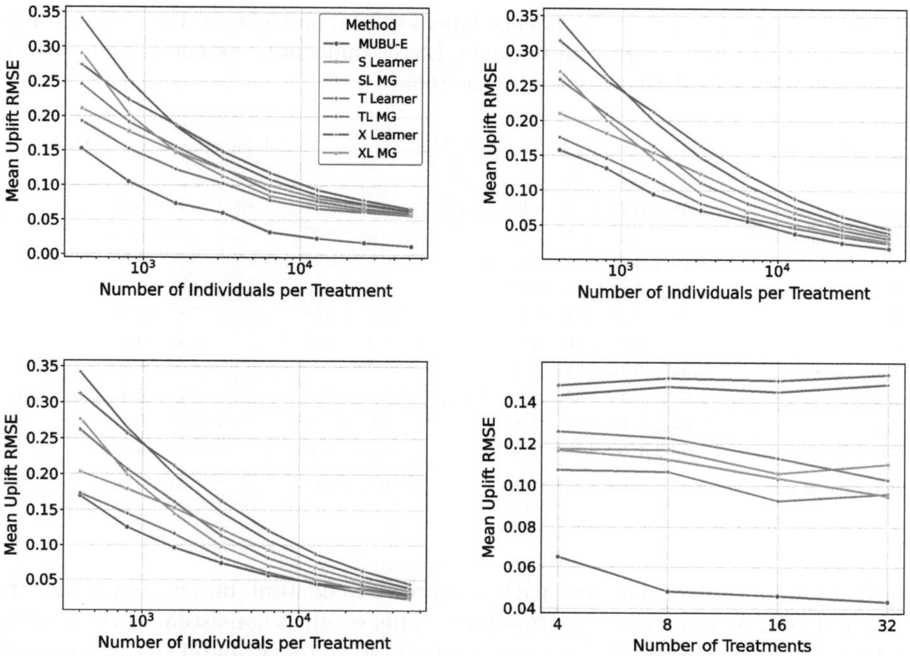

Fig. 3. The plots report the Uplift RMSE (lower is better) for different experimental settings: (Top-Left) *Crenels*; (Top-Right) *Scissors discrete*; (Bottom-Left) *Scissors continuous*; (Bottom-Right) Impact of the number of treatments on the *Random* dataset. The first three plots show performance as a function of sample size (x-axis, base-10 log scale), while the bottom-right plot shows performance as a function of the number of treatments $T \in \{4, 8, 16, 32\}$ (x-axis, base-2 log scale).

process relies on the generative models exemplified in Fig. 2. With $X \sim \mathcal{U}([0,1])$ and T sampled uniformly, the outcome Y is drawn from a Bernoulli distribution parameterized by the ground-truth probability $Y \sim \mathcal{B}(P(Y=1 \mid T, X))$.

5.4 Experimental Results

Section 5 presents our experimental results. Table 1 summarizes the mean ranks (based on RMSE) for each method across the 10 different datasets, averaged over all sample sizes and folds. Figure 3 illustrates the impact of sample size on model performance for three of these datasets, as well as the performance relative to the number of treatments on the *Random* dataset. In these figures, S Learner, T Learner and X Learner enhanced by MUBU-G are abbreviated as SL-MG, TL-MG, and XL-MG, respectively.

RQ1: How does MUBU-E perform as a direct uplift estimator compared to other methods?

As evidenced in Table 1, MUBU-E achieves superior performance compared to the SOTA methods on the majority of the datasets. This dominance is partic-

Table 1. Mean Rank ± 95% Confidence Interval based on Uplift RMSE. The last row shows the average rank across all datasets. Lower ranks indicate better performance (green and italics if MUBU-G improves the rank).

Method Dataset	MUBU-E	SL	SL MG	TL	TL MG	XL	XL MG
Scissors discrete	**1.55 ± 0.13**	4.31 ± 0.10	*1.83 ± 0.05*	6.62 ± 0.07	*4.19 ± 0.06*	6.13 ± 0.05	*3.37 ± 0.10*
Scissors continuous	2.17 ± 0.16	4.25 ± 0.10	***1.59 ± 0.05***	6.60 ± 0.07	*4.08 ± 0.07*	6.11 ± 0.05	*3.19 ± 0.11*
Crenels	**1.15 ± 0.05**	4.52 ± 0.11	*2.74 ± 0.12*	6.09 ± 0.11	*4.19 ± 0.12*	5.56 ± 0.15	*3.76 ± 0.16*
Crenels no group	**1.15 ± 0.08**	3.48 ± 0.15	*3.41 ± 0.15*	5.58 ± 0.13	*5.41 ± 0.12*	4.61 ± 0.15	*4.37 ± 0.14*
Sinus	3.34 ± 0.23	3.98 ± 0.11	***1.92 ± 0.07***	6.54 ± 0.06	*3.89 ± 0.09*	5.82 ± 0.09	*2.51 ± 0.14*
Sinus no group	5.83 ± 0.28	**2.13 ± 0.12**	2.31 ± 0.13	5.47 ± 0.11	*5.17 ± 0.10*	3.73 ± 0.22	*3.35 ± 0.21*
Random sinus	**1.58 ± 0.19**	3.68 ± 0.24	*3.61 ± 0.26*	5.70 ± 0.20	*3.97 ± 0.28*	4.72 ± 0.26	*2.50 ± 0.38*
Random sinus no group	**1.77 ± 0.19**	2.63 ± 0.26	4.53 ± 0.27	5.09 ± 0.21	5.84 ± 0.18	3.79 ± 0.32	4.35 ± 0.19
Random	**1.03 ± 0.06**	3.95 ± 0.18	*2.64 ± 0.20*	6.65 ± 0.10	*4.83 ± 0.17*	5.77 ± 0.19	*3.12 ± 0.15*
Random no group	**1.00 ± 0.00**	4.56 ± 0.23	*3.54 ± 0.32*	6.09 ± 0.20	*4.70 ± 0.19*	5.04 ± 0.30	*3.07 ± 0.18*
Average Rank	**2.12 ± 0.07**	3.85 ± 0.05	*2.56 ± 0.05*	6.13 ± 0.04	*4.53 ± 0.04*	5.31 ± 0.06	*3.43 ± 0.05*

ularly pronounced on datasets with a piecewise-constant underlying structure (e.g., *Crenels*, *Scissors discrete*, *Random*). This result is consistent with the theoretical design of MUBU-E: since the method estimates probabilities by interval, it is more easily aligned with the data-generating process in these scenarios.

More importantly, MUBU-E proves to be highly efficient on datasets generated from continuous functions (e.g., *Sinus*, *Scissors continuous*), where the discretization assumption is theoretically less advantageous. Despite this, it outperforms XGBoost-based meta-learners in most continuous scenarios. This performance extends to high-dimensional treatment spaces. As illustrated in Fig. 3 (bottom-right), MUBU-E turns high dimensionality into an asset: unlike standard learners whose performance deteriorates with added complexity, it exploits the larger number of treatments (from 4 to 32) to increase meaningful aggregation opportunities, effectively enriching the data to even improve performance. Crucially, the method excels in low-data settings, with the error margin relative to other models increasing as the number of individuals per treatment drops. Furthermore, the method maintains high performance even on datasets where treatments are distinct and no natural grouping exists, demonstrating that the criterion successfully avoids forcing artificial groupings.

RQ2: Does using MUBU-G as a model-agnostic preprocessing strategy enhance the performance of existing meta-learners?
The results demonstrate that using MUBU-G as a preprocessing strategy consistently enhances the estimation accuracy of state-of-the-art meta-learners. This improvement is observed across diverse structural configurations: notably in challenging continuous scenarios like *Scissors continuous* and *Sinus*, as well as in piecewise settings like *Crenels* and *Scissors discrete*.

This confirms that pooling information from locally similar treatments effectively improves uplift estimation. It is particularly evident when the treatment space becomes high-dimensional. As illustrated in Fig. 3 (bottom-right), while the task difficulty inherently increases with the number of treatments, the performance of methods enhanced by MUBU-G remains stable as the number of treatments increases from 4 to 32. Crucially, the approach remains safe in "worst-case" scenarios: even when treatment effects are distinct or non-beneficial (where no grouping should theoretically occur), the performance remains at least comparable to the state-of-the-art methods, ensuring that MUBU-G does not negatively impact the models when the underlying assumption of treatment similarity is not met.

6 Conclusion and Future Work

In this paper, we address the significant challenge of data sparsity and imbalance in multi-treatment uplift modeling. We propose a novel approach that leverages information pooling. Our work introduces a novel criterion that simultaneously discretizes the feature space and performs local treatment grouping.

Our primary contribution is MUBU, a parameter-free Bayesian criterion based on the MODL methodology. This criterion effectively identifies and groups treatments that exhibit similar causal effects, adapting locally to different subpopulations based on their features. Experimental results show the effectiveness of our approach. When used as a model-agnostic preprocessing strategy, MUBU-G significantly enhances the performance of existing meta-learners. Secondly, our standalone model MUBU-E consistently outperforms all uplift methods.

A direction for future work is to extend our approach to a multi-dimensional feature space, allowing the local groupings to be conditioned on multiple variables. Additionally, we plan to apply our method to large-scale, real-world datasets from marketing or personalized medicine to further validate its practical impact.

Disclosure of Interests. Nathan Le Boudec reports financial support was provided by Orange Innovation Lannion. Nicolas Voisine reports financial support was provided by Orange Innovation Lannion. Bruno Cremilleux reports financial support was provided by University of Caen Normandy.

References

1. Aoki, R., Chen, Y., Ester, M.: Multi-treatment effect estimation from biomedical data. In: Pacific Symposium on Biocomputing, vol. 27, pp. 415–426 (2022)
2. Athey, S., Imbens, G.W.: Machine learning methods for estimating heterogeneous causal effects. stat **5**(1050), 1–26 (2015)
3. Boullé, M.: A grouping method for categorical attributes having very large number of values. In: Machine Learning and Data Mining in Pattern Recognition. Lecture Notes in Computer Science, vol. 3587, pp. 228–242. Springer, Cham (2005)
4. Boullé, M.: Modl: a bayes optimal discretization method for continuous attributes. Mach. Learn. **65**(1), 131–165 (2006)

5. Chawla, N.V., Bowyer, K.W., Hall, L.O., Kegelmeyer, W.P.: Smote: synthetic minority over-sampling technique. J. Artif. Intell. Res. **16**, 321–357 (2002). https://doi.org/10.1613/jair.953
6. Fayyad, U.M., Irani, K.B.: On the handling of continuous-valued attributes in decision tree generation. Mach. Learn. **8**(1), 87–102 (1992)
7. Guelman, L., Guillén, M., Pérez-Marinín, A.M.: Uplift random forests. Cybern. Syst. **46**(3–4), 230–248 (2015)
8. Hansotia, B., Rukstales, B.: Incremental value modeling. J. Interact. Mark. **16**, 35–46 (2002)
9. Kaddour, J., Zhu, Y., Liu, Q., Kusner, M.J., Silva, R.: Causal effect inference for structured treatments. Adv. Neural. Inf. Process. Syst. **34**, 23906–23919 (2021)
10. Künzel, S.R., Sekhon, J.S., Bickel, P.J., Yu, B.: Metalearners for estimating heterogeneous treatment effects using machine learning. Proc. Natl. Acad. Sci. **116**(10), 4156–4165 (2019)
11. Le Boudec, N., Voisine, N., Crémilleux, B.: Multi-treatment uplift evaluation on non-random assignment biased data. Data Know. Eng. **163**, 102565 (2026). https://doi.org/10.1016/j.datak.2026.102565
12. Olaya, D., Coussement, K., Verbeke, W.: A survey and benchmarking study of multitreatment uplift modeling. Data Min. Knowl. Disc. **34**, 273–308 (2020)
13. Piskorz, J., Huynh, N., Berrevoets, J., Luyten, M.R., van der Schaar, M.: Enhancing treatment effect estimation with generation-driven data augmentation (2025). https://openreview.net/forum?id=K0WzGPTGwx
14. Radcliffe, N.J., Surry, P.D.: Real-world uplift modelling with significance-based uplift trees. White Paper TR-2011-1, Stochastic Solutions, pp. 1–33 (2011)
15. Rafla, M., Voisine, N., Crémilleux, B.: A non-parametric Bayesian approach for uplift discretization and feature selection. In: ECML-PKDD (2022)
16. Rubin, D.: Using propensity scores to help design observational studies: application to the tobacco litigation. Health Serv. Outcomes Res. Methodol. **2**, 169–188 (2001). https://doi.org/10.1023/A:1020363010465
17. Rubin, D.B.: Estimating causal effects of treatments in randomized and nonrandomized studies. J. Educ. Psychol. **66**(5), 688 (1974)
18. Rzepakowski, P., Jaroszewicz, S.: Decision trees for uplift modeling with single and multiple treatments. Knowl. Inf. Syst. **32**(2), 303–327 (2012)
19. Zhang, W., Li, J., Liu, L.: A unified survey of treatment effect heterogeneity modeling and uplift modeling (2021). https://arxiv.org/abs/2007.12769
20. Zhao, Y., Zhang, H., Lyu, S., Jiang, R., Gu, J., Zhang, G.: Multiple instance learning for uplift modeling. In: Proceedings of the 32nd ACM International Conference on Information and Knowledge Management, pp. 3605–3614. ACM (2023)

Exceptional Model Residual Mining, and Three Richer EMM Description Languages

Aniket Mishra, Cristiana Cărbunaru, and Wouter Duivesteijn(✉)

Technische Universiteit Eindhoven, Eindhoven, the Netherlands
{a.mishra3,c.carbunaru}@student.tue.nl, w.duivesteijn@tue.nl

Abstract. Exceptional Model Mining (EMM) seeks subgroups of the dataset that are interpretable, and display exceptional behavior. We make contributions to both of these EMM properties. To behavior, we contribute Exceptional Model Residual Mining (EMRM), gauging exceptionality of subgroup behavior on the residuals of a regression model. Existing versions of EMM with regression models focus on exceptionality of regression parameters themselves; EMRM instead discovers subgroups explaining systematic model under- or overperformance. To interpretability, we contribute three description languages. Existing work almost always restricts the subgroup search to those subsets that belong to the Conjunctive description language; we propose Polynomial Predicates, Shallow Decision Trees, and Symbolic Expressions as alternatives. Experiments on five UCI datasets illustrate what kind of subgroups EMRM can find, when optimizing for high (underperformance) and low (overperformance) residuals. We also compare the four languages, by inspecting how they fare in trading off subgroup quality and interpretability: Symbolic Expressions achieve the best trade-off, capturing highly exceptional regions with minimal complexity, while Conjunctions achieve the highest peak quality across the majority of datasets.

Keywords: Exceptional Model Mining · Subgroup Discovery · Residual Analysis · Interpretability · Description Languages

1 Introduction

Exceptional Model Mining (EMM) [8,17] seeks subgroups of a dataset that satisfy two properties. On the one hand, they need to *be interpretable*: we need to be able to concisely summarize what defines a subgroup, in language that is understandable for anyone familiar with the dataset. On the other hand, they need to *display exceptional behavior*: some kind of modeling over some target subspace of the data space must exhibit exceptional parameter values. This paper makes contributions to both properties of EMM subgroups.

Regarding exceptional behavior, we introduce *Exceptional Model Residual Mining* (EMRM), where the residuals of a black-box regression model become

M. Baratchi et al. (Eds.): IDA 2026, LNCS 16513, pp. 269–282, 2026.
https://doi.org/10.1007/978-3-032-23833-7_20

the statistics on which exceptionality of subgroup behavior is gauged. In real-world forecasting applications, model failures are often limited to specific, complex feature interactions. As an example, in an industrial solar power plant model, high residuals may occur only when a conjunction of high ambient temperature (say, $> 60°C$) and low humidity is present. Similarly, a wind turbine downtime predictor may fail systematically for a subgroup defined by a highly specific rule, such as $\texttt{Gen Speed} > 300 \wedge \texttt{Wind Speed} \geq 3 \wedge \texttt{Wind Speed} \leqslant 25 \wedge \texttt{Pitch Angle} \geq 75$. Motivated by real-world instances of localized model failure, mining for exceptional residuals with EMRM provides a direct means of discovering interpretable subgroups that explain systematic model underperformance or bias. This work is most related to other regression-based EMM papers: [17, Section 3.2] and [5,7,21].

Interpretability is typically enforced by restricting the EMM search to only those subgroups that belong to a *description language* $\mathcal{L}$. While the setting of $\mathcal{L}$ is notionally a parameter choice of EMM, this choice is almost always set to some variant of Conjunctions ($\mathcal{L}_{\text{conj}}$) of conditions on single input attributes of the dataset, often with an upper bound on the number of conjuncts. There is some variety in allowed operators *within conjuncts*: in the context of itemset and association rule mining, a line of work involves itself with negations [2,4]; in the context of subgroup discovery, subset conditions for categorical attributes and interval conditions for numerical attributes exist [19]. Such more complex conditions are typically subsequently conjoined with other conditions anyway. A rare alternative example includes disjunctions [15]: these tend to harm more than help[1] in the presence of a richer input attribute space. We introduce three new description languages for EMM: Polynomial Predicates ($\mathcal{L}_{\text{poly}}$), Shallow Decision Trees ($\mathcal{L}_{\text{tree}}$), and Symbolic Expressions ($\mathcal{L}_{\text{symb}}$).

We show what subgroups EMRM discovers on five datasets with a regression target, separately for each of the four description languages $\mathcal{L}_{\text{conj}}$, $\mathcal{L}_{\text{poly}}$, $\mathcal{L}_{\text{tree}}$, and $\mathcal{L}_{\text{symb}}$. We then compare the four languages in terms of the trade-off between subgroup quality and interpretability. We find that highly compact, algebraically rich expressions as provided by $\mathcal{L}_{\text{symb}}$ offer the most diagnostic insight.

2 Exceptional Model Residual Mining: Preliminaries

Assume a trained black-box model $f : \mathcal{X} \to \mathbb{R}$ and a dataset $D = ((x_i, y_i))_{i=1}^{n}$. The domain $\mathcal{X}$ is a Cartesian product of domains of individual input *attributes*: $\mathcal{X} = \times_{i=1}^{k} \mathcal{X}_i$. In both Subgroup Discovery (SD) [11,16,26] and Exceptional Model Mining (EMM) [8,17], we seek *subgroups* which are induced by *descriptions* on this dataset. Formally, a *description* is a function $d : \mathcal{X} \to \{0, 1\}$. This function determines for each record in the dataset whether it is *covered* by the

[1] Disjunctive descriptions suffer from repetitive results. Especially when many numerical or high-cardinality categorical attributes are present, many near-copies of high-quality subgroups can be created by disjoining conditions that cover a tiny number of records. Thus, disjunctive descriptions pollute the search space with many repetitions of the same underlying concepts, causing trouble for many search algorithms.

description; the *subgroup* S_d induced by d is the set of records covered by d: $S_d = \{(x_i, y_i) \in D \mid d(x_i) = 1\}$. We restrict the SD/EMM search algorithm to consider only those descriptions encompassed by a *description language* $\mathcal{L}$.

Many task formulations of SD/EMM exist; the most common is to report the top-k best subgroups visited by the selected search algorithm. "Best" is governed by the definition of a *Quality Measure* (QM) $q : \mathcal{L} \to \mathbb{R}$. The QM is formally defined to assess the quality of a description, because subgroup induction from a description is not an injective function: if the QM input would be defined as the subgroup, the QM may not use properties from the description since it may not be unique; if the QM input is defined as the description, this induces a unique subgroup and so the QM may use properties of both description and subgroup.

SD and EMM are differentiated by their assessment of subgroup exceptionality. SD assesses the extent to which the *distribution* of the target variable y_i differs from that same distribution on the full dataset; the target is typically a single dimension of the dataset. EMM assesses the extent to which *some modeling* over the target space y_i has *exceptional parameters*; the target space typically spans more than one dimension, and the modeling typically evaluates some interaction between these dimensions. In this paper, we find ourselves straddling SD and EMM. We are looking at regression models with only a single output variable y_i, which looks like the single-target setting which is typical of SD. But we are not interested in the distribution of y_i itself; instead, we are building a regression model and then we employ the residuals from that model for assessing exceptionality of subgroups, which looks like the exceptionality of parameters of some modeling over a target space which is typical of EMM.

3 Related Work

Using a black-box model's residuals as an exceptional concept is not a new idea: this transition from diagnosing the underlying data to diagnosing the model's behavior is a current practice for modern performance analysis of regression models [12,22,23]. Interpretable subgroups can be identified where the model consistently exhibits over- or under-fitting, revealing systematic biases or a combination of features that cause poor predictions [12,23]. The utility of this is demonstrated by work that applies SD to output, such as providing interpretable summaries of black-box incident triage decisions [23], or utilizing residual network architectures to enhance interpretability [12].

Work on improving description languages for SD/EMM mainly operates inside the standard conjunctive template, by enriching the atomic predicates, for example, optimal numeric intervals and nominal value sets [19]. In contrast, richer logical description languages $\mathcal{L}$ that can represent interactions more directly are also considered, including shallow decision trees, low-degree polynomial predicates, and compact symbolic expressions [3,18]. Because numeric handling is central to these constructions, a local discretization strategy guided by a systematic SD framework is adopted [20]. This study aims to discover patterns that are both expressive and sparse, by explicitly constraining model depth/degree/length, and penalizing complexity to preserve interpretability [3], which is particularly important for fairness-aware subgroup analyses [13] and highlighting heterogeneous causal structure where richer rules are needed [18].

The application of EMM to regression models requires careful choice of the exceptionality metric. The first regression model class for EMM [17, Section 3.2] allowed for only one input and one output variable in the regression model. The second [7] generalized this to any number of input variables for linear regression. A recent extension incorporates mixed models [21]. Various Goodness-of-Fit metrics have also been explored within regression-EMM [5].

4 Three Richer EMM Description Languages

We discuss four description languages, three of which are new to EMM. We impose some bounds and constraints on each language, to ensure that the resulting subgroups remain human-interpretable. For ease of comparison, we start by formalizing the existing conjunctive language. The Conjunctive description language $\mathcal{L}_{\text{conj}}$ uses conjunctions of attribute-value conditions. It is the most interpretable form, expressing subgroups as a set of jointly satisfied predicates.

$$\mathcal{L}_{\text{conj}} = \{x_i\,\theta_i\,v_i \ \wedge\ \cdots\ \wedge\ x_j\,\theta_j\,v_j\}$$

We set an upper bound of 5 on the number of conjuncts. Each value v_i is taken from the domain $\mathcal{X}_i$ of their corresponding attribute x_i; this domain also governs which operators θ_i are allowed ($\leqslant$ and $\geq$ for numeric attributes; $=$, $\neq$, and $\in$ for categorical attributes, ...). An example description (based on the Boston Housing dataset) identifies a subgroup of expensive houses with many rooms and low lower-status population:

$$(\texttt{RM} > 7) \wedge (\texttt{LSTAT} < 5)$$

4.1 Polynomial Predicates

The Polynomial Predicates description language $\mathcal{L}_{\text{poly}}$ captures nonlinear relationships through polynomial combinations of numeric features [6]. It allows expressing curved decision boundaries or interaction effects.

$$\mathcal{L}_{\text{poly}} = \left\{ a_0 + \sum_i a_i f_i + \sum_{i,j} a_{ij} f_i f_j \ \leqslant\ \tau \right\}$$

To maintain tractability and interpretability, polynomial predicates are restricted to degree at most two. Coefficients are estimated via least squares over candidate feature pairs, and thresholds are quantized to a fixed grid. This limits the hypothesis space to $\mathcal{O}(p^2)$ candidate monomials for p features, avoiding combinatorial explosion. An example description identifies non-linear relation conditions between room count and status ratio that predicts higher prices:

$$0.6 \cdot \texttt{RM}^2 - 0.4 \cdot \texttt{LSTAT} < 30$$

4.2 Shallow Decision Trees

The Shallow Decision Trees description language $\mathcal{L}_{\text{tree}}$ represents subgroups as a decision tree of limited depth. Each path corresponds to a conjunction of conditions learned from data, offering a balance between interpretability and expressiveness [9].

$$\mathcal{L}_{\text{tree}} = \{\text{shallow trees with depth } \leqslant 3\}$$

We set an upper bound of 3 to the depth of the tree. The search integrates the `lnca` approach to locally discretize numeric descriptors from [20]. An example path defines a subgroup where the model tends to under-predict for affluent neighborhoods with large houses:

$$\texttt{LSTAT} < 10 \rightarrow \texttt{RM} > 6 \rightarrow \texttt{Residual} > 2$$

4.3 Symbolic Expressions

The Symbolic Expressions description language $\mathcal{L}_{\text{symb}}$ allows flexible symbolic expressions (e.g., ratios, differences, aggregates) generated through symbolic regression [1], providing interpretable yet expressive formulas.

$$\mathcal{L}_{\text{symb}} = \{\text{combinations of operators } +, -, \times, \div, \leqslant, >, =\}$$

We set an upper bound of 3 on the number of operators, and an upper bound of 2 on the number of distinct features. Simple ratio or difference formulas are allowed, such as $\frac{f_i}{f_j} \leqslant \tau$ or $(f_i - f_j) > \tau$, which cover many interpretable relationships (e.g., efficiency ratios or deviations). Thresholds τ are quantized to a predefined resolution. This keeps the search space finite and the resulting expressions human-readable. An example description indicates houses where the ratio of room count to lower-status percentage is high, showing systematic over-prediction:

$$\frac{\texttt{RM}}{\texttt{LSTAT}} > 1.5$$

4.4 Assessing Interpretability Across Languages

The more complex we allow a description language to be, the more likely it is to discover high-quality subgroups. Such an optimization would ignore a core property of subgroups, which is their inherent interpretability. Instead of ignoring this dimension, ideally we would want to trade off subgroup quality and interpretability. To do so, we must have a formal definition of interpretability that can function across the four description languages in this paper. We start by inverting interpretability into complexity:

$$I(d) = \frac{1}{1 + \mathcal{C}(d)}$$

The complexity function aggregates measurable aspects of the descriptions:

$$\begin{aligned}
\mathcal{L}_{\text{conj}} &: \mathcal{C}(d) = \#\text{predicates}(d) \\
\mathcal{L}_{\text{poly}} &: \mathcal{C}(d) = \#\text{terms}(d) + \text{degree}(d) \\
\mathcal{L}_{\text{tree}} &: \mathcal{C}(d) = \text{depth}(d) + \#\text{splits}(d) \\
\mathcal{L}_{\text{symb}} &: \mathcal{C}(d) = \#\text{tokens}(d) \text{ (total count of operands + operators)}
\end{aligned}$$

Alternatively, one could formally analyze the complexity of descriptions from all languages using the Minimum Description Length (MDL) principle [10,24]. For a set of discovered descriptions, applying the MDL principle requires deriving the length in bits of the descriptions, and the length in bits of the data when compressed using the descriptions. This principle has proven to be very effective in discovering the set of itemsets that together optimally describe the data [25]. However, here, we are not concerned with itemset mining: we introduce three languages for an EMM task with exceptional regression residuals. Design choices must be made to enable application of MDL on individual descriptions from these languages, while doing justice to what the subgroups represent in target space. We do not think that this is impossible, but for the purposes of the comparison we are making in this paper, the cruder complexity function defined above will suffice.

5 Exceptional Model Residual Mining: Continued

Given the model f and dataset D, let the residuals be defined by $r_i = (y_i - f(x_i))^2$, capturing pointwise prediction error magnitude for each instance. As QM we define the residual exceptionality score, $q_{\text{res}}(d)$, which measures the deviation of the subgroup's mean residual from the global mean, normalized by global variance and subgroup size, similar to a standard error:

$$q_{\text{res}}(d) = \frac{|\mathbb{E}_{x \in S_d}[r(x)] - \mathbb{E}_{x \in D}[r(x)]|}{\sqrt{\text{Var}_{x \in D}[r(x)]/|S_d|}}$$

This metric behaves similarly to a z-score, highlighting subgroups where residuals deviate substantially beyond what is expected by chance. A high $q_{\text{res}}(d)$ shows a subgroup where the model systematically underperforms (large residuals) or overperforms (small residuals), showing bias (underfitting) or overfitting.

We evaluate candidate subgroups in terms of $q(d)$ and $I(d)$. Pareto-optimal solutions are retained to visualize the interpretability-quality frontier (cf. Fig. 1; circle sizes are commensurate with the corresponding subgroup size). This frontier illustrates the set of non-dominated solutions, showing the maximum achievable subgroup quality for any given level of description complexity. The resulting plot clearly visualizes the trade-off of each language, illustrating, for instance, how $\mathcal{L}_{\text{symb}}$ clusters toward the lowest complexity while $\mathcal{L}_{\text{tree}}$ explores a wider complexity range.

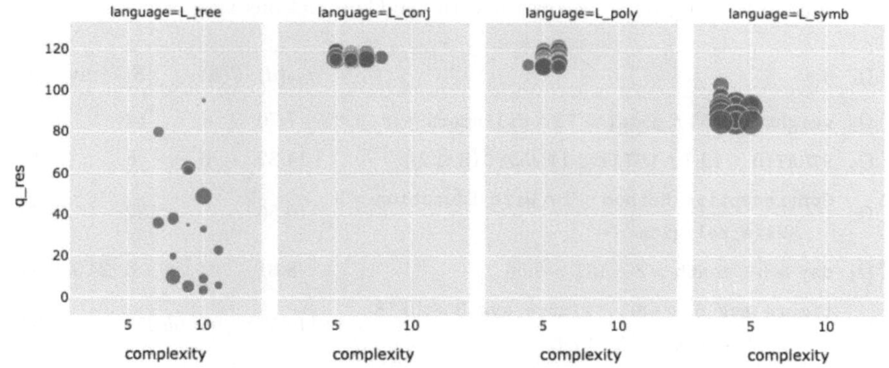

Fig. 1. Quality vs. complexity tradeoff: Pareto-optimal (Blue) and other (red) subgroups, for (left-to-right) $\mathcal{L}_{\text{tree}}$, $\mathcal{L}_{\text{conj}}$, $\mathcal{L}_{\text{poly}}$, and $\mathcal{L}_{\text{symb}}$. (Color figure online)

Table 1. Dataset metadata.

Ω_i	**Name**	n	k	**Target**	**Attributes**
Ω_1	Auto-mpg	398	9	`class`	numeric
Ω_2	Boston Housing	506	14	`MEDV`	numeric
Ω_3	Contraceptive Method Choice	1 473	10	`Wife_Age`	mixed
Ω_4	Forest Fires	517	13	`area`	numeric
Ω_5	Year Prediction MSD	515 345	90	`year`	numeric

6 Experimental Setup

We run experiments on five regression datasets, selected to vary systematically in terms of dimensionality, size, and feature interpretability. All datasets can be found in the UCI ML repository [14], except for Boston Housing which used to be in that repository but can now be found in scikit-learn, on Kaggle, and elsewhere online. Metadata on the five datasets can be found in Table 1.

We pre-train a Random Forest as black-box regressor on each dataset. Its pointwise residuals $r_i = (y_i - f(x_i))^2$ are then treated as model-induced numeric targets, which are fed to the QM q_{res}. Candidate subgroup descriptions $d \in \{\mathcal{L}_{\text{conj}}, \mathcal{L}_{\text{poly}}, \mathcal{L}_{\text{tree}}, \mathcal{L}_{\text{symb}}\}$ are jointly assessed on subgroup quality $q(d)$ and interpretability $I(d)$. The resulting Pareto set $\{(q(d), I(d))\}$ visualizes the interpretability-quality frontier, highlighting both over-performing (low-residual) and under-performing (high-residual) subgroups.

For each EMM run, we gather the top-25 (by quality) non-dominated (by quality and interpretability) subgroups discovered by beam search [8, Algorithm 1]. We set the beam width to a generous $w = 200$ to promote an extensive exploration of the search space. We locally discretize the numeric descriptors with the `lnca` strategy from [20] with 5 cut points.

Table 2. Subgroups discovered with the Conjunctions language $\mathcal{L}_{\text{conj}}$.

Ω_i	d	$q_{\text{res}}(d)$	$\mathcal{C}(d)$	$\lvert S \rvert$	$\text{avg}_S\, r_i$
Ω_1	weight $> 2155 \wedge$ model $> 79 \wedge$ cylinders > 4	7.70	4	14	6.99
Ω_2	PTRATIO $> 19.7 \wedge$ LSTAT $\leqslant 11.45 \wedge$ DIS $\leqslant 2.1$	14.82	6	8	24.47
Ω_3	Contraceptive_Method $\leqslant 1 \wedge$ Wife_Education > 3 $\wedge$Wife_religion > 0	10.56	5	132	37.72
Ω_4	day $> 6 \wedge$ month $> 8 \wedge$ ISI > 8.56	8.81	4	6	24 018.55
Ω_5	timbre_avg_6 $> -8.1 \wedge$ timbre_avg_3 > 16.18 $\wedge$timbre_cov_3_3 > 19.42	119.57	5	65 956	24.63

7 Experimental Results

In the remainder of this paper, you will find tables of results, where for each combination of $\mathcal{L}$ and D we display one hand-picked representative subgroup from the top-25 non-dominated discovered subgroups. A comprehensive overview of all top-25 s of discovered non-dominated subgroups, for each $\mathcal{L}$-D-combination, can be found in our GitHub repository[2], along with all datasets and code.

7.1 Exceptional Subgroups per Description Language

We let beam search determine the top Pareto-optimal subgroups when the QM q_{res} is to be maximized, that is: high positive residuals, systematic underperformance of the regressor. Results are listed in Tables 2-5.

Conjunctions: $\mathcal{L}_{\text{conj}}$ (Baseline). Table 2 lists the top Pareto-optimal subgroups discovered with the Conjunctive language $\mathcal{L}_{\text{conj}}$. Highly exceptional failure regions are effectively detected, with interpretable subgroups found in low-dimensional datasets like Boston Housing and complex ones like Forest Fires (i.e., late weekend fires during dry season).

Polynomial Predicates: $\mathcal{L}_{\text{poly}}$. Table 3 lists the subgroups discovered with the Polynomial Predicates language $\mathcal{L}_{\text{poly}}$, which aimed to introduce non-linear interaction terms. The subgroup for Forest Fires shows a simple conjunctive structure being preferred by the algorithm, while the Year Prediction MSD rule demonstrates the language's core advantage by capturing a non-linear interaction term: timbre_avg_3 $\times$ timbre_cov_3_3 $> 1758 \wedge$ timbre_avg_1 $\times$ timbre_avg_6 > -409.9, achieving the highest peak quality of $q_{\text{res}} = 121.73$ for this complex domain.

[2] https://github.com/Aniket-Mishra/EMRM/tree/IDA_2026_Submission.

Table 3. Subgroups discovered with the Polynomial Predicates language $\mathcal{L}_{\text{poly}}$.

Ω_i	d	$q_{\text{res}}(d)$	$\mathcal{C}(d)$	$\lvert S\rvert$	$\text{avg}_S\, r_i$
Ω_1	$\texttt{horsepower} \times \texttt{weight} > 2.281 \cdot 10^5 \wedge \texttt{model} > 79$	5.82	5	30	4.13
Ω_2	$\texttt{DIS} \leqslant 1.951 \wedge \texttt{LSTAT} \leqslant 13.66$	10.36	5	21	11.38
Ω_3	$\texttt{Children} \times \texttt{Husband\_Occupation} \leqslant 2$ $\wedge \texttt{Contraceptive\_Method} \times \texttt{Husband\_Occupation} \leqslant 1$	9.35	9	95	38.82
Ω_4	$\texttt{day} > 6 \wedge \texttt{temp} > 24.1$	5.25	3	16	9 131.29
Ω_5	$\texttt{timbre\_avg\_3} \times \texttt{timbre\_cov\_3\_3} > 1758$ $\wedge \texttt{timbre\_avg\_1} \times \texttt{timbre\_avg\_6} > -409.9$	121.73	6	61 841	25.30

Shallow Decision Trees: $\mathcal{L}_{\text{tree}}$. Table 4 lists the subgroups discovered with the Shallow Decision Trees language $\mathcal{L}_{\text{tree}}$, which captures sequential and hierarchical relationships. The subgroup for Forest Fires is highly specific to a narrow temperature band, showing the tree's ability to partition data based on thresholds, achieving a q_{res} competitive with $\mathcal{L}_{\text{conj}}$. Notably, $\mathcal{L}_{\text{tree}}$ achieved the highest peak quality on the Year Prediction MSD dataset, with $q_{\text{res}} = 95.25$.

Symbolic Expressions: $\mathcal{L}_{\text{symb}}$. Table 5 lists the subgroups discovered with the Symbolic Expressions language $\mathcal{L}_{\text{symb}}$, providing the most compact and highly effective subgroups, demonstrating an excellent trade-off between quality and complexity, particularly on the complex Year Prediction MSD dataset. The Forest Fires subgroup is the simplest, with $q_{\text{res}} = 3.84$, using a single feature predicate ($\texttt{temp} > 24.6$), demonstrating $\mathcal{L}_{\text{symb}}$'s preference for maximum simplicity.

7.2 Exceptional Subgroups with Inverted QM

Alternatively, we can tell beam search to determine the top Pareto-optimal subgroups when the QM q_{res} is to be minimized: low positive residuals, systematic overperformance of the regressor. We do so for the Shallow Decision Trees language $\mathcal{L}_{\text{tree}}$; results are listed in Table 6. The consistent discovery of simple subgroups with substantial negative residual deviation demonstrates discovery of regions where the model is highly accurate, often corresponding to the most common or easily characterized segments of the input space.

7.3 Comparing the Languages

Table 7 summarizes conclusions of the competition between the four languages $\mathcal{L}_{\text{conj}}$, $\mathcal{L}_{\text{poly}}$, $\mathcal{L}_{\text{tree}}$, and $\mathcal{L}_{\text{symb}}$, on all five datasets, in qualitative (subgroup quality $q_{\text{res}}(d)$ and interpretability $I(d)$) and quantitative (structural insight) terms.

Before running the experiments, we expected that shallow trees or low-degree polynomials would offer the best balance between performance and interpretability. Our hypothesis was refuted by the dominance of Symbolic Expressions

Table 4. Subgroups discovered with the Shallow Decision Trees language $\mathcal{L}_{\text{tree}}$.

(a) Metadata.

Ω_i	$q_{\text{res}}(d)$	$\mathcal{C}(d)$	$\lvert S\rvert$	$\text{avg}_S\, r_i$
Ω_1	7.41	3	6	9.77
Ω_2	8.05	3	10	12.63
Ω_3	10.99	11	55	53.01
Ω_4	8.55	4	6	23 331.97
Ω_5	95.25	10	7 055	43.39

(b) Ω_1 subgroup.

```
|--- model <= 79.50
|   |--- displacement <= 90.50
|   |   |--- model <= 77.50
|   |   |   |--- weight <= 2091.00
|   |   |   |   |--- value: [0.87]
|   |   |   |--- weight >  2091.00
|   |   |   |   |--- value: [3.20]
|   |   |--- model >  77.50
|   |   |   |--- value: [3.80]
|   |--- displacement >  90.50
|   |   |--- weight <= 3199.00
|   |   |   |--- acceleration <= 13.35
|   |   |   |   |--- value: [1.24]
|   |   |   |--- acceleration >  13.35
|   |   |   |   |--- value: [0.57]
|   |   |--- weight >  3199.00
|   |   |   |--- weight <= 4057.50
|   |   |   |   |--- value: [0.35]
|   |   |   |--- weight >  4057.50
|   |   |   |   |--- value: [0.18]
|--- model >  79.50
|   |--- displacement <= 212.50
|   |   |--- acceleration <= 13.85
|   |   |   |--- value: [7.81]
|   |   |--- acceleration >  13.85
|   |   |   |--- displacement <= 99.00
|   |   |   |   |--- value: [2.64]
|   |   |   |--- displacement >  99.00
|   |   |   |   |--- value: [0.86]
|   |--- displacement >  212.50
|   |   |--- value: [9.77]
```

(c) Ω_2 subgroup.

```
|--- DIS <= 1.34
|   |--- value: [12.63]
|--- DIS >  1.34
|   |--- RM <= 8.28
|   |   |--- RM <= 6.77
|   |   |   |--- DIS <= 1.42
|   |   |   |   |--- value: [4.18]
|   |   |   |--- DIS >  1.42
|   |   |   |   |--- value: [0.79]
|   |   |--- RM >  6.77
|   |   |   |--- INDUS <= 6.66
|   |   |   |   |--- value: [1.42]
|   |   |   |--- INDUS >  6.66
|   |   |   |   |--- value: [4.02]
|   |--- RM >  8.28
|   |   |--- value: [12.44]
```

(d) Ω_3 subgroup.

```
|--- Husband_Occupation <= 1.50
|   |--- Contraceptive_Method <= 1.50
|   |   |--- Children <= 1.50
|   |   |   |--- SOLI <= 2.50
|   |   |   |   |--- value: [9.10]
|   |   |   |--- SOLI >  2.50
|   |   |   |   |--- value: [53.01]
|   |   |--- Children >  1.50
|   |   |   |--- Children <= 5.50
|   |   |   |   |--- value: [25.79]
|   |   |   |--- Children >  5.50
|   |   |   |   |--- value: [4.04]
|   |--- Contraceptive_Method >  1.50
|   |   |--- Children <= 5.50
|   |   |   |--- SOLI <= 3.50
|   |   |   |   |--- value: [8.77]
|   |   |   |--- SOLI >  3.50
|   |   |   |   |--- value: [16.66]
|   |   |--- Children >  5.50
|   |   |   |--- Children <= 6.50
|   |   |   |   |--- value: [1.99]
|   |   |   |--- Children >  6.50
|   |   |   |   |--- value: [5.45]
|--- Husband_Occupation >  1.50
|   |--- Contraceptive_Method <= 1.50
|   |   |--- SOLI <= 2.50
|   |   |   |--- Children <= 5.50
|   |   |   |   |--- value: [8.15]
|   |   |   |--- Children >  5.50
|   |   |   |   |--- value: [3.41]
|   |   |--- SOLI >  2.50
|   |   |   |--- Wife_Education <= 3.50
|   |   |   |   |--- value: [13.64]
|   |   |   |--- Wife_Education >  3.50
|   |   |   |   |--- value: [24.27]
|   |--- Contraceptive_Method >  1.50
|   |   |--- SOLI <= 2.50
|   |   |   |--- Husband_Education <= 3.50
|   |   |   |   |--- value: [3.66]
|   |   |   |--- Husband_Education >  3.50
|   |   |   |   |--- value: [6.20]
|   |   |--- SOLI >  2.50
|   |   |   |--- Children <= 3.50
|   |   |   |   |--- value: [10.03]
|   |   |   |--- Children >  3.50
|   |   |   |   |--- value: [6.33]
```

(e) Ω_4 subgroup.

```
|--- temp <= 25.05
|   |--- DMC <= 103.55
|   |   |--- temp <= 24.15
|   |   |   |--- ISI <= 11.05
|   |   |   |   |--- value: [20.08]
|   |   |   |--- ISI >  11.05
|   |   |   |   |--- value: [64.27]
|   |   |--- temp >  24.15
|   |   |   |--- value: [233.29]
|   |--- DMC >  103.55
|   |   |--- DC <= 434.65
|   |   |   |--- value: [1228.09]
|   |   |--- DC >  434.65
|   |   |   |--- Y <= 2.50
|   |   |   |   |--- value: [409.17]
|   |   |   |--- Y >  2.50
|   |   |   |   |--- value: [83.46]
|--- temp >  25.05
|   |--- temp <= 25.45
|   |   |--- value: [23331.97]
|   |--- temp >  25.45
|   |   |--- DMC <= 220.05
|   |   |   |--- DMC <= 186.25
|   |   |   |   |--- value: [246.53]
|   |   |   |--- DMC >  186.25
|   |   |   |   |--- value: [1322.82]
|   |   |--- DMC >  220.05
|   |   |   |--- value: [14905.53]
```

(f) Ω_5 subgroup.

```
|--- timbre_avg_6 <= -8.76
|   |--- timbre_avg_1 <= 42.61
|   |   |--- timbre_cov_7_10 <= -23.93
|   |   |   |--- timbre_cov_6_10 <= 15.76
|   |   |   |   |--- value: [12.76]
|   |   |   |--- timbre_cov_6_10 >  15.76
|   |   |   |   |--- value: [18.77]
|   |   |--- timbre_cov_7_10 >  -23.93
|   |   |   |--- timbre_avg_3 <= -7.14
|   |   |   |   |--- value: [7.75]
|   |   |   |--- timbre_avg_3 >  -7.14
|   |   |   |   |--- value: [12.16]
|   |--- timbre_avg_1 >  42.61
|   |   |--- timbre_avg_6 <= -15.95
|   |   |   |--- timbre_avg_1 <= 49.36
|   |   |   |   |--- value: [6.51]
|   |   |   |--- timbre_avg_1 >  49.36
|   |   |   |   |--- value: [4.10]
|   |   |--- timbre_avg_6 >  -15.95
|   |   |   |--- timbre_cov_1_11 <= 214.01
|   |   |   |   |--- value: [10.85]
|   |   |   |--- timbre_cov_1_11 >  214.01
|   |   |   |   |--- value: [6.31]
|--- timbre_avg_6 >  -8.76
|   |--- timbre_avg_3 <= 26.79
|   |   |--- timbre_avg_1 <= 40.38
|   |   |   |--- timbre_avg_3 <= -16.76
|   |   |   |   |--- value: [13.09]
|   |   |   |--- timbre_avg_3 >  -16.76
|   |   |   |   |--- value: [21.70]
|   |   |--- timbre_avg_1 >  40.38
|   |   |   |--- timbre_cov_1_8 <= 683.75
|   |   |   |   |--- value: [12.68]
|   |   |   |--- timbre_cov_1_8 >  683.75
|   |   |   |   |--- value: [6.43]
|   |--- timbre_avg_3 >  26.79
|   |   |--- timbre_avg_2 <= -54.36
|   |   |   |--- timbre_cov_7_10 <= -104.34
|   |   |   |   |--- value: [43.39]
|   |   |   |--- timbre_cov_7_10 >  -104.34
|   |   |   |   |--- value: [24.19]
|   |   |--- timbre_avg_2 >  -54.36
|   |   |   |--- timbre_cov_3_3 <= 37.20
|   |   |   |   |--- value: [15.84]
|   |   |   |--- timbre_cov_3_3 >  37.20
|   |   |   |   |--- value: [24.13]
```

Table 5. Subgroups discovered with the Symbolic Expressions language $\mathcal{L}_{\text{symb}}$. For CMC, both rules are equivalent Pareto solutions producing the same results.

Ω_i	d	$q_{\text{res}}(d)$	$\mathcal{C}(d)$	$\lvert S\rvert$	$\text{avg}_S\, r_i$
Ω_1	(model + origin) > 80	5.55	3	103	2.65
Ω_2	(CHAS + RAD) > 24	8.30	3	8	14.34
Ω_3	Wife_Education − Contraceptive_Method > 2 Wife_Education ÷ Contraceptive_Method > 3	8.75	3	175	30.73
Ω_4	temp > 24.6	3.84	2	71	3 537.61
Ω_5	(timbre_avg_6 + timbre_avg_3) > 33.77	103.25	3	73 605	22.24

Table 6. Subgroups discovered with the Shallow Decision Trees language $\mathcal{L}_{\text{tree}}$, when minimizing rather than maximizing q_{res}.

Ω_i	d	$q_{\text{res}}(d)$	$\text{avg}_S\, r_i$	Interpretation
Ω_1	model ⩽ 79.5 ∧ displacement > 90.5 ∧ weight ⩽ 3199.0 ∧ acceleration ⩽ 13.35	0.18	1.24	Older, mid-to-heavy cars. The model confidently predicts low mileage.
Ω_2	DIS > 1.34 ∧ RM ⩽ 8.28 ∧ RM > 6.77 ∧ INDUS ⩽ 6.66	0.04	1.42	Residential area with moderate rooms, far from employment.
Ω_3	Husband_Occupation ⩽ 1.5 ∧ Contraceptive_Method ⩽ 1.5 ∧ Children ⩽ 1.5 ∧ SOLI ⩽ 2.5	0.38	9.09	Married women with high-mid occupation/method use.
Ω_4	temp ⩽ 25.05 ∧ DMC > 103.55 ∧ DC > 434.65 ∧ Y ⩽ 2.5	0.12	409.17	Fires occurring under specific low temperature/dry fuel conditions.
Ω_5	timbre_avg_6 ⩽ −8.76 ∧ timbre_avg_1 ⩽ 42.61 ∧ timbre_cov_7_10 > −23.93 ∧ timbre_avg_3 > −7.14	3.40	12.16	Highly constrained, specific area of the timbre space.

($\mathcal{L}_{\text{symb}}$) at the lowest complexity levels. This language achieves high q_{res} scores with minimal complexity (e.g., $q_{\text{res}} = 103.25$ with $\mathcal{C}(d) = 3$ on Year Prediction MSD in Table 5). This shows that a constrained search over fundamental arithmetic expressions yields superior results.

The efficacy of each language is domain-dependent. $\mathcal{L}_{\text{conj}}$ remains the most robust generalist, succeeding where model errors align with orthogonal features, achieving the highest peak quality on three out of five datasets: Auto-mpg, Boston Housing, and Forest Fires. $\mathcal{L}_{\text{poly}}$ achieves the highest peak quality of

Table 7. Winners and outcomes of quantitative ($q_{\text{res}}(d)$ and $I(d)$) and qualitative (structural insight) competition between the four languages. Values represent peak performance achieved within each category/dataset.

Ω_i	$q_{\text{res}}(d)$	$\mathcal{L}$	$I(d)$	$\mathcal{L}$	Structural Insight	$\mathcal{L}$
Ω_1	7.70	$\mathcal{L}_{\text{conj}}$	1.16	$\mathcal{L}_{\text{symb}}$	Hierarchical, mechanical conditions	$\mathcal{L}_{\text{conj}}$
Ω_2	14.82	$\mathcal{L}_{\text{conj}}$	1.01	$\mathcal{L}_{\text{poly}}$ / $\mathcal{L}_{\text{symb}}$	Spatial socioeconomic topology	$\mathcal{L}_{\text{conj}}$
Ω_3	10.99	$\mathcal{L}_{\text{tree}}$	1.04	$\mathcal{L}_{\text{symb}}$	Hierarchical socioeconomic conditions	$\mathcal{L}_{\text{tree}}$
Ω_4	8.81	$\mathcal{L}_{\text{conj}}$	0.97	$\mathcal{L}_{\text{symb}}$	Seasonal and environmental structures	$\mathcal{L}_{\text{conj}}$
Ω_5	121.73	$\mathcal{L}_{\text{poly}}$	3.56	$\mathcal{L}_{\text{symb}}$	Non-linear dependency in bands	$\mathcal{L}_{\text{poly}}$

$q_{\text{res}} = 121.73$ on the complex Year Prediction MSD dataset, highlighting its ability to use non-linear interaction terms to capture specific, high-magnitude errors. $\mathcal{L}_{\text{symb}}$ provides the greatest overall gain in insight and performance at high interpretability levels, capturing succinct non-linear functional relationships (e.g., ratios, sums, differences). $\mathcal{L}_{\text{poly}}$ and $\mathcal{L}_{\text{tree}}$ introduce structural complexity that often limits the search algorithm's ability to find the absolute exceptionality peak, making them less competitive for this specific residual-based EMM task under tight complexity constraints, even though $\mathcal{L}_{\text{tree}}$ does achieve the peak quality on the CMC dataset ($q_{\text{res}} = 10.99$).

8 Conclusions

We introduce the task of Exceptional Model Residual Mining (EMRM): finding interpretable subgroups in a dataset with a regression task, where the regression residuals are exceptionally high or low. Previous work on Exceptional Model Mining (EMM) with regression models focused on exceptional parameters of the regression model itself. EMRM instead focuses on exceptionality of the regression *residuals*, steering the search towards interpretable subgroups of the input space where the regressor performs particularly well or badly. Moreover, this paper provides new answers to the question what exactly makes a subgroup interpretable, by introduce three description languages for EMM: Polynomial Predicates ($\mathcal{L}_{\text{poly}}$), Shallow Decision Trees ($\mathcal{L}_{\text{tree}}$), and Symbolic Expressions ($\mathcal{L}_{\text{symb}}$). On five UCI datasets with a regression target, we illustrate the kinds of subgroups these languages can find, as well as the standard Conjunctive baseline $\mathcal{L}_{\text{conj}}$. The experimental results demonstrate that $\mathcal{L}_{\text{symb}}$ offers the most effective balance, achieving peak quality at minimal interpretability cost.

The comparative analysis provides that while the Conjunctive language ($\mathcal{L}_{conj}$) achieves the highest peak q_{res} on the majority of datasets, the most effective language for balancing exceptional subgroup quality (q_{res}) and human interpretability ($I(d)$) is the Symbolic Expressions language ($\mathcal{L}_{\text{symb}}$). The power of $\mathcal{L}_{\text{symb}}$ lies in its ability to achieve strong performance (e.g., $q_{\text{res}} = 103.25$ on the Year Prediction MSD dataset, reported in Table 5) with minimal complexity, favoring simple arithmetic relationships that are often more insightful than complex conjunctive rules. This result pushes the state-of-the-art by demonstrating that maximal diagnostic insight is achieved not through the complexity of the logical structure (as in $\mathcal{L}_{\text{tree}}$ or $\mathcal{L}_{\text{conj}}$), but through constrained algebraic conciseness. While the Conjunctive language ($\mathcal{L}_{\text{conj}}$) proves to be competitive on lower-dimensional data, $\mathcal{L}_{\text{symb}}$ succeeds in revealing fundamental functional relationships such as ratios, sums, or deviations that capture the underlying non-linear error topology far more efficiently than complex Boolean conjunctions or fixed-degree polynomials. This conciseness allows the diagnostic tool to extract actionable, foundational knowledge from model failures, addressing the existing gap in the literature regarding the optimal expressive language for model-agnostic error diagnosis.

Disclosure of Interests. The authors have no competing interests to declare that are relevant to the content of this article.

References

1. Aldeia, G.S.I., de Franca, F.O.: Interpretability in symbolic regression: a benchmark of explanatory methods using the Feynman data set. Genet. Program Evolvable Mach. **23**, 309–349 (2022)
2. Antonie, M., Zaïane, O.R.: Mining positive and negative association rules: an approach for confined rules. In: Proceedings of PKDD, pp. 27–38 (2004)
3. Bach, J.: Subgroup discovery with small and alternative feature sets. Proc. ACM Manag. Data **3**(3) (2025)
4. Boulicaut, J.F., Bykowski, A., Jeudy, B.: Mining association rules with negations. Technical report, INSA Lyon (2000)
5. Branderhorst, N.M.: Goodness-of-Fit in Exceptional Regression Model Mining. Master's thesis, Eindhoven University of Technology (2021)
6. Buzmakov, A.: Towards polynomial subgroup discovery by means of FCA. In: Proceedings of FCA4AI, pp. 57–68 (2020)
7. Duivesteijn, W., Feelders, A., Knobbe, A.: Different slopes for different folks: mining for exceptional regression models with cook's distance. In: Proceedings of KDD, pp. 868–876 (2012)
8. Duivesteijn, W., Feelders, A.J., Knobbe, A.: Exceptional model mining. Data Min. Knowl. Disc. **30**(1), 47–98 (2016)
9. Good, J.H., Kovach, T., Miller, K., Dubrawski, A.: Feature learning for interpretable, performant decision trees. In: Proceedings of NeurIPS, pp. 66571–66582 (2023)
10. Grünwald, P.: The Minimum Description Length Principle. MIT Press, Cambridge (2007)
11. Herrera, F., Carmona, C.J., González, P., Del Jesus, M.J.: An overview on subgroup discovery: foundations and applications. Knowl. Inf. Syst. **29**(3), 495–525 (2011)
12. Kamal, K., Farooq, B.: Ordinal-reslogit: interpretable deep residual neural networks for ordered choices. J. Choice Model. **50**, 100454 (2024)
13. Kearns, M.J., Neel, S., Roth, A., Wu, Z.S.: An empirical study of rich subgroup fairness for machine learning. In: Proceedings of FAT*, pp. 100–109 (2019)

14. Kelly, M., Longjohn, R., Nottingham, K.: The UCI machine learning repository. https://archive.ics.uci.edu
15. Kim, S., Ryu, K.: Subgroup discovery method with internal disjunctive expression. J. Korea Soc. Comput. Inf. **22**, 23–32 (2017)
16. Klösgen, W.: Explora: a multipattern and multistrategy discovery assistant. In: Fayyad, U.M., Piatetsky-Shapiro, G., Smyth, P., Uthurusamy, R. (eds.) Advances in Knowledge Discovery and Data Mining, pp. 249–271. AAAI/MIT Press (1996)
17. Leman, D., Feelders, A., Knobbe, A.: Exceptional model mining. In: Proceedings of ECMLPKDD, pp. 1–16 (2008)
18. Liu, L., Tang, Y., Zhang, K., Sun, Q.: Causal learning for heterogeneous subgroups based on nonlinear causal kernel clustering. In: Proceedings of IJCNN, pp. 1–8 (2024)
19. Mampaey, M., Nijssen, S., Feelders, A., Knobbe, A.: Efficient algorithms for finding richer subgroup descriptions in numeric and nominal data. In: Proceedings of ICDM, pp. 499–508 (2012)
20. Meeng, M., Knobbe, A.: For real: a thorough look at numeric attributes in subgroup discovery. Data Min. Knowl. Disc. **35**(1), 158–212 (2021)
21. Mulders, P.J.A.M., van den Heuvel, E.R., Reidsma, P., Duivesteijn, W.: Introducing exceptional growth mining – analyzing the impact of soil characteristics on on-farm crop growth and yield variability. PLoS ONE **19**(1), 1–26 (2024)
22. Pimentel, J., Azevedo, P.J., Torgo, L.: Subgroup mining for performance analysis of regression models. Expert Syst. J. Knowl. Eng. **40**(1) (2023)
23. Remil, Y., Bendimerad, A., Plantevit, M., Robardet, C., Kaytoue, M.: Interpretable summaries of black box incident triaging with subgroup discovery. In: Proceedings of DSAA, pp. 1–10 (2021)
24. Rissanen, J.: Modeling by shortest data description. Automatica **14**(5), 465–471 (1978)
25. Vreeken, J., van Leeuwen, M., Siebes, A.: Krimp: mining itemsets that compress. Data Min. Knowl. Discov. **23**(1), 169–214 (2011)
26. Wrobel, S.: An algorithm for multi-relational discovery of subgroups. In: Proceedings of PKDD, pp. 78–87 (1997)

Heterogeneous Pattern Sampling According to Frequency

Rayane Lachache[1(✉)], Djawad Bekkoucha[1,3], Abdelkader Ouali[1], Bruno Crémilleux[1], Thi-Bich-Hanh Dao[2], and Christel Vrain[2]

[1] University of Caen Normandy, CNRS, UMR 6072 GREYC, 14032 Caen, France
{rayane.lachache,djawad.bekkoucha,abdelkader.ouali, bruno.cremilleux}@unicaen.fr

[2] University of Orléans, INSA Centre Val de Loire, LIFO EA 4022, Orléans, France
{thi-bich-hanh.dao,christel.vrain}@univ-orleans.fr

[3] LISN, CNRS (UMR 9015), Paris Saclay University, Gif-sur-Yvette, France
djawad.bekkoucha@lisn.fr

Abstract. Pattern sampling has emerged as an efficient paradigm for extracting knowledge from large-scale tabular datasets. By stochastically drawing patterns in proportion to a specified interestingness measure, such as frequency, these methods provide an anytime mechanism able to supply users with representative patterns. Existing approaches, however, operate only on homogeneous data structures such as sequential, numerical, or transactional data. In contrast, many real-world applications produce datasets that combine several heterogeneous types of information. In this article, we introduce the first frequency-based sampling method designed specifically for heterogeneous tabular datasets. Our method employs a multi-step decomposition of the sampling process and tackles the central challenge of precisely computing, for each data instance, the exact number of heterogeneous patterns that cover it. We provide a formal proof establishing that the resulting sample is proportional to pattern frequency. A comprehensive experimental study shows the quality of the drawn patterns using several indicators.

Keywords: Pattern Sampling · Data Mining · Heterogeneous tabular Data

1 Introduction

Many works in pattern mining are devoted on a single type of data such as Boolean data, sequential data or graph data [1]. However it is common to cope with heterogeneous data where data come from different type sources. For example, when studying radiologists' behaviour by analysing their sequences of actions over time, it is valuable to enrich this analysis with additional contextual information about the radiologists and their working situation. These pieces of information, which can be categorical, numerical and sequential lead to an heterogeneous tabular database representation. In order to enable the development

M. Baratchi et al. (Eds.): IDA 2026, LNCS 16513, pp. 283–297, 2026.
https://doi.org/10.1007/978-3-032-23833-7_21

of user-centered methods that address, in particular, the problem of the huge number of patterns produced by exhaustive methods, output pattern sampling is an efficient approach that instantly returns patterns thus enabling to produce pattern-based models at any time [2,5]. It consists in randomly selecting a sample of patterns with a probability proportional to an interestingness measure among the patterns that would have been mined from the complete dataset. For example, a pattern X_1 that is twice as frequent as a pattern X_2 will be twice as likely to be selected. A weighted draw based on the interestingness measure on the whole set of patterns is not possible due to the large size of the search space. The first pattern sampling approach was introduced by Hasan and Zaki [2] and it uses a Markov chain Monte Carlo algorithm. It handles a graph pattern language and is biased towards the frequency measure. Boley et al. [5] defined an approach based on two successive sampling steps. It consists in drawing an instance of the dataset and then drawing a pattern contained in this instance. By judiciously selecting the two draw distributions, this strategy *formally* guarantees an exact sampling according to the distribution induced by the interestingness measure. There are a couple of works following this strategy and addressing several pattern languages such as Boolean data [5], sequences [7] or numerical data [3,11]. However, these pattern languages are considered separately. To the best of our knowledge, there is no work addressing pattern sampling in heterogeneous data. In this paper, we propose SEHP, the first method for sampling heterogeneous patterns. SEHP is based on a two-step method [5] and instantly returns heterogeneous patterns sampled proportional to frequency. We formally prove that SEHP draw patterns proportionally to their frequency. The main challenge lies in computing the exact number of heterogeneous patterns that cover a given object in order to find an appropriate two-step decomposition to obtain the desired distribution. A large set of experimental results according to several criteria (frequency, impact of the long-tail phenomenon, diversity of the sampled patterns, and overlap of their coverages) assess the quality of sampled patterns. Section 2 introduces the preliminaries and the problem statement. Section 3 deals with related work. In Sect. 4, we present our SEHP method, and Sect. 5 provides a comprehensive set of experiments.

2 Preliminaries

Heterogeneous Database. A heterogeneous database is defined as a pair $\mathcal{H} = (\mathcal{G}, \mathcal{M})$, where $\mathcal{G}$ denotes the set of objects and $\mathcal{M} = \{M_1, \ldots, M_n\}$ denotes the set of attributes describing these objects. The set of attributes $\mathcal{M}$ is made of three disjoint subsets. First, the binary attributes subset, denoted $\mathcal{M}_B = \{M_1, \ldots, M_i\}$ where $i < n$, consists of attributes $b \in \mathcal{M}_B$ that take their values from the domain $\mathcal{H}_B = \{0, 1\}$. Next is the subset of numerical attributes, $\mathcal{M}_N = \{M_{i+1}, \ldots, M_{n-1}\}$ where each attribute $f \in \mathcal{M}_N$ takes its values from a finite domain $\mathcal{H}_N$, which is defined as the set union of all values assigned to $\mathcal{M}_N$. The last subset is the single sequential attribute $\mathcal{M}_S = \{M_n\}$, whose domain is denoted by $\mathcal{H}_S$ and defined over a set of literals $\mathcal{I}$. An itemset s is defined as a subset of $\mathcal{I}$. A sequence $s = \langle s_1, \ldots, s_c \rangle$ is an ordered list of non-empty itemsets, where $s_i \subseteq \mathcal{I}$ for $1 \leq i \leq c$, and its size is denoted by $|s| = c$. Each object $g \in \mathcal{G}$ is represented by a vector $\langle v_{g,m} \rangle_{m \in \mathcal{M}}$, where each component $v_{g,m}$ belongs to its corresponding domain $\mathcal{H}_m$.

Example 1. Let Table 1 be our running example of a heterogeneous database containing 4 objects $\mathcal{G} = \{g_1, g_2, g_3, g_4\}$, described over three binary attributes $\mathcal{M}_B = \{b_1, b_2, b_3\}$, two numerical attributes $\mathcal{M}_N = \{f_1, f_2\}$, and one sequential attribute $\mathcal{M}_S = \{S\}$.

Table 1. Example of a heterogeneous database H.

	$\mathcal{M}_B$			$\mathcal{M}_N$		$\mathcal{M}_S$
$\mathcal{G}$	b_1	b_2	b_3	f_1	f_2	S
g_1	1	0	1	7	12	$\langle (ab), (c) \rangle$
g_2	0	1	0	4	9	$\langle (ab), (c), (ac) \rangle$
g_3	0	0	1	5	11	$\langle (ab), (cd) \rangle$
g_4	1	0	1	6	15	$\langle (a), (abd), (abc), (bd) \rangle$

Heterogeneous Pattern. A pattern extracted from the dataset $\mathcal{H}$ is called a heterogeneous pattern and is represented as a triple $X_{\mathrm{H}} = \langle X_{\mathrm{B}}, X_{\mathrm{N}}, X_{\mathrm{S}} \rangle$, where each component corresponds to a specific pattern type. X_{B} is a subset of the binary attributes $\mathcal{M}_B$. X_{N} is an interval pattern defined as a vector of intervals $X_{\mathrm{N}} = \langle [a_f, b_f] \rangle_{\forall f \in \mathcal{M}_N}$, i.e. such that $a_f, b_f \in \mathcal{H}_N$. Each interval of the vector corresponds to a numerical attribute, following a canonical order on the set $\mathcal{M}_N$. X_{S} is a sequential pattern extracted over the set of sequences in $\mathcal{H}$. A sequential pattern $X_{\mathrm{S}} = \langle s'_1, \ldots, s'_m \rangle$ is said to be a subsequence of $s = \langle s_1, \ldots, s_c \rangle$, denoted $X_{\mathrm{S}} \preceq s$, if there exists an index sequence $1 \leq i_1 < i_2 < \cdots < i_m \leq c$ such that $s'_j \subseteq s_{i_j}$ for all $j \in [1..m]$, with $m \leq c$.

Projection Operators. A set of projection operators is defined over the disjoint attribute subsets $(\mathcal{M}_B, \mathcal{M}_N, \mathcal{M}_S)$. For an object $g \in \mathcal{G}$, each operator associates the most specific pattern type that describes g according to the corresponding attribute subset.

$$\begin{cases} g[\mathcal{M}_B] = \{b \in \mathcal{M}_B \mid v_{g,b} = 1\}, \\ g[\mathcal{M}_N] = \langle [v_{g,f}, v_{g,f}] \rangle_{f \in \mathcal{M}_N}, \\ g[\mathcal{M}_S] = v_{g,S}, \; S \in \mathcal{M}_S. \end{cases}$$

The projection $g[\mathcal{M}_B]$ on the set of binary attributes $\mathcal{M}_B$ returns the finite set of attributes present in g, i.e. such that $v_{g,b} = 1$. The projection $g[\mathcal{M}_N]$ represents the interval vector restricted to the values of the numerical attributes $\mathcal{M}_N$. The projection $g[\mathcal{M}_S]$ returns the sequence $v_{g,S}$ associated with g.

Example 2. Applying the projection operators to g_1 results in the following:

$$g_1[\mathcal{M}_B] = \{b_1, b_3\}, \quad g_1[\mathcal{M}_N] = \langle [7,7], [12,12] \rangle, \quad g_1[\mathcal{M}_S] = \langle (ab), (c) \rangle.$$

Coverage and Frequency. An object $g \in \mathcal{G}$ is said to be in the coverage of a heterogeneous pattern X_{H}, denoted by $g \models X_{\mathrm{H}}$, if and only if the following conjunctive conditions are satisfied: $X_{\mathrm{B}} \subseteq g[\mathcal{M}_B] \ \wedge \ X_{\mathrm{N}} \sqsubseteq g[\mathcal{M}_N] \ \wedge \ X_{\mathrm{S}} \preceq g[\mathcal{M}_S]$. More precisely, X_{B} must be a subset of the set $g[\mathcal{M}_B]$, and each interval in $g[\mathcal{M}_N]$ must be included in the corresponding interval of X_{N}. Moreover, the sequential pattern X_{S} must be a subsequence of the sequence associated with g.

Example 3. Let $X_{\mathrm{H}} = \big(\{b_3\}, \langle [6,8], [12,15] \rangle, \langle (a)(c) \rangle\big)$ be a sampled pattern from $\mathcal{H}$. The object g_1 is in the cover of X_{H}, since it satisfies all the three conditions: $(g1 \models X_{\mathrm{H}}) \equiv \{b_3\} \subseteq \{b_1, b_3\} \wedge [6,8] \supseteq ([v_{1,1}, v_{1,1}] = [7,7]) \wedge [12,15] \supseteq ([v_{1,2}, v_{1,2}] = [12,12]) \wedge \langle (a)(c) \rangle \preceq \langle (ab), (c) \rangle$.

The set of all objects in $\mathcal{G}$ that satisfy the heterogeneous pattern X_{H} is defined as its cover in the dataset $\mathcal{H}$: $\mathrm{cover}(X_{\mathrm{H}}, \mathcal{H}) = \{\, g \in \mathcal{G} \mid g \models X_{\mathrm{H}} \,\}$. The cardinality of this set corresponds to the frequency of X_{H} in the dataset $\mathcal{H}$, that is, $\mathrm{freq}(X_{\mathrm{H}}, \mathcal{H}) \ = \ |\mathrm{cover}(X_{\mathrm{H}}, \mathcal{H})|$.

Problem Statement. Let Ω be a population and $f : \Omega \longrightarrow [0,1]$ a measure. The notation $x \sim f(\Omega)$ states that the element x is drawn randomly from Ω with a probability distribution $\pi(x) = f(x)/\sum_{x' \in \Omega} f(x')$. In our work we focus on the language of heterogeneous patterns $\mathcal{L}_H$ and the frequency measure. Given a heterogeneous database $\mathcal{H}$ and $k \in \mathbb{N}$, the problem of sampling heterogeneous patterns consists on extracting k patterns $X_{\mathrm{H}}^1, \ldots, X_{\mathrm{H}}^k$ from $\mathcal{L}_H$, where each pattern X_{H}^i is randomly drawn, with replacement, with a probability proportional to its frequency value $\mathrm{freq}(X_{\mathrm{H}}^i, \mathcal{H})$, $i \in \{1, \ldots, k\}$.

3 Related Work

There are a few works on mining patterns from different type sources. In the context of Relational Concept Analysis, [6] introduced heterogeneous pattern structures as a model to describe objects in a combination of spaces represented in a hybrid formal context such as numerical and Boolean data. Heterogeneity has been studied in multidimensional sequence mining, where sequences are enriched with categorical dimensions [13]. This has been extended to handle multiple levels of granularity by M3SP [14] and it has been further refined by explicitly taking data types and external taxonomies into account [10]. All of these approaches rely on exhaustive pattern extraction over the entire search space and there is no sampling. The literature on pattern output sampling methods can be broadly categorized into three families. The first family is based on stochastic methods such as Markov chain Monte Carlo algorithms. This family was introduced by [2] for sampling graph patterns according to the frequency. In this line of research, Boley et al. [4] handle itemsets and address positive interestingness measures. Despite their accuracy, stochastic approaches often suffer from slow convergence. The second family is based on declarative paradigms. Dzyuba et al. [8] uses a SAT solver to sample itemsets. Several measures are tackled. However, modeling other pattern languages and therefore heterogeneous patterns is challenging as the encoding needs to be adapted for each language. The last family is composed of multi-step methods. The pioneering work was carried out by Boley at al. [5], using a two-step draw to sample patterns according to the frequency. In short, these methods draw an object of the dataset and then draw a pattern supported by this object. By finding an appropriate decomposition to obtain the desired distribution – which remains challenging – these methods provide formal guarantees on the sampling. This line of research has been extended to other pattern languages such as sequences [7], numerical data [3,11] or richer patterns such as subgroups [12]. However, there is no work on sampling that addresses several pattern languages simultaneously, which is the aim of our work.

4 Two-Step Heterogeneous Pattern Sampling Approach

This section introduces SEHP, our sampling method on heterogeneous patterns (cf. Sect. 2).

Key Ideas. In order to draw a pattern proportionally to its frequency, it is necessary to compute the sum of the frequencies of all patterns in the solution space, i.e. $Z = \sum_{X \in \mathcal{L}_H} \text{freq}(X, \mathcal{H})$. However, due to the size of this space, it is infeasible to compute this sum by an exhaustive enumeration. By designing an appropriate two-step sampling procedure [5], such an enumeration can be avoided. The key idea is to compute the weight of each object g which is the number of heterogeneous patterns covering g. Indeed, we observe that $\sum_{X \in \mathcal{L}_H} \text{freq}(X, \mathcal{H}) = \sum_{X \in \mathcal{L}_H} \sum_{g \in \mathcal{G}} 1_{g \models X_{\text{H}}} = \sum_{g \in \mathcal{G}} \sum_{X \in \mathcal{L}_H} 1_{g \models X_{\text{H}}}$. This equality enables us to reduce the problem to computing the number of patterns that cover the object. Then, knowing the weight of each object makes it possible to efficiently compute Z without exploring the entire pattern space. The principle is then as follows: (1) sample an object proportionally to its weight, and (2) select uniformly a pattern among those that cover it. The main challenge lies in computing the weight of each object. We address this issue by introducing the NHP function.

Number of Heterogeneous Patterns (NHP). The function NHP computes for each object $g \in \mathcal{G}$ the number of heterogeneous patterns covering g. To this end, we first need to determine, for each object, how many sub-patterns of each type cover it. Formally, NHP is defined as follows:

$$\text{NHP}(g) = \text{NI}(g) \times \text{NIP}(g) \times \text{NSP}(g). \tag{I}$$

For each object g, NHP first performs a projection $g[\mathcal{M}_B]$ onto the binary dimension. The function NI(g) then computes the exact number of attribute-subset patterns X_{B} covering this projection, as in [5]. A second projection onto the numerical attributes, $g[\mathcal{M}_N]$, allows NIP(g) to compute the exact number of interval patterns X_{N} covering g, by reusing the NIP function proposed in [3]. The final projection, $g[\mathcal{M}_S]$, concerns the sequential dimension and enables NSP(g) to compute the exact number of distinct subsequences X_{S} over the sequence component of g. This computation follows the procedure described in [9]. By distinct subsequences, we mean that the same subsequence X_{S} may occur several times within a single sequence. However, in order to compute Z over $\mathcal{M}_S$, NSP must count each subsequence pattern at most once per sequence, as discussed in [7]. Example 4 illustrates the computation of the number of heterogeneous patterns covering a given object.

Example 4. Consider the object $g_1 \in \mathcal{G}$,

$$g_1[\mathcal{M}_B] = \{b_1, b_3\}, \quad g_1[\mathcal{M}_N] = \langle [7,7], [12,12] \rangle, \quad g_1[\mathcal{M}_S] = \langle (ab), (c) \rangle.$$

(1) *Computation of $NI(g_1)$*. The present binary attributes are b_1 and b_3, hence $\text{NI}(g_1) = 2^{|\{b_1,b_3\}|} = 2^2 = 4$. (2) *Computation of $NIP(g_1)$*. For $f_1 = 7$ with $\mathcal{H}_{f_1} = \{4,5,6,7\}$, let $\text{NIA}(g_1, f_1) = |\{v \in \mathcal{H}_{f_1} \mid v \leq 7\}| \cdot |\{v \in \mathcal{H}_{f_1} \mid v \geq 7\}| = 4 \times 1 = 4$. For $f_2 = 12$ with $\mathcal{H}_{f_2} = \{9, 11, 12, 15\}$, we obtain $\text{NIA}(g_1, f_2) = |\{v \in \mathcal{H}_{f_2} \mid v \leq 12\}| \cdot |\{v \in \mathcal{H}_{f_2} \mid v \geq 12\}| = 3 \times 2 = 6$. Thus, $\text{NIP}(g_1) = 4 \times 6 = 24$. (3) *Computation of $NSP(g_1)$*. The sequence is $\langle (ab), (c) \rangle$. Applying the function NSP yields $\text{NSP}(g_1) = 8$. (4) *Final result.* $\text{NHP}(g_1) = \text{NI}(g_1) \times \text{NIP}(g_1) \times \text{NSP}(g_1) = 4 \times 24 \times 8 = 768$. Hence, the object g_1 is covered by exactly 768 heterogeneous patterns.

NHP Complexity. The NHP worst case time complexity is given by:

$$\mathcal{O}\big(|\mathcal{M}_B| + |\mathcal{G}| \cdot |\mathcal{M}_N| + L \cdot 2^L \cdot T\big)$$

The term $|\mathcal{M}_B|$ comes from NI complexity, which scans the number of binary attributes of each object; the term $|\mathcal{G}| \cdot |\mathcal{M}_N|$ corresponds to NIP complexity in the worst case, where $|\mathcal{G}|$ corresponds to the number of distinct values in the worst case of an attribute; the remaining term $L \cdot 2^L \cdot T$ is due to NSP complexity obtained by adapting the complexity analysis reported in [7], where L is the maximum number of itemsets in a sequence and T the maximum itemset size.

SEHP Algorithm. This section presents the sampling algorithm SEHP, described in Algorithm 1. Given a heterogeneous database $\mathcal{H}$, SEHP samples a heterogeneous pattern X_{H} proportionally to its frequency of occurrences in $\mathcal{H}$. In a preprocessing phase, each object $g \in \mathcal{G}$ is assigned a weight $w(g)$, defined as the number of heterogeneous patterns covering g, computed through the NHP function (line 1). The sampling procedure then proceeds in two steps: (i) an object g is drawn with probability proportional to $w(g)$ (line 2). (ii) given the selected

Algorithm 1. Sampling an heterogeneous pattern proportionally to its frequency (SEHP)

Require: A heterogeneous database $\mathcal{H}$
Ensure: A heterogeneous pattern X_{H} sampled proportionally to its frequency

Preprocessing: For each object $g \in \mathcal{G}$, compute $w(g) = \mathrm{NHP}(g)$.

Step 1: Draw an object $g \sim w(g)$.
Step 2: Uniformly draw a pattern X_{H} among those covering g.
  $X_{\mathrm{B}} \leftarrow$ `DrawUniformItemset`$(\mathcal{M}_B, g[\mathcal{M}_B])$.
  $X_{\mathrm{N}} \leftarrow$ `DrawUniformIntervalPattern`$(\mathcal{M}_N, g[\mathcal{M}_N])$.
  $X_{\mathrm{S}} \leftarrow$ `DrawUniformSubSequence`$(\mathcal{M}_S, g[\mathcal{M}_S])$.
return $X_{\mathrm{H}} \leftarrow \langle X_{\mathrm{B}}, X_{\mathrm{N}}, X_{\mathrm{S}} \rangle$

function DrawUniformItemset$(\mathcal{M}_B, g[\mathcal{M}_B])$
  Initialize $X_{\mathrm{B}} \leftarrow \emptyset$.
  for all present attributes $i \in g[\mathcal{M}_B]$ **do**
    Include i in X_{B} with probability 1/2.
  end for
  return X_{B}
end function

function DrawUniformIntervalPattern$(\mathcal{M}_N, g[\mathcal{M}_N])$
  Initialize $X_{\mathrm{N}} \leftarrow \langle\ \rangle$.
  for all numerical attributes $f \in \mathcal{M}_N$ **do.**
    Let $v_{g,f}$ be the value of feature f from $g[\mathcal{M}_N]$.
    Let $A_f = \{v \in \mathcal{H}_f \mid v \leq v_{g,f}\}$ and $B_f = \{v \in \mathcal{H}_f \mid v \geq v_{g,f}\}$.
    Draw a_f uniformly from A_f and b_f uniformly from B_f.
    Append the interval $[a_f, b_f]$ to X_{N}.
  end for
  return X_{N}
end function

function DrawUniformSubSequence$(\mathcal{M}_S, g[\mathcal{M}_S])$
  Let F be the flatten list of all items in the sequence $g[\mathcal{M}_S]$.
  repeat
    Initialize a set of positions $P \leftarrow \emptyset$.
    for all positions $j \in \{1, \ldots, |F|\}$ **do**
      Include j in P with probability 1/2.
    end for
    $X_{\mathrm{S}} \leftarrow$ Group items at positions in P which belong to the same itemset in $g[\mathcal{M}_S]$.
  until the sub-sequence X_{S} starts at the leftmost occurrence in $g[\mathcal{M}_S]$.
  return X_{S}
end function

g, one heterogeneous pattern X_{H} is drawn uniformly at random among those covering g (line 3). Repeating these two steps independently k times yields a sample of k heterogeneous patterns from $\mathcal{H}$.

Proof. We prove that Algorithm 1 performs sampling proportional to the frequency of patterns. Let $\mathcal{L}_H$ denote the space of all heterogeneous patterns, and let Z be the normalization constant defined as $Z = \sum_{\forall g \in \mathcal{G}} \mathrm{NHP}(g)$. Consider a pattern $X_{\mathrm{H}} \in \mathcal{L}_H$. In Step 1, an object $g \in \mathcal{G}$ is selected with probability proportional to its weight, namely $\pi(g) = \frac{\mathrm{NHP}(g)}{Z}$. In Step 2, if X_{H} covers the object g, then X_{H} is chosen uniformly among the $\mathrm{NHP}(g)$ patterns covering g. In this case, $\pi(X_{\mathrm{H}} \mid g) = \frac{1}{\mathrm{NHP}(g)}$. In the other case, $\pi(X_{\mathrm{H}} \mid g) = 0$. By combining the two steps, we obtain $\pi(X_{\mathrm{H}}) = \sum_{g \in \mathcal{G}} \pi(g)\, \pi(X_{\mathrm{H}} \mid g) = \sum_{g \in \mathrm{cover}(X_{\mathrm{H}}, \mathcal{H})} \frac{\mathrm{NHP}(g)}{Z} \times \frac{1}{\mathrm{NHP}(g)}$. The terms simplify, leading to $\pi(X_{\mathrm{H}}) = \sum_{g \in \mathrm{cover}(X_{\mathrm{H}}, \mathcal{H})} \frac{1}{Z} = \frac{|\mathrm{cover}(X_{\mathrm{H}}, \mathcal{H})|}{Z}$. Since $|\mathrm{cover}(X_{\mathrm{H}}, \mathcal{H})|$ is precisely the frequency of X_{H} in $\mathcal{H}$, it follows that $\pi(X_{\mathrm{H}}) = \frac{\mathrm{freq}(X_{\mathrm{H}}, \mathcal{H})}{Z}$. Therefore, the probability of drawing a pattern is proportional to its frequency.

5 Experiments

The experimental study aims to evaluate the efficiency of SEHP by addressing the following questions.[1]

1. What is the quality of the heterogeneous patterns sampled with respect to frequency, and how does the long-tail phenomenon affect SEHP?
2. What is the diversity of the patterns sampled by SEHP?
3. To what extent do the sampled pattern's overlap ?
4. What is the efficiency of SEHP in terms of CPU time?

5.1 Synthetic Datasets

In order to assess the performance of our algorithm, we use a configurable generator of heterogeneous tabular datasets that we have developed for the evaluation. Given the number of objects $|\mathcal{G}|$, the generator creates instances $g \in \mathcal{G}$ based on the following parameters. For a given number of binary attributes, a relative density parameter $\rho_B \in [0, 1]$ is used to specify the total number of randomly selected pairs $(g \in \mathcal{G}, b \in \mathcal{M}_B)$ whose value is equal to 1 in the dataset. All the non-selected pairs are equal to zero. For a given number of numerical attributes, a parameter $\rho_N \in [0, 1]$ is used to specify the number of distinct numerical values used in the dataset proportional to the total number of pairs $(g \in \mathcal{G}, f \in \mathcal{M}_N)$. Each pair is assigned to a value drawn uniformly from the given set of distinct

[1] All source code (including the SEHP implementation, the synthetic data generator, and the baseline approach), as well as the full set of experimental results, are available at https://github.com/lacray/SEHP-Sampling-heterogeneous-pattern-proportionally-to-its-frequency.git.

Table 2. Characteristics of the 10 synthetic datasets used in the experiments.

Dataset	$\|\mathcal{G}\|$	$\|\mathcal{M}_B\|$	ρ_B	$\mathcal{M}_N$	ρ_N	$\|\mathcal{I}\|$	$\|X_S\|_{max}$	$\|X_S\|_{mean}$	$\|X_S\|_\sigma$
test1	10000	13	80	18	18000	15	103	40.32	11.65
test2	250	4	80	4	20	4	15	8.88	2.27
test3	500	4	80	4	20	4	16	8.84	2.27
test4	250	9	80	6	15	4	21	12.28	3.25
test5	850	3	35	5	85	7	20	9.28	3.19
test6	1500	2	80	3	18	5	18	8.61	3.26
test7	800	6	20	4	19	9	35	12.21	5.17
test8	800	6	70	8	1216	6	18	8.54	2.84
test9	900	7	70	5	22	20	53	20.45	7.67
test10	750	6	70	6	540	7	30	14.0	4.56

values. For the sequential part, the generator considers a set of literals $\mathcal{I}$. Each literal $i \in \mathcal{I}$ is attributed an occurrence rate ρ_i, which is uniformly drawn from the given interval $[\rho_{\min}, \rho_{\max}]$. This rate ρ_i then determines the proportion of objects associated with the literal i. Each itemset in the sequence is constituted by a set of literals which are uniformly drawn from those associated to g. These literals are selected without replacement throughout the sequence generation, ensuring that the selected literal is used exactly once in the current itemset. For each object g, the length of its sequence, in terms of the number of itemsets, is drawn uniformly between two bounds $\eta_{\min}$ and $\eta_{\max}$.

By varying the parameters $(\rho_B, \rho_N, |\mathcal{I}|, \rho_{\min}, \rho_{\max}, \eta_{\min}, \eta_{\max})$, we can generate datasets with different characteristics across the binary, the numerical, and the sequential attributes. Moreover, we do not enforce any explicit similarity structure between heterogeneous objects, which makes the sampling of frequent patterns more challenging. Table 2 summarises all generated datasets and their main characteristics.

5.2 Independent Sampling Approach (I-SEHP)

As there is no existing approach addressing the sampling of heterogeneous patterns, we therefore introduce I-SEHP as an intuitive comparative baseline. Given a heterogeneous database, I-SEHP independently samples one pattern from each component using the corresponding state-of-the-art algorithm, and then aggregates the three resulting sub-patterns to construct a heterogeneous pattern X_{H}. However, we observe that a substantial proportion of the sampled patterns have an empty coverage, as illustrated in Table 3. This situation arises when no object simultaneously belongs to the coverages of the three sub-patterns. Since such patterns are of limited practical interest for the user, we incorporate a rejection mechanism to address this issue. More precisely, whenever the coverage of a sampled pattern X_{H} is empty, the pattern is rejected and the sampling procedure is repeated until a pattern covering at least one object is obtained (cf Algorithm 2). This mechanism preserves frequency-proportional sampling within each component while providing a meaningful and relevant baseline for comparison.

Table 3. Percentage of patterns sampled by I-SEHP having empty coverage (50 000 sampled patterns).

Databases	test1	test2	test3	test4	test5	test6	test7	test8	test9	test10
% empty-coverage	99.99	57.95	51.96	65.03	69.56	62.36	67.37	71.12	74.70	77.37

Algorithm 2. I-SEHP

Require: Heterogeneous database $\mathcal{H}$; flag `ALLOWSEmptyCovers` $\in \{$`true`, `false`$\}$
Ensure: A heterogeneous pattern X_{H}

Preprocessing: define weights $w_{\mathcal{D}_B}(g) = \mathrm{NI}(g)$, $w_{\mathcal{D}_N}(g) = \mathrm{NIP}(g)$, $w_{\mathcal{D}_S}(g) = \mathrm{NSP}(g)$.
repeat
  Step 1 — Separate two-step sampling in each sub-database:
  Sample X_{B} from $\mathcal{D}_B$, X_{N} from $\mathcal{D}_N$, and X_{S} from $\mathcal{D}_S$
  Step 2 — Aggregation: $X_{\mathrm{H}} \leftarrow (X_{\mathrm{B}}, X_{\mathrm{N}}, X_{\mathrm{S}})$
  Coverage: $\mathrm{cover}(X_{\mathrm{H}}, \mathcal{H}) \leftarrow \mathrm{cover}(X_{\mathrm{B}}, \mathcal{D}_B) \cap \mathrm{cover}(X_{\mathrm{N}}, \mathcal{D}_N) \cap \mathrm{cover}(X_{\mathrm{S}}, \mathcal{D}_S)$
  if `ALLOWSEmptyCovers` = **true then**
    break ▷ accept the pattern even if $\mathrm{cover}(X_{\mathrm{H}}, \mathcal{H}) = \varnothing$
  end if
until $\mathrm{cover}(X_{\mathrm{H}}, \mathcal{H}) \neq \varnothing$
return X_{H}

5.3 Pattern Frequency and the Impact of the Long Tail

Figure 1 presents the frequencies of 500 patterns sampled with SEHP and I-SEHP. The patterns are sorted in descending order of frequency. We observe that I-SEHP samples patterns with a higher frequency across all datasets. This is largely a consequence of the sampling procedure's reliance on stochastic independence, where each draw of a sub-pattern is carried using only a subset of attributes. This process, particularly when subjected to rejection, tends to select frequent sub-patterns which leads to increased overlap in their coverage of objects when such overlap exists. It should be noted, however, that the rejection mechanism could be time consuming, as discussed in CPU-time subsection.

In pattern sampling, when an interestingness measure proportional to frequency is applied, it is well known that methods suffer from the curse of the long tail. This phenomenon is characterized by a highly asymmetrical distribution, where a large number of patterns presenting a low interest value form

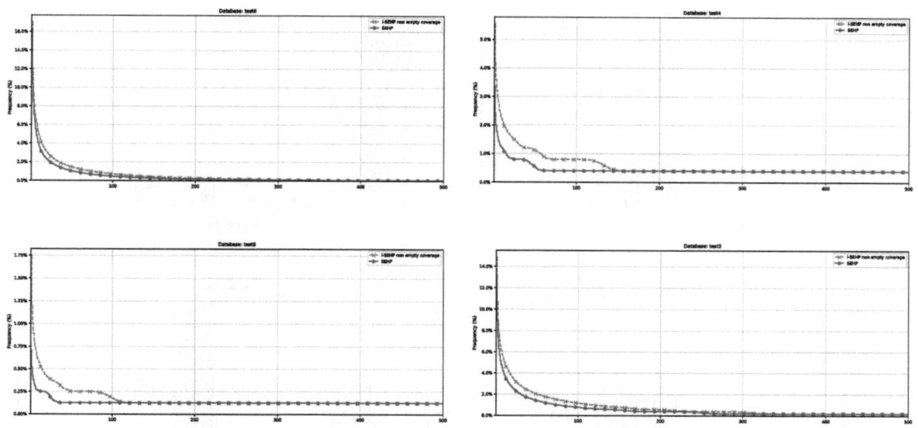

Fig. 1. Frequency evaluation of 500 patterns sampled by SEHP and I-SEHP methods.

the tail, while a small number of patterns with a high interest value form the head. Assuming the long tail is associated with patterns having a frequency lower than 1%, our two methods do not escape from this curse. More precisely, for the test6 dataset, 15% of I-SEHP sampled patterns are non-tail (vs. 10.2% for SEHP), and on the imbalanced test8 dataset, I-SEHP only manages 0.6% non-tail patterns, whereas SEHP patterns are completely in the long-tail. Furthermore, test4 being a variant of test3 having more attributes but fewer objects, a significant decrease in performance is observed. More precisely, the non-tail rate of SEHP drops from 14.8% on test3 to 2.8% on test4, and the rate of I-SEHP falls from 22.8% to 10.8%, showing that increased dimensionality has an effect on the efficient sampling of frequent pattern, even when the overall object count is reduced. We subsequently evaluated the sampled heterogeneous patterns through their sub-patterns. More precisely, we consider a heterogeneous pattern to be more relevant for the analyst when its sub-patterns do not exhibit a full coverage of the objects in the dataset. Therefore, for each dataset, we report for the 100 most frequent patterns sampled by both methods the number of sub-patterns having a full coverage of all the objects in the dataset. The threshold of 100 frequent patterns is determined by observing the Area Under the Curve in Fig. 1 which becomes significantly larger starting from the top 100 patterns. The results obtained are illustrated in Fig. 2. We observe that SEHP exhibits slightly better sub-patterns than I-SEHP on relatively balanced datasets such as test3 and test6. However, SEHP yields significantly superior results on the more imbalanced datasets, namely test5 and test7. Extracting respectively 44 and 50 patterns carrying better sub-patterns compared to only 26 and 17 for I-SEHP.

5.4 CPU Time

Figure 3a shows the cumulative CPU time required for sampling 500 patterns using SEHP and I-SEHP. We observe that SEHP is faster on the majority of

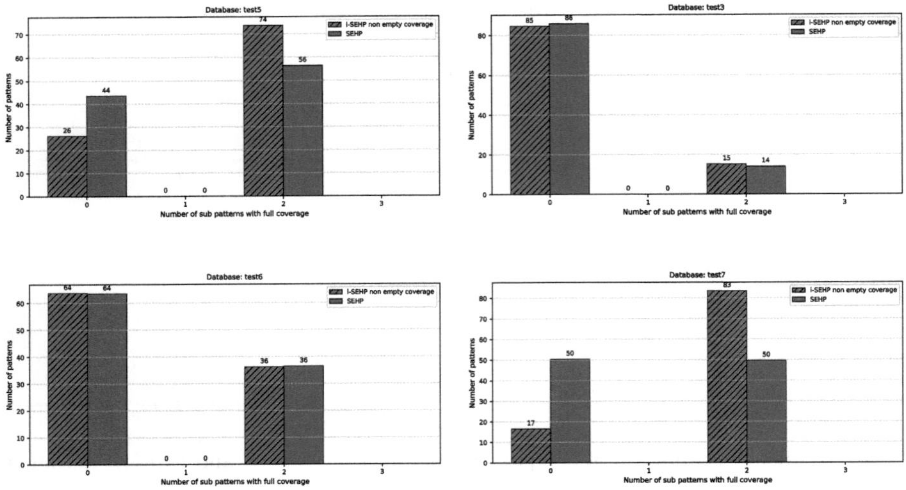

Fig. 2. The number of sub patterns with full coverage among the top 100 frequent sampled heterogeneous Patterns per dataset (SEHP vs. ISEHP).

datasets, providing the required sample in a near-instantaneous time compared to I-SEHP. This difference in performance is caused by the rejection mechanism employed by I-SEHP to ensure a set of 500 patterns with non-empty coverage, as illustrated in Fig. 3b, which shows the necessary number of rejections. Conversely, SEHP guarantees by design through step 1 a coverage of at least one object. We also note that I-SEHP on the test1 dataset results with 0 pattern drawn, which is due to the fact that the total sampling timeout, fixed to 20 min, was reached.

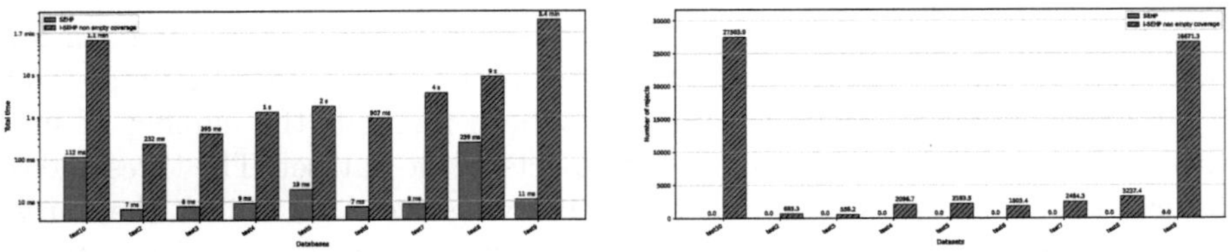

a Cumulative CPU time required by SEHP and I-SEHP for sampling 500 patterns across all datasets.

b Number of rejections performed by I-SEHP to sample 500 patterns on each dataset.

Fig. 3. (a) Cumulative CPU time required by SEHP and I-SEHP for sampling 500 patterns across all datasets. (b) Number of rejections performed by I-SEHP to sample 500 patterns on each dataset.

5.5 Diversity

Sampling diverse patterns is critical for the exploration of the solution space in an interactive mining process. We quantify this aspect by using an extension of the diversity metric introduced in [11] for heterogeneous patterns. Diversity$(K,\mathcal{H}) = \frac{|\{\,\text{cover}(X_{\text{H}}^{1},\mathcal{H}),\ldots,\text{cover}(X_{\text{H}}^{|K|},\mathcal{H})\,\}|}{|K|}$, where $|K|$ is the total number of sampled patterns, and for each $i \in \{1,\ldots,|K|\}$, cover$(X_{\text{H}}^{i},\mathcal{H})$ is the cover of pattern X_{H}^{i} in $\mathcal{H}$. Figure 4a reports the diversity obtained over 50,000 patterns. As I-SEHP samples more frequent patterns than SEHP, it leads to a more varied set of covers and, consequently, higher diversity. While both methods achieve a moderate level of diversity across most datasets, they show notably low values for test7 and test9. This is attributed to the characteristics of these datasets, which exhibit a low number of distinct values alongside longer sequence lengths, thereby forcing the sampling of highly similar patterns. We complete the diversity assessment by calculating the mean pairwise Jaccard distance between the covers of the $|K|$ sampled patterns.

Diversity$_J(K,\mathcal{H}) = \frac{2}{|K|(|K|-1)}\sum_{1\leq i<\,j\leq|K|}\left(1-\frac{|\,\text{cover}(X_{\text{H}}^{i},\mathcal{H})\cap\text{cover}(X_{\text{H}}^{j},\mathcal{H})\,|}{|\,\text{cover}(X_{\text{H}}^{i},\mathcal{H})\cup\text{cover}(X_{\text{H}}^{j},\mathcal{H})\,|}\right)$. Figure 4b reports results over 50,000 sampled patterns, showing that the performances of I-SEHP and SEHP are nearly identical across all datasets, although low values are observed for test7 and test9, confirming that highly similar patterns are drawn from these specific datasets. This low diversity for SEHP is explained by the fact that the same object with a high NHP value is frequently drawn in step 1, leading to similar pattern coverage.

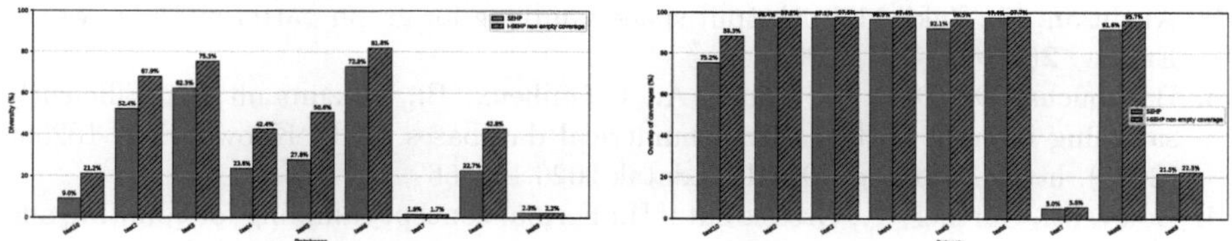

a Diversity of 50,000 sampled patterns across all datasets.

b Mean overlap evaluation (SEHP vs. I-SEHP; 50,000 sampled patterns).

Fig. 4. (a) Diversity of 50,000 sampled patterns across all datasets. (b) Mean overlap evaluation (SEHP vs. I-SEHP; 50,000 sampled patterns).

6 Conclusion and Perspectives

In this paper, we introduced SEHP, the first sampling method to tackle heterogeneous patterns. We formally proved that SEHP samples patterns proportionally to their frequency. SEHP instantly returns heterogeneous patterns, enabling tight coupling between the system and the user and improving knowledge discovery. As future work, we plan to address alternative interestingness measures tailored specifically to heterogeneous patterns and to enrich SEHP with constraints to combine the advantages of the worlds of sampling and constraint-based pattern mining [1]. Another direction is to evaluate the usefulness of sampled patterns for designing supervised or unsupervised models. In the context of classification, as discussed in [7], sampled patterns can be used to create features that add information to the original data. It is hoped that classification models built on data enriched by these new features will perform better than models trained on the original data. Finally, we also aim to integrate SEHP into applications implementing interactive pattern mining processes.

Acknowledgement. This work was supported by the French National Research Agency (ANR) through the project FIDD (grant agreement ANR-24-CE23-0711). The second author was supported by the French National Research Agency (ANR) and the Normandie Region through the HAISCoDe project (ANR-20-THIA-0021), and by the DeMik project under the UPS Excellences/Springboard Program (grant SPATTR_2025-12). The work of the fourth author is partly supported the Pandora project ANR-24-CE23-0950.

Disclosure of Interests. The authors have no competing interests to declare that are relevant to the content of this article.

References

1. Aggarwal, C.C., Jiawei, H.: Frequent Pattern Mining. Springer (2014)
2. Al Hasan, M., Zaki, M.J.: Output space sampling for graph patterns. Proc. VLDB Endow. **2**(1), 730–741 (2009)
3. Bekkoucha, D., Diop, L., Ouali, A., Crémilleux, B., Boizumault, P.: Efficiently sampling interval patterns from numerical databases. Data Knowl. Eng. 102566 (2026). https://doi.org/10.1016/j.datak.2026.102566
4. Boley, M., Gärtner, T., Grosskreutz, H.: Formal concept sampling for counting and threshold-free local pattern mining. In: Proceedings of the 2010 SIAM International Conference on Data Mining, pp. 177–188. SIAM (2010). https://doi.org/10.1137/1.9781611972801.16
5. Boley, M., Lucchese, C., Paurat, D., Gärtner, T.: Direct local pattern sampling by efficient two-step random procedures. In: Proceedings of the 17th ACM SIGKDD International Conference on Knowledge Discovery and Data Mining, pp. 582–590 (2011). https://doi.org/10.1145/2020408.2020500
6. Codocedo, V., Napoli, A.: A proposition for combining pattern structures and relational concept analysis. In: International Conference on Formal Concept Analysis, pp. 96–111. Springer (2014). https://doi.org/10.1007/978-3-319-07248-7_8
7. Diop, L., Diop, C.T., Giacometti, A., Li, D., Soulet, A.: Sequential pattern sampling with norm-based utility. Knowl. Inf. Syst. **62**(5), 2029–2065 (2020). https://doi.org/10.1007/s10115-019-01417-3
8. Dzyuba, V., van Leeuwen, M., De Raedt, L.: Flexible constrained sampling with guarantees for pattern mining. Data Min. Knowl. Disc. **31**(5), 1266–1293 (2017). https://doi.org/10.1007/s10618-017-0501-6

9. Egho, E., Raïssi, C., Calders, T., Jay, N., Napoli, A.: On measuring similarity for sequences of itemsets. Data Min. Knowl. Disc. **29**(3), 732–764 (2015). https://doi.org/10.1007/s10618-014-0362-1
10. Egho, E., Raïssi, C., Jay, N., Napoli, A.: Mining heterogeneous multidimensional sequential patterns. In: ECAI 2014, pp. 279–284. IOS Press (2014). https://doi.org/10.3233/978-1-61499-419-0-279
11. Giacometti, A., Soulet, A.: Dense neighborhood pattern sampling in numerical data. In: Proceedings of the 2018 SIAM International Conference on Data Mining, pp. 756–764. SIAM (2018). https://doi.org/10.1137/1.9781611975321.85
12. Moens, S., Boley, M.: Instant exceptional model mining using weighted controlled pattern sampling. In: International Symposium on Intelligent Data Analysis, pp. 203–214. Springer (2014). https://doi.org/10.1007/978-3-319-12571-8_18
13. Pinto, H., Han, J., Pei, J., Wang, K., Chen, Q., Dayal, U.: Multi-dimensional sequential pattern mining. In: Proceedings of the Tenth International Conference on Information and Knowledge Management, pp. 81–88 (2001). https://doi.org/10.1145/502585.502600
14. Plantevit, M., Laurent, A., Laurent, D., Teisseire, M., Choong, Y.W.: Mining multidimensional and multilevel sequential patterns. ACM Trans. Knowl. Discov. Data (TKDD) **4**(1), 1–37 (2010). https://doi.org/10.1145/1644873.1644877

Drop the Mask! GAMM - A Taxonomy for Graph Attributes Missing Mechanisms

Richard Serrano[1(✉)], Baptiste Jeudy[1], Charlotte Laclau[2], and Christine Largeron[1]

[1] Laboratoire Hubert Curien, Saint-Étienne 42000, France
{richard.serrano,baptiste.jeudy,christine.largeron}@univ-st-etienne.fr
[2] Télécom Paris, Institut Polytechnique de Paris, Paris, France
charlotte.laclau@telecom-paris.fr

Abstract. Exploring missing data in attributed graphs introduces unique challenges beyond those found in tabular datasets. In this work, we extend the taxonomy for missing data mechanisms to attributed graphs by proposing GAMM (Graph Attributes Missing Mechanisms), a framework that systematically links missingness probability to both node attributes and the underlying graph structure. Our taxonomy enriches the conventional definitions of masking mechanisms by introducing graph-specific dependencies.

We empirically demonstrate that state-of-the-art imputation methods, while effective on traditional masks, significantly struggle when confronted with these more realistic graph-aware missingness scenarios.

Keywords: Attributed Graph · Missingness Mechanisms · Masking Taxonomy · Missing Values Imputation

1 Introduction

Attributed graphs provide a powerful framework for modeling complex systems, from social networks to biological interactions. Unlike tabular data, graph-structured data captures the inherent connections between entities, where a node is defined by both its attributes and its relational context. However, this structure creates unique challenges when dealing with missing data. In practice, node attributes are often partially unobserved, posing a significant problem for methods like Graph Neural Networks (GNNs) that usually require fully observed feature matrices. For instance, in social networks, users may omit sensitive profile information depending not only on their own characteristics but also on their position within the network or the attributes of their friends. A user with many connections might be more cautious about sharing personal data, or missingness may cluster within communities due to shared privacy norms or platform features. Such dependencies between missingness and network structure are not captured by classical missing data models, which assume independence from relational context. Accurately modeling these effects is crucial for fair and effective

M. Baratchi et al. (Eds.): IDA 2026, LNCS 16513, pp. 298–311, 2026.
https://doi.org/10.1007/978-3-032-23833-7_22

inference on graph data. This need has led to substantial research into missing attribute imputation [16,18,23]. However, a critical issue is that these methods are often evaluated using simplistic masking mechanisms, e.g., MCAR, MAR, MNAR [7], which provides a crucial foundation, but do not explicitly account for structural dependencies. In real-world graphs, missingness may be influenced by a node's connectivity or the characteristics of its neighborhood.

Contributions. We introduce the Graph Attributes Missing Mechanisms (GAMM) taxonomy, which extends traditional missingness frameworks by incorporating structural dependencies and classifying missingness into four sources: (A) node attributes, (S) structural properties, (N) neighbors' attributes, and (G) their combinations. Focusing solely on missing node attributes under a known topology, we show that graph-aware mechanisms substantially degrade the performance of state-of-the-art imputers relative to attribute-only missingness. To structure the investigation, we examine how (Research Question 1) S- and N-based mechanisms affect reconstruction performance across graphs and missing rates, (RQ2) homophily/heterophily modulates the gap between attribute- and neighborhood-based MAR, (RQ3) neighbor-dependent MNAR compares to attribute-based MNAR and in which settings it becomes most severe, and (RQ4) whether imputers preserve feature distributions under these structurally informed conditions. An open-source implementation of all mechanisms and protocols is provided.[1]

This paper is structured as follows. Section 2 outlines existing masking strategies and discusses their limits. Section 3 presents GAMM; Sect. 4 details experiments and results. Finally, Sect. 5 concludes.

2 Limits of Traditional Missing Data Mechanisms

2.1 Traditional Missing Data Mechanisms

The traditional taxonomy identifies three missingness mechanisms [9,15]: (1) **Missing Completely At Random (MCAR)**, where missingness is independent of all data; (2) **Missing At Random (MAR)**, where it is influenced only by observed data; and (3) **Missing Not At Random (MNAR)**, where it depends on unobserved values. For instance, in a corporate network, missing employee *Salary* data would be MCAR if it resulted from a randomly occurring glitch in the employee database, MAR if junior employees (i.e., observed *Experience* attribute) are less likely to report it, and MNAR if employees with higher salaries (the unobserved feature itself) are the ones who mostly withhold this information. Most imputation methods assume MCAR or MAR conditions [8], while MNAR requires specialized approaches [3]. Existing extensions to this taxonomy, including graphical models [11] and conditional independence approaches [2], enhance the framework for tabular data but fail to address network-specific challenges where structural relationships directly influence missingness patterns.

[1] Code and Supplementary Material: https://github.com/RichardSrn/GAMM.

Similarly, recent work advocating for explicit measurement process modeling [6] acknowledges the limitations of the rigid MCAR/MAR/MNAR classification in complex dependencies, yet still overlooks how network topology itself can determine whether attributes are observed. While these foundational missingness mechanisms provide a basis for our graph-based taxonomy, they require significant extensions to address the unique complexities of network structure.

2.2 Taxonomies and Extensions for Graph-Based Missingness

When translating missingness concepts to graphs, structural relationships between entities add complexity beyond tabular settings. Missingness can depend on node attributes, topological properties, or relational influences [4,5].

Limitations. We claim that traditional definitions of missingness mechanisms are inadequate for graph-structured data, where missingness may depend on structural properties. The following two examples illustrate this argument:

1. **Structurally-Induced Missingness**: Nodes with specific structural properties (e.g., low degree) may exhibit missing attributes due to their position in the network. In social networks, peripheral users may be less likely to report certain attributes, like alcohol consumption [4]. This represents a MAR mechanism dependent on graph topology rather than attributes values; a distinction classical frameworks fail to characterize.
2. **Neighbor-Influenced Missingness**: In social networks, individuals might withhold information (e.g., salary) not based on their own values, but influenced by their friends' values. This creates dependencies unexplainable without incorporating graph structure [5]. This scenario represents a graph-specific form of MNAR because the missingness of a node's attribute is conditioned on the unobserved attributes of its neighbors. Here, traditional MNAR frameworks are inadequate, as they are unable to model dependencies that arise from the underlying graph topology. This concept of leveraging network information can also be applied to the MAR framework, for instance, by conditioning the missingness on the *observed* features of neighboring nodes.

In the following, we propose a new taxonomy for attributed graphs, providing a basis for clear methodology and consistent evaluation of graph-imputation methods under more realistic missingness conditions.

3 Taxonomy for Graph Attributes Missing Mechanisms (GAMM)

In this section, we introduce GAMM (Graph Attributes Missing Mechanisms), a novel taxonomy for understanding and generating missing data masks in attributed graphs. GAMM extends traditional missingness definitions by explicitly incorporating the interplay between node attributes and the graph structure.

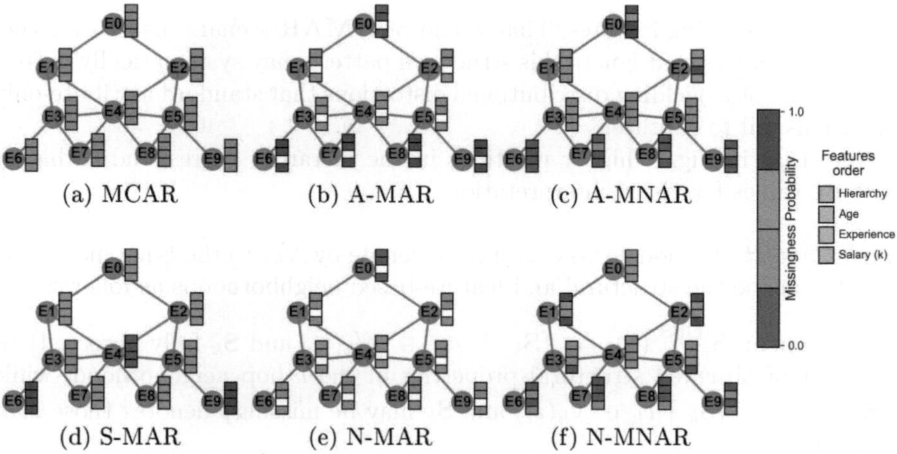

Fig. 1. Traditional (a–c) and graph-aware (d–f) missingness mechanisms in GAMM. Each node shows four attribute slots (top-to-down: Hierarchy, Salary, Experience, Age); red shades indicate higher missingness probability, while blue indicates lower probability (white = observed). Consistent with the example description in paragraph *Example*, higher positions in the hierarchy correspond to higher attribute values.(Color figure online)

To illustrate these mechanisms, we use a toy example of a corporate email network (Fig. 1). Each node represents an employee, and edges denote communication. Node attributes include *Hierarchy*, *Salary*, *Experience*, and *Age*, with some features correlated to aid visual intuition. To ensure consistent interpretation across panels, each node displays four attribute slots in the fixed top-to-down order *[Hierarchy, Salary, Experience, Age]*. We encode the missingness probability p_{miss} using a color scale: *white* denotes observed entries ($p_{\text{miss}} = 0$), followed by *blue* for $p_{\text{miss}} \in (0, 0.25]$, *green* for $p_{\text{miss}} \in (0.25, 0.50]$, *orange* for $p_{\text{miss}} \in (0.50, 0.75]$, and *red* for $p_{\text{miss}} \in (0.75, 1.0]$.

We consider an undirected attributed graph $G = (\mathcal{V}, \mathcal{E}, F)$ with n nodes, edges $\mathcal{E} \subset \mathcal{V} \times \mathcal{V}$, and feature matrix $F \in \mathbb{R}^{n \times d}$ where $F_{i\cdot}$ is the feature vector of node v_i and $\mathbf{S}_i$ are its structural properties (degree, ...). Missing entries are indicated by a binary mask $\Omega \in \{0,1\}^{n \times d}$, with $\Omega_{ij} = 1$ when feature j of v_i is observed. We decompose F into $F^{(\text{OBS})}$ (observed features) and $F^{(\text{MIS})}$ (features subject to missingness).

Example. Consider a corporate email network. In a **MAR** scenario, suppose *Experience* is fully observed while *Hierarchy*, *Age* and *Salary* may be missing features. This partitions the features into $F^{(\text{OBS})} = [\textit{Experience}]$ and $F^{(\text{MIS})} = [\textit{Hierarchy, Age, Salary}]$. In contrast, in a **MNAR** scenario, we may have no fully observed features ($F^{(\text{OBS})} = \emptyset$), meaning all attributes fall into $F^{(\text{MIS})}$.

Crucially, missingness in graphs often correlates with structure. For instance, junior employees, often located in peripheral regions of the graph, may be more likely to have missing features. That would be a **MAR** scenario based on structure, and imputers that ignore this structural pattern may systematically overestimate their roles, yielding distributional distortions that standard attribute-only evaluations fail to capture.

Note that in Fig. 1, higher positions in the hierarchy correspond to higher attribute values for visual interpretation.

Graph Context. To model this context, we denote by $\mathcal{N}_h(v_i)$ the h-hop neighbors of v_i. We define the structural and feature-based neighborhoods as follows:

- **Structure:** $\mathbf{S}_i^{(\text{OBS})}(h) = \{\mathbf{S}_k \mid v_k \in \mathcal{N}_h(v_i) \text{ and } \mathbf{S}_k \text{ fully observed}\}$ is the set of observed structural properties in the h-hop neighborhood, while $\mathbf{S}_i^{(\text{MIS})}(h) = \{\mathbf{S}_k \mid v_k \in \mathcal{N}_h(v_i) \text{ and } \mathbf{S}_k \text{ may be missing}\}$ denotes those that may be missing.
- **Features:** Similarly, $\mathbf{N}_i^{(\text{OBS})}(h) = \{F_{k\cdot}^{(\text{OBS})} \mid v_k \in \mathcal{N}_h(v_i)\}$ represents the observed features of neighbors, and $\mathbf{N}_i^{(\text{MIS})}(h) = \{F_{k\cdot}^{(\text{MIS})} \mid v_k \in \mathcal{N}_h(v_i)\}$ represents the potentially missing features.

For $h = 0$, $\mathbf{S}_i^{(\text{OBS})}(0)$ and $\mathbf{S}_i^{(\text{MIS})}(0)$ correspond to the properties of v_i itself, abbreviated as $\mathbf{S}_i^{(\text{OBS})}$ and $\mathbf{S}_i^{(\text{MIS})}$. A full notation list is provided in Section A.3 of the Supplementary Material.

3.1 Proposed Mechanisms

We are now ready to present GAMM, a Graph-Aware Missingness Mechanism framework, under the assumption that the graph structure $(\mathcal{V}, \mathcal{E})$ is fully observed. The probability that an attribute F_{ij} is missing ($P(\Omega_{ij} = 0)$) is governed by a function $g(.)$ over relevant inputs:

$$P(\Omega_{ij} = 0 \mid G) = P(\Omega_{ij} = 0 \mid \cdot) = g(\cdot). \tag{1}$$

The nature of the inputs to $g(.)$ determines the missingness mechanism. Table 1 summarizes our taxonomy, and illustrative examples are shown in Fig. 1. In addition to the MCAR setting, we distinguish four key families of mechanisms that capture different sources of dependency in graphs.

*Missing Completely At Random (**MCAR**).* Missingness occurs independently of both node attributes and the graph structure: the missingness probability is the same constant for all entries of $F^{(\text{MIS})}$, as shown in Fig. 1(a).

*Attribute-based Missingness (**A-**).* In this setting, missingness for a node v_i depends only on its own attributes, mirroring standard tabular assumptions. **A-MAR** arises when the probability of missingness depends on observed attributes of v_i (e.g., in Fig. 1(b), low observed *Experience* correlates with more missing values). Conversely, under **A-MNAR**, missingness is driven by the node's unobserved attributes (Fig. 1(c) illustrates higher true *Salary* leading to missing entries).

Table 1. GAMM Taxonomy: Characterization of graph-aware missingness mechanisms based on the inputs to the probability function $g(\cdot)$. ✓ indicates an input is potentially used by $g(\cdot)$. A stands for Attribute-based, S for Structure, N for Neighborhood, and G for Generic.

Cat.	Mech.	Formal Definition of $P(\Omega_{ij} = 0 \mid G)$	Possible dependencies for $g(\cdot)$					
			$F_{i\cdot}^{(\text{OBS})}$	$F_{i\cdot}^{(\text{MIS})}$	$\mathbf{S}_i^{(\text{OBS})}(h)$	$\mathbf{S}_i^{(\text{MIS})}(h)$	$\mathbf{N}_i^{(\text{OBS})}(h)$	$\mathbf{N}_i^{(\text{MIS})}(h)$
	MCAR	$P(\Omega_{ij} = 0)$						
A-	MAR	$P(\Omega_{ij} = 0 \mid F_{i\cdot})$	✓					
	MNAR		✓	✓				
S-	MAR	$P(\Omega_{ij} = 0 \mid \mathbf{S}_i)$			✓			
	MNAR				✓	✓		
N-	MAR	$P(\Omega_{ij} = 0 \mid \mathbf{N}_i(h))$					✓	
	MNAR						✓	✓
G-	MAR	$P(\Omega_{ij} = 0 \mid F_{i\cdot}, \mathbf{S}_i, \mathbf{N}_i(h))$	✓		✓		✓	
	MNAR		✓	✓	✓	✓	✓	✓

*Structural Properties-aware Missingness (**S-**).* Here, missingness depends on the structural characteristics of v_i (degree, centrality, etc.). Under **S-MAR**, it is governed by observed structural features (Fig. 1(d) shows low-degree nodes having more missing data). Under **S-MNAR**, missingness depends on missing structural characteristics; for instance, nodes with missing edges in the graph may exhibit more missing values on features too.

*Neighborhood-aware Missingness (**N-**).* In this case, missingness of v_i depends on attributes of its h-hop neighbors (where the neighborhood is defined by the graph topology and not from the feature-space). Under **N-MAR**, it depends only on neighbors' observed attributes (e.g., Fig. 1(e), neighbors' observed *Experience*). Under **N-MNAR**, it depends on neighbors' unobserved attributes, as in Fig. 1(f).

*Generic Combination Missingness (**G-**).* This category captures the most general and realistic forms of missingness. It allows dependencies across a node's own attributes (both observed and unobserved), structural properties, and neighborhood information. Such cases reflect complex real-world patterns where missingness cannot be attributed to a single source but instead arises from intertwined local and structural factors. These scenarios often blur the line between MAR and MNAR, making the masking mechanism particularly hard to disentangle without additional assumptions or annotations.

Remark On the Identifiability of Missingness Mechanisms. A central challenge in modeling missing data is distinguishing MAR from MNAR mechanisms, a task that is fundamentally unidentifiable without additional assumptions or external supervision. This difficulty arises because the same observed missingness patterns can often be explained by either observed or unobserved variables.

Consider a feature with missing entries, $F_{\cdot j}$, that is correlated with a fully observed feature $F_{\cdot k}$. If missingness depends on $F_{\cdot k}$, it falls under MAR; but if it instead depends on the unobserved values of $F_{\cdot j}$, the mechanism is MNAR. In practice, these explanations are statistically indistinguishable from observational data alone leading to intrinsic ambiguity in determining the true source of missingness. In the tabular settings, identifiability is often assumed under the i.i.d. sampling of rows (i.e., independent individuals) [21], though this assumption is stronger than strictly necessary [2]. For graphs, this assumption breaks down: nodes are not independent. However, for identifying the masking mechanism, we do not require node-level independence. Instead, it suffices to assume independence across features, i.e., columns of F are uncorrelated.

While this assumption is rarely satisfied in real-world data (whether graph-based or tabular), it enables clear identifiability in controlled experiments. By constructing synthetic data under known and independent feature distributions, we can isolate and verify the missingness mechanism. This is crucial when benchmarking models or validating imputation strategies under known ground truth.

Summary. GAMM provides a formal framework to characterize and simulate diverse missingness mechanisms in graph data, capturing dependencies on node attributes, structural properties, and neighborhood information. This enables the creation of realistic missing data scenarios and systematic evaluation of their impact on graph learning methods. In the next section, we use GAMM to assess the robustness of state-of-the-art imputer models across these mechanisms. Our results show that performance and reliability vary significantly with the missing data type, highlighting the importance of accounting for these distinctions in graph algorithm design and evaluation.

4 Experiments

4.1 Experimental Protocol

Our experimental design directly addresses **RQ1**–**RQ4** by systematically varying missingness mechanisms (traditional vs. graph-aware), graph homophily, and missing rates, and evaluating tabular and graph-based imputers using MAE/RMSE and distributional analysis.

Datasets. We select twelve benchmark real-world graph datasets [1,14,17,22,25] (detailed in Section A.1 of the Supplementary Material) with diverse network properties, notably varying levels of homophily, which is crucial for evaluating our proposed mechanisms.

Table 2 summarizes their characteristics and groups them into Homophilic, Neutral, and Heterophilic categories based on their adjusted homophily (H_{adj}) [13].

Table 2. Presentation of the benchmark datasets, categorized by their adjusted homophily (H_{adj}).

	Dataset	#Nodes	#Edges	#Classes	Features	H_{adj}
Homophilic	Cora	2708	5278	7	$\{0,1\}^{1433}$	0.768
	PubMed	19717	44324	3	$[0,2]^{500}$	0.693
	CiteSeer	3327	4552	6	$\{0,1\}^{3702}$	0.678
Neutral	Chameleon	2277	18050	5	$\{0,1\}^{2325}$	0.042
	Squirrel	5201	108536	5	$\{0,1\}^{2089}$	0.030
	Minesweeper	10000	39402	2	$\{0,1\}^{7}$	0.009
	Actor	7600	15009	5	$\{0,1\}^{932}$	0.007
	Roman-empire	22662	32927	18	$[0,1]^{300}$	-0.045
Heterophilic	Tolokers	11758	519000	2	$[0,1]^{10}$	-0.188
	Wisconsin	251	257	5	$\{0,1\}^{1702}$	-0.189
	Cornell	183	149	5	$\{0,1\}^{1702}$	-0.219
	Texas	183	162	5	$\{0,1\}^{1702}$	-0.425

Missingness Mask Generation. We generate masks at overall rates of $p_{\text{miss}} \in \{20\%, 50\%, 80\%\}$ using distinct mechanisms from our taxonomy. As introduced in Sect. 3, we define the function $g(\cdot)$ that maps specific informational inputs to a missingness probability, $P(\Omega_{ij} = 0) = g(\cdot)$. The specific ways in which $g(\cdot)$ is determined, which instantiate mechanisms based on a node's own attributes (A-), its structure (S-), or its neighbors' attributes (N-), are defined as follows:

$$
\begin{aligned}
&\text{MCAR: } g() = p_{\text{miss}} \\
&\text{A-MAR/MNAR: } g(F_{i\cdot}) = \begin{cases} \sigma(\omega_1 F_{i\cdot}^{(\text{OBS})} + b_1), & \text{A-MAR}, \\ \sigma(\omega_2 F_{ij}^{(\text{MIS})} + b_2), & \text{A-MNAR}, \end{cases} \\
&\text{S-MAR/MNAR: } g(\mathbf{S}_i) = \begin{cases} \sigma(\omega_3 \deg(\mathbf{S}_i) + b_3), & \text{S-MAR}, \\ \sigma(\omega_3 \mathbf{S}_i^{(\text{MIS})}(h) + b_3), & \text{S-MNAR}, \end{cases} \\
&\text{N-MAR/MNAR: } g(\mathbf{N}_i(h)) = \begin{cases} \sigma(\omega_4 \mathbf{N}_i^{(\text{OBS})}(h) + b_4), & \text{N-MAR}, \\ \sigma(\omega_5 \mathbf{N}_i^{(\text{MIS})}(h) + b_5), & \text{N-MNAR}. \end{cases}
\end{aligned}
\tag{2}
$$

where, $\sigma(\cdot)$ is the sigmoid function, and deg stands for *degree*. The parameters ω_k and b_k control the dependency strength and are calibrated to achieve the target probability p_{miss}. A detailed description of this calibration is provided in Section A.2 of the Supplementary Material.

Remark on S-MNAR. While GAMM defines formally S-MNAR (see Eq. 2), we exclude it from experiments as we assume a fully observed topology. Treating edges as missing shifts the problem to *link prediction*. Furthermore, without edge masks to distinguish unobserved links from non-edges, the structural set $\mathbf{S}_i^{(\mathrm{MIS})}(h)$ remains empty in our setting.

Imputation Methods. We evaluate a suite of state-of-the-art methods, focusing on those that represent current best practices. These methods have been selected because they demonstrate superior performance compared to older or less advanced baselines, thus we do not report standard baselines such as MICE [24].

Tabular Imputers. process node attributes as a standard feature matrix, inherently ignoring graph structure: TABULAR AVERAGE which imputes missing values with the global mean of each feature; OT-TAB [12] that leverages optimal transport theory to minimize distributional discrepancies between observed and imputed data.

Graph Imputers. explicitly exploit network topology for enhanced reconstruction. Notably we assess: GRAPH AVERAGE, a direct neighborhood average; FP [16] that propagates known values as a heat diffusion process along edges; PCFI [23] which introduces a novel channel-wise confidence metric for node features and utilizes shortest path distances for feature propagation and refinement; and finally GRIOT [18] which combines optimal transport with graph neural networks, to incorporates both node features and graph structure for robust imputation of missing attribute values.

Evaluation. We assess imputation quality using Mean Absolute Error (MAE) and Root Mean Square Error (RMSE). Since our primary contribution is a new missingness-generation taxonomy, these reconstruction metrics provide the most direct and controlled evaluation of how methods respond to the proposed mechanisms, avoiding confounding effects from downstream tasks. The relationship between reconstruction fidelity and downstream performance is known to be complex: downstream models may tolerate substantial missingness [18], and strong downstream accuracy does not necessarily imply reliable interpretability or high-quality imputation [19,20]. By focusing on reconstruction, we obtain a rigorous and targeted validation of our core contribution. All experiments are repeated eight times, and we use a Wilcoxon–Mann–Whitney U test [10], with a significance level $p < 0.05$ to assess statistical significance, testing the null hypothesis H_0 that the performance distributions of the compared methods are identical.

4.2 Results and Analysis

We evaluate state-of-the-art imputers under our taxonomy of missingness mechanisms, with a focus on the most challenging MNAR variants (Table 3), and summarise all 2,304 configurations in Table 4.

Table 3. Percentage degradation in MAE of N-MNAR over A-MNAR ($p_{\text{miss}} = 20\%$) across dataset homophily groups. Negative results indicate deterioration in MAE with the N-MNAR mechanism compared to A-MNAR. (*) indicates statistical significance (Mann–Whitney U, $p < 0.05$).

a) HOMOPHILIC datasets

Imputer	CiteSeer	Cora	PubMed
Tabular_Avg	-50.6*	-48.7*	-9.1*
OT-tab	-10.3*	-14.0*	-2.1*
FP	-72.9*	-68.9*	-15.5*
GRIOT	-12.5	-17.1*	-5.1*
Graph_avg	-70.1*	-64.4*	-12.8*
PCFI	-55.6*	-55.5*	-13.0*

b) HETEROPHILIC datasets

Imputer	Tolokers	Cornell	Texas	Wisconsin
Tabular_Avg	-9.0*	-47.5*	-46.5*	-47.2*
OT-tab	-7.6*	-30.8*	-25.5*	-31.2*
FP	-35.3*	-46.1*	-67.7*	-54.4
GRIOT	+0.9	-77.1*	-72.3*	-72.4*
Graph_avg	-5.0*	-53.1*	-62.5*	-54.3*
PCFI	-4.0*	-49.4*	-38.3*	-48.9*

c) NEUTRAL datasets

Imputer	Actor	Chameleon	Squirrel	Minesweeper	Roman_Empire
Tabular_Avg	-53.1*	-38.8*	-34.5*	-23.1*	+11.8*
OT-tab	-6.4*	-6.2*	-7.8*	-20.1*	+3.1*
FP	-73.3*	-86.8*	-78.6*	-27.4*	-0.6
GRIOT	-14.3*	-11.2	-14.1	-0.0*	+3.3*
Graph_avg	-63.5*	-38.8*	-36.0*	-26.4*	+6.2*
PCFI	-51.6*	-17.2*	-17.4*	-25.5*	+6.5*

RQ1: Impact of Graph-Aware Mechanisms. Across configurations, **graph-aware missingness markedly increases difficulty**. As captured in Table 4, moving from attribute-only to structurally informed masks leads to significant degradation in 46.6% of cases, consistently across tabular and graph-based imputers. These results indicate that conventional attribute-only masks provide an overly optimistic view of robustness and that integrating structure in mask generation exposes realistic failure modes.

Table 4. Percentage Summary of Statistical Outcomes by mechanism. Total: 100% = 2,304 experiments.

Type	Sign. Degradation	No Sign. Change	Sign. Improvement	**Total**
A-MAR v.s. N-MAR	8.0%	32.2%	2.9%	**43.1%**
A-MNAR v.s. N-MNAR	38.6%	9.7%	8.6%	**56.9%**
Total	**46.6%**	**41.9%**	**11.5%**	**100.0%**

RQ2: Role of Homophily Under MAR. Under MAR, the contrast between attribute-based (A-MAR) and neighborhood-based (N-MAR) mechanisms depends strongly on network homophily. In homophilic or neutral graphs, both mechanisms yield similar performance, as neighbors are either similar or redundant (Supplementary Material Section C). In heterophilic graphs, however, switching from A-MAR to N-MAR leads to a significant degradation for most approaches. This degradation arises because aggregating dissimilar neighbors injects misleading signals, highlighting the fragility of methods that implicitly rely on homophily.

RQ3: MNAR with Neighborhood Dependence. The MNAR setting amplifies these effects. Missingness that depends on unobserved values interacts with topology in ways that make reconstruction substantially harder.

Homophilic and Neutral Graphs. In these graphs, N-MNAR consistently outperforms A-MNAR in difficulty: Table 3(a,c) shows widespread and significant degradation when moving from A-MNAR to N-MNAR. For example, on Cora, GRIOT and FP incur a deterioration of 17.1% and 68.9%, respectively, for the MAE (both significant). Although the homophilic structure typically helps imputation, N-MNAR introduces dependencies between a node's missingness and its neighbors' unobserved values, making the reconstruction problem inherently more entangled than in A-MNAR.

Heterophilic Graphs. The effect is strongest in heterophilic graphs (Table 3(b)). For instance, FP presents a decrease of 35.3% in MAE on Tolokers, and GRIOT's performance decrease by 77.1% on Cornell ($p < 0.05$). Here, dependence on unobserved values compounds with misleading neighbor information, creating a worst-case regime where all core assumptions of existing imputers are violated. This heterophilic N-MNAR combination emerges as a universal Achilles' heel across methods and datasets.

RQ4: Distributional Fidelity. Pointwise errors do not guarantee correct distributional reconstruction. Figure 2 illustrates distortions in binary attribute distributions on Tolokers (heterophilic) with 20% missing values: imputers underestimate the mass at 0 in N-MNAR, and at 1 in A-MNAR, flattening the bimodal structure. This skew is visible under A-MAR but intensifies under N-MAR and becomes severe under N-MNAR. In these conditions, imputers face biased supervision (missingness depends on unobserved values) and misleading structural cues (dissimilar neighbors), jointly causing systematic underestimation and flattening.

Summary of Findings. We synthesise the main insights across RQs:

1. Graph-aware missingness markedly degrades imputation performance in nearly half of all configurations.

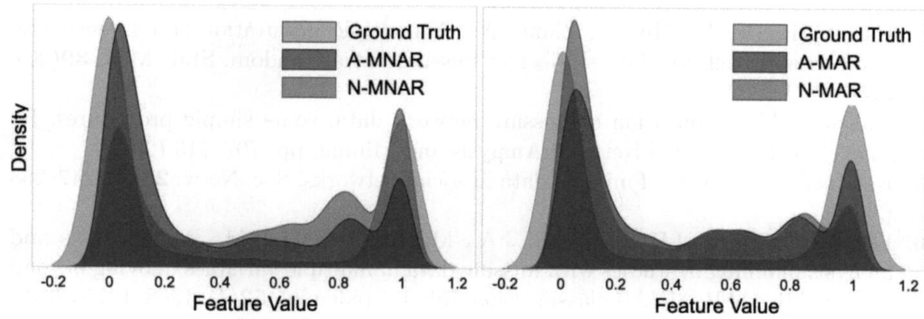

Fig. 2. Density (KDE) plots of imputed features on Tolokers with 20% of missing values. MNAR (left); MAR (right).

2. Heterophily drives the MAR gap, with significant drops from A-MAR to N-MAR.
3. Under MNAR, N-MNAR is consistently the hardest regime, especially on heterophilic graphs.
4. Graph-aware mechanisms break distributional fidelity, introducing biases even when MAE/RMSE appear acceptable.

5 Conclusion and Future Directions

We introduced **GAMM**, a taxonomy extending traditional missing-data definitions to account for graph structure. Our experiments show that graph-aware missingness, especially neighbor-aware MNAR in heterophilic networks, can severely degrade imputation performance, and current state-of-the-art methods, designed for classical settings, struggle under these conditions. Although some downstream tasks remain robust, many rely on accurate feature reconstruction. We highlight the need for imputation methods resilient to these challenges, and for future work on missing structural data. By providing a theoretical framework, empirical evidence, and open-source code (see footnote 1), we aim to support more rigorous evaluation and development of imputation approaches for graphs.

Acknowledgment. This work has been partly funded by a public grant from the French National Research Agency (ANR) under the "France 2030" investment plan, which has the reference EUR MANUTECH SLEIGHT - ANR-17-EURE-0026.

References

1. Craven, M., et al.: Learning to extract symbolic knowledge from the World Wide Web. AAAI/IAAI **3**(3.6), 2 (1998)
2. Doretti, M., Geneletti, S., Stanghellini, E.: Missing data: a unified taxonomy guided by conditional independence. Int. Stat. Rev. **86**(2), 189–204 (2018)

3. Hsu, C.H., He, Y., Hu, C., Zhou, W.: A multiple imputation-based sensitivity analysis approach for data subject to missing not at random. Stat. Med. **39**(26), 3756–3771 (2020)
4. Huisman, M.: Imputation of missing network data: some simple procedures. In: Encyclopedia of Social Network Analysis and Mining, pp. 707–715 (2014)
5. Kossinets, G.: Effects of missing data in social networks. Soc. Netw. **28**(3), 247–268 (2006)
6. Lee, K.J., Carlin, J.B., Simpson, J.A., Moreno-Betancur, M.: Assumptions and analysis planning in studies with missing data in multiple variables: moving beyond the MCAR/MAR/MNAR classification. Int. J. Epidemiol. **52**(4), 1268–1275 (2023)
7. Li, C.: Little's test of missing completely at random. Stata J. (2013)
8. Little, R.J., Carpenter, J.R., Lee, K.J.: A comparison of three popular methods for handling missing data: complete-case analysis, inverse probability weighting, and multiple imputation. Soc. Methods Res. **53**(3), 1105–1135 (2024)
9. Little, R.J.A.: A test of missing completely at random for multivariate data with missing values. J. Am. Stat. Association **83**(404), 1198–1202 (1988)
10. Mann, H.B., Whitney, D.R.: On a test of whether one of two random variables is stochastically larger than the other. Annals Math. Stat. 50–60 (1947)
11. Mohan, K., Pearl, J.: Graphical models for processing missing data. J. Am. Stat. Assoc. **116**(534), 1023–1037 (2021)
12. Muzellec, B., Josse, J., Boyer, C., Cuturi, M.: Missing data imputation using optimal transport. In: ICML, pp. 7130–7140 (2020)
13. Pei, H., Wei, B., Chang, K.C.C., Lei, Y., Yang, B.: Geom-GCN: Geometric Graph Convolutional Networks. In: ICLR (2020)
14. Platonov, O., Kuznedelev, D., Diskin, M., Babenko, A., Prokhorenkova, L.: A critical look at evaluation of GNNs under heterophily: are we really making progress? In: The Eleventh International Conference on Learning Representations (2023)
15. Rebin, D.B.: Inference and missing data. Biometrika **63**(3), 581–592 (1976)
16. Rossi, E., et al.: On the unreasonable effectiveness of feature propagation in learning on graphs with missing node features. In: Learning on Graphs Conference (2022)
17. Rozemberczki, B., Allen, C., Sarkar, R.: Multi-scale attributed node embedding. J. Complex Netw. **9**(2) (2021)
18. Serrano, R., Laclau, C., Jeudy, B., Largeron, C.: Reconstructing the unseen: griot for attributed graph imputation with optimal transport. In: ECML-PKDD, pp. 269–286. Springer (2024)
19. Shadbahr, T., et al.: Classification of datasets with imputed missing values: Does imputation quality matter? arXiv:2206.08478 (2022)
20. Shadbahr, T., et al.: The impact of imputation quality on machine learning classifiers for datasets with missing values. Commun. Med. **3**(1), 139 (2023)
21. Sportisse, A.: Handling heterogeneous and MNAR missing data in statistical learning frameworks : imputation based on low-rank models, online linear regression with SGD, and model-based clustering. Ph.D. thesis, Sorbonne Université (2021)
22. Tang, J., Sun, J., Wang, C., Yang, Z.: Social influence analysis in large-scale networks. In: Proceedings of the 15th ACM SIGKDD international conference on Knowledge discovery and data mining, pp. 807–816 (2009)
23. Um, D., Park, J., Park, S., young Choi, J.: Confidence-based feature imputation for graphs with partially known features. In: ICLR (2023)

24. Van Buuren, S., Groothuis-Oudshoorn, K.: mice: Multivariate imputation by chained equations in R. J. Stat. Softw. **45**, 1–67 (2011)
25. Yang, Z., Cohen, W., Salakhudinov, R.: Revisiting semi-supervised learning with graph embeddings. In: ICML, pp. 40–48 (2016)

CARTGen-IR: Synthetic Tabular Data Generation for Imbalanced Regression

António Pedro Pinheiro[1,2(✉)] and Rita P. Ribeiro[1,2]

[1] University of Porto, 4169-007 Porto, Portugal
[2] INESC TEC, 4200-465 Porto, Portugal
up201704931@up.pt

Abstract. Handling imbalanced target distributions in regression poses a persistent challenge, as the underrepresentation of relevant target values can significantly hinder model performance. Existing data-level solutions often adapt classification-oriented techniques, introducing arbitrary thresholds over the continuous target and leading to artificial and potentially misleading problem formulations. Deep generative models offer flexible sample synthesis but are computationally intensive and difficult to interpret. We propose a CART-based synthetic sampling method specifically designed for imbalanced regression on tabular data. The method integrates relevance- and density-guided sampling to address sparse target regions without thresholding, and employs a feature-driven tree structure to generate realistic tabular samples across heterogeneous features and non-linear interactions. Experiments on benchmark datasets for extreme-value prediction show that the proposed approach is competitive with state-of-the-art resampling and generative methods while offering faster execution and greater transparency. These results highlight its potential as a scalable and interpretable data-level strategy for improving regression models in imbalanced domains.

Keywords: Imbalanced Regression · Data-level Strategies · Sampling · Synthetic Data Generation · Extreme Value Prediction

1 Introduction

Research on imbalanced domain learning for tabular data has predominantly centered on classification tasks [4,22], with a strong emphasis on improving predictions for minority classes. Yet, many real-world problems involve continuous targets, where similar challenges arise in regression settings. These include forecasting extreme weather events [23], predicting exceptionally high sea surface temperatures [2], identifying unusually low drug responses in cancer cells that signal increased sensitivity [16], and detecting major financial frauds. Unlike classification, regression tasks introduce an additional difficulty: determining which regions of the continuous target space should be considered rare, relevant, and worthy of focused modeling.

M. Baratchi et al. (Eds.): IDA 2026, LNCS 16513, pp. 312–326, 2026.
https://doi.org/10.1007/978-3-032-23833-7_23

Data-level approaches remain among the most widely used strategies for addressing imbalance: they modify the training distribution to emphasize the most relevant cases, offer considerable flexibility, and can be applied during preprocessing, thereby extending the applicability of standard machine learning algorithms [4].

In imbalanced regression, several data-level techniques have been proposed [3,5,6,10,11,17,25–27]. However, many of these proposals exhibit intrinsic limitations. Methods that discretize the continuous target variable through artificial thresholds often compromise interpretability and transparency, effectively transforming the approach into a black-box. Moreover, overly simplistic data-generation mechanisms can increase the risk of overfitting, and numerous strategies have limited ability to handle categorical features or missing values.

To address these issues, this paper adapts a CART-based data augmentation method [18], originally designed for classification, to the context of imbalanced regression. Our approach avoids user-defined thresholds on the target variable, thereby eliminating the arbitrariness associated with domain discretization in existing methods [5,6,28]. Moreover, because the technique leverages decision trees for synthetic data generation, it inherits the transparency and interpretability characteristic of CART models, while naturally supporting numeric and categorical features as well as missing values.

The remainder of this paper is structured as follows: Sect. 2 delineates the problem definition and provides a concise overview of the pertinent literature, including existing oversampling and data augmentation techniques. Section 3 presents our proposed solutions, while Sect. 4 discusses the outcomes of an extensive experimental evaluation. Concluding remarks are provided in Sect. 5.

2 Related Work

Supervised learning aims to approximate an unknown function $Y = f(X_1, \ldots, X_p)$ using a model h trained on data $D = \{\langle \mathrm{x}_i, y_i \rangle\}_{i=1}^{n}$, with tasks framed as classification or regression depending on the nature of target variable Y, and optimized through criteria such as error rate or squared error [4,22]. In many regression tasks, the most relevant cases correspond to extreme target values. Their underrepresentation biases models toward predicting average outcomes, causing them to overlook rare yet critical extremes. Addressing such imbalanced regression problems [4,21] involves (i) identifying key target ranges, (ii) assessing performance in these areas, and (iii) biasing learning toward rare instances [22]. Existing research tackles these issues from different perspectives [3,5,6,10,11,17,22,26–28]. This paper focuses on (i), adopts suitable metrics for (ii), by selecting specially suited metrics for imbalanced regression tasks, instead of commonly used standard error metrics, and proposes a new data-level method for (iii).

A central challenge in imbalanced regression is defining non-uniform preferences over continuous domains. While full domain knowledge would ideally guide this process, according to the user needs and preferences, it is rarely available, particularly for infinite target domains. Two main approaches address this issue.

The first, proposed by [21,22], approximates a relevance function $\phi() \in [0,1]$ via interpolation of domain-based control points; when knowledge is lacking, it employs a non-parametric, data-driven procedure that assumes extreme values are most important and derives control points from adjusted boxplot statistics [14]. The second, DenseWeight [24], assigns weights inversely proportional to the estimated density of target values using Kernel Density Estimation (KDE), emphasizing underrepresented outcomes by prioritizing low-density regions; all weights are positive and normalized to ensure stable gradient descent.

Data-level strategies address imbalanced regression by modifying data distributions to better represent relevant instances, using sampling and data augmentation techniques [4]. Examples include Random Undersampling (RU), Random Oversampling (RO), and WEighted Relevance Combination Strategy (WERCS) [6], which employ relevance functions for guided sampling. Gaussian Noise (GN) [6] augments data by perturbing relevant cases while undersampling common ones. Additionally, SMOTER [27] adapts SMOTE for regression by interpolating synthetic samples based on relevance, while SMOGN [5] combines undersampling with SMOTER and GN to balance synthetic data fidelity and variability. Furthermore, G-SMOTER [10] extends G-SMOTE using geometric transformations to diversify generated samples.

However, many of these methods stem from classification settings, often relying on arbitrary thresholds to partition continuous target spaces, which is ill-suited for regression tasks due to the introduction of artificial discretization in the target domain. This is particularly detrimental to the nature of a continuous target variable. By introducing a threshold, the user, for instance, stipulates that any value exceeding the threshold is significant and valuable, while any sample below the threshold is deemed common and insignificant. Consequently, if the threshold is, for example, $x = 5$, then 5.1 is considered important, whereas 4.9 is not. Intuitively, this does not make much sense and creates an abrupt division within the domain. Furthermore, it implies that, for instance, a sample with a target value of 1.2 and another sample with 4.9 are both categorized as normal, common cases, but the magnitude of the target variable value is disregarded. Consequently, artificial thresholds end up categorizing an infinite domain and eliminating the inherent value of a continuous target variable.

WSMOTER [11] mitigates this by integrating DenseWeight with SMOTE, using probabilistic weighting to focus on sparse regions. KNNOR-REG [3] enhances SMOTE with k-NN filtering to identify representative minority points, addressing intra-domain imbalance and noise. Deep learning models like GANs and VAEs have also been explored for synthetic data generation, with some especially developed in the context of imbalanced regression tasks. Models like TVAE [29], CTGAN [29], CopulaGAN [19] and diffusion-based TabDDPM [15] have been widely studied for synthetic case generation but lack specific focus on imbalanced regression tasks. DIRVAE [26] uses a dual-model GAN framework to improve generative performance on sparse regression data. IRGAN [17] integrates generation, correction, discrimination, and regression modules for synthetic sample creation. DAVID [25] combines regression training with a β-VAE architecture. Despite their flexibility, these generative models are computationally intensive and offer limited interpretability due to their black-box nature.

In [20], the author introduced the use of the CART algorithm [8] for synthetic data generation to protect sensitive microdata, emphasizing its advantages over parametric models in handling unknown distributions, complex interactions, and missing values without explicit imputation. Following the same idea, [1] applied CART to generate synthetic patient data for survival analysis in oncology trials, where it outperformed methods such as Random Forests, Bayesian Networks, and CTGAN—particularly in low-data settings. Building further on the ability of CART to model complex relationships, [18] evaluated a CART-based approach for synthetic tabular data generation in the context of class imbalance and fairness, showing strong performance compared with traditional techniques such as SMOTE and GANs. Overall, prior research demonstrates that CART-based data generation effectively captures complex dependencies in the original data while avoiding the distortions often introduced by SMOTE and GANs. Further expanding on these tree-based mechanisms, [9] proposes a random forest approach for regression augmentation. By sampling from the terminal nodes of the ensemble, they approximate the underlying conditional distributions and effectively model non-linear dependencies.

3 Our Proposal

Building on the good results of CART-based synthetic data generation reported in the literature [1,18,20], we propose CARTGen-IR, a method that uses CART to generate synthetic tabular data for imbalanced regression. Our approach removes the need for arbitrary user-defined thresholds when identifying relevant or rare cases, an issue that conflicts with the continuous nature of regression targets. By avoiding crisp partitions of the target domain, CARTGen-IR prevents the domain discretization required by methods such as SMOTER [27] and SMOGN [5]. Since the method builds on decision trees, it samples according to the conditional distributions estimated by CART while also retaining the algorithm's inherent transparency and interpretability. The recursive partitioning process provides an auditable view of how synthetic data is generated. In addition, CARTGen-IR naturally handles numeric and categorical variables, as well as missing values, making it broadly applicable to tabular data.

The overall procedure of CARTGen-IR is summarized in Algorithm 1. The method begins by weighting target values according to their rarity or relevance, assigning higher weights to rare cases. Rarity is estimated using either the DenseWeight method [24] or the relevance function [22], selected through the ρ weight scheme hyperparameter. The former uses probabilistic estimation, which may not match user preferences when density imbalance doesn't reflect their priorities, unlike the relevance-based approach. These scores are adjusted with a rarity exponent α, normalized, and used to resample the dataset with replacement so that rare instances are more likely to be selected.

Rarity-based weighting is also employed in methods such as WERCS [6] and G-SMOTER [10], but CARTGen-IR distinguishes itself by resampling the original dataset in a way that largely excludes frequent target values from augmentation. This produces synthetic data that concentrates more effectively on the rare and relevant regions of the target space. Although we implement DenseWeight and the relevance function, the method supports any rarity estimation mechanism without requiring changes to the overall framework, ensuring adaptability to future developments.

The extent of resampling is controlled by the hyperparameter $\eta \in [0, 1]$, which determines the proportion of synthetic samples to be generated. Following established practice in synthetic data generation [12,13], each selected instance serves as the basis for generating multiple synthetic samples.

Because resampling can create duplicate instances, an optional noise mechanism controlled by δ can be applied to numeric features to reduce overfitting. The resampled dataset, enriched with rare cases, is then used to generate synthetic data through a CART-based, sequential, attribute-wise data generation procedure via sampling in the leaves, described in [18,20], where a CART model is iteratively fitted for each attribute using the remaining, previous attributes as predictors. The synthesis follows a three-step protocol. First, several CART trees are trained by considering each variable as the target and using only the preceding variables as predictors (e.g., the model for X_3 is trained using X_1 and X_2), through the `FitCARTModels` function. Second, the `GenSynthetic` function generates data for a new synthetic case by iteratively selecting the appropriate terminal/leaf node for each variable based on the values of the preceding variables already generated for that case. For the first variable in the sequence, where no prior values exist for the synthetic sample, the process defaults to a root node, and the entire pool of values for that variable is used to select a random value. Third, a value is then randomly drawn from the members of that node and the observed value is taken as the synthetic output for that variable. Furthermore, for continuous variables, instead of sampling directly from the discrete values in a leaf node, the method fits a Gaussian kernel density estimator to the values in that leaf and samples from that smooth distribution. This improves generalization by generating plausible values that fit the statistical profile, including sparse tails, without being limited to the exact points found in the training set. This process is repeated for each synthetic case, always using the same trees fitted in the `FitCARTModels` stage, derived solely from the original data, and is not affected by the new synthetic data generated. This process is illustrated in Fig. 1.

Figure 2 shows a simple example with two features and target values labeled as common or rare. This setup shows how CARTGen-IR handles imbalanced regression differently from interpolation-based methods. While SMOTER creates unrealistic synthetic points and KNNOR-REG fails to generate diverse feature values and only performs augmentation in one rare cluster, CARTGen-IR uses tree-based partitioning to generate samples locally within rare regions, preserving the original data structure.

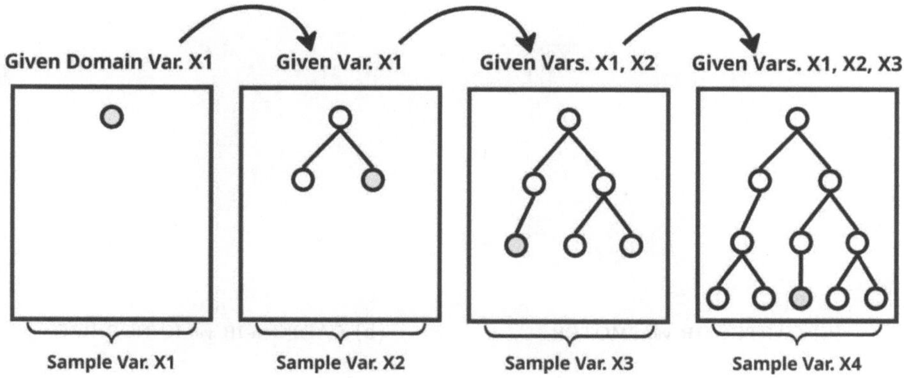

Fig. 1. CART consecutive column-wise data generation.

Algorithm 1: CARTGen-IR Algorithm

```
function CARTGen-IR(⟨X, Y⟩, ρ: weight scheme, α: exponent, η: sampling prop., δ: noise):
    w ← RarityWeights(Y, ρ, α)
    N ← ⌊η · |Y|⌋                                  // nr. cases to generate
    N_re ← ⌊N/5⌋                                   // nr. cases to resample [12,13]
    ⟨X_new, Y_new⟩ ← sample(⟨X, Y⟩, N_re, w, repl =T)
    if δ > 0 then
        X_new ← JitterDuplicates(X_new, δ)         // add Gaussian noise to duplicates
    Trees ← FitCARTModels(⟨X_new, Y_new⟩)          // one CART tree per attribute as target
    ⟨X_synth, Y_synth⟩ ← GenSynthetic(Trees, N)    // attribute-wise generation of synthetic data
    return ⟨X, Y⟩ ∪ ⟨X_synth, Y_synth⟩
function RarityWeights(Y, ρ, α):
    if ρ = denseweight then
        w ←DenseWeight(Y)                          // algorithm presented in [24]
    else if ρ = relevance then
        w ← φ(Y)                                   // algorithm presented in [22]
    w ← w^α
    return w/∑w
```

4 Experimental Study

Our primary objective is to evaluate CARTGen-IR performance and capabilities within the broader context of a comparative study against other state-of-the-art data-level strategies for imbalanced regression tasks. With the experimental study, we aim to answer the following research questions: (**RQ1**) Is CARTGen-IR effective for imbalanced regression scenarios, both in terms of standard error metrics and specially-suited error metrics? (**RQ2**) How does it compare to the state-of-the-art data-level methods proposed for imbalanced regression tasks for tabular data, in both rare and common target subdomains?; and, finally (**RQ3**) What tradeoff do these methods offer regarding predictive performance and execution time?

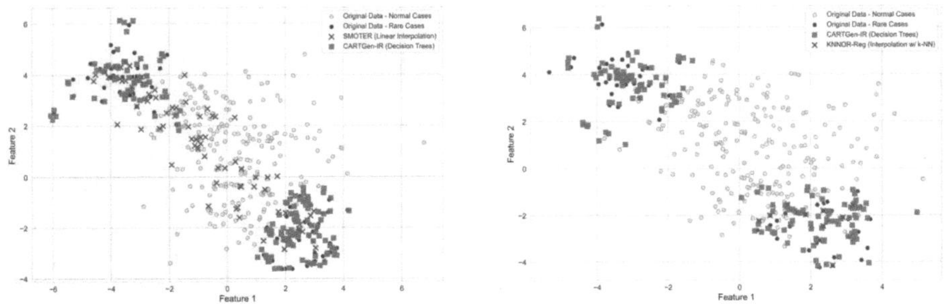

(a) CARTGen-IR vs. SMOTER (b) CARTGen-IR vs. KNNOR-REG

Fig. 2. Synthetic sample generation comparison.

Table 1. Datasets used in the experiments, with nr. of cases, nominal and numerical features, extreme-value type, and rarity statistics for a threshold of 0.8.

Dataset	#cases	#NomFeat	#NumFeat	ExtrType	# Rare	% Rare
strikes	625	0	6	High	15	2.40
forestFires	517	0	12	High	15	2.90
ele-1	495	0	2	High	21	4.24
cpuSm	8,192	0	12	Low	371	4.53
airfoil	1,503	0	5	High	80	5.32
fuelConsumption	1,764	12	25	Both	167	9.47
heat	7,400	3	8	Both	833	11.26
sensory	576	0	11	Both	69	11.98
mortgage	1,049	0	15	Low	133	12.68
maxTorque	1,802	13	19	Both	235	13.04
treasury	1,049	0	15	Low	137	13.06
availablePower	1,802	7	8	Both	305	16.93
housingBoston	506	0	13	Both	105	20.75
abalone	4,177	1	7	Both	1033	24.73
servo	167	4	0	Both	59	35.33

4.1 Experimental Setup

In this study, we used 15 widely used regression datasets. The key attributes of these datasets are summarized in Table 1, which also reports the absolute and relative frequencies of rare instances, as well as the types of extremes, defined according to a relevance threshold of 0.8. To accomplish this, we derived a relevance function for each dataset using the automated approach outlined in [22].

The chosen datasets present a broad range of characteristics such as numeric and nominal features, instances, types of extremes, and rare occurrences.

We evaluated a comprehensive set of preprocessing strategies to address data imbalance in regression tasks. The approaches considered include RU, RO, WERCS and GN from [6], SMOTER [27], SMOGN [5], WSMOTER [11], G-SMOTER [10], DAVID [25], KNNOR-REG [3] and CARTGen-IR, which were specifically developed for imbalanced regression tasks. Furthermore, we included other state-of-the-art deep learning techniques which, although not developed purposefully to tackle imbalanced regression problems, are still suitable for these scenarios: (i) VAE-based method: TVAE [29], (ii) GAN-based methods: CTGAN [29] and CopulaGAN [19]; 3) diffusion-based method: TabDDPM [15]. A complete overview of the 56 resampling configurations is presented in Table 2. The hyperparameter values used for each preprocessing strategy were selected based on those used in the literature where they were proposed or introduced. For CARTGen-IR, the hyperparameter search space was defined by synthesizing established values from related literature and refining them through preliminary empirical analysis. It is important to note that a direct comparison with the DIRVAE approach proposed by [26] as well as the IRGAN approach proposed by [17] was not possible, as both source codes are not publicly available.

Table 2. Data-level strategies and their hyperparameter settings

Strategy	Hyperparameters
RU [6]	%u = {balance, extreme}
RO [6]	%o = {balance, extreme}
WERCS [6]	%u/%o = {0.5, 0.75}
GN [6]	%u/%o = {balance, extreme}, δ = {0.05, 0.1, 0.5}
SMOTER [27]	%u/%o = {balance, extreme}
SMOGN [5]	%u/%o = {balance, extreme}, δ = {0.05, 0.1, 0.5}
WSMOTER [11]	ratio = {1.5, 1.75}, β = {1, 2}
G-SMOTER [10]	strategy = {minority, majority, combined}, $\xi = 0.7$, $\eta = 0.75$, θ = {-0.5, 0.5}, $k = 5$
DAVID [25]	α = {1, 2}
TVAE [29]	–
CTGAN [29]	–
CopulaGAN [19]	–
TabDDPM [15]	–
KNNOR-REG [3]	–
CARTGen-IR	ρ = {denseweight, relevance}, α = {1.0, 1.5, 2.0}, η = {0.5, 0.75}, δ = {0.001, 0}

To assess the effectiveness of these data-level strategies, we paired them with three algorithms: Random Forest (RF), Support Vector Regressor (SVR), and XGBoost (XGB). The experimental setup had 14 hyperparameter combinations for tuning the learning models: RF with 'n_estimators' (100, 200) and 'max_features' ('sqrt', 'log2'), SVR with 'rbf' kernel, 'C' (1, 10, 100), and 'epsilon' (0.1, 0.5), and XGB with 'n_estimators' (100, 200) and 'max_depth' (3, 6). Each model was evaluated across 15 regression datasets under all 56 preprocessing conditions, resulting in a total of 11.760 experiments ($14 \times 15 \times 56$).

In this study, we used both the Squared Error–Relevance Area (SERA) [22] and the relevance weighted RMSE (RW-RMSE), proposed by [7] as evaluation metrics to better assess model performance under imbalanced conditions. We also included RMSE as a standard regression error metric.

All evaluation metrics were computed using a stratified, repeated 2×5-fold cross-validation procedure to ensure robust and reliable performance estimates, coupled with a nested grid search to simultaneously tune the regressor hyperparameters and the sampling strategy configurations. Data augmentation was applied strictly within the cross-validation loop on the training folds only, ensuring no synthetic data leaked to the test sets. The stratification procedure was conducted considering the quantiles of the target variable.

4.2 Obtained Results

Figure 3 provides an overview of the wins and losses of each data augmentation method across all evaluation metrics, highlighting statistically significant differences determined using the Wilcoxon signed-rank test at the 95% confidence level. For readability, the figure reports only the six best-performing configurations of CARTGen-IR (out of 16 tested), together with all competing approaches. The purpose of this test is to enable a comparative analysis among the methods without an absolute pre-determined threshold for distinguishing between good and poor performances.

From this analysis, WSMOTER emerges as the most consistent performer across metrics, followed by KNNOR-REG and G-SMOTER. CARTGen-IR ranks fourth in overall consistency, with a level of robustness not observed in other methods. For instance, WERCS performs well in its oversampling-dominant versions but declines when undersampling becomes more prominent. Notably, although CARTGen-IR is not the most frequent winner, it exhibits a superior significant win-to-loss ratio compared to similarly performing methods, indicating that its victories are generally more meaningful. Additionally, the strongest CARTGen-IR variants exhibit a consistent set of characteristics: they introduce Gaussian noise to numerical attributes and perform well under both density-weighting schemes and rarity exponents of 1.5 and 2.0. The proportion of synthetic samples has little influence, as both tested values produced similar results.

Following the observation of strong performance from both CARTGen-IR and WSMOTER from previous tests, a Bayesian signed-rank test was conducted to compare them, utilizing a Region of Practical Equivalence (ROPE) of $[-1\%, 1\%]$. The results indicated that CARTGen-IR consistently outperformed WSMOTER

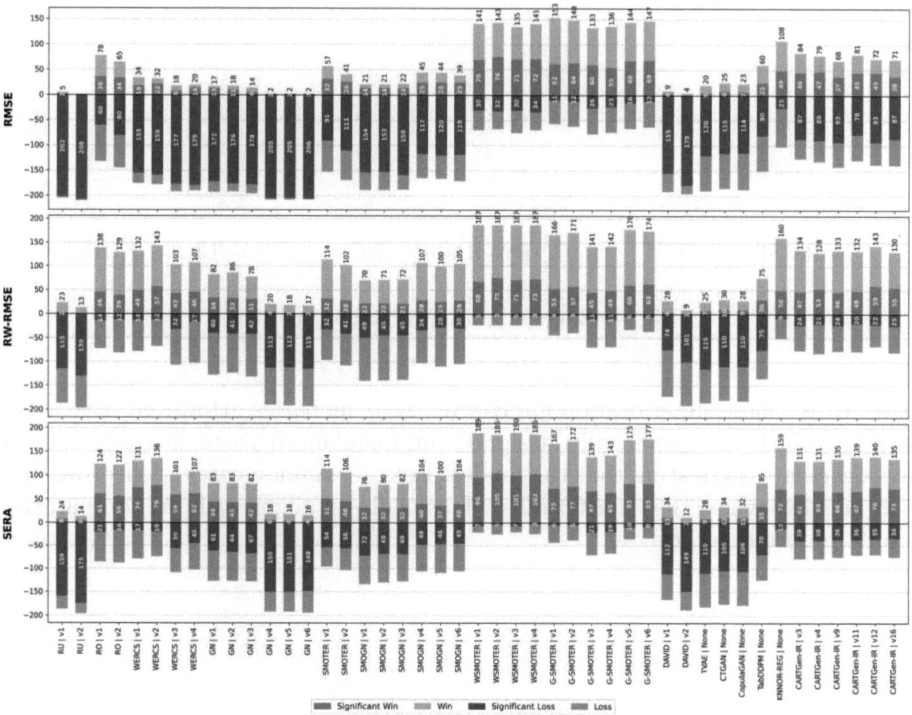

Fig. 3. Comparison of wins and losses, including significant outcomes at the 95% confidence level, for each data-level strategy against a no-preprocessing baseline, across different learners, datasets, and evaluation metrics.

across the evaluated models. Specifically, for RF, CARTGen-IR demonstrated a 99% probability of superiority across all metrics - Fig. 4b. While SVR performance was balanced regarding RMSE and RW-RMSE (Fig. 4a), CARTGen-IR dominated the SERA metric with probabilities exceeding 90%. For XGBoost, CARTGen-IR showed a probability of superiority above 70% for most metrics - Fig. 4c. Overall, the analysis suggests that CARTGen-IR is generally the superior method.

We also conducted a hyperparameter sensitivity analysis of CARTGen-IR, examining the effects of the density scheme ρ, exponent α, sampling proportion η, and noise level δ on performance. Results were obtained using a Random Forest model on a representative regression dataset and are summarized in Fig. 5.

The *relevance* weighting mechanism consistently led to better performance than *denseweight*. For α, values between 1.5 and 2.5 proved most effective with *relevance*, whereas $\alpha = 1$ was preferable for *denseweight*. Increasing η benefited *relevance* but had the opposite effect for *denseweight*. Noise had asymmetric effects as well: additional jitter improved *denseweight* yet degraded *relevance*. RMSE exhibited a distinct pattern. Since it is not tailored to imbalanced regression and penalizes all deviations uniformly, emphasizing rare cases naturally

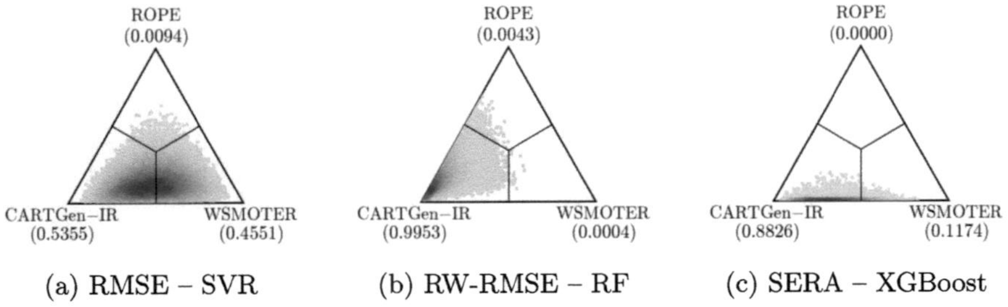

Fig. 4. Bayesian posterior ternary plots.

leads to a slight deterioration in RMSE as η increases. However, the other metrics, which are specifically designed for imbalanced regression, show clear improvements, indicating that the modest loss on common cases is offset by substantial gains where they matter most. This trade-off aligns with the comparative results reported earlier.

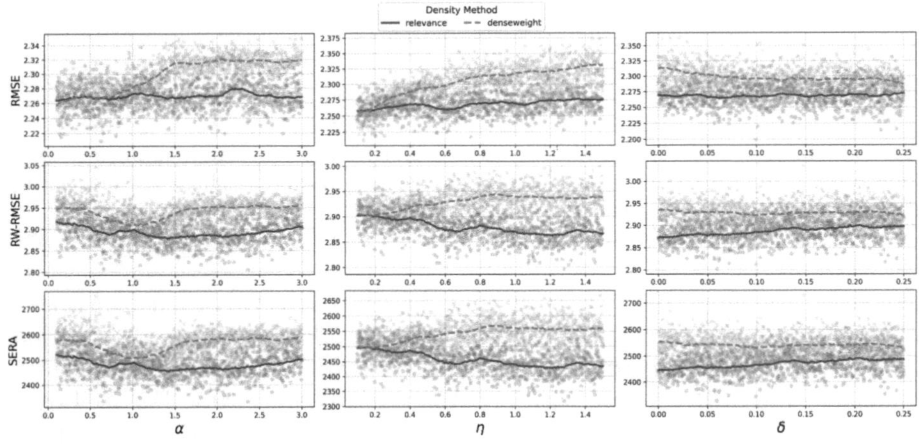

Fig. 5. Sensitivity analysis of CARTGen-IR with respect to the density weighting scheme (ρ), exponent (α), sampling proportion (η), and noise level (δ).

To conclude our experimental study, we performed a runtime comparison across all data augmentation techniques. For fairness, we measured execution times exclusively for the data augmentation procedures under identical parallelization conditions. Figure 6 presents the runtime values (in seconds) for each strategy, in a logaritmic scale. CARTGen-IR stands out as one of the fastest techniques among those generating synthetic data. Sampling-based methods such as RU, RO, and WERCS exhibit the lowest runtimes, as they do not synthesize new data. Among augmentation methods that create new synthetic data, only

KNNOR-REG surpasses CARTGen-IR in speed, though CARTGen-IR demonstrates lower standard deviation, indicating greater consistency. WSMOTER and G-SMOTER have runtimes close to those of CARTGen-IR, while other SMOTER-based methods are significantly slower. Deep learning based methods are the slowest, taking on average 131 times longer to run than CARTGen-IR.

The code used in this study, along with all experimental results, is available at SynthTabularDataGeneration-IR

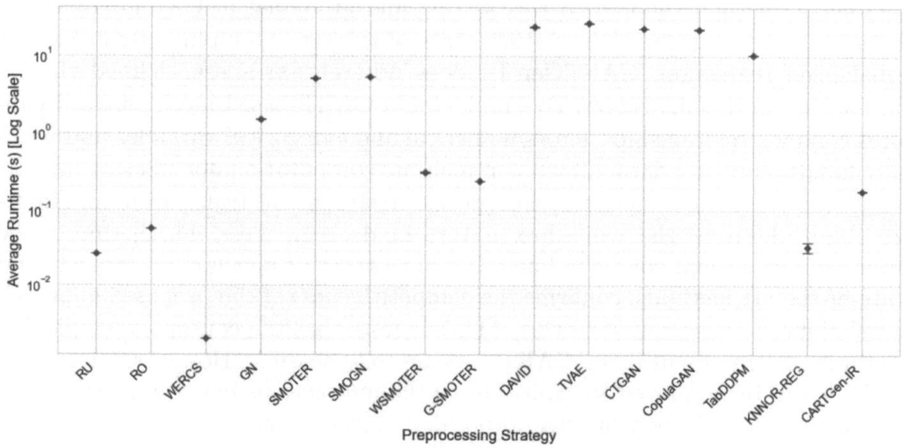

Fig. 6. Aggregated average runtime per data-level strategy.

4.3 Discussion

In response to **RQ1** and **RQ2**, CARTGen-IR has proven to be effective for imbalanced regression tasks. It consistently achieves strong performance across all datasets and ranks the highest overall. When compared to the leading state-of-the-art resampling strategies, CARTGen-IR either improves upon or matches these methods, particularly in balancing the focus between rare-valued and common-valued cases. For the RW-RMSE and SERA metrics, which are specifically designed to assess performance in imbalanced regression tasks, WSMOTER, KNNOR-REG, and CARTGen-IR emerge as clear winners, also demonstrating solid performance for RMSE. This indicates that CARTGen-IR generalizes well across the entire domain without compromising the overall predictive performance. Furthermore, it exhibits a significantly superior win-to-loss ratio compared to the other two methods. Concerning **RQ3**, when analyzing the characteristics of each strategy, we can categorize the methods into sampling and augmentation techniques. Sampling techniques, due to their simplicity in implementation, result in the lowest execution times; however, they also yield inconsistent and lower-ranked scores. Among the methods that generate synthetic samples, KNNOR-REG is the fastest, although it performs poorly on rare target values. CARTGen-IR is the second fastest and offers a strong balance between efficiency and predictive performance.

5 Conclusions

This work addresses the problem of imbalanced regression, where the goal is to predict rare and atypical values of a continuous target, a task that remains challenging for many learning algorithms. We introduced CARTGen-IR, a non-parametric method tailored to this setting and grounded in CART-based synthetic data generation. Unlike other state-of-the-art resampling strategies for imbalanced regression, CARTGen-IR does not rely on a user-defined threshold for the continuous target. By employing a CART-based mechanism, it captures complex relationships across numerical and categorical variables and naturally handles missing data, all while remaining computationally efficient. Importantly, the synthetic data generation process itself remains transparent and interpretable, inheriting the white-box nature of decision trees. Furthermore, the empirical evaluation, conducted across a diverse set of benchmark datasets and state-of-the-art methods, confirms the competitiveness of the proposed approach in imbalanced regression scenarios. This suggests that the concept of employing a data-level strategy utilizing CART appears to be feasible. However, alternative methods can be employed to capitalize on the interactions between features and the framework laid upon by this proposal, as future work.

These findings indicate that a CART-based data-level strategy is a viable approach for imbalanced regression and point to several promising directions for future work. These include expanding the experimental evaluation to a broader range of datasets, particularly those with non-extreme rare intervals; exploring alternative decision-tree learners within the proposed framework to capture more complex feature interactions; incorporating cost-sensitive learning approaches tailored to imbalanced regression metrics such as SERA; and analysing how varying the number of synthetic samples generated per resampled instance affects overall performance.

Acknowledgments. This article is co-funded by the European Regional Development Fund (ERDF) through the Innovation and Digital Transition Programme (COMPETE 2030) under Portugal 2030 and by National Funds through the FCT - Fundação para a Ciência e a Tecnologia, I.P. (Portuguese Foundation for Science and Technology) within project FOXPM, with reference 14731 (COMPETE2030-FEDER-00923300).

Disclosure of Interests. The authors have no competing interests to declare that are relevant to the content of this article.

References

1. Akiya, I., Ishihara, T., Yamamoto, K.: Comparison of synthetic data generation techniques for control group survival data in oncology clinical trials: simulation study. JMIR Med. Inform. **12**, e55118 (2024)

2. Alerskans, E., Zinck, A., Nielsen-Englyst, P., Høyer, J.: Exploring machine learning techniques to retrieve sea surface temperatures from passive microwave measurements. Remote Sens. Environ. **281**, 113220 (2022)
3. Belhaouari, S., Islam, A., Kassoul, K., Al-Fuqaha, A., Bouzerdoum, A.: Oversampling techniques for imbalanced data in regression. Expert Syst. Appl. **252**, 124118 (2024)
4. Branco, P., Torgo, L., Ribeiro, R.: A survey of predictive modeling on imbalanced domains. ACM Comput. Surv. **49**, Article 31 (2016)
5. Branco, P., Torgo, L., Ribeiro, R.: SMOGN: a pre-processing approach for imbalanced regression. In: Proceedings of the 1st International Workshop on Learning with Imbalanced Domains: Theory and Applications, pp. 36–50 (2017)
6. Branco, P., Torgo, L., Ribeiro, R.: Pre-processing approaches for imbalanced distributions in regression. Neurocomputing **343**, 76–99 (2019)
7. Branco, P., Tulon, S.: imbalance-metrics: a Python package for evaluating imbalanced datasets. Version 0.1.6, available via PyPI (2025)
8. Breiman, L., Friedman, J., Olshen, R., Stone, C.: Classification and Regression Trees. Chapman, New York (1984)
9. Caiola, G., Reiter, J.P.: Random forests for generating partially synthetic. Categorical Data. Trans. Data Privacy **3**(1), 27–42 (2010)
10. Camacho, L., Douzas, G., Bacao, F.: Geometric SMOTE for regression. Expert Syst. Appl. **193**, 116387 (2022)
11. Camacho, L., Bacao, F.: WSMOTER: a novel approach for imbalanced regression. Appl. Intell. **54**, 8789–8799 (2024)
12. Chawla, N., Bowyer, K., Hall, L., Kegelmeyer, W.: SMOTE: Synthetic Minority Over-sampling Technique. JAIR **16**, 321–357 (2002)
13. He, H., Bai, Y., Garcia, E., Li, S.: ADASYN: Adaptive synthetic sampling approach for imbalanced learning. In: IEEE International Joint Conference on Neural Networks, pp. 1322–1328 (2008)
14. Hubert, M., Vandervieren, E.: An adjusted boxplot for skewed distributions. Comput. Stat. Data Anal. **52**, 5186–5201 (2008)
15. Kotelnikov, A., Baranchuk, D., Rubachev, I., Babenko, A.: TabDDPM: modelling tabular data with diffusion models. In: Proceedings of the 40th International Conference on Machine Learning (ICML'23), Article No. 725, pp. 1–16 (2023)
16. Lenhof, K., Eckhart, L., Gerstner, N., Kehl, T., Lenhof, H.: Simultaneous regression and classification for drug sensitivity prediction using an advanced random forest method. Sci. Rep. **12** (2022)
17. Liu, X., Tian, H.: Research on imbalanced data regression based on confrontation. Processes **12**, 375 (2024)
18. Panagiotou, E., Roy, A., Ntoutsi, E.: Synthetic tabular data generation for class imbalance and fairness: a comparative study. arXiv:2409.05215 (2024)
19. Patki, N., Wedge, R., Veeramachaneni, K.: The synthetic data vault. In: 2016 IEEE International Conference on Data Science and Advanced Analytics (DSAA), pp. 399–410 (2016)
20. Reiter, J.: Using CART to generate partially synthetic, public use microdata. J. Official Stat. **21** (2005)
21. Ribeiro, R.: Utility-based Regression. Faculty of Sciences, University of Porto, Department of Computer Science (2011)
22. Ribeiro, R., Moniz, N.: Imbalanced regression and extreme value prediction. Mach. Learn. **109**, 1803–1835 (2020)
23. Schultz, M., et al.: Can deep learning beat numerical weather prediction?. Philos. Trans. Roy. Soc. Math. Phys. Eng. Sci. **379** (2021)

24. Steininger, M., Kobs, K., Davidson, P., Krause, A., Hotho, A.: Density-based weighting for imbalanced regression. Mach. Learn. **110**, 2187–2211 (2021)
25. Stocksieker, S., Pommeret, D., Charpentier, A.: Data augmentation with variational autoencoder for imbalanced dataset. arXiv preprint arXiv:2412.07039 (2024)
26. Tian, H., Tian, C., Li, K., Jia, W.: Unbalanced regression sample generation algorithm based on confrontation. Inf. Sci. **642**, 119157 (2023)
27. Torgo, L., Ribeiro, R., Pfahringer, B., Branco, P.: SMOTE for regression. In: Progress in Artificial Intelligence, pp. 378–389 (2013)
28. Torgo, L., Branco, P., Ribeiro, R., Pfahringer, B.: Resampling strategies for regression. Expert. Syst. **32**, 465–476 (2015)
29. Xu, L., Skoularidou, M., Cuesta-Infante, A., Veeramachaneni, K.: Modeling tabular data using conditional GAN. arXiv preprint arXiv:1907.00503 (2019)

Conditional Motif-Based Graph Convolutional Network for Anomaly Detection in the Waste Management Network

Sara Oliveira[1(✉)], Shazia Tabassum[1], João Gama[1], Ana Garcia[2], and Pedro Santana[2]

[1] INESC TEC Porto, 4200-465 Porto, Portugal
sara.l.oliveira@inesctec.pt
[2] IGAMAOT, 1200-433 Lisboa, Portugal
https://www.inesctec.pt, https://www.igamaot.gov.pt

Abstract. Illicit activities in the waste management network, such as waste laundering, misreporting, or trade of stolen waste pose serious environmental and regulatory challenges. Detecting these behaviours is challenging, because they often emerge from higher-order interactions among multiple entities, and are not continuous over time. Furthermore, these activities often manifest as triangles in the network, and the participation of individuals in these waste transfer structures is additionally suspicious. Traditional anomaly detection methods, which rely on first-order relationships or static analyses, struggle to capture these complex, temporally dynamic patterns. To address this challenge, we propose a Conditional Motif-Based Graph Convolutional Network (CM-GCN) that integrates condition-driven triangular motifs directly into the GCN message-passing mechanism. The CM–GCN learns structural embeddings that encode both local graph topology and node attributes–based connectivity to triangular motifs. To detect sudden or sporadic changes, these weekly embeddings are processed by a Long Short–Term Memory Variational Autoencoder (LSTM–VAE), which models temporal behaviour and identifies anomalies through spikes in reconstruction error. Experiments on one year of Portuguese waste transport data demonstrate that the proposed approach effectively highlights companies with known illicit behaviour. The CM–GCN–LSTM–VAE outperformed a standard GCN–LSTM–VAE that ignores motif structure. Results are comparable to, and slightly improve upon, an LSTM–VAE trained on a manually engineered triangle–based feature. This demonstrates that higher–order structural representations learned by the model provide a more informative signal, while simple pairwise relationships contribute little to the detection of complex behaviours.

Keywords: Graph Convolutional Network · Autoencoder · Waste Management

M. Baratchi et al. (Eds.): IDA 2026, LNCS 16513, pp. 327–339, 2026.
https://doi.org/10.1007/978-3-032-23833-7_24

1 Introduction

Waste transportation plays a vital role in the broader waste management system. Waste produced by industrial, commercial, and household activities must be transferred to facilities where they can be treated, recycled, or disposed of in accordance with environmental standards. The proper functioning of this system has direct implications for environmental protection, resource efficiency, and public health. The management of waste transportation in Portugal is governed by strict environmental regulations which are designed to minimize pollution, prevent illegal dumping and trade of stolen waste, and overall promote accountability among the entities that are involved in the waste chain.

Despite this structured regulatory environment, waste transport monitoring remains a challenging task. Waste management operations involve numerous stakeholders, producers, transporters, intermediaries, and recipients. Some entities may engage in falsification, misreporting, or omission of waste transport data to circumvent regulatory obligations, reduce costs, or avoid taxes associated with proper waste disposal. In more serious cases, these actions develop into organized crimes involving the theft and illegal trade of valuable materials such as scrap metal or car parts. Others take advantage of regulatory gaps between countries, engaging in cross–border waste trafficking that shifts environmental burdens to less regulated destinations.

To address these risks, Portuguese authority IGAMAOT (The General Inspectorate of Agriculture, Sea, Environment and Spatial Planning) [9], carries out inspections and audits to detect irregularities and enforce compliance. However, the volume of data generated through e–GAR [7], the mandatory electronic documentation system for waste shipments, is huge. This makes manual analysis increasingly impractical.

Thus, an efficient, data–driven tool is needed to detect waste operating companies engaging in potentially illicit activities, to be flagged for inspection. However, these anomalies can be difficult for traditional fraud detection methods to detect, given that they often emerge from the collusive relation of several entities in the network and that they typically happen sporadically, among periods of normal activity. Thus, they require an approach that can capture the higher–order network topology and its temporal dynamics. Additionally, illicit activities in the waste management network have been confirmed by the domain experts to frequently manifest as triangle formations, and individual waste senders are associated with informal waste trades and thus suggestive of suspicious activities.

Therefore, we propose a time–aware anomaly detection framework with a novel conditional motif–based Graph Convolutional Network (CM–GCN). CM–GCN integrates condition–driven triangular motifs into its message–passing mechanism. By conditioning propagation on the presence of specific motif patterns, CM–GCN embeds higher–order connectivity patterns directly into node representations. Across time, the CM–GCN produces conditional motif–informed dynamic embedding timeseries for each node, which are then processed by a Long Short–Term Memory Variational Autoencoder (LSTM–VAE) to detect temporal anomalies through periods of high reconstruction error. This approach is compared to a standard GCN–LSTM–VAE model, and to a LSTM–VAE model that replaces graph embeddings with manually engineered graph theoretical time-series feature as input.

2 Literature Review

Graph anomaly detection identifies unusual patterns in network data like nodes, edges, or subgraphs that deviate from normal behavior, with applications in fraud detection, cybersecurity, and social networks. Graph Neural Networks (GNNs) have seen growing research interest, due to their ability to learn graph representations through their message passing mechanism, that aggregates the feature information from the neighbours of a given node [10]. Therefore, they have been used for several tasks, including anomaly detection. Graph Convolutional Networks (GCNs) are one of the most widely used for anomaly detection, particularly in unsupervised settings [14]. They learn graph data by aggregating and transforming node features through multiple layers of convolutions, similar to traditional convolutional networks but adapted for irregular graph structures [5,13,22]. Other notable GNN architectures include Graph Attention Networks (GATs) [6,21], which use attention mechanisms to learn different levels of importance of the neighbors when aggregating information; and GraphSAGE [4,8], which uses neighbour sampling and learns flexible neighborhood aggregation functions. While GNNs effectively learn meaningful graph representations, they are typically combined with other deep learning architectures, to capture other types of information, such as temporal patterns [10]. These additional architectures include Transformers [13], Gated Recurring Units [22] and Long Short–Term Memory networks [18].

Traditional GNN–based anomaly detection methods typically rely on pairwise relationships between first–order neighbors, and are limited in expressiveness to what the 1–dimensional Weisfeiler–Lehman (1–WL) graph isomorphism test can distinguish [15]. Therefore, these standard GNN architectures may be unable to capture anomalies associated with complex higher–order structures. This has lead to the development of motif–aware approaches. Motifs refer to any recurrent subgraph patterns, such as wedges, triangles, or more complex structures, that represent interaction patterns beyond simple edges. Instead of aggregating information solely from immediate neighbors, these approaches embed motif–induced adjacency matrices [11,12] or aggregate information from motif–derived "supernodes" [19,20] into the GNN architecture, to enhance node representations and node–level anomaly detection, or to detect anomalous motif instances. However, the dependency of motifs on specific contextual conditions or attribute–driven categories have not been explored until now. Furthermore, integrating such conditional motif reasoning within a sequential, time–aware framework for dynamic graph anomaly detection introduces an additional layer of novelty, enabling the identification of evolving structural irregularities over time. Other approaches have also been developed to capture higher–order structures in GNNs. k–WL based approaches [1,15] aim to overcome the expressive limitations of standard GNNs by aligning architectures with the k–WL hierarchy of graph isomorphism tests. These k–dimensional GNNs (k–GNNs) run message passing through higher–order graph structures (k–tuples of nodes). While theoretically powerful, these methods face significant computational overhead, limiting applicability. Message Passing Simplicial Networks (MPSNs) [3] instead perform message passing on simplicial complexes. CW networks [2] generalize this idea further through more flexible cell complexes, allowing for multi–dimensional message passing on a broader class of higher–order cells, rather than focusing specifically on predefined motifs.

In the context of waste management, literature on anomaly detection remains sparse. A notable exception [17] proposed a graph–based framework for analyzing the e–GAR waste transportation data. The authors combine network–based analysis and machine learning to detect anomalies in the waste transportation data. Their method integrates sub–graph detection, community structure analysis, and centrality measures to extract structural features, and machine learning techniques, including clustering, autoencoders, and density–based methods, are then applied for detecting anomalies. Our previous work [16] extended this approach by introducing temporal analysis through change detection methods, the Page–Hinkley and CUSUM tests, and deep learning, through a LSTM–VAE model, on features created using Social Network Analysis. The results suggested that the feature representing triangles with certain node types provided the strongest signal of anomalous activity. In this work, we will further build on this approach, leveraging the informative power of attribute–based triangle motifs suggested by the results obtained in [16], with the ability of Graph Neural Networks to learn meaningful representations of connectivity patterns.

3 Dataset

The dataset analyzed in this study consists of a comprehensive collection of anonymized electronic Waste Transport records (e–GAR), documenting the movement of waste materials across Portugal during the year 2022. Initially, the raw data comprised over 4.5 million transport records, which underwent extensive preprocessing and cleaning to remove inconsistencies, incomplete entries, and redundant information. After this process, the final dataset contained approximately 3.5 million valid transfer records. These correspond to transactions involving more than 111,000 unique waste producers (senders) and 1,591 distinct authorized receiver companies (which may also send waste themselves). Each record includes details about the sender and receiver, the type and quantity of waste, timestamp of each operation, the operation codes, type of material and more. While this data is rich in information, it also possesses an inherent network structure from the relationships based on transfers. These relationships are evolving over time.

The dataset is unlabeled without the information on fraudulent activities or users in the network. However, a very small subset of four entities is provided by the client who have been confirmed to have engaged in suspicious or fraudulent activities. Due to the extreme scarcity of these labeled nodes, they are not used for training of a supervised framework. Rather, they serve as reference points for evaluating model performance in detecting anomalous behavior, and for parameter tuning.

4 Methodology

The adopted methodology consists of a sequential processing framework that processes weekly transaction graphs to detect anomalous company behaviour. After data preprocessing, each weekly graph is encoded using a Conditional Motif–Based GCN, which integrates both pairwise edges and attribute–centered triangle motifs to produce structural embeddings. These embeddings are then passed through an LSTM to model temporal dynamics across weeks, and a variational autoencoder (VAE) reconstructs the temporal embeddings and assigns an anomaly score based on reconstruction error, indicating how strongly a company's behaviour deviates from learned normal patterns. The overall workflow is illustrated in Fig. 1.

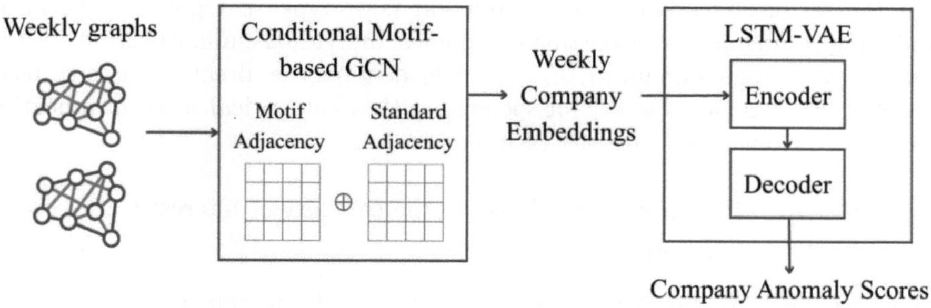

Fig. 1. Overview of the Anomaly Detection model.

4.1 Data Preprocessing

Before conducting our analysis, we performed several data preprocessing steps to ensure the quality of the data. Firstly, we removed all e–GAR documents that had invalid states, such as 'Nullified', 'Rejected', or 'Deleted'. Additionally, we removed residual transfers from 2021 and 2023, restricting our analysis to waste transfers taking place in 2022. Operations related to civil construction were also excluded from the dataset, based on guidance from inspectors who indicated that these records were not relevant for the objectives of our study. For the temporal aspects of our analysis, the date corresponding to the beginning of the transport was selected, as it most accurately reflects the actual activity of the waste operators. After these cleaning steps, the resulting dataset contained over 3.5 million transfer records, with more than 111,000 unique senders and

1,591 unique waste receiving companies, which are the target of our anomaly detection.

Using this cleaned data, we constructed weekly graphs composed of the waste transfers taking place between the entities during each given week. The nodes correspond to the senders and receivers of waste, which can be individuals, waste recycling companies or other organizations. The directed edges in the network represent the flow of waste from sender to receiver, and are attributed with the date of transfer. Nodes are further attributed with the type of entity. Specifically, each weekly graph is composed of the target companies and their 2–hop neighbourhoods, i.e., their direct neighbors, and the neighbors of those neighbors.

4.2 Problem Formulation

The dynamic waste–transfer system is modeled as a discrete temporal directed graph comprising heterogeneous node types

$$\mathcal{G} = \{G_1, G_2, \ldots, G_T\},$$

where each snapshot $G_t = (V_t, E_t, X_t)$ represents the network at time t. Here, V_t is the set of nodes belonging to different node types/categories (e.g., waste producing companies, transporting companies, individuals, unauthorized entities or waste recycling companies), $E_t \subseteq V_t \times V_t$ denotes the directed waste transfer relations, Each node $v_i \in V_t$ is associated with a categorical feature indicating whether it belongs to a category of interest.

$$x_i^{(t)} = \begin{cases} 1, & \text{if node } i \text{ belongs to the category of interest } C, \\ 0, & \text{otherwise.} \end{cases}$$

Thus, a node belonging to a category of interest C is considered a node–of–interest. In our case, the category of interest is individuals, as they are typically associated with informal waste trading. Note that nodes–of-interest are not prelabeled fraudulent nodes and are not subject to prior inspection, and their probability of exhibiting fraudulent behaviour is assumed to be the same as any other node in the network. However, the participation of waste receiving companies, the target of anomaly detection, in triangles with individuals is considered highly suspicious by the domain experts, and is therefore the target of our analysis.

In addition, more features from data (for example: type of operation, the type of waste material handled, hazardous materials) can be incorporated into the feature matrix to provide richer contextual information and weighting the conditional representations learned by the model.

4.3 Conditional Motif-Based Graph Convolutional Network

A graph convolutional network (GCN) is a neural network architecture specifically designed to operate on graph-structured data. GCNs extend the notion of convolution from regular grids to irregular domains such as networks. In a traditional GCN, each node updates its representation by aggregating the features of its neighbors through a propagation function. This allows the network to learn latent embeddings that reflect structural proximity: nodes that are connected or share many neighbors tend to obtain similar representations. However, traditional GCNs rely exclusively on pairwise edges and therefore cannot explicitly capture higher-order connectivity patterns, such as triangles or other network structures. To model such patterns, we present CM-GCN where we augment the GCN with a conditional triangle motif adjacency matrix, aiming to capture irregular activity with nodes-of-interest participating in triangles, which are highly suggestive of suspicious activity.

For computing the motif adjacency matrix, for each pair of nodes (i, j), we count the number of triangles (i, j, k) where one of the nodes is a node-of-interest (individual waste sender). Each triangle is assigned a weight of 1. The resulting motif adjacency is $M_{ij} = \sum_{k \in \Gamma(i) \cap \Gamma(j)} 1$ if triangle (i, j, k) contains atleast one node-of-interest, and 0 otherwise, where $\Gamma(i)$ denotes the neighbors of node i. Therefore, M assigns higher weight to node pairs that jointly participate in many triangles including individual waste senders.

On the other hand, the standard adjacency matrix A encodes the observed pairwise relationships between nodes, that is $A_{ij} = 1$ if there is an edge between i and j, and 0 otherwise. Both A and M are symmetrically normalized with added self-loops using

$$\hat{A} = D_A^{-\frac{1}{2}}(A + I)D_A^{-\frac{1}{2}} \quad \text{and} \quad \hat{M} = D_M^{-\frac{1}{2}}(M + I)D_M^{-\frac{1}{2}},$$

where D_A and D_M denote the corresponding degree matrices.

The matrices are then processed in parallel during training using the ReLU activation function:

$$H_A = \text{ReLU}(\hat{A}XW_A), \quad H_M = \text{ReLU}(\hat{M}XW_M), \tag{1}$$

The final embedding is a combination of the two representations:

$$H = \alpha \odot H_A + (1 - \alpha) \odot H_M, \tag{2}$$

where $\alpha \in (0, 1)$ is a learned coefficient that adaptively determines how much each node should rely on local pairwise structure versus motif-induced higher-order structure.

The model is trained across all weeks and target nodes, excluding the four known anomalous nodes, to minimize a loss that combines the reconstruction loss of the motif adjacency matrix with a temporal smoothing term:

$$\mathcal{L} = \mathcal{L}_{\text{rec}} + \lambda_{\text{temp}} \mathcal{L}_{\text{temp}} = \|HH^\top - \hat{M}\|_2^2 + \lambda_{\text{temp}} \|H_c^{(t)} - H_c^{(t-1)}\|_2^2, \tag{3}$$

where $HH^\top$ is the similarity matrix from the embeddings, and λ_{temp} controls the strength of temporal smoothing. This encourages the model to preserve the structure of node-of-interest centered triangle motifs while maintaining temporal consistency of the embeddings.

After training, the best model is re-run on each weekly graph to extract the embeddings for all target nodes, which are then passed to the LSTM-VAE model.

4.4 LSTM Sequence Autoencoding for Temporal Anomaly Detection

The node embeddings obtained by the Conditional motif-aware GCN are then passed into a Long Short-Term Memory Variational Autoencoder (LSTM-VAE) model to detect anomalies over time. The LSTM-VAE is a deep generative model that combines the ability of LSTM networks to model sequential data with the probabilistic latent variable framework of Variational Autoencoders. The LSTM component captures long-term dependencies and temporal dynamics in the companies' behavior, while the Variational Autoencoder provides the training objective of learning a compressed, probabilistic representation of normal activity that is robust to noise and variability in the data. The model consists of an encoder LSTM that processes input windows of shape (t, n), where t is the window size and n is the number of features. The encoder outputs two vectors, the mean and the log-variance, which define the approximate posterior distribution over the latent variables. A latent vector is then sampled from this distribution using the reparameterization trick. This sampled latent vector is passed to a decoder LSTM, which reconstructs the original input sequence. The model is trained to minimize a loss function that combines two terms: the reconstruction loss and the Kullback–Leibler divergence between the approximate posterior $q(z \mid X)$ and a standard normal prior $p(z) = \mathcal{N}(0, I)$, where z denotes the latent variable. The impact of the KL divergence term on the total loss is controlled by the parameter β, which regulates how much weight is given to the KL divergence term relative to the reconstruction loss. The reconstruction loss is defined as the mean squared error between an input sequence X and its reconstruction $\hat{X}$.

$$\mathcal{L} = \|X - \hat{X}\|_2^2 \;+\; \beta\, D_{\mathrm{KL}}(q(z \mid X) \,\|\, p(z)) \,, \tag{4}$$

High reconstruction errors indicate that the observed sequence deviates from the patterns seen during training, and can therefore be used for anomaly detection.

The model was implemented using the `Keras` library in Python, using sliding windows constructed over each company's time series of embeddings. After training, the model achieving the minimum training loss is used to reconstruct the sequences, and the anomaly score for each company is defined as the mean of its 10 highest reconstruction errors. The top 10 errors were selected instead of the mean of all errors because the objective is to find spikes in periods where it was especially difficult for the model to reconstruct the pattern, as the premise of our approach is that suspicious activities happen sporadically rather than continuously. On the other hand, using the mean reconstruction error would dilute the signal from isolated suspicious spikes.

5 Experiments and Results Discussion

The Conditional motif-based GCN-LSTM-VAE model was applied to the full dataset, consisting of 1,591 target companies and 52 weeks of data. The GCN module used two graph-convolutional layers with hidden dimension 16, producing node embeddings of size 8, and was trained for 15 epochs with learning rate 5×10^{-4}, weight decay 10^{-6}, and no dropout.

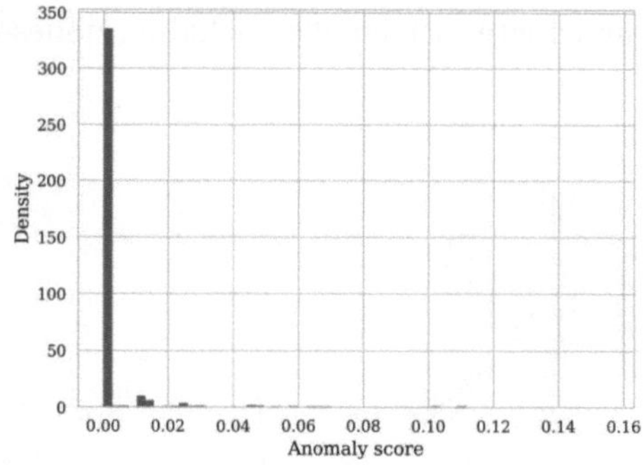

Fig. 2. Distribution of anomaly scores across all companies.

Table 1. Anomaly score percentiles of the four anomalous companies.

Company	Percentile (%)
A	99.50
B	91.95
C	80.26
D	99.56

The weekly embeddings were then arranged into temporal windows of length 8 and fed into the LSTM-VAE for anomaly detection. The encoder contained an LSTM with 32 hidden units followed by two dense layers producing an 8-dimensional latent mean and log-variance. The latent space was regularized through a β-weighted KL divergence with $\beta = 0.5$. The LSTM-VAE was trained for 15 epochs with learning rate 5×10^{-4}, batch size 32, and dropout 0.1. Hyperparameters were tuned by defining ranges for each hyperparameter, testing 15 random combinations, and selecting the combination achieving the best performance on the four anomalous companies.

After running the full pipeline, each target company obtained an anomaly score, reflecting whether it presented periods of unusual connectivity Fig. 2 shows the resulting distribution of anomaly scores.

The obtained distribution is highly positively skewed, with most companies receiving a near-zero anomaly score, consistent with the expectation that most companies exhibit normal behavior. The anomaly score percentiles of the four anonymized anomalous companies are presented in Table 1. Companies A and

D were strongly detected, lying in the top 0.5% of anomaly scores, followed by Company B in the top 8.05%. Company C was weakly detected, appearing in the top 19.74%. Figure 3 shows the static network of three of the pre-labeled anomalous nodes A, B and C only (as the network of all the nodes in the network is infeasible to visualize due to its size). Consistent with the adopted methodology, these nodes jointly participate in triangles including nodes-of-interest (individuals).

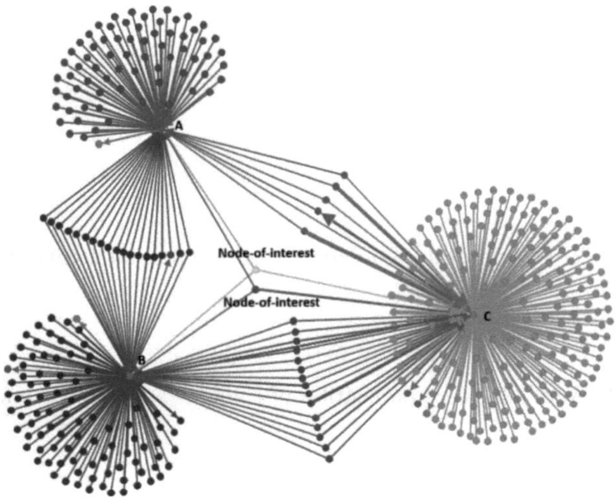

Fig. 3. Network of detected anomalous nodes A, B and C. The node colors in the network are influenced by the color of their neighbors, analogous to message passing in GCN.

5.1 Ablation Study

The performance of the proposed Conditional Motif-based GCN (CM-GCN-LSTM-VAE) model was assessed by comparison against two baseline approaches. The first baseline was a standard GCN-LSTM-VAE relying only on pairwise edges rather than conditional motifs, to detect deviations in companies' overall network connectivity. The second baseline (F-LSTM-VAE) replaced the GCN embeddings with a simple Graph theoretical time-series input feature: for each company and week, the feature consisted of the total number of triangles involving nodes-of-interest connected to that company. An LSTM-VAE was then trained on this sequence to identify anomalous deviations in the company's connectivity patterns. To maintain consistency, all models were trained using the same hyperparameter configuration as the CM-GCN-LSTM-VAE. The resulting anomaly scores for the four anomalous companies under each method are presented in Fig. 4.

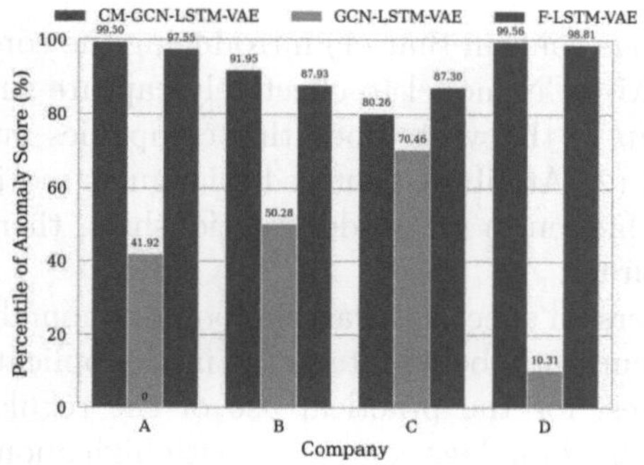

Fig. 4. Comparison of Conditional Motif-based GCN-LSTM-VAE with two baselines.

Firstly, the GCN-LSTM-VAE did not effectively detect any of the four anomalies, assigning them anomaly scores in very low percentiles, which suggests that pairwise connectivity alone is not sufficient to capture anomalous behaviour in the waste management network. Therefore, augmenting the GCN with a category-centered triangle connectivity contributed positively to the detection performance of the GCN-LSTM-VAE model. On the other hand, the F-LSTM-VAE model obtained similar detection performance to the CM-GCN-LSTM-VAE, with the Conditional Motif-based model obtaining a higher anomaly score percentile for three of the four anomalous companies. This result indicates that the connectivity with nodes-of-interest participating in triangles provides an informative signal for anomaly detection in the waste management network.

6 Conclusions

This work proposes a Conditional Motif-based Graph Convolutional Network for detecting anomalous activity in waste management networks, by explicitly encoding higher-order connectivity patterns based on other node attributes. Unlike standard GCNs, which rely solely on pairwise relationships, the CM-GCN integrates conditional motif-induced adjacency into its message-passing mechanism. This allows the model to embed structural patterns that are strongly associated with illicit behaviour, that is, interaction with nodes-of-interest (individuals) participating in triangles.

Combined with an LSTM-VAE model, the proposed framework successfully identifies periods in which companies exhibit unusual motif-based connectivity, producing anomaly scores that highlight entities whose behaviour deviates from learned normal patterns. Experimental results on a year of Portuguese waste transport data (e-GAR) show that the CM-GCN–LSTM-VAE substantially outperforms a standard GCN baseline without motif adjacency, which fails to detect any of the known anomalous companies. The method performs marginally better than an LSTM-VAE using a manually engineered triangle-based feature, demonstrating that the CM-GCN is able to learn this structure directly from the graph.

Overall, the findings confirm that (1) introducing the conditional motif adjacency enabled the CM-GCN model to effectively capture suspicious patterns in the network behaviour of the waste operating companies much better than the motif-less GCN, and (2) Attribute-centered triangular motifs provide powerful insights that remain hidden in first-order relationships, thereby contributing to higher detection accuracy.

This approach offers an effective, scalable tool for anomaly detection in waste management, and is currently being integrated in an application, in combination with other approaches, for the practical use of the regulatory authorities of IGAMAOT. This application flags companies with high anomaly scores, to guide and prioritize the routine inspections, and therefore enhance the oversight of waste management operations.

Acknowledgements. This work was co-funded by national funds for Project EnSafe through FCT - Fundação para a Ciência e a Tecnologia, I.P., under the support 2024.07677.IACDC, and by the European Regional Development Fund (ERDF) through the Innovation and Digital Transition Programme (COMPETE 2030) under Portugal 2030 and by National Funds through the FCT - Fundação para a Ciência e a Tecnologia, I.P. (Portuguese Foundation for Science and Technology) within project FOXPM, with reference 14731 (COMPETE2030-FEDER-00923300).

References

1. Bevilacqua, B., et al.: Equivariant subgraph aggregation networks. ArXiv **abs/2110.02910** (2021). https://api.semanticscholar.org/CorpusID:238407774
2. Bodnar, C., et al.: Weisfeiler and lehman go cellular: CW networks (2022). https://arxiv.org/abs/2106.12575
3. Bodnar, C., et al.: Weisfeiler and lehman go topological: message passing simplicial networks (2021). https://arxiv.org/abs/2103.03212
4. Caville, E., Lo, W.W., Layeghy, S., Portmann, M.: Anomal-e: a self-supervised network intrusion detection system based on graph neural networks. Knowl. Based Syst. **258**, 110030 (2022)
5. Ding, K., Li, J., Bhanushali, R., Liu, H.: Deep anomaly detection on attributed networks. In: Proceedings of the 2019 SIAM international conference on data mining, pp. 594–602. SIAM (2019)
6. Fan, H., Zhang, F., Li, Z.: Anomalydae: dual autoencoder for anomaly detection on attributed networks. In: ICASSP 2020-2020 IEEE International Conference on Acoustics, Speech and Signal Processing (ICASSP), pp. 5685–5689. IEEE (2020)
7. e GAR: electronic transport document (2025). https://apoiosiliamb.apambiente.pt/temas/e-gar
8. Hamilton, W., Ying, Z., Leskovec, J.: Inductive representation learning on large graphs. Adv. Neural Inf. Process. Syst. **30** (2017)
9. IGAMAOT: inspeção-geral da agricultura, do mar, do ambiente e do ordenamento do território (2025). https://www.igamaot.gov.pt/
10. Kim, H., Lee, B.S., Shin, W.Y., Lim, S.: Graph anomaly detection with graph neural networks: Current status and challenges. IEEe Access **10**, 111820–111829 (2022)

11. Lee, J.B., Rossi, R.A., Kong, X., Kim, S., Koh, E., Rao, A.: Graph convolutional networks with motif-based attention. In: Proceedings of the 28th ACM International Conference On Information And Knowledge Management, pp. 499–508 (2019)
12. Li, S., Zhou, J., Mo, C., Li, J., Tso, G.K., Tian, Y.: Motif-aware temporal GCN for fraud detection in signed cryptocurrency trust networks. arXiv preprint arXiv:2211.13123 (2022)
13. Li, Y., et al.: Glad: content-aware dynamic graphs for log anomaly detection. In: 2023 IEEE International Conference on Knowledge Graph (ICKG), pp. 9–18. IEEE (2023)
14. Mesgaran, M., Hamza, A.B.: A graph encoder-decoder network for unsupervised anomaly detection. Neural Comput. Appl. **35**(32), 23521–23535 (2023)
15. Morris, C., et al.: Weisfeiler and leman go neural: higher-order graph neural networks. Proc. AAAI Conf. Artif. Intell. **33**(01), 4602–4609 (2019). https://doi.org/10.1609/aaai.v33i01.33014602, https://ojs.aaai.org/index.php/AAAI/article/view/4384
16. Oliveira, S., Tabassum, S., Gama, J., Santana, P.: Detecting suspicious activities in waste transport data via temporal and network-aware change detection. In: Machine Learning and Knowledge Discovery in Databases. Workshops Track: European Conference, ECML PKDD 2025, Porto, Portugal, September 15–19. vol. in press. Springer (2025)
17. Shaji, N., Tabassum, S., Ribeiro, R.P., Gama, J., Santana, P., Garcia, A.: Network-based anomaly detection in waste transportation data. In: International Conference on Complex Networks and Their Applications, pp. 273–284. Springer (2024)
18. Tian, Z., Zhuo, M., Liu, L., Chen, J., Zhou, S.: Anomaly detection using spatial and temporal information in multivariate time series. Sci. Rep. **13** (2023). https://api.semanticscholar.org/CorpusID:257560549
19. Yuan, Z., Shao, M., Yan, Q.: Motif-level anomaly detection in dynamic graphs. IEEE Trans. Inf. Forensics Secur. **18**, 2870–2882 (2023)
20. Zhang, Z., Song, L., Bao, E., Lv, X., Wang, X.: Atm-gad: adaptive temporal motif graph anomaly detection for financial transaction networks. arXiv preprint arXiv:2508.20829 (2025)
21. Zhao, H., et al.: Multivariate time-series anomaly detection via graph attention network. In: 2020 IEEE International Conference On Data Mining (ICDM), pp. 841–850. IEEE (2020)
22. Zheng, L., Li, Z., Li, J., Li, Z., Gao, J.: Addgraph: anomaly detection in dynamic graph using attention-based temporal GCN. In: IJCAI. vol. 3, p. 7 (2019)

Fair Multilayer Community Detection: A Research Agenda

Georgios Panayiotou(✉) and Matteo Magnani

InfoLab, Department of Information Technology, Uppsala University, Uppsala, Sweden
{georgios.panayiotou,matteo.magnani}@it.uu.se

Abstract. Fairness-aware community detection has recently emerged as an important direction within fair social network analysis. While existing work has begun to integrate fairness constraints into traditional community detection methods, these efforts are limited to single-layer networks, despite many real-world social systems being inherently multilayered. Before extending such methods to multilayer settings, it is critical to clarify how the multilayer network structure affects the feasibility and interpretation of fairness objectives. In this paper, we identify four interrelated dimensions along which fairness-relevant considerations arise: the networked system and its intrinsic or modeling-induced biases, the types of multilayer communities that different methods can detect, the fairness definitions and how the various layers contribute to them, and algorithmic factors that can potentially impact fairness outcomes. By examining how these aspects interact, we provide a foundation for principled method design, and highlight key open challenges, most notably the need for multilayer-specific fairness definitions and comprehensive benchmarking.

Keywords: Fair community detection · Multilayer networks · Fair social network analysis · Algorithmic fairness

1 Introduction

Research on algorithmic fairness has recently expanded to include social network analysis. Several studies have shown that applying fairness-oblivious methods to network data can reinforce latent power asymmetries and demographic biases present in the network [2,9,14,15,22,23,28,36]. These findings have motivated the emergence of fair social network analysis, one of the central goals being the design of methods mitigating such inequalities, and ensuring that algorithmic outcomes do not disproportionately disadvantage specific demographic groups [35].

Within this context, community detection stands out as a fundamental social network analysis task. Community detection is central to understanding the social and functional organization of networks, and the structure of detected

M. Baratchi et al. (Eds.): IDA 2026, LNCS 16513, pp. 340–353, 2026.
https://doi.org/10.1007/978-3-032-23833-7_25

communities can affect downstream tasks such as recommendation, information diffusion, and influence maximization [10,20]. Ensuring fairness in community detection is therefore vital, as biased community structures may reinforce segregation or marginalization of certain groups, especially with the use of network data to represent social networks in very large platforms, including online social media and online retail systems.

Consider, for example, an online social network platform detecting communities, which are then used to drive personalization, a typical practice in online social networks such as Facebook [26] and Twitter [39]: users in the same community receive similar follow recommendations and are exposed to similar content, and information diffusion campaigns are seeded by selecting influential users within each community. If traditional clustering methods are applied behind recommendation systems, users can be clustered almost entirely by homophilous interaction patterns, causing communities to align strongly with protected demographic attributes (e.g., gender or ethnicity). In turn, community-based recommendation may inadvertently become a mechanism for segregation in opportunity and information access, repeatedly routing users back to same-group connections and content. At the same time, diffusion-based tasks can also lead to inequitable outcomes; as influence is measured relative to the community structure, groups appearing structurally smaller or less dense can be treated as low-priority targets. These concerns become still more pronounced in realistic settings, where user ties stem from multiple contexts of interaction, as community structure and, in turn, recommendations, can be strongly affected by the type of interaction between two users [37].

Despite its importance, fair community detection has only been explored within single-layer network settings. Recent works have audited traditional community detection algorithms [6,7,21,30], showing that they can introduce or amplify demographic disparities. A parallel line of work has developed fairness-aware variants of modularity-based community detection [11,29,32], demonstrating that fairness constraints can be integrated into traditional objective functions.

However, many real-world social systems are inherently multilayered, comprising multiple types of ties and interactions, such as friendship, communication, and collaboration, between the same set of individuals. Multilayer networks provide a flexible framework for modeling such systems, as they enable the simultaneous representation and analysis of multiple network layers [1,8,17]. Yet, fairness-aware social network analysis for multilayer networks remains unexplored: there are no fairness definitions nor extensions of fairness-aware methods to multilayer settings, and no guidance on how multilayer-specific modeling choices and structural properties interact with fairness objectives [35]. Before new fairness-aware algorithms can be extended to multilayer network settings, it is critical to identify potential sources of bias that multilayer networks can introduce, including heterogeneity and demographic imbalances across layers, and representational biases induced by modeling choices.

The goal of this paper is to fill this gap by examining the key factors that should guide the development of fairness-aware community detection methods for multilayer networks. We primarily focus on the special case of multiplex networks; that is, single-dimension multilayer networks, where the same actors can participate in different interaction contexts, represented as network layers. Still, these considerations apply to most types of complex networks that can be modeled as a multilayer network [17]. Our key contributions are twofold. First, we identify the major dimensions that can affect detecting fair communities in multilayer networks, and examine their potential impact. We organize these into four interrelated classes: (i) the representation of a social system as a multilayer network, including its structural biases and modeling choices, (ii) the community structures that methods can detect, (iii) the concepts of fairness that can be applied across layers, and (iv) the algorithmic aspects of community detection methods that can influence fairness outcomes. Second, we compile a list of directions for future work using this analysis, proposing future developments.

Overall, we provide a conceptual foundation that highlights the interdependencies between multilayer network community detection and fairness considerations, and supports the principled development of future fairness-aware multilayer community detection methods.

1.1 Related Work

Research on algorithmic fairness in community detection, while relatively limited, has gained attention recently. Several studies have shown that fairness-oblivious community detection algorithms can produce biased partitions [6,7, 21,25]. In response, a number of fairness-aware community detection algorithms have been proposed over the last two years, incorporating group fairness constraints to modularity-based community detection [11,29,32]. While these methods represent important progress, they are designed for single-layer networks, and rely on fairness definitions that do not generalize to settings with multiple interaction types or cross-layer dependencies.

In contrast, the literature on multilayer networks is extensive, including multiple surveys on their modeling and structure [1,3,8,17], and community detection specifically [13,16,19]. However, we are not aware of existing work addressing concepts of fairness in multilayer networks, nor examining how multilayer-specific properties interact with fairness objectives.

2 Fairness Aspects in Multilayer Community Detection

Fairness in multilayer community detection cannot be treated as an isolated algorithmic constraint. Instead, it emerges from a complex interplay between the underlying system represented as a network, the types of communities permitted by the representation, the desired fairness objectives, and the algorithmic mechanisms used to optimize or enforce these constraints. Each of these dimensions can potentially introduce or amplify biases, shape the space of feasible partitions, and ultimately influence the fairness outcomes that can be achieved. These aspects are deeply interconnected, as choices made in one of these dimensions can restrict the possible design and solution space in others.

2.1 Multilayer Network Representation

Multilayer community detection methods operate on a network representation of a social system, which is shaped by two distinct forces: the structure of the social system itself, representing how people interact in each relational context, and the modeling decisions made during network construction. Both sources can encode latent biases in the input, before any algorithm is applied. Structural properties, such as degree imbalances, homophily [9,14,15], and the presence of multiple network hubs [27], can introduce representational disparities and propagate unfairness in the network. At the same time, practical modeling choices, such as layer selection, definition of protected attributes, and weighting of inter-layer coupling edges, can further shape how actors and their relationships are represented across layers. These decisions may implicitly prioritize certain community structures over others, influencing which fairness objectives are meaningful or even feasible within communities.

Intrinsic Biases. A first step in fairness-aware community detection is understanding how the underlying social system may encode social or demographic biases, even before network construction. Many network systems exhibit phenomena such as homophily, where actors with similar attributes preferentially connect [9,14,24,34]; rich-get-richer effects, where well-connected nodes accumulate even more connections over time [9,38]; and degree imbalance correlated with demographic groups, where actors from privileged groups (e.g., individuals with higher income) tend to receive more connections [4,5,33], while also likely occupying structurally central positions. These phenomena often manifest in the network as homogeneous modules, densely connected regions dominated by a single demographic group, or majority-group hubs that serve as bridges between communities. In multilayer networks, such patterns can differ across layers, as each layer reflects a distinct interaction type or relational context. Layers with high assortativity, skewed degree distributions, or sparse connectivity can embed demographic imbalances that remain hidden unless examined.

These effects have direct implications for the fairness and quality of communities that can be discovered, even when the community detection algorithm is designed to enforce fairness constraints. Homophilous structures naturally promote communities that reinforce demographic segregation, making it difficult for algorithms to identify partitions that are both demographically balanced and structurally coherent. The presence of strongly connected homogeneous modules can force fairness-aware methods into trade-offs: achieving representational balance may require splitting strongly connected groups, which lowers structural quality. Moreover, when such groups are present in the network, it can be hard to discover communities (or dense subgraphs in general) that are both balanced with respect to the demographic groups and high in quality, even when the algorithm is specifically designed to promote fairness outcomes [12,32]. Similarly, when high-degree hubs disproportionately belong to majority groups, community detection methods centering communities around the denser network core may amplify the representation of privileged groups.

In the multilayer context, these challenges compound. Structural biases may differ substantially between layers, meaning that groups that appear fair in one layer may be highly imbalanced in another. Layers with strong assortativity or sparse connectivity can further reinforce segregation or disproportionately isolate certain demographic groups, and later constrain the types of communities that algorithms can identify. Fairness-aware algorithm design, therefore, requires sensitivity to the structural heterogeneity across layers, ensuring that differences in network topology do not translate into unfairness in the resulting communities.

Modeling Choices. Modeling decisions made during the construction of a multilayer network representation can also amplify biases. These decisions include defining which relational contexts and demographic information are to be represented as layers and sensitive attributes respectively, whether node attributes apply across all layers simultaneously, and specifying how layers are coupled through inter-layer edges. Each choice potentially constrains what types of fair communities can be discovered.

A first concern is class imbalance, arising when demographic groups are unequally represented within layers. Such imbalances may stem from data collection or sampling practices [15], but also from explicit choices behind what should be considered as a sensitive attribute, particularly when generating networks from secondary data sources. This poses challenges even in single-layer settings, with recent works showing that class imbalances can affect the interpretation of fairness scores if not taken into account [30,32]. Multilayer networks introduce additional complexity, as demographic distributions can differ across layers. A group underrepresented in one relational context may be overrepresented in another, yielding conflicting fairness outcomes across layers. These imbalances can also constrain the types of communities that can be identified, especially in cases where different layers contain metadata for distinct demographic attributes, for example when we are interested in the interactions between intersectional identities of actors in the network [22]. As a result, fairness metrics ignoring group distribution, such as simple group balance within communities, may lead to limited ability in detecting fair communities.

A second modeling choice is the definition and weighting of inter-layer coupling edges, linking an actor's representations across layers, as coupling strength practically controls what types of community structures can be detected. Strong coupling weights enforce cross-layer consistency, likely leading to so-called pillar communities, aligning actors across all layers. While this can simplify enforcing fairness constraints, it may also lead to representational inequalities across different layers. Weak coupling, by contrast, enables identifying non-pillar communities, but may render certain fairness objectives infeasible.

These modeling decisions can thus largely affect fairness-aware community detection. Demographic underrepresentation can force algorithms into trade-offs, sacrificing structural quality at the expense of achieving high fairness scores, while methods aiming to balance demographic representation can also inadvertently reproduce structural inequalities. Coupling edge choices, on the other hand, can determine the space of possible multilayer community structures, which in turn influence how fairness constraints are feasible across different layers. Designing and evaluating fairness-aware methods thus requires careful attention to how modeling choices impact both the resulting community structure and the fairness objectives.

2.2 Community Structure

The second set of considerations concerns the types of communities that a fairness-aware multilayer community detection method is designed to identify. Multilayer networks admit a range of community structure types combining multiple layers [19], and choices about which community types are permitted directly shape fairness outcomes. In particular, the distinction between pillar and non-pillar communities, along with the presence of overlapping or optional community memberships, constrain the expressive power of the algorithm and determine how demographic representation manifests in the detected communities.

Pillar Communities. A central question in multilayer network community detection is whether the method should identify so-called pillar communities, comprised of the same set of actors across all layers, or non-pillar communities, where different layers may involve different subsets of actors [19]. Pillar communities impose strong inter-layer consistency constraints, assuming that actors participate in structurally similar communities regardless of the relational context captured by each layer. In contrast, non-pillar communities reflect variation across layers, allowing actors to belong to distinct communities depending on the interaction type or context represented.

This distinction has important implications for fairness outcomes. Pillar communities may obscure encoded biases on individual layers; for example, minority groups that are structurally marginalized in only one of the layers in the network may not be represented adequately in pillar communities because of the enforced alignment across layers. In contrast, while non-pillar community structures do not necessarily suffer from this constraint, algorithmic approaches prioritizing fairness scores over community quality can yield weakly connected non-pillar communities that artificially raise partition fairness scores. To avoid such effects, it is critical to consider mechanisms that are able to adjust between community quality and partition fairness when designing fairness-aware methods.

Overlapping Communities. Another key design choice concerns whether communities are overlapping and, if so, what type of overlap is permitted. In multilayer settings, overlap can occur in three ways: (i) actor-overlapping communities, where the same actor may belong to multiple communities across layers; (ii) node-overlapping communities, where the same actor can participate in multiple communities on the same layer; and (iii) non-overlapping communities, where each actor is assigned to exactly one multilayer community across the entire network [19].

These distinctions, again, have consequences for fairness outcomes. Overlapping community structures offer more flexible support of fairness objectives; for instance, allowing minority-group actors to participate in multiple communities simultaneously can reduce representational disparities. However, overlapping community membership can also introduce potential pitfalls, as fairness-aware algorithms may exploit overlapping memberships to artificially inflate fairness scores, for example by placing minority-group hub nodes into many communities to balance group proportions. Designing fairness-aware methods therefore requires a clear specification of which types of overlap are permissible and how overlapping memberships should contribute to fairness metrics. Otherwise, fairness constraints may become trivial to satisfy, and fail to represent genuine equity in the underlying network structure.

Community Completeness. Lastly, we should consider whether the method enforces community completeness, that is, whether all actors must be assigned to communities. Methods that leave sparsely connected or structurally peripheral nodes unassigned may also disadvantage groups already underrepresented in the network, especially when such nodes correspond to minority-group actors. Excluding them from community assignments can further distort demographic proportions and reduce visibility of minority groups in communities. Conversely, methods enforcing strict completeness introduce their own challenges. Such methods might assign poorly connected nodes to arbitrary communities simply to maintain demographic balance, which can produce communities that artificially inflate fairness scores without reflecting meaningful structural relationships.

Yet again, these issues are amplified when we consider the structural and representational heterogeneities of multilayer networks: nodes may be peripheral in only a subset of the network layers, and in turn completeness constraints may force inconsistent community assignments across layers.

2.3 Fairness Outcomes

A third set of considerations concerns the fairness outcomes that a multilayer community detection method is expected to achieve. Here, we broadly refer to fairness as the desire that algorithmic outcomes (in this context, community assignments) do not systematically disadvantage any particular individual or group. However, fairness is not a single, universally agreed-upon objective; instead, the literature proposes multiple, often incompatible, metrics and definitions, that reflect various normative assumptions about what it means for a partition to be fair. In multilayer settings, these nuances become even more pronounced, because of the complexity imposed by the model and its flexibility in discovering various community structures. Therefore, we need to carefully consider how fairness is conceptualized, both at the individual community level and the multilayer network as a whole, and how these settings can impact the detected communities.

Community-level Fairness. A fundamental design choice lies in selecting the type of fairness outcomes the method should pursue. Recent work on fairness-aware community detection has proposed various definitions, including statistical parity-style group balance and structural fairness constraints, each capturing different goals of equity within a community. The majority of the literature has focused on group-based metrics, expressing fairness with respect to demographic groups defined by one or more sensitive attributes [11,21,29,30]. Even within this category, there is still an important distinction to be made: should a fair partition consider only the demographic groups in each community, or also the position of the groups with respect to the edges?

Node-based fairness metrics evaluate a community assignment purely in terms of the demographic composition of communities. Recent examples of node-based metrics aim to balance demographic groups within a community [29,30,32]. Such criteria are generally easier to extend to settings with multiple sensitive groups and attributes, and later to multilayer networks, because they depend only on the group memberships of nodes. However, node-based fairness metrics are agnostic to structural variations between the layers.

Conversely, edge-based fairness metrics consider not only demographic groups, but also their positions in the network topology. Such criteria can capture structural intra- and inter-group connectivity patterns [11,21]. Edge-based fairness objectives are conceptually well-aligned with multilayer networks, where connectivity between actors can vary across layers. However, this expressiveness comes at a cost: edge-based metrics are more complex to formalize, both with respect to multiple sensitive groups and when considering multilayer networks, and can introduce large computational overhead.

Beyond sensitive attribute-based metrics, other types of fairness constraints consider the communities' structural properties [6,7]. While such metrics are also relatively easy to extend to a multilayer network setting, they ignore demographic information for actors in the network. In any case, the chosen metric behind the design of a fairness-aware method shapes both the types of partitions that can be discovered and the types of biases a method can address, as well as the computational complexity of the method.

Multilayer Network Fairness. One should also consider whether the desired concept of fairness should apply to each layer individually, or the multilayer network as a whole. This aspect is particularly important, as it largely affects what types of fair multilayer partitions can be discovered. To illustrate these two approaches, we take group balance-based fairness constraints as an example, where the goal is to obtain communities whose nodes are balanced with respect to the various demographic groups in the dataset.

To express this fairness metric over multilayer communities, a simple approach would be to ensure that the nodes in each layer are group-balanced. Such an approach, however, can limit the types of community structures that can be detected; in cases where sensitive attributes are aligned across layers, enforcing layer-by-layer balance is likely to favor pillar communities. Moreover, enforcing group balance on each layer while also aiming for partitions of high quality, can be difficult and in some cases impossible, especially in datasets with stark demographic imbalances. However, individual-layer approaches can be useful when enforcing constraints expressed in terms of unfairness towards a group, such as the proposed modularity-based community fairness metric [21], or even necessary when actors can belong to groups in multiple sensitive demographics.

In contrast, multi-layer group balance constraints can ensure that the overall demographic balance in the entire community, spanning all layers simultaneously, remains high. Such constraints can be seen as a relaxation of the strict approach above, and can be less complex to implement, as they do not need to take individual layer community structure into account. Additionally, they can be ideal when discovering pillar communities, as the same actors participate in the same community across all layers. However, for non-pillar communities, such metrics can receive high scores when merging highly homogeneous well-connected groups from different layers under a single community.

We can similarly consider methods blending the previous two approaches; for example, enforcing multi-layer group balance on a subset of the layers, and individual-layer group balance on the rest, or applying different fairness constraints between layers. Such methods carry the benefits and drawbacks of the aforementioned approaches.

2.4 Algorithmic Aspects

The final set of considerations pertains to the community detection algorithm itself. Various methods for multilayer network community detection have been introduced in recent years, but many are constrained to community quality objectives prioritizing pillar and non-overlapping community structures [19], or specific types of multilayer networks [13]. Here, quality refers to what is considered a good community structure, typically representing how well the partition captures meaningful structural regularities in the network, such as high connectivity between nodes in a community and low connectivity between nodes in different communities. Thus, the choice of algorithmic approach implicitly limits both the types of network structures that are supported, and to what extent fairness outcomes can be achieved. At the same time, how fairness objectives are integrated into the algorithm can introduce a tension between the quality of a community assignment and its fairness outcomes. Therefore, we should also consider to what extent we should sacrifice partition coherence to detect fair communities.

A core driver of the community detection method is its objective function, defining what constitutes a good community structure. Multiple multilayer community detection methods with varying quality objectives have been proposed, including multilayer modularity, random walker, and stochastic block model-based methods [13,19]. Each method embeds implicit assumptions about desirable structural patterns, for instance favoring dense intra-layer connectivity, or penalizing community fragmentation.

A related challenge is how the fairness constraints should be integrated into the objective function, resulting in a tension between balancing partition quality and fairness. Algorithmic strategies differ in how they navigate this trade-off. Some proposed methods explicitly include a user-input weight balancing this trade-off directly into the optimization function [29,32]. Others adopt constrained update rules, restricting node movements between communities that would reduce fairness, even if such a move would increase partition quality [11]. Each approach has varying consequences: strategies integrating fairness constraints directly into the optimization process may allow fairness violations when structurally convenient, while stricter constraints may prevent the algorithm from escaping low-quality partitions [29].

These choices restrict the size and types of communities that the algorithm can feasibly discover, which in turn can potentially affect fairness outcomes [6,7,30]. For example, modularity-based methods tend to favor large, densely connected communities, which may overlook low-degree nodes, often corresponding to minority groups [25]. Similarly, objective functions that reward strong inter-layer alignment may bias the method toward pillar communities, practically restricting the types of communities that can be discovered. Moreover, algorithms relying on greedy local search may also become trapped in regions of the solution space where fairness constraints are difficult to satisfy [29,32]. Similarly, any mechanisms prioritizing partition quality over fairness can determine what types of fair multilayer communities can be discovered, and whether fairness constraints are feasible at all, especially when prioritizing high quality partitions. Algorithmic choices can also lead to scalability limitations, restricting the size of networks that can be analyzed in practice, as the computational cost of multilayer tasks poses significant challenges for large-scale analyses [31].

3 Discussion

In this paper, we examine the challenges that arise when bringing fairness considerations into multilayer community detection. We argue that fairness concerns emerge from every stage of the analysis pipeline: the structure of the social system and its network representation, our choices behind the types of community structures and the fairness objectives, and the algorithmic mechanisms used to optimize them. Each of these aspects introduces potential constraints or sources of bias, and as they are highly interdependent, fairness in multilayer communities cannot be addressed solely as an algorithmic problem.

We highlight several aspects specific to fair community detection in multilayer networks, as opposed to its single-layer variant. These include the need to reason about fairness objectives in multiple dimensions, the influence of layer heterogeneity on the feasibility of certain fairness objectives, and the way coupling edge choices or community types desired implicitly constrain the kinds of fair community structures that can be detected. Moreover, multilayer settings introduce additional tensions between partition quality and fairness; for example, communities that are fair in one layer may be structurally incompatible in another, while fairness scores can be increased artificially, given less strict constraints on the community structures that should be discovered.

Based on these insights, we identify several opportunities for future research. First, there is a need for multilayer-specific fairness definitions that capture how fairness objectives should be expressed when communities span multiple interaction contexts. Such definitions must account for structural and demographic variation across layers, as well as the possibility that the fairness objectives they quantify may be incompatible across different layers.

Second, drawing on these definitions, future work should develop new algorithmic approaches that balance partition quality and fairness in multilayer networks. Existing fairness-aware community detection methods are designed for single-layer settings and do not address challenges such as layer heterogeneity, cross-layer dependencies, or the risk of enforcing fairness in one layer at the expense of another. At the same time, method scalability to large-scale networks remains a practical limitation that should be addressed [31], considering the computational complexity imposed by both the multilayer network model, and community detection itself.

Finally, we argue that future method development requires systematic benchmarking. Different network taxonomies can lead to vastly different, and in some cases impossible, fairness outcomes, even for single-layer networks [12,32]. Method evaluation should, therefore, explicitly account for the diverse sources of bias discussed in this paper. In this context, synthetic multilayer network models allowing for layer-specific community settings [18] are particularly useful, as they enable controlled variation of network and community structures, and can potentially isolate the impact of small design choices on fairness outcomes. Such systematic benchmarking ensures that any fairness-aware methods are evaluated realistically with respect to the underlying data, and that their limitations are properly understood.

Together, these directions form a basis for principled advances in fairness-aware multilayer community detection. We hope that this work provides a starting point guiding the development of future methods that reflect the complexity of multilayer data, and offer transparent and robust tools for analyzing networked systems.

Acknowledgments. This work has been partly funded by eSSENCE, an e-Science collaboration funded as a strategic research area of Sweden, and Swedish Research Council grant 2025–04504.

Disclosure of Interests. The authors declare no competing interests.

References

1. Aleta, A., et al.: Multilayer network science: theory, methods, and applications (2026). https://doi.org/10.48550/arXiv.2511.23371
2. Avin, C., Lotker, Z., Nahum, Y., Peleg, D.: Modeling and analysis of glass ceiling and power inequality in Bi-populated societies. In: Shmueli, E., Barzel, B., Puzis, R. (eds.) 3rd International Winter School and Conference on Network Science, pp. 61–73. Springer International Publishing, Cham (2017). https://doi.org/10.1007/978-3-319-55471-6_6

3. Boccaletti, S., et al.: The structure and dynamics of multilayer networks. Phys. Rep. **544**(1), 1–122 (2014). https://doi.org/10.1016/j.physrep.2014.07.001
4. Bokányi, E., Heemskerk, E.M., Takes, F.W.: The anatomy of a population-scale social network. Sci. Rep. **13**(1), 9209 (2023). https://doi.org/10.1038/s41598-023-36324-9
5. Cremers, J., et al.: Unveiling the social fabric through a temporal, nation-scale social network and its characteristics. Sci. Rep. **15**(1), 18383 (2025). https://doi.org/10.1038/s41598-025-98072-2
6. de Vink, E., Saxena, A.: Group fairness metrics for community detection methods in social networks. In: Cherifi, H., Donduran, M., Rocha, L.M., Cherifi, C., Varol, O. (eds.) Complex Networks & Their Applications XIII, pp. 43–56. Springer Nature Switzerland, Cham (2025). https://doi.org/10.1007/978-3-031-82435-7_4
7. de Vink, E., Takes, F.W., Saxena, A.: Measuring group fairness in community detection. PLOS ONE **20**(11), e0336212 (2025). https://doi.org/10.1371/journal.pone.0336212
8. Dickison, M.E., Magnani, M., Rossi, L.: Multilayer social networks. Cambridge University Press, Cambridge (2016). https://doi.org/10.1017/CBO9781139941907
9. Espín-Noboa, L., Wagner, C., Strohmaier, M., Karimi, F.: Inequality and inequity in network-based ranking and recommendation algorithms. Sci. Rep. **12**(1), 2012 (2022). https://doi.org/10.1038/s41598-022-05434-1
10. Fortunato, S.: Community detection in graphs. Phys. Rep. **486**(3), 75–174 (2010). https://doi.org/10.1016/j.physrep.2009.11.002
11. Gkartzios, C., Pitoura, E., Tsaparas, P.: Fair network communities through group modularity. In: Proceedings of the ACM on Web Conference 2025, pp. 506–517. Association for Computing Machinery, New York, NY, USA (2025). https://doi.org/10.1145/3696410.3714625
12. Honcharov, D., Soundarajan, S.: Fair dense subgraph discovery: algorithms and analysis. Soc. Netw. Anal. Min. **16**(1), 4 (2025). https://doi.org/10.1007/s13278-025-01543-7
13. Huang, X., Chen, D., Ren, T., Wang, D.: A survey of community detection methods in multilayer networks. Data Min. Knowl. Disc. **35**(1), 1–45 (2021). https://doi.org/10.1007/s10618-020-00716-6
14. Karimi, F., Génois, M., Wagner, C., Singer, P., Strohmaier, M.: Homophily influences ranking of minorities in social networks. Sci. Rep. **8**(1), 11077 (2018). https://doi.org/10.1038/s41598-018-29405-7
15. Karimi, F., Oliveira, M., Strohmaier, M.: Minorities in networks and algorithms (2022). https://doi.org/10.48550/arXiv.2206.07113
16. Kim, J., Lee, J.G.: Community detection in multi-layer graphs: a survey. ACM SIGMOD Rec. **44**(3), 37–48 (2015). https://doi.org/10.1145/2854006.2854013
17. Kivela, M., Arenas, A., Barthelemy, M., Gleeson, J.P., Moreno, Y., Porter, M.A.: Multilayer networks. J. Complex Netw. **2**(3), 203–271 (2014). https://doi.org/10.1093/comnet/cnu016
18. Kraiński, Ł., Czuba, M., Bródka, P., Prałat, P., Kamiński, B., Théberge, F.: Multilayer artificial benchmark for community detection (mABCD). Expert Syst. Appl. **307**, 130920 (2026). https://doi.org/10.1016/j.eswa.2025.130920
19. Magnani, M., Hanteer, O., Interdonato, R., Rossi, L., Tagarelli, A.: Community detection in multiplex networks. ACM Comput. Surv. **54**(3) (2021). https://doi.org/10.1145/3444688
20. Malliaros, F.D., Vazirgiannis, M.: Clustering and community detection in directed networks: a survey. Phys. Rep. **533**(4), 95–142 (2013). https://doi.org/10.1016/j.physrep.2013.08.002

21. Manolis, K., Pitoura, E.: Modularity-based fairness in community detection. In: Proceedings of the International Conference on Advances in Social Networks Analysis and Mining, pp. 126–130. ACM, Kusadasi Turkiye (2023). https://doi.org/10.1145/3625007.3627518
22. Martin-Gutierrez, S., Cartier van Dissel, M.N., Karimi, F.: Intersectional inequalities in social ties. Sci. Adv. **11**(45), eadu9025 (2025). https://doi.org/10.1126/sciadv.adu9025
23. Masrour, F., Wilson, T., Yan, H., Tan, P.N., Esfahanian, A.: Bursting the filter bubble: fairness-aware network link prediction. Proc. AAAI Conf. Artifi. Intell. **34**(01), 841–848 (2020). https://doi.org/10.1609/aaai.v34i01.5429
24. McPherson, M., Smith-Lovin, L., Cook, J.M.: Birds of a feather: homophily in social networks. Ann. Rev. Soc. **27**(27), 415–444 (2001). https://doi.org/10.1146/annurev.soc.27.1.415
25. Mehrabi, N., Morstatter, F., Peng, N., Galstyan, A.: Debiasing community detection: the importance of lowly connected nodes. In: Proceedings of the 2019 IEEE/ACM International Conference on Advances in Social Networks Analysis and Mining, pp. 509–512. ASONAM '19, Association for Computing Machinery, New York, NY, USA (2020). https://doi.org/10.1145/3341161.3342915
26. Meta: The AI behind unconnected content recommendations on facebook and instagram (2023). https://ai.meta.com/blog/ai-unconnected-content-recommendations-facebook-instagram/
27. Mosleh, M., Heydari, B.: Fair topologies: community structures and network hubs drive emergence of fairness norms. Sci. Rep. **7**(1), 2686 (2017). https://doi.org/10.1038/s41598-017-01876-0
28. Oliveira, M., Neuhauser, L., Karimi, F.: Stronger together? The homophily trap in networks (2024). https://doi.org/10.48550/arXiv.2412.20158
29. Panayiotou, G., Magnani, M.: Fair-mod: fair modular community detection. In: Cherifi, H., Donduran, M., Rocha, L.M., Cherifi, C., Varol, O. (eds.) Complex Networks & Their Applications XIII, pp. 91–102. Springer Nature Switzerland, Cham (2025). https://doi.org/10.1007/978-3-031-82435-7_8
30. Panayiotou, G., Magnani, M., Calikus, E.: Towards intersectional fairness in community detection. In: Workshop Proceedings of the 19th International AAAI Conference on Web and Social Media 2025, 40 (2025). https://doi.org/10.36190/2025.40
31. Panayiotou, G., Magnani, M., Pinaud, B.: Current challenges in multilayer network engineering. Appl. Netw. Sci. **9**(1), 1–23 (2024). https://doi.org/10.1007/s41109-024-00686-4
32. Panayiotou, G., Simon, A.M.M., Magnani, M., Calikus, E.: MOUFLON: multi-group modularity-based fairness-aware community detection (2025). https://doi.org/10.48550/arXiv.2510.12348
33. Panayiotou, G., Wohlert, I.K., Bask, M., Bask, M., Magnani, M., Mäkinen, I.H.: Anatomy of a swedish population-scale network. Sci. Rep. **15**(1), 30300 (2025). https://doi.org/10.1038/s41598-025-15966-x
34. Rizi, A.K., Michielan, R., Stegehuis, C., Kivelä, M.: Homophily within and across groups. Nat. Commun. **16**(1), 11351 (2025). https://doi.org/10.1038/s41467-025-66415-2
35. Saxena, A., Fletcher, G., Pechenizkiy, M.: FairSNA: algorithmic fairness in social network analysis. ACM Comput. Surv. **56**(8), 213:1–213:45 (2024https://doi.org/10.1145/3653711

36. Sekara, V., Dotu, I., Cebrian, M., Moro, E., Garcia-Herranz, M.: Detecting bias in algorithms used to disseminate information in social networks and mitigating it using multiobjective optimization. PNAS Nexus **4**(10), pgaf291 (2025). https://doi.org/10.1093/pnasnexus/pgaf291
37. Steindl, B., Kolb, T.E., Neidhardt, J.: Beyond demographics: evaluating news recommender systems fairness through behavioural communities. In: Adjunct Proceedings of the 33rd ACM Conference on User Modeling, Adaptation and Personalization, pp. 13–17. UMAP Adjunct '25, Association for Computing Machinery, New York, NY, USA (2025). https://doi.org/10.1145/3708319.3733694
38. Subramonian, A., Sagun, L., Sun, Y.: Networked inequality: preferential attachment bias in graph neural network link prediction (2024). https://doi.org/10.48550/arXiv.2309.17417
39. Twitter: twitter's recommendation algorithm (2023). https://blog.x.com/engineering/en_us/topics/open-source/2023/twitter-recommendation-algorithm

Section: Temporal Data

An End-to-End Framework for Measuring Product Cannibalization Using Multivariate Time Series Forecasting

Daniela Martins[1] and Luís Miguel Matos[1,2](✉)

[1] ALGORITMI/LASI Center – Dep. Information Systems, Escola de Engenharia, Universidade do Minho, Campus de Azurém, Guimarães, 4800-058 Braga, Portugal
luis.matos@dsi.uminho.pt

[2] EPMQ, CCG ZGDV Institute, Campus de Azurém, Guimarães, 4804-533 Braga, Portugal

Abstract. When launching new products, firms must determine whether the new offering will cannibalize existing sales or expand overall portfolio performance. This requires distinguishing genuine cannibalization from coincidental shifts caused by external market conditions. This study presents a three-phase framework that combines multivariate time series forecasting with counterfactual analysis to quantify cannibalization in a rigorous way. Phase 1 identifies key external drivers such as macroeconomic indicators, market trends and competitive actions using correlation and multicollinearity analysis. Phase 2 builds counterfactual forecasts based on pre-launch data to estimate how the incumbent product would have performed in the absence of the new product. Phase 3 converts forecast deviations into business metrics such as cannibalization rate, displaced revenue and net portfolio impact. Five forecasting methods (SARIMAX, Prophet, Random Forest, XGBoost and LightGBM) were evaluated using rolling-window and growing-window validation. In the medical devices case study, the SARIMAX model with rolling-window validation showed the greatest consistency. Results indicate that Product B cannibalized 28% of Product A's sales while still generating a positive net revenue effect, which supports the launch decision. Overall, this framework provides a robust and replicable approach for assessing product interactions and enabling evidence-based decisions in multi-product portfolios.

Keywords: Time Series · Multivariate forecasting · Machine Learning · Product Cannibalization

1 Introduction

Product cannibalization arises when a newly introduced item reduces the sales of an existing product within the same portfolio. Correctly estimating this effect is essential for pricing, marketing, and portfolio planning [1], yet traditional before-and-after comparisons often fail to account for seasonality, macroeconomic shifts, or evolving consumer behavior, producing biased conclusions.

M. Baratchi et al. (Eds.): IDA 2026, LNCS 16513, pp. 357–370, 2026.
https://doi.org/10.1007/978-3-032-23833-7_26

In dynamic markets such as healthcare and consumer goods, where product launches are frequent and target audiences overlap, organizations require methods that distinguish true substitution from broader market changes. This study proposes a multivariate, data-driven time-series framework to quantify product cannibalization by constructing a counterfactual scenario for the incumbent item. The approach is applied to Company X, a medical devices manufacturer in Spain, to assess how the launch of a lower-priced product (Product B) affected an existing one (Product A) over a six-month period. Two questions guide the analysis: To what extent did Product B displace Product A, and what was the net revenue effect for the portfolio?

The framework leverages statistical models (SARIMAX, Prophet) and machine-learning models (Random Forest, XGBoost, LightGBM), each evaluated using rolling and growing temporal validation. Final forecasts are translated into business metrics, including unit substitution and net revenue impact.

The remainder of the paper is organized as follows. Section 2 reviews prior work and outlines methodological challenges. Section 3 presents the forecasting and counterfactual estimation framework. Section 4 reports predictive results and quantifies cannibalization. Section 5 discusses implications, limitations, and avenues for future research. Overall, this work contributes a replicable methodology that integrates temporal validation and external drivers to provide reliable cannibalization estimates for real-world portfolio decisions.

2 Related Work

2.1 Product Cannibalization and Measurement Challenges

Product cannibalization refers to the decline in sales of an existing product following the launch of a related new offering [2, 3]. Although common in innovation-driven markets [4], its accurate measurement remains difficult. Traditional approaches, typically simple before-and-after comparisons, treat pre-launch sales as stable baselines, overlooking seasonality, macroeconomic shifts, and changes in consumer attention [5]. The central methodological challenge is constructing a credible counterfactual reflecting how the original product would have performed absent the new launch.

Research on cannibalization measurement has evolved substantially. Early studies used static sales comparisons, attributing deviations from baseline directly to cannibalization while disregarding dynamic market patterns [2, 6, 7]; while later work used consumer panel data to capture switching behavior [8]. Although richer in behavioral insights, these later approaches depend on granular panel data that are rarely available outside fast-moving consumer goods (FMCG) contexts.

Econometric models, including Vector autoregressive (VAR) models [9], allow dynamic substitution but assume structural stability and require extensive data. More recent machine-learning methods, such as neural forecasting models [10] and multivariate Hawkes processes with inhibition [4], capture complex dependencies but reduce interpretability, limiting their usefulness for business decision-making.

These limitations have prompted a shift toward framing cannibalization estimation as a causal inference problem. Bayesian structural time-series models, as in the

CausalImpact framework [11], estimate post-intervention effects by constructing a synthetic counterfactual integrating trend, seasonality, and exogenous drivers. This approach has been successfully applied to short-term promotional cannibalization in grocery retail contexts [12], analyzing daily sales across multiple stores during discrete promotional windows. While effective for temporary promotional effects spanning days to weeks, permanent product launches require different considerations, including longer evaluation horizons and gradual rather than temporary market adaptation. Additionally, recent studies advocate hybrid approaches combining machine learning with multivariate forecasting to incorporate competitive, macroeconomic, or behavioral signals [4]. These developments highlight the need for frameworks that balance predictive accuracy, interpretability, and practical applicability across industries.

2.2 Research Gap and Contribution

Despite these methodological advances, significant gaps remain. Existing approaches often emphasize either predictive accuracy (e.g., machine learning methods) or causal interpretability (e.g., Bayesian structural models) but rarely integrate both within a systematic comparison framework. Most published research focuses on temporary promotional effects in fast-moving consumer goods [12], leaving permanent product launches in durable goods contexts underserved. Furthermore, few studies provide end-to-end operational frameworks that translate from data preparation through model validation to actionable business metrics, offering theoretical contributions but limiting practical guidance. These limitations hinder practical implementation in business contexts, where portfolio strategy depends on reliable cannibalization measurement.

This study introduces a multivariate forecasting framework that extends causal inference principles to permanent product launches. Unlike promotional analysis where intervention timing is precisely defined and effects are temporary [12], permanent launches require longer evaluation horizons and must account for gradual market adaptation. The framework operationalizes counterfactual reasoning through systematic comparison of SARIMAX, Prophet, Random Forest, XGBoost, and LightGBM models under rolling and growing windows. By integrating both internal sales behavior and external contextual factors, the framework produces transparent and actionable cannibalization estimates, suitable for industries where granular behavioral data may be scarce.

3 Materials and Methods

3.1 Proposed Framework

The proposed framework estimates cannibalization by forecasting a counterfactual—how Product A would have performed had Product B not launched—and comparing this baseline with observed post-launch sales. The methodology unfolds in three conceptually distinct phases that map to the subsequent methodological sections:

Phase 1 – Data and Market Understanding (Sects. 3.2–3.5). This phase establishes the analytical foundation by defining the business context and gathering internal and external data. A comprehensive dataset is constructed integrating sales records with macroeconomic indicators, and consumer interest proxies. The temporal experimental design ensures strict separation between training and evaluation periods to prevent data leakage. Exploratory analysis identifies temporal patterns and seasonal effects, while correlation and multicollinearity analyses guide selection of meaningful predictors that are both statistically valid and practically forecastable. The output is a validated feature set balancing statistical rigor, business interpretability, and forecasting practicality.

Phase 2 – Counterfactual Forecasting (Sects. 3.6–3.8). The causal question is translated into a multivariate forecasting task using complementary model families (SARIMAX, Prophet, Random Forest, XGBoost, LightGBM) to approximate the sales trajectory of the existing product. Model robustness is assessed through rolling- and growing-window validation to reflect real-world temporal variation. Accuracy is evaluated using Root Mean Squared Error, Mean Absolute Error, and Mean Absolute Percentage Error, enabling selection of the best-performing model. Feature importance is assessed through SHAP analysis to ensure interpretability and validate feature selection.

Phase 3 – Cannibalization Quantification (Sect. 3.8). The best model is used to forecast the no-launch trajectory during the post-launch period. The gap between predicted and observed sales quantifies the cannibalized volume. Results are translated into actionable business metrics, including total units cannibalized, lost revenue, and net portfolio impact accounting for the new product's sales.

3.2 Data Description

A comprehensive monthly dataset was constructed by integrating internal sales data with external indicators capturing macroeconomic conditions, consumer interest, market dynamics, and customer sentiment. All variables span January 2018 to March 2025 and were aligned to a monthly frequency to ensure consistency across analyses.

The primary data source consists of internal monthly sales records for Product A and Product B obtained from the internal reporting environment. Product A serves as the dependent variable for forecasting and for constructing the counterfactual trajectory used in the cannibalization estimation.

To account for broader economic influences, several macroeconomic indicators were incorporated. Four Harmonized Index of Consumer Prices (HICP) series were retrieved from the national statistical authority, covering overall inflation (all items) and category-specific inflation in healthcare, recreation and culture, and hospitality. These indicators act as proxies for cost and demand pressures that may affect purchasing behavior independently of portfolio dynamics.

Given limited internal visibility into competitive activity and consumer attention, Google Trends data was used to approximate market dynamics within the Spanish medical devices sector. Three indicators were collected: search interest for Company X, search interest for three competitor brands, and broader category interest based on terms associated with invisible orthodontics. These variables provide a proxy for shifts in consumer awareness that may correlate with product demand.

Customer sentiment was incorporated through the organization's monthly Net Promoter Score (NPS), capturing customer satisfaction over time, obtained from an internal platform.

All variables were synchronized by date and merged into a unified analytical dataset, forming the basis for the subsequent sections and exercises.

3.3 Temporal Experimental Design

To ensure rigorous counterfactual estimation and prevent data leakage, a strict temporal separation was enforced between model development and evaluation phases:

Pre-Product B Period (Training/Validation): January 2018–September 2024. This period was dedicated to model development, encompassing exploratory analysis, feature selection, and hyperparameter optimization. All validation exercises, including rolling and growing windows, were restricted to this timeframe to prevent look-ahead bias in model specification.

Post-Product B Period (Test/Evaluation): October 2024–March 2025. The six-month period following Product B's launch served as a strict holdout set, excluded from all training and tuning phases. It was utilized solely for the final counterfactual forecast and cannibalization analysis, ensuring an out-of-sample prediction based exclusively on pre-launch dynamics.

3.4 Data Pre-processing and Exploratory Data Analysis

An extensive exploratory data analysis (EDA) was conducted to understand temporal patterns and seasonal effects. This stage included visualization, time series decomposition, and derivation of additional features.

The sales data for Product A and Product B exhibit distinct trends and seasonal patterns. Product A demonstrated a strong upward trajectory from January 2018 until the introduction of Product B in October 2024, with recurring peaks at the end of each quarter and consistent annual lows in August. Following the launch of Product B, Product A sales experienced a decline; however, the aggregated sales of both products showed a net increase, suggesting an overall market expansion.

The four HICP series (overall consumer inflation, healthcare, recreation, and hospitality) showed generally upward trends with varying magnitudes. Healthcare inflation remained relatively stable, while recreation and hospitality demonstrated pronounced seasonal patterns, with peaks in summer months corresponding to higher leisure demand. Derived ratio features were introduced to capture relative price developments, such as HICP health divided by the remaining HICP series, reflecting the affordability of healthcare relative to other sectors. These ratios exhibited a declining trend, indicating that healthcare costs grew more slowly than other categories over time, which may influence demand for healthcare-related products.

Google Trends data showed that Company X's search volume rose steadily until mid-2023 before plateauing, with seasonal peaks early and late in the year. Competitor interest increased noticeably after mid-2023, while overall category interest trended upward, accelerating from early 2022 with regular seasonal fluctuations. To quantify brand relative performance, two additional ratios were created: Company X searches relative to total category interest, and Company X searches relative to competitor searches. These metrics highlighted periods of brand dominance and decline relative to competitors and the overall market.

The NPS series reflected high customer satisfaction from 2018 to mid-2022, followed by a decline from mid-2023 onward.

3.5 Correlation, Multicollinearity and Variable Selection

Cross-correlation and Pearson correlation coefficients (denoted by r) were computed for the pre-Product B period (January 2018-September 2024) to quantify relationships between Product A sales and independent variables. Significant correlations were observed across macroeconomic, consumer interest, and derived ratio variables. Lag structures varied from immediate effects, such as Company X searches and overall HICP, to multi-month lags for relative inflation ratios and category-level interest; reflecting delayed influences on Product A sales.

Multicollinearity among independent variables was assessed using Variance Inflation Factors (VIFs), which quantify how much a predictor's variance is inflated due to correlation with other predictors. VIF is calculated as $VIF_j = \frac{1}{1-R_j^2}$ QUOTE , where R_j^2 is obtained by regressing predictor j on all other predictors, with VIF > 10 indicating problematic multicollinearity that can destabilize coefficient estimates and reduce model interpretability. High VIF values were observed for several macroeconomic indicators and derived ratios due to compositional relationships (e.g., HICP Health per Restaurants). Derived ratio features generally exhibited weak correlations with other variables. NPS displayed weak correlations across all series, highlighting limited contribution. Pairwise correlation analysis further confirmed strong interdependencies among macroeconomic indicators and moderate correlations between brand and competitor search volumes.

Considering these insights, a variable selection process was applied to reduce redundancy and enhance model interpretability. Variables with strong to very strong correlations ($r \geq 0.6$) with one another were excluded. Category Interest at lag 3 was therefore selected as the sole exogenous regressor, capturing the strongest independent association with Product A sales ($r = 0.695$) while avoiding estimation instability from remaining correlated variables. The lag-3 specification also ensures genuine ex-ante forecastability. To ensure methodological rigor and prevent data leakage, the framework employs an unconditional forecasting approach restricted to information observable at time t. Specifically, the model utilizes Category Interest with a three-month lag ($t-3$ QUOTE), ensuring that for a six-month horizon, all exogenous inputs precede the forecast window. By maintaining this temporal precedence across all rolling and growing-window folds, the cross-validation framework provides an authentic assessment of predictive performance.

This targeted selection ensures that the multivariate forecasting model includes variables that are both meaningful and interpretable, while adhering to the information availability constraints necessary for credible counterfactual estimation. These features provide a robust foundation for the subsequent modeling stage.

3.6 Modeling and Hyperparameter Optimization with Rolling and Growing Windows

To forecast Product A sales, SARIMAX, Prophet, Random Forest, XGBoost, and LightGBM were employed to capture linear and nonlinear dynamics, seasonal patterns, and the influence of external variables. Hyperparameter optimization was conducted under both rolling and growing window schemes to evaluate model stability across different temporal conditions.

SARIMAX (Seasonal ARIMA with eXogenous variables) incorporates seasonal components and external predictors, providing interpretable forecasts for seasonal [13]. Prophet captures multiple seasonalities, holiday effects, and structural trend changes, offering flexible forecasts for business applications [14]. Random Forest models nonlinear relationships through ensembles of decision trees, providing robustness and resistance to overfitting [15]. XGBoost and LightGBM are gradient boosting frameworks that iteratively improve accuracy, with XGBoost prioritizing robustness and LightGBM focusing on computational efficiency and scalability [16, 17].

Two complementary windowing approaches were employed to assess performance: rolling and growing window. The rolling window repeatedly trains the model on a fixed-length segment of recent data that moves forward in time, capturing short-term dynamics. The growing window expands the training set sequentially, leveraging the full historical record to improve long-term stability. Together, they provide a comprehensive evaluation of model generalization [18]. In this study, the rolling window used a fixed 36-month training set that slid forward monthly, forecasting a 6-month horizon for each fold; while the growing window began with a 36-month training set that expanded sequentially, forecasting the same 6-month horizon.

Hyperparameter tuning was performed via grid search with RMSE as the objective function. Each hyperparameter configuration was evaluated across all folds within both windowing strategies, and average RMSE per configuration was used to identify the optimal specification. Grid search was conducted over the following parameter spaces. For SARIMAX, all combinations of non-seasonal components $p \in \{0, 1, 2\}$, $d \in \{0, 1, 2\}$, $q \in \{0, 1\}$ and seasonal components $P \in \{0, 1\}$, $D \in \{0, 1\}$, $Q \in \{0, 1\}$ (with fixed seasonal period $m = 12$) were evaluated. Prophet's flexibility parameters included changepoint_prior_scale $\in \{0.001, 0.01, 0.1, 0.5\}$ and seasonality_prior_scale $\in \{0.01, 0.1, 1.0, 10.0\}$. Random Forest explored ntrees $\in \{200, 300, 500\}$, mtry $\in \{2, 3, 4, 5\}$, min_n $\in \{3, 5, 8, 12\}$, sample_size $\in \{0.8, 0.9, 1.0\}$, and max_depth $\in \{15, 25, 35\}$. XGBoost parameters comprised nrounds $\in \{100, 200, 300\}$, max_depth $\in \{3, 5, 7\}$, eta $\in \{0.01, 0.1, 0.2\}$, subsample $\in \{0.6, 0.8, 1.0\}$, and colsample_bytree $\in \{0.6, 0.8, 1.0\}$. LightGBM's search space included num_leaves $\in \{31, 100, 200\}$, learning_rate $\in \{0.05, 0.1, 0.2\}$, min_data_in_leaf $\in \{10, 20, 50\}$, feature_fraction $\in \{0.8, 0.9, 1.0\}$, and lambda_l2 $\in \{0, 0.1, 1.0\}$. Each configuration was evaluated across all cross-validation folds within both windowing strategies, with average RMSE determining the optimal parameters. This procedure ensures that the final model settings balance responsiveness to emerging patterns with stability across longer historical contexts. This process can be observed in Fig. 1, and the selected optimal parameters are summarized in Table 1.

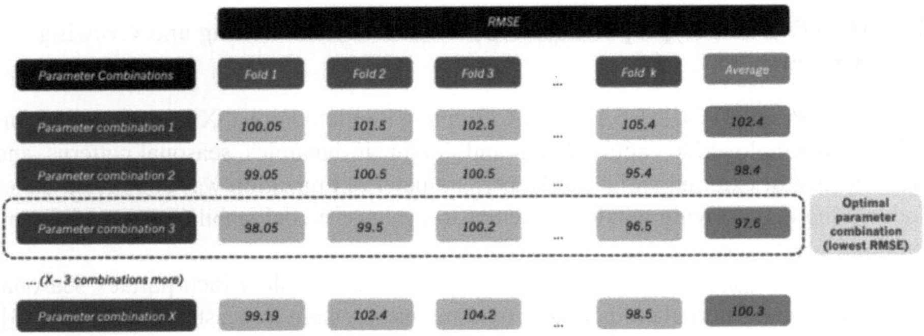

Fig. 1. Process of Hyperparameter Optimization with Grid Search

Table 1. Optimal Parameters for models under Rolling and Growing window approaches

Model	Validation Strategy	Parameters
SARIMAX	Rolling	p = 1, d = 0, q = 0, P = 0, D = 1, Q = 0
SARIMAX	Growing	p = 1, d = 0, q = 0, P = 0, D = 1, Q = 0
Prophet	Rolling	changepoint_prior_scale = 0.1, seasonality_prior_scale = 0.1
Prophet	Growing	changepoint_prior_scale = 0.5, seasonality_prior_scale = 0.1
RF	Rolling	ntrees = 200, mtry = 5, min_n = 3, sample_size = 1, max_depth = 35
RF	Growing	ntrees = 200, mtry = 5, min_n = 3, sample_size = 1, max_depth = 25
XGBoost	Rolling	nrounds = 200, max_depth = 5, eta = 0.1, subsample = 0.6, colsample_bytree = 1
XGBoost	Growing	nrounds = 200, max_depth = 5, eta = 0.1, subsample = 0.8, colsample_bytree = 0.8
LightGBM	Rolling	num_leaves = 31, learning_rate = 0.1, min_data_in_leaf = 10, feature_fraction = 0.8, lambda_l2 = 0.1
LightGBM	Growing	num_leaves = 200, learning_rate = 0.2, min_data_in_leaf = 10, feature_fraction = 0.8, lambda_l2 = 0

3.7 Model Evaluation and Interpretability

Forecast accuracy was evaluated using three complementary metrics. Root Mean Squared Error (RMSE) measures the average magnitude of forecast errors, giving greater weight to larger deviations. Mean Absolute Error (MAE) captures the average absolute difference between predicted and actual values, providing a robust measure less sensitive to outliers. Finally, Mean Absolute Percentage Error (MAPE) expresses errors relative to actual values, allowing for comparison across different scales. While RMSE and MAE are scale-dependent, they are appropriate for this analysis where all models predict the same target variable in identical units. RMSE's sensitivity to large errors is particularly valuable for counterfactual estimation, where substantial forecast deviations would compromise cannibalization quantification. MAPE provides scale-independent evaluation, enabling interpretation relative to sales levels and facilitating potential cross-product comparisons in future applications.

Models used the optimal parameters based on the hyperparameter optimization exercise for both rolling and growing window strategies. Errors were calculated across both frameworks, averaging results over multiple windows to reflect performance across varying training and testing periods.

To enhance model interpretability, SHAP (SHapley Additive exPlanations) analysis was employed to quantify feature contributions to forecasts. Traditional statistical models, like SARIMAX, encode temporal relationships implicitly within model parameters rather than through explicit feature representations, making them incompatible with standard SHAP computation methods designed for machine learning architectures. To address this limitation, a surrogate modeling approach was implemented [19].

For each of the last six windows, a feature matrix was constructed containing 12 lagged values of Product A sales (lags 1–12) and Category Interest at lag 3, transforming the time series into a cross-sectional format suitable for interpretability analysis. All features were standardized using Z-score normalization to ensure comparability across different numerical scales and prevent scale-induced biases in importance measures.

A linear regression surrogate model was then trained using these standardized features as inputs and the statistical model's forecasts as target outputs. This surrogate approximates the input-output behavior of the statistical model while providing the explicit linear structure required for efficient SHAP computation. SHAP values were calculated on the resulting surrogate models using the FastSHAP algorithm, revealing the contribution of each lagged variable and independent variable to the forecasts. The complete feature importance workflow is illustrated in Fig. 2.

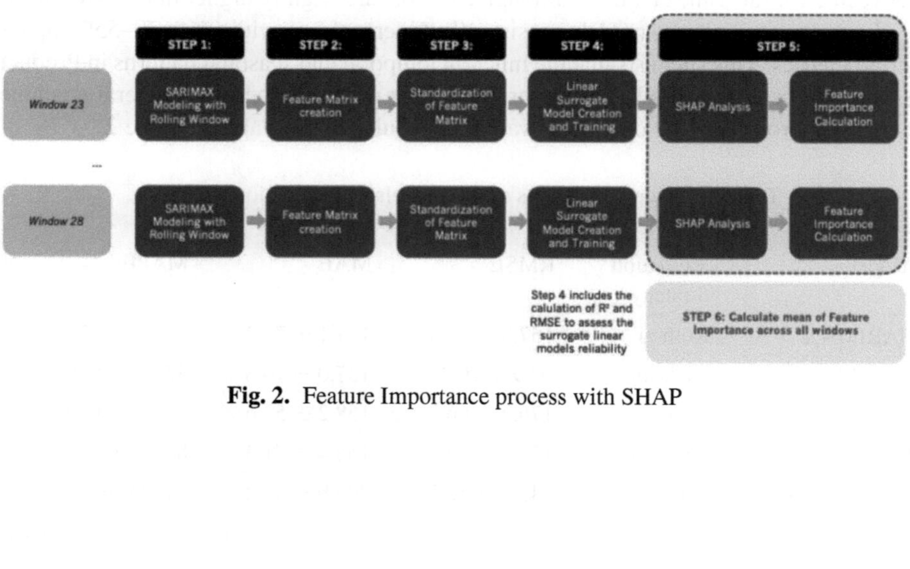

Fig. 2. Feature Importance process with SHAP

3.8 Cannibalization Quantification

The best model's forecasts represent the counterfactual sales of Product A from October 2024 to March 2025, had Product B not launched. Comparing these forecasts with actual sales quantifies cannibalization, which can be expressed as a rate relative to Product B's sales.

$$Sales\,Lost\,from\,Existing\,Product(\#, e) = Forecast(\#, e) - Actuals(\#, e) \quad (1)$$

The cannibalization rate is then calculated as the proportion of lost sales relative to the new product's sales, providing a standardized measure of market displacement.

$$Cannibalization\ Rate(\%) = \frac{Sales\ Lost\ from\ Existing\ Product(\#,e)}{Sales\ of\ New\ Product(\#,e)} \times 100\% \quad (2)$$

This methodology can be applied to both units and revenue, with revenue-based estimates obtained by multiplying the forecasted units by the average selling price. The approach ensures that the estimation of cannibalization is methodologically consistent, statistically grounded, and reproducible, maintaining alignment with the rigorous data preparation, modeling, and validation processes established in the earlier phases.

4 Results

4.1 Predictive Results

Results show that SARIMAX consistently delivers the most accurate forecasts, outperforming all other models across RMSE, MAE, and MAPE. Its performance remains stable across both windowing strategies, with minimal differences between rolling and growing windows, highlighting its suitability for forecasting Product A sales. Prophet emerges as a strong alternative, offering moderate accuracy with a favorable balance between error rates and stability, although its errors are slightly higher than SARIMAX.

In contrast, tree-based models consistently underperform, with errors 30–55% higher across metrics. This suggests that the inherent temporal and seasonal patterns in Product A sales are better captured by time series-specific models rather than general machine learning approaches. The complete evaluation results are available in Table 2.

Table 2. Evaluation results across models and validation strategies

Model	Validation Strategy	RMSE	MAE	MAPE
SARIMAX	Rolling	157.9 ± 82.3	137.1 ± 73.2	7.7% ± 3.7%
SARIMAX	Growing	159 ± 79.2	137.0 ± 70.6	7.7% ± 3.6%
Prophet	Rolling	176.7 ± 66.3	158.2 ± 57.6	8.7% ± 3.5%
Prophet	Growing	179.1 ± 82.3	155.4 ± 71.3	8.4% ± 3.8%
Random Forest	Rolling	245.7 ± 52.7	209.9 ± 46.2	11.3% ± 3.2%

(continued)

Table 2. (*continued*)

Model	Validation Strategy	RMSE	MAE	MAPE
Random Forest	Growing	241.5 ± 53.1	206.3 ± 44.7	11.1% ± 3.0%
XGBoost	Rolling	211.1 ± 63.3	178.3 ± 55.6	9.5% ± 3.1%
XGBoost	Growing	214.9 ± 67.8	176.6 ± 58.0	9.3% ± 3.0%
LightGBM	Rolling	238.6 ± 63.3	199.2 ± 64.6	10.3% ± 3.8%
LightGBM	Growing	240.0 ± 59.3	195.9 ± 49.1	10.2% ± 2.9%

The superior performance of SARIMAX and Prophet over tree-based models reflects the dataset's limited temporal dimension rather than algorithmic weaknesses. With only 80 monthly observations in the training period, gradient boosting and ensemble methods lack sufficient data to learn complex patterns that might improve upon well-specified statistical models. Tree-based approaches typically require hundreds to thousands of observations to outperform traditional time series methods, particularly for data with strong seasonal structure that SARIMAX and Prophet explicitly model. This finding highlights that model selection should prioritize dataset characteristics over algorithmic complexity: statistical methods remain optimal for moderately sized temporal datasets with clear patterns, while machine learning approaches require larger samples to demonstrate their advantages.

To enhance interpretability, SHAP analysis was conducted using surrogate models trained to approximate SARIMAX predictions. These surrogate models replicated SARIMAX behavior with high fidelity, achieving an average R2 of 0.911 and RMSE of 0.292, indicating that they captured more than 90% of the underlying predictive structure. SHAP results show that the 12-month seasonal lag and Category Interest (lag 3) are the dominant drivers of Product A forecasts, confirming the relevance of the selected features and providing clear, business-oriented insights into the model's behavior.

4.2 Forecasting Product a Sales

Using the validated SARIMAX model under the rolling-window strategy, Product A sales were forecasted for the post-launch period (October 2024-March 2025, as defined in Sect. 3.3) using the most recent 36 months of pre-launch data (October 2021-September 2024) and Category Interest at lag 3. This forecast represents the expected sales trajectory absent Product B's launch. Confidence intervals (80% and 95%) quantify predictive uncertainty. The model projects cumulative sales of approximately 12,921 units, serving as the baseline for cannibalization estimation. The monthly forecast is presented in Fig. 3.

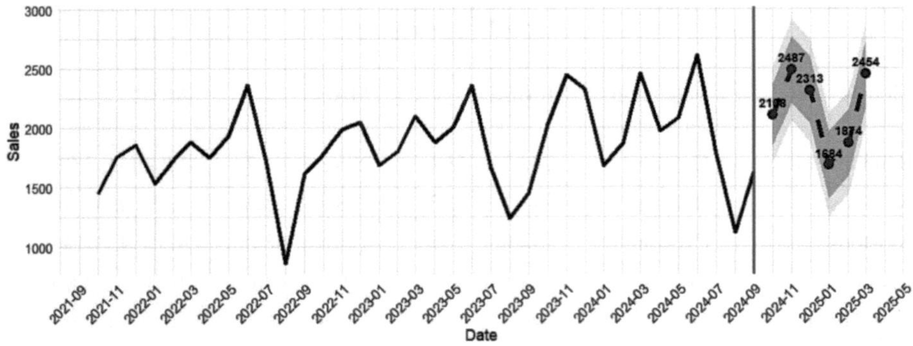

Fig. 3. Product A Forecast using SARIMAX with Rolling Window

4.3 Cannibalization Analysis

Compared to actual post-launch sales of 10,517 units, the forecast indicates a loss of 2,404 units and a baseline cannibalization rate of 28%. This suggests that 28% of Product B's demand directly substituted Product A, while 72% represented incremental growth, signaling that the launch expanded the overall market rather than solely shifting internal demand.

Cannibalization peaked in November 2024 (46%), coinciding with Product B's initial marketing push, before moderating to around 13% by January 2025 as the market stabilized. Cumulatively, Product B generated 8,715 units sold, leading to a net portfolio gain of 6,311 units over the six-month period.

Revenue analysis confirmed a positive net portfolio. While Product A experienced an estimated loss of €6.15 million, Product B contributed €15.25 million, yielding a net revenue increase of €9.1 million. This demonstrates that, despite measurable cannibalization, the new product introduction was strategically beneficial, enhancing both sales volume and financial performance across the portfolio. This analysis is presented in Table 3.

Table 3. Cannibalization Results in Units and Revenue (€)

Month	Product A Actual Units	Product A Forecast Units	Units Lost	Product B Units	Cannibalization Rate (%)	Net Unit Gain	Product A Revenue	Product B Revenue	Net Revenue Gain
Oct-24	2,006	2,108	102	288	36%	186	5,135,360	504,000	241,651
Nov-24	1,910	2,487	577	1,266	46%	689	4,889,600	2,215,500	737,458
Dec-24	1,843	2,313	470	1,663	28%	1,193	4,718,080	2,910,250	1,707,997
Jan-25	1,493	1,684	191	1,504	13%	1,313	3,822,080	2,632,000	2,142,528
Feb-25	1,487	1,874	387	1,741	22%	1,354	3,806,720	3,046,750	2,055,748
Mar-25	1,778	2,454	676	2,253	30%	1,577	4,551,680	3,942,750	2,212,881
Total	**10,517**	**12,921**	**2,404**	**8,715**	**28%**	**6,311**	**26,923,520**	**15,251,250**	**9,098,264**

5 Conclusion and Future Work

This study developed a multivariate time series framework for quantifying product cannibalization through counterfactual forecasting. The framework systematically compares statistical and machine learning models under rolling and growing window validation to identify the most reliable approach for each context, translating forecasts into actionable business metrics.

Although demonstrated in the medical devices industry, the methodology is designed for cross-industry applicability. Its core components of external driver's identification, counterfactual forecasting, and translation to tangible business metrics, rely on standard time series structures transferable across sectors. Successful implementation requires quality historical sales data and access to relevant external indicators that influence demand patterns, prerequisites that organizations in consumer goods, telecommunications, financial services, and other industries typically possess. The framework's logic and analytical steps can be adapted to any context where product launches may affect existing portfolio performance.

Applied to Company X's product launch, SARIMAX effectively captured temporal dynamics and enabled robust cannibalization estimation. Results confirmed that while meaningful substitution occurred, the new product generated substantial incremental demand, yielding a net positive portfolio impact that supported the strategic launch decision.

This study has limitations that suggest future research directions. First, the analysis relies on a single medical devices dataset; cross-industry validation would strengthen generalizability. Second, the absence of competitor sales data constrained full market decomposition, limiting separation of market share redistribution from category expansion. Third, the six-month evaluation window captures initial patterns but may miss long-term trends; extending to 12- and 18-month horizons would reveal whether cannibalization rates remain stable or evolve. Finally, the counterfactual assumes external drivers would have evolved similarly regardless of the launch – an assumption requiring context-specific validation.

Future research could incorporate competitor data and validate the framework across industries with varying demand patterns. This work demonstrates that rigorous counterfactual modeling can transform cannibalization analysis from retrospective estimation into a generalizable tool for portfolio strategy across sectors.

Acknowledgments. This work has been supported by FCT – Fundação para a Ciência e Tecnologia within the R&D Unit Project Scope UID/00319/2025 - Centro ALGORITMI (ALGORITMI/UM).

Disclosure of Interests. The authors have no competing interests to declare that are relevant to the content of this article.

References

1. Raghavan Srinivasan, S., Ramakrishnan, S., Grasman, S.E.: Identifying the effects of cannibalization on the product portfolio. Mark. Intell. Plan. **23**(4), 359–371 (2005). https://doi.org/10.1108/02634500510603465

2. Copulsky, W.: Cannibalism in the Marketplace. J. Mark. **40**(4), 103–105 (1976). https://doi.org/10.1177/002224297604000415
3. Kotler, P., Keller, K.: Marketing Management. Prentice Hall (2012)
4. Deutsch, I., Ross, G.J.: Estimating Product Cannibalisation in Wholesale using Multivariate Hawkes Processes with Inhibition (2023), arXiv: arXiv:2201.05009. https://doi.org/10.48550/arXiv.2201.05009
5. Okorie, O., Obi, M., Russell, J., Charnley, F., Salonitis, K.: A triple bottom line examination of product cannibalisation and remanufacturing: a review and research agenda. Sustain. Prod. Consum. **27**, 958–974 (2021). https://doi.org/10.1016/j.spc.2021.02.013
6. Traylor, M.B.: Cannibalism in multibrand firms. J. Consum. Mark. **3**(2), 69–75 (1986). https://doi.org/10.1108/eb008164
7. Mason, C.H., Milne, G.R.: An approach for identifying cannibalization within product line extensions and multi-brand strategies. J. Bus. Res. **31**(2–3), 163–170 (1994). https://doi.org/10.1016/0148-2963(94)90080-9
8. Lomax, W., Hammond, K., East, R., Clemente, M.: The measurement of cannibalization, vol. 14, no.7. Marketing Intelligence & Planning (1996)
9. McColl, R., Macgilchrist, R., Rafiq, S.: Estimating cannibalizing effects of sales promotions: the impact of price cuts and store type. J. Retail. Consum. Serv. **53**, 101982 (2020). https://doi.org/10.1016/j.jretconser.2019.101982
10. Garnier, R.: Concurrent Neural Network: a model of competition between times series (2021), arXiv: arXiv:2009.14610. https://doi.org/10.48550/arXiv.2009.14610
11. Brodersen, K.H., Gallusser, F., Koehler, J., Remy, N., Scott, S.L.: Inferring causal impact using Bayesian structural time-series models. Ann. Appl. Stat. **9**(1) (2015). https://doi.org/10.1214/14-AOAS788
12. Aguilar-Palacios, C., Munoz-Romero, S., Rojo-Alvarez, J.L.: Causal quantification of cannibalization during promotional sales in grocery retail. IEEE Access **9**, 34078–34089 (2021). https://doi.org/10.1109/ACCESS.2021.3062222
13. Hyndman, R.J., Athanasopoulos, G.: Forecasting: Principles and Practice, 3rd edn. OTexts (2021). https://otexts.org/fpp2/
14. Taylor, S.J., Letham, B.: Forecasting at scale. Am. Stat. **72**(1), 37–45 (2018). https://doi.org/10.1080/00031305.2017.1380080
15. Breiman, L.: Random forests. Mach. Learn. **45**(1), 5–32 (2001). https://doi.org/10.1023/A:1010933404324
16. Chen, T., Guestrin, C.: XGBoost: A Scalable Tree Boosting System (2016). https://doi.org/10.48550/ARXIV.1603.02754
17. Ke, G., et al.: LightGBM: a highly efficient gradient boosting decision tree. In: Guyon, I., et al. (eds.) Advances in Neural Information Processing Systems. Curran Associates, Inc. (2017). https://proceedings.neurips.cc/paper_files/paper/2017/file/6449f44a102fde848669bdd9eb6b76fa-Paper.pdf
18. Cerqueira, V., Torgo, L., Mozetic, I.: Evaluating time series forecasting models: an empirical study on performance estimation methods (2019). https://doi.org/10.48550/ARXIV.1905.11744
19. Guidotti, R., Monreale, A., Ruggieri, S., Turini, F., Giannotti, F., Pedreschi, D.: A survey of methods for explaining black box models. ACM Comput. Surv. **51**(5), 1–42 (2019). https://doi.org/10.1145/3236009

The Window Dilemma: Why Concept Drift Detection is Ill-Posed

Brandon Gower-Winter, Misja Groen, and Georg Krempl(✉)

Utrecht University, Heidelberglaan 8, 3584 CS Utrecht, The Netherlands
g.m.krempl@uu.nl
https://www.uu.nl/en

Abstract. Non-stationarity of an underlying data generating process that leads to distributional changes over time is a key characteristic of Data Streams. This phenomenon, commonly referred to as Concept Drift, has been intensively studied, and Concept Drift Detectors have been established as a class of methods for detecting such changes (drifts). For the most part, Drift Detectors compare regions (windows) of the data stream and detect drift if those windows are sufficiently dissimilar.

In this work, we introduce the Window Dilemma, an observation that perceived drift is a product of windowing and not necessarily the underlying data generating process. Additionally, we highlight that drift detection is ill-posed, primarily because verification of drift events are implausible in practice. We demonstrate these contributions first by an illustrative example, followed by empirical comparisons of drift detectors against a variety of alternative adaptation strategies. Our main finding is that traditional batch learning techniques often perform better than their drift-aware counterparts further bringing into question the purpose of detectors in Stream Classification.

Keywords: Concept Drift · Change Detection · Data Stream

1 Introduction

Many real-world machine learning tasks face the challenge that the underlying data generating process is non-stationary, in particular when data arrives over time in so-called Data Streams [33]. A common strategy to address this problem of Concept Drift (distributional changes over time) is to use only a subset of the data for building (or updating) a model. Typically, this is a window of instances sampled consecutively within an interval. Various Change (or Drift) Detection approaches have been proposed, most of which rely on windows to detect changes between concepts.

The contributions of this paper are threefold: First, in Sect. 3 we show that there is a dilemma in choosing the correct instance selection procedure. Instance selection innately determines the perception of concepts and drifts, which might not match the actual ones. Second, in Sect. 4 we illustrate why drift detection is

M. Baratchi et al. (Eds.): IDA 2026, LNCS 16513, pp. 371–384, 2026.
https://doi.org/10.1007/978-3-032-23833-7_27

ill-posed. It depends itself on the selected sample to be an accurate proxy of the current state (concept) of the underlying data generating process. To support this claim, we integrate several well known limitations of drift detection such as the inability to verify all drift in practice. Third, our empirical evaluation (Sect. 5) of various drift-aware adaptive classifiers and drift detectors against drift-unaware classifiers indicates that the type of classifier is often more important than drift-awareness. We conclude by providing directions for future research, in particular by moving towards techniques that allow identifying change patterns that allow specific actions and a more stringent evaluation.

An Extended Version of this paper can be found on `arXiv.org`. The source code is available at https://github.com/BrandonGower-Winter/TheWindowDilemma.

2 Background and Related Work

A data stream may be defined as a mode of access to a potentially infinite sequence of instances generated by some concept and delivered to a learning algorithm by an instance-delivery-process (IDP) [32]. Streaming data exists in many real world settings such as sensor, stock market and financial data [33].

In traditional batch machine learning, we assume the underlying data generating process $\mathcal{D}$ remains constant throughout the training, testing and deployment of a model [20]. That is to say that n i.i.d. samples are taken from a stationary distribution, that is, $X_0, X_1, ..., X_n \sim \mathcal{D}$.[1]

On the other hand the data stream in a stream learning task is not time-invariant, so each sample is potentially sampled from a unique distribution $X_i \sim \mathcal{D}_i$. Concept drift, a change in the underlying data generating process through time, occurs when for two timesteps i, j, $\mathcal{D}_i \neq \mathcal{D}_j$. Given a stream learning task $\mathcal{D}_i = P(X, Y)_i$, concept drift often occurs in the class conditional probabilities $P(Y|X)$ or the covariates $P(X)$. Concept drift may be categorized by its observed behavior: If a concept changes suddenly, it is known as abrupt drift. If it changes slowly over time, it is known as incremental or gradual drift. If these changes are repetitive (e.g. public transportation usage patterns over workweeks compared to weekends), the drift is known as reoccurring drift [27].

2.1 Detection of Drift with Change Detectors

Naturally, if Concept Drift describes the change in a distribution generating process $\mathcal{D}_t$ over time t, then a Change (or Drift) Detector is a tool for its detection. Informally, a drift detector is a binary classifier that observes some number of samples, and predicts if the underlying data distribution that produced said samples changed through time.

The drift detection process typically consists of 4 stages [20,27]. The first stage is data acquisition which refers to the selecting of instances to pair with

[1] In the Supervised Learning setting X_i may be written as (X_i, Y_i).

a drift detector. The most common solutions are window-based instance selection with the sliding window technique being the most popular [27]. Alternative options including forgetting mechanisms [32].

The second stage is the data modeling stage whereby descriptors are built. This stage is optional and only relevant if some kind of preprocessing is necessary for the drift detector to function. Examples of such techniques include dimensionality reduction and binning of numerical values [20].

In Stage 3, the detector computes the dissimilarity of the descriptors, and Stage 4 calibrates the dissimilarity which identifies drift in the selected instances. We are intentionally vague about the last two stages because they are not the focus of this paper. A typical drift detector will either compute some magnitude of dissimilarity [14,21] or conduct a statistical test [28].

Interestingly, few research endeavors focus on Stage 1 (Instance Selection) despite it being one of the most important factors in detecting concept drift [18,24,29]. In window-based methods, this often comes down to choosing the appropriate window size. Small windows offer reactivity, while larger windows offer stability. Furthermore, the speed of the concept drift in a data stream will influence the window size. Fast drift needs to be detected quickly (small window), while slow drift requires larger windows [18].

Drift Detection research that consider window sizes are Lazarescu et al. [25] who keep multiple reference windows of different sizes. The detectors ADWIN [3], and PUDD [28] use adaptive windowing techniques to be robust to concept drift at different time scales. It is worth noting though that the vast majority of drift detectors simply have the window size as a tunable parameter. It should be emphasized that any approach that relies on a sample of instances for change detection (or classification) inherently depends on the window in which the corresponding instances were observed.

Drift detectors can be categorized in several ways. First, some drift detectors are designed for Supervised or Unsupervised Learning tasks. DDM [12], ADWIN [3], and PUDD [28] are examples of the former, and D3 [16], IBDD [35] are examples of the latter. Unsupervised Drift Detectors can be used in both Supervised and Unsupervised Learning tasks with some detectors, such as DAWIDD [18], being designed with both settings in mind.

Both Supervised and Unsupervised techniques are limited in the types of drift they can detect [39]. Supervised (performance-based) detectors often rely on the presence of labeled data which may not be readily available. Some techniques such as STUDD [8] and PUDD [28] aim to work around this limitation, but a performance-based drift detector is only reliable if changes in the performance of the deployed predictor reliably indicate drift [19,24,32]. Conversely, Unsupervised (distribution-based) detectors can only detect changes in the features $P(X)$ and not in the class-conditional probabilities $P(Y|X)$. For further reading, we refer the reader to the following survey papers [1,13,20,27,29,32].

2.2 Adaptive Models

Drift detectors are often used to create adaptive models [27]. The exact method for imbuing a model with adaptivity is both model and detector dependent, but the general process is that when a drift event is detected, the deployed model will reset itself in some capacity. This can be through a complete model reset (start training from scratch) or by intelligently altering some aspects of the model's internal structure as is the case with many of the Hoeffding Tree [22] variants. For example, the Hoeffding Adaptive Tree [4] will use a detector to monitor the performance of different subtrees and replace them if necessary.

There also exist ensemble methods such as Adaptive Random Forests [15] and the drift aware variants of Bagging and Boosting [5,31] which will replace poorly performing base learners. It is also possible to have adaptive models that do not utilize drift detectors [32] such as the Aggregated Mondrian Forest [30]. Adaptivity may also be added by employing simple reset or retraining regimes [2].

3 The Window Dilemma

The first stage of Drift Detection is instance selection for which windowing methods are the most popular [27,29]. This Section introduces the Window Dilemma, an observation that illustrates why the act of selecting any finite set of past instances (windowing), at best, provides a surrogate representation of the current concept. We also highlight that, in practice, Concept Drift is usually undetectable from the observation of the data stream alone.

Recall that the generating distribution of a data stream is not time-invariant, so each sample is potentially sampled from a unique distribution $X_i \sim \mathcal{D}_i$. Concept drift occurs when for two timesteps i, j, $\mathcal{D}_i \neq \mathcal{D}_j$.

When attempting to detect if $\mathcal{D}_i \neq \mathcal{D}_j$, a time window $W(i,j) = \{X_i, ..., X_j\}$ captures the instances between timesteps i and j. Ideally, we want to select some intermediate timestep $i < k < j$ such that $\mathcal{D}_{W(i,k)} \approx \mathcal{D}_i$ and $\mathcal{D}_{W(k,j)} \approx \mathcal{D}_j$ which leads to $\mathcal{S}(W(i,k), W(k,j)) \approx \mathcal{S}(\mathcal{D}_i, \mathcal{D}_j)$ for some dissimilarity metric $\mathcal{S}$.

Herein lies the essence of the Window Dilemma: selecting the right k is unlikely without access to the underlying data generating processes $\mathcal{D}_i$ and $\mathcal{D}_j$. In practice, a window is often simply subdivided in half or an adaptive windowing technique is used to evaluate the window over multiple partitions [3].

Given this, any conclusions drawn from drift detection or analysis are conditioned on the instances selected and how they are partitioned, not necessarily the underlying data generating process.

Consider the following example where we have a data stream containing a single random variable that is sampled from a Gaussian such that $X_i \sim \mathcal{D}_i = \mathcal{N}(\mu_i, \sigma^2)$ where $\sigma = 1$. We induce incremental concept drift by altering μ_i every 10 units of time such that $\mu_i = \frac{i}{10}$. Figure 1 illustrates how various window partitions generate different approximations of distributions $\mathcal{D}_i$ and $\mathcal{D}_j$. When $k \to j$, the approximated distribution $\mathcal{D}_{W(k,j)} \to \mathcal{D}_j$. Conversely, when $k \to i$, the approximated distribution $\mathcal{D}_{W(i,k)} \rightarrow \mathcal{D}_i$. Here we see the Window Dilemma in action, as the choice of window partitioning affects the accuracy of the approximated distributions at timesteps i and j. One can construct such scenarios for a myriad of drift types. In the abrupt drift setting, the accuracy at which one approximates distributions $\mathcal{D}_i$ and $\mathcal{D}_j$ is entirely dependent on how precisely the partition point k approximates the moment at which the abrupt drift occurs.

The Window Dilemma is inherent to instance selection over a non-stationary distribution. Thus, it also applies to models that do not use windows explicitly, but implicitly through their instance selection mechanism. We are intentionally vague about the number of samples considered in these streams, because it affects how accurately the generating distributions can be represented. We are also not commenting on whether a drift detector would detect drift in these cases. Even the approximated distributions in Fig. 1 are clearly different and it is possible that a drift detector would pick up on this. Whether that is useful is a different matter which we address in the next Section.

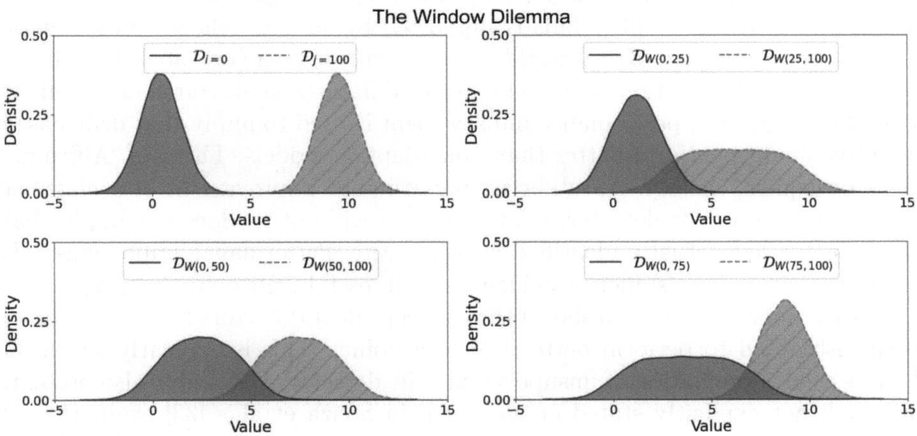

Fig. 1. An illustrative example of the Window Dilemma. In the top left we have two distributions $\mathcal{D}_i$ and $\mathcal{D}_j$ for which we would like to detect Concept Drift. In practice, we have to achieve this by observing samples generated over the timesteps $i = 0$ and $j = 100$ and keeping them in a Window $W(i, j)$. By choosing some intermediate point k, we can construct two sub-windows which we would use to detect drift. The Window Dilemma shows that, depending on our choice of k, we will observe varying distributions, none of which are the distributions we are actually trying to compare. In this Incremental Drift setting, choosing a k close to either i or j will approximate the generating distributions of either $\mathcal{D}_i$ (top right figure) or $\mathcal{D}_j$ (bottom right figure) accurately, at the cost of poorly approximating the other. In the case where k is chosen as the mid-point (bottom left figure), as often done in practice, we see that we get the right shape for each distribution, but the overlap between them isn't representative of the overlap between the actual distributions $\mathcal{D}_i$ and $\mathcal{D}_j$.

4 Concept Drift Detection Is Ill-Posed

This Section aims to demonstrate why concept drift detection is ill-posed. Formally, a Concept Drift Detector $\mathcal{M}$ may be defined as mapping a window of instances $W(i,j)$ over timesteps i and j to an outcome: $\mathcal{M} : W(i,j) \mapsto \{0,1\}$. In essence, a drift detector is seen as a binary classification task where the goal is to correctly identify whether concept drift has occurred at a given timestep. This claim is further supported by the use of Accuracy, AUC and Confusion Matrices when reporting on the performance of drift detectors [18,20].

As shown in Sect. 3, instance selection affects the approximated distributions that are used to detect concept drift. This means that any detection event raised by a drift detector is, at best, an indication that the partitioning of the selected instances exhibited what appears to be drift, not that concept drift has actually occurred.

It is possible that the detected drift event corresponds to some real change in concept, but it is often infeasible to verify this. Despite being a classification task, we often do not know when real drift events happen in practice (i.e. we don't have drift labels) [33]. This is supported by an over-reliance on synthetic data and the imputing of real-world datasets with known (i.e. labeled) drift.

Because drift detection on *actual* real-world datastreams is often an Unsupervised Learning task, performance improvement is used to imply that drift-based adaptive models perform better than non-adaptive models. This is a "Affirming the Consequent" fallacy. An observed performance improvement of a classifier equipped with a drift detector, relative to one without it, does not imply that the detector has actually identified concept drift. Performance improvement is neither necessary nor sufficient evidence of successful drift detection [32].

Furthermore, it has been shown that concept drift detectors that detect more frequently, tend to perform better (up to a point). This is explicitly shown in Lukats et al.'s evaluation of unsupervised drift detectors [29] and it also present, although not explicitly stated or analyzed, in Souza et al.'s [33] evaluation of performance-based detectors. It would appear that drift detectors that detect frequently are able to more frequently retrain classifiers on more recent data. Given the prominence of autocorrelation and time-dependency present in several data streams [38], it is perhaps not surprising that frequent retraining results in better performing models. This has also been validated empirically with simple retraining regimes that do not make use of drift detectors [2].

The effect of retraining a classifier depends on both the start and the end point of the training window. A frequent update of the end point allows timely incorporation of new observations. In contrast, updating the starting point triggers forgetting, which positively removes outdated/irrelevant instances at the risk of negatively excluding still relevant instances.

Another limiting aspect of drift detection that is not often discussed is what a "drift detection" means for the overall system. Liu et al. [27] call this the "When" of drift detection, noting that few drift detectors go beyond that to determine "Where" and "How" the drift occurred. Webb et al. [36] provide several measures for quantifying the amount of drift a system experiences: (1) Drift Duration (how long is drift), (2) Drift Magnitude (how much drift), (3) Path Length (how much total drift), (4) Drift Rate (how frequently does drift occur) but these measures themselves are susceptible to the Window Dilemma.

In terms of detectability, it may also be useful to think of: (1) Persistence (duration for which a Concept remains stable) (may also be called Concept Stability [36]), (2) Sample Size (number of samples from a persistent concept), (3) Sample Rate (the rate at which samples arrive from a persistent concept) (i.e. Sample Rate = Sample Size/Persistence). From a theoretical perspective, a concept with a low sample rate may be fundamentally undetectable (this is known as blip drift [36]), and a concept with a high sample rate is easier to detect provided the appropriate windows are chosen. Practically, it is not clear how useful such ideas are without access to the ground truth drift labels.

Overall, our point is that simply "detecting drift" is insufficient, because (1) it is often impractical to verify whether these drift events happened, and (2) a drift event often tells very little about how the overall system or data stream itself is changing. Some drift may be irrelevant, some may be important, neither of which can be determined by viewing drift detection as binary (Yes/No) task, which we show empirically in the next Section.

5 Experimental Design and Results

In this Section we empirically evaluate the usefulness of drift-aware classifiers across a variety of benchmarking datasets. To achieve this, we will make use of two supervised drift detectors: (1) DDM [12] and (2) ADWIN [3] and two unsupervised drift detectors: (1) D3 [16] and (2) IBDD [35]. We chose ADWIN and DDM because of their popularity in literature, and D3 and IBDD because they were the best performing detectors in a recent benchmark evaluation [29]. For the D3 detector, we create two variants which use (1) Logistic Regression (D3-LR) and a (2) Hoeffding Tree (D3-HT) as the discriminatory classifier respectively. This was done because the choice of discriminator can have an impact of the performance of the detector.

For our base model, we make use of the popular (1) Naive Bayes (NB) and (2) Hoeffding Tree (HT) classifiers, which we also use as baselines. For each drift detector, we use each base model to create two adaptive-models per detector variant. For example, the DDM detector is paired with both the Naive Bayes (DDM-NB) and the Hoeffding Tree (DDM-HT) for a total of 10 detector-model pairs. When one of these detector-model pairs detects drift, the base model is reset and will start learning from scratch. We make also use of a periodic resetting technique as described by Bifet [2]. Every N instances, the base model is simply reset. We do this for both base models (R-NB) and (R-HT).

We extended our comparison to include also more sophisticated adaptive online learners. To that end, we also evaluate the Hoeffding Adaptive Tree [4] (HAT), Adaptive Random Forests [15] (ARF) and Aggregated Mondrian Forests [30] (AMF). HAT and ARF have built-in mechanisms for utilizing concept drift detectors to adapt their internal structure (replacing subtrees in HAT, replacing base learners in ARF). AMF is a purely online technique that does not utilize concept drift detection.

We also include traditional batch learning techniques in the comparison with the aforementioned stream learning methods. We create three variants of the popular Random Forests [7] (RF): (1) a static Random Forest (S-RF) which is trained on the first $N = 100$ instances of a stream and then never trained again, (2) a resetting Random Forest (R-RF) which is retrained every $N = 50$ iterations on the past $N = 50$ instances, and (3) an incremental Random Forest (I-RF) which is retrained every $N = 50$ instances on all seen data.

Lastly, we use of two baseline predictors: (1) Last Class (LC) and (2) Majority Class, which predict the last label seen and the majority class seen respectively. We include these baselines as it will reveal class imbalances and autocorrelation which are present in several data streams [2,37].

For evaluation, we use prequential evaluation (test-then-train) whereby a model will first make a prediction on an instance X_i before it receives the label Y_i to train on. We record the accuracy after each prediction and report the mean accuracy in this report. We acknowledge that accuracy is a potentially misleading metric in Drift Research [2,8] and include the mean Cohen's Kappa statistic in the Extended Version of this paper. It does not change the overall message of our paper so we show accuracy for interpretability. We also use a psuedo-random number generator to initialize each model, ensuring reproducibility by providing the seeds. The list of the parameter settings used are included in the Extended Version of this paper.

5.1 Datasets

Using a subset of the USP Data Stream Repository [33], we evaluate each model across 11 binary and multiclass data streams ranging over a variety of problem domains. These datasets are: Electricity (EL) [17], Forest Covertype (FC) [6], Insects-Abrupt (balanced) (IA) [33], Insects-Incremental (balanced) (II) [33], Keystroke (KS) [34], Luxembourg (LX) [37], MIRS (MR) [9], NOAA Weather (NW) [11], Ozone (OZ) [10], Rialto (RT) [26], and Yoga (YG) [9]. Note that we do not make use of any purely synthetic datasets because we believe it diminishes from the message of the paper which is concerned about drift detection in the practical (real/semi-real) setting. We include a short description of each dataset in the Extended Version of this paper.

5.2 Results

Table 1 reports the results from our experiments. Note that while we report the median rank, the observed performance of a particular classifier is often data stream specific. For example, the ARF classifier is the most consistent classifier whereas the R-RF and I-RF classifiers perform exceptionally on some datasets, while performing comparatively poor on others (increasing their median rank). Nevertheless, we can still make several observations from these results.

Table 1. The Average Accuracy for each technique across all benchmark data steams investigated in this work. Bold values indicate the highest mean accuracy across the entire data stream. The median rank (MedRank, the lower the better) of each technique is also reported.

	EL	FC	IA	II	KS	LX	MR	NW	OZ	RT	YG	MedRank
Baselines												
LC	84,8	**91,4**	25,0	15,9	6,7	48,0	49,8	68,1	90,2	0,0	48,7	22
MC	57,4	56,6	17,7	11,0	6,8	52,3	54,1	69,8	92,3	10,0	52,0	21
Naive Bayes												
NB	76,3	56,6	53,6	46,5	69,9	55,6	63,8	69,6	73,8	18,8	54,3	17
DDM-NB	82,4	79,3	61,2	52,5	70,5	55,6	64,0	71,1	89,9	25,0	54,3	14
ADWIN-NB	82,3	71,4	63,2	52,6	69,9	55,6	64,2	70,7	85,7	30,3	55,1	16
R-NB	86,3	81,5	54,2	39,2	72,8	56,3	65,9	73,2	91,9	34,9	54,6	12
D3-LR-NB	82,3	70,0	58,6	44,4	75,6	55,6	66,9	72,7	89,6	34,6	54,6	15
D3-HT-NB	86,6	76,2	63,7	52,4	75,6	61,2	65,0	73,0	88,8	37,9	55,9	9
IBDD-NB	86,7	82,0	59,8	46,4	72,5	56,6	66,1	71,0	88,1	23,6	56,8	10
Hoeffding Trees												
HT	80,2	79,7	54,9	48,1	70,0	90,2	63,6	72,6	92,3	27,3	54,6	13
DDM-HT	86,1	85,3	60,5	52,5	70,6	90,2	63,7	71,4	92,1	41,5	54,6	9
ADWIN-HT	83,9	79,0	61,3	52,8	70,0	85,0	63,8	70,6	92,2	44,5	54,8	12
R-HT	86,0	87,3	40,0	33,3	69,7	64,0	63,0	72,6	**92,8**	53,9	52,5	13
D3-LR-HT	84,3	80,4	46,7	41,0	77,5	90,2	65,3	71,8	92,4	51,0	53,2	10
D3-HT-HT	86,4	85,7	59,4	51,8	74,9	87,5	64,5	72,2	92,4	55,8	55,4	9
IBDD-HT	86,1	87,0	50,7	44,0	71,8	85,5	65,2	70,5	92,0	36,1	55,8	11
HAT	84,6	79,7	57,7	49,4	70,5	91,9	63,7	73,8	92,3	29,3	55,1	11
Online Random Forests												
AMF	85,0	90,5	67,2	57,5	**88,7**	**95,4**	73,7	76,4	92,0	**82,8**	75,5	3
ARF	**88,3**	89,4	**71,4**	**58,7**	87,4	94,8	76,5	78,1	92,3	72,8	68,8	2
Batch Random Forests												
S-RF	62,6	67,3	55,8	19,3	51,9	91,0	69,7	72,8	92,3	14,2	60,0	14
R-RF	81,9	78,1	69,8	58,4	81,7	93,9	73,9	72,1	91,6	80,3	62,9	4
I-RF	78,2	78,7	63,2	55,0	82,3	94,3	**78,9**	**79,1**	92,6	81,3	**79,0**	3

Table 2. Experiments comparing the batch Reset and Incremental Random Forests (R-RF and I-RF) to the online Aggregated Mondrian Forest (AMF) and Adaptive Random Forest (ARF). These results show how AMF and ARF lose some of their efficacy when they follow the same label-delayed training regime as R-RF and I-RF.

	EL	FC	IA	II	KS	LX	MR	NW	OZ	RT	YG
AMF	76,9	75,6	63,3	55,7	78,8	91,6	73,1	76,5	92,0	74,7	75,2
ARF	81,1	69,7	66,7	56,8	77,0	93,7	74,1	77,2	92,2	63.3	67,8
R-RF	**81,9**	78,1	**69,8**	**58,4**	81,7	93,9	73,9	72,1	91,6	80,3	62,9
I-RF	78,2	**78,7**	63,2	55,0	**82,3**	**94,3**	**78,9**	**79,1**	**92,6**	**81,3**	**79,0**

On average, our results suggest that it is often better to have some form of retraining or resetting when a data stream is involved. This is identifiable by the poor performance of the non-adaptive Naive Bayes (NB), Hoeffding Tree (HT) and static Random Forest (S-RF).

In terms of the adaptive-classifiers, it is unclear which adaptive-technique is most appropriate. For the Naive Bayes, simple resetting (R-NB) works quite well for some datasets, but noticeably poor on both the Insect data streams (IA and II). For the Hoeffding Tree variants, resetting (R-HT) is less consistent, which may reflect in the HT's sensitivity to the data windows it is trained on.

Of the drift detectors, D3-HT performed the best for both the NB and HT base models. This is in line with the findings of Lukats et al. [29]. The HAT classifier performed worse than expected. Recall that HAT replaces poorly performing subtrees. This adaptive strategy does not seem to consistently outperform simply resetting the entire Hoeffding Tree when a drift event is detected.

However, the classification strategy had the greatest impact on performance. More specifically, Adaptive Forest (tree-ensemble) techniques outperformed the single learner NB and HT variants. Interestingly, the drift-unaware Aggregated Mondrian Forest (AMF) performed competitively with the drift-aware Adaptive Random Forest (ARF). As noted before, the ARF classifier was the most consistent classifier we evaluated. We attribute this to ARF's ability to replace base learners that are no longer performing well.

Interestingly, the resetting Random Forest (R-RF) and incremental Random Forest (I-RF) performed competitively on most of the datasets.[2] This result is surprising because R-RF and I-RF are drift-unaware and for all intents and purposes, are batch learners applied in a stream learning context. It is often assumed that previous instances should be forgotten when drift is detected, our results suggest that keeping past data is often useful and periodically retraining a model on increasingly large training sets is often an effective strategy. There

[2] The Last Class classifier performed best on ForestCoverType because it abuses the known autocorrelation in the data stream. If we were to exclude this result, drift-ignorant classifiers would have performed the best on all but three of the data streams.

are some settings where this is not the case, but the performance of the R-RF classifier suggests that simple reset retraining on recent data is also viable.

Given that the R-RF and I-RF classifiers are at a disadvantage because they are only updated every N instances whereas the stream learning classifiers could update themselves after every instance, we reran our experiments with the AMF and ARF models following the update regime used by R-RF and I-RF. The results reported in Table 2 are quite telling. Both AMF and ARF's consistency diminishes across many of the datasets. In fact, these results suggest that drift-aware stream learners are often inferior to simple batch learning procedures provided an appropriate classifier is chosen. Furthermore, the performance degradation of the AMF classifier suggests than periodically retraining your entire classifier from scratch has benefits over obeying the strict stream learning paradigm of learning from instances in the order they arrive. Overall, we believe these results reinforce our central claim that concept drift detection is broken. If simply keeping data to retrain on is the optimal strategy (as is the case for many of the data streams investigated), what is the purpose of a drift detector?

6 Conclusions and Future Work

This research highlighted two observations regarding Concept Drift Detection:

(1) We showed how the process of selecting instances for drift detection innately determines if drift is perceived. This perceived drift may also be unreflective of a data stream at it's current point in time. The process of choosing the correct instance selection procedure is what we call the Window Dilemma.

(2) Due to our inability to verify the accuracy to which our instance selection procedure approximates a concept at some time point, we argued that drift detection is ill-posed. We support this claim by synthesizing several well known limitations of drift detection such as our inability to verify all drift in practice.

We then empirically investigated a number of drift-aware adaptive classifiers (using various drift detectors) and compared them to simple online and drift-unaware adaptive classifiers across several data streams. Our results showed that the type of classifier used is often more important than drift-awareness.

We want to make it clear that we are not suggesting that any of theoretical discoveries surrounding concept drift are invalid, rather we wished to illustrate the apparent disconnect between the theory, and what we see in practice. We empirically showed that a state-of-the-art learning model (in combination with a little bit of domain knowledge) will often outperform a drift-aware model. This complements traditional data stream research methods, which in the past mainly focused on efficient retraining under limited computational resources. Our findings do not mean that drift detection is pointless, rather that its current purpose is misplaced. We believe our results are a call to action. It is not sufficient to reduce drift detection to a binary "When", and we should move towards developing drift understanding ("Where" and "How") [27], so that we may better exploit the data to create even better adaptive models. While the

Window Dilemma itself is inherent to any choice of training window, insights from Ensemble Learning might offer some mitigation. Further integration with areas such as Transfer Learning, Domain adaptation and Change Mining have potential [23]. In short, we believe learning how to exploit drift patterns (e.g. incremental) may be more useful than a simple drift detector could ever be.

Acknowledgments. We are thankful for the financial support of the ICAI lab AI4Oversight. We thank Bernhard Pfahringer and members of the AI4Oversight Lab for early discussions, and the anonymous reviewers for their comments.

References

1. Bayram, F., Ahmed, B.S., Kassler, A.: From concept drift to model degradation: an overview on performance-aware drift detectors. Knowl. Based Syst. **245**, 108632 (2022)
2. Bifet, A.: Classifier concept drift detection and the illusion of progress. In: Int. Conf. on Artificial Intelligence and Soft Computing, pp. 715–725. Springer (2017)
3. Bifet, A., Gavalda, R.: Learning from time-changing data with adaptive windowing. In: Proc. of the 2007 SIAM Int. Conf. on Data Mining, pp. 443–448. SIAM (2007)
4. Bifet, A., Gavalda, R.: Adaptive learning from evolving data streams. In: Int. Symposium on Intelligent Data Analysis, pp. 249–260. Springer (2009)
5. Bifet, A., Holmes, G., Pfahringer, B., Kirkby, R., Gavalda, R.: New ensemble methods for evolving data streams. In: Proc. of the 15th ACM SIGKDD Int. Conf. on Knowledge Discovery and Data Mining, pp. 139–148 (2009)
6. Blackard, J.A., Dean, D.J.: Comparative accuracies of artificial neural networks and discriminant analysis in predicting forest cover types from cartographic variables. Comput. Electron. Agricul. **24**(3), 131–151 (1999)
7. Breiman, L.: Random forests. Mach. Learn. **45**(1), 5–32 (2001)
8. Cerqueira, V., Gomes, H.M., Bifet, A., Torgo, L.: STUDD: a student-teacher method for unsupervised concept drift detection. Mach. Learn. **112**(11), 4351–4378 (2023)
9. Dau, H.A., et al.: Hexagon-ML: The UCR Time Series Classification Archive (2018)
10. Dheeru, D., Taniskidou, E.K.: UCI Machine Learning Repository (2017)
11. Ditzler, G., Polikar, R.: Incremental learning of concept drift from streaming imbalanced data. IEEE Trans. Knowl. Data Eng. **25**(10), 2283–2301 (2012)
12. Gama, J., Medas, P., Castillo, G., Rodrigues, P.: Learning with drift detection. In: Brazilian Symposium on Artificial Intelligence, pp. 286–295. Springer (2004)
13. Gama, J., Žliobaitė, I., Bifet, A., Pechenizkiy, M., Bouchachia, A.: A survey on concept drift adaptation. ACM Comput. Surv. (CSUR) **46**(4), 1–37 (2014)
14. Goldenberg, I., Webb, G.I.: Survey of distance measures for quantifying concept drift and shift in numeric data. Knowl. Inf. Syst. **60**(2), 591–615 (2019)
15. Gomes, H.M., et al.: Adaptive random forests for evolving data stream classification. Mach. Learn. **106**(9), 1469–1495 (2017)
16. Gözüaçık, Ö., Büyükçakır, A., Bonab, H., Can, F.: Unsupervised concept drift detection with a discriminative classifier. In: Proc. of the 28th ACM Int. Conf. on Information and Knowledge Management, pp. 2365–2368 (2019)
17. Harries, M.: Splice-2 comparative evaluation: electricity pricing (1999)

18. Hinder, F., Artelt, A., Hammer, B.: Towards non-parametric drift detection via dynamic adapting window independence drift detection (DAWIDD). In: Int. Conf. on Machine Learning, pp. 4249–4259. PMLR (2020)
19. Hinder, F., Vaquet, V., Brinkrolf, J., Hammer, B.: On the change of decision boundary and loss in learning with concept drift. In: Int. Symposium on Intelligent Data Analysis, pp. 182–194. Springer (2023)
20. Hinder, F., Vaquet, V., Hammer, B.: One or two things we know about concept drift—a survey on monitoring in evolving environments. part a: detecting concept drift. Front. Artif. Intell. **7**, 1330257 (2024)
21. Hoens, T.R., Chawla, N.V., Polikar, R.: Heuristic updatable weighted random subspaces for non-stationary environments. In: 2011 IEEE 11th Int. Conf. on Data Mining, pp. 241–250. IEEE (2011)
22. Hulten, G., Spencer, L., Domingos, P.: Mining time-changing data streams. In: Proc. of the seventh ACM SIGKDD Int. Conf. on Knowledge Discovery and Data Mining, pp. 97–106 (2001)
23. Krempl, G., Hofer, V., Webb, G., Hüllermeier, E.: Beyond adaptation: understanding distributional changes (Dagstuhl Seminar 20372). In: Dagstuhl Reports. vol. 10, pp. 1–36. Schloss Dagstuhl-Leibniz-Zentrum für Informatik (2021)
24. Kuncheva, L.I., Žliobaitė, I.: On the window size for classification in changing environments. Intell. Data Anal. **13**(6), 861–872 (2009)
25. Lazarescu, M.M., Venkatesh, S., Bui, H.H.: Using multiple windows to track concept drift. Intell. Data Anal. **8**(1), 29–59 (2004)
26. Losing, V., Hammer, B., Wersing, H.: Knn classifier with self adjusting memory for heterogeneous concept drift. In: IEEE 16th Int. Conf. on Data Mining (ICDM), pp. 291–300. IEEE (2016)
27. Lu, J., Liu, A., Dong, F., Gu, F., Gama, J., Zhang, G.: Learning under concept drift: a review. IEEE Trans. Knowl. Data Eng. **31**(12), 2346–2363 (2018)
28. Lu, P., Lu, J., Liu, A., Zhang, G.: Early concept drift detection via prediction uncertainty. In: Proc. of the AAAI Conf. on Artificial Intelligence. vol. 39, pp. 19124–19132 (2025)
29. Lukats, D., Zielinski, O., Hahn, A., Stahl, F.: A benchmark and survey of fully unsupervised concept drift detectors on real-world data streams. Int. J. Data Sci. Anal. **19**(1), 1–31 (2025)
30. Mourtada, J., Gaïffas, S., Scornet, E.: AMF: aggregated mondrian forests for online learning. J. Royal Stat. Soc. Seri. B Stat. Methodol. **83**(3), 505–533 (2021)
31. Oza, N.C., Russell, S.J.: Online bagging and boosting. In: Int. Workshop on Artificial Intelligence and Statistics, pp. 229–236. PMLR (2001)
32. Read, J., Žliobaitė, I.: Supervised learning from data streams: an overview and update. ACM Comput. Surv. (2025)
33. Souza, V.M.A., Reis, D.M.d., Maletzke, A.G., Batista, G.E.A.P.A.: Challenges in benchmarking stream learning algorithms with real-world data. Data Mining Knowl. Discov. **34**, 1805–1858 (2020)
34. Souza, V.M.A., Silva, D.F., Gama, J., Batista, G.E.A.P.A.: Data stream classification guided by clustering on nonstationary environments and extreme verification latency, pp. 873–881 (2015)
35. Souza, V.M., Parmezan, A.R., Chowdhury, F.A., Mueen, A.: Efficient unsupervised drift detector for fast and high-dimensional data streams. Knowl. Inf. Syst. **63**(6), 1497–1527 (2021)
36. Webb, G.I., Hyde, R., Cao, H., Nguyen, H.L., Petitjean, F.: Characterizing concept drift. Data Mining Knowl. Discov. **30**(4), 964–994 (2016)

37. Žliobaitė, I.: Combining similarity in time and space for training set formation under concept drift. Intell. Data Anal. **15**(4), 589–611 (2011)
38. Zliobaite, I.: How good is the electricity benchmark for evaluating concept drift adaptation. arXiv preprint arXiv:1301.3524 (2013)
39. Zliobaitė, I.: Change with delayed labeling: when is it detectable? In: IEEE Int. Conf. on Data Mining Workshops (ICDMW 2010), pp. 843 – 850 (2010)

Toward Improved Time-Series Explanations for Federated Learning in Healthcare

Christoph Düsing(✉) and Philipp Cimiano

CITEC, Bielefeld University, Bielefeld, Germany
{cduesing,cimiano}@techfak.uni-bielefeld.de

Abstract. *Federated Learning* (FL) for time-series data has emerged as a promising paradigm to train *Machine Learning* (ML) models across institutions without sharing data, which is essential in sensitive healthcare settings. Unfortunately, federated models share the black-box characteristics of conventional ML models. While methods such as *Time-SHAP* provide feature-, event-, and cell-wise explanations for centralized time-series models, their use in FL is limited because they depend on a background dataset that reflects the joint training distribution. As data in FL is typically not identically distributed, clients do not hold such representative datasets, ultimately limiting the quality of explanations they can produce locally. To overcome this limitation, we introduce *TimeSHAP–FL*, which facilitates a federated, differentially private *Generative Adversarial Network* (GAN) to synthesize a representative background dataset, which is subsequently incorporated during client-side explanation generation. Our results (1) demonstrate the applicability of *TimeSHAP–FL* for the task of sepsis onset prediction, (2) indicate improved explanation quality (measured as Spearman Correlation Score) through the combination of local and synthesized background datasets, and (3) account for possible privacy-leakage through the GAN-based data synthesis.

Keywords: Federated Learning · Explainability · SHAP · Time-Series

1 Introduction

Federated Learning (FL) [19] has emerged as a promising paradigm for training *Machine Learning* (ML) models across multiple institutions – so-called *clients* – without sharing raw data. This makes FL particularly attractive for healthcare, where sensitive patient information cannot be exchanged between hospitals due to ethical, legal, and practical constraints. By distributing model training, FL enables healthcare institutions to collaboratively improve predictive performance while preserving data locality. Despite these benefits, FL inherits the black-box problem common to modern ML models, where complex architectures hinder transparency and raise concerns about reliability and trustworthiness [24]. In

M. Baratchi et al. (Eds.): IDA 2026, LNCS 16513, pp. 385–397, 2026.
https://doi.org/10.1007/978-3-032-23833-7_28

high-stakes domains such as intensive care medicine, clinicians often require accurate explanations to understand and validate model predictions [24]. Among the various approaches in *Explainable Artificial Intelligence* (XAI), *feature importance* methods are especially valuable in healthcare due to their intuitive nature and their ability to describe both individual and global model behavior [3]. However, adapting XAI techniques to FL introduces unique challenges. Since hospitals typically hold small and non-identically distributed datasets [7], they cannot generate explanations that reliably reflect the global model's reasoning. This discrepancy can lead to inconsistent or misleading explanations across institutions [21], underscoring the need for FL-specific XAI solutions.

These challenges become even more significant for time-series data. Methods such as *TimeSHAP* [2] and *WindowSHAP* [20] provide fine-grained feature-, event-, and cell-wise explanations by removing or perturbing temporal segments and quantifying their influence on the model's prediction. However, these methods – much like regular *SHAP* [18] – critically depend on a representative background dataset that reflects the joint training distribution. While such a dataset is readily available in centralized settings, this assumption fundamentally breaks down in FL, where each institution observes only a local, and often highly biased, subset of the full data distribution [7]. As a result, the lack of a globally representative background distribution emerges as a critical bottleneck for time-series explainability in FL, leading to explanations that vary significantly across clients and fail to reflect the temporal dependencies learned by the global model.

To address this limitation, we introduce *TimeSHAP–FL*, a framework that explicitly targets the absence of a representative background distribution for time-series explanations in FL. The key idea is to synthesize such a background dataset through a federated, differentially private *Generative Adversarial Network* (GAN) [17], thereby decoupling explanation quality from any single institution's local data. Instead of relying on locally observed samples, the GAN learns the global temporal structure of the federation during training and subsequently generates synthetic time-series data that can be shared across clients. These samples preserve the temporal coherence required for *TimeSHAP*, while maintaining privacy guarantees through federated and differentially private training.

Ultimately, *TimeSHAP–FL* provides two practical benefits for federated time-series explanation: First, the synthetic background dataset enables clients to compute feature-, event-, and cell-wise explanations that more accurately reflect the behavior of the global model, even when their local data distribution differs substantially from the federation. Second, incorporating a mixture of synthetic and local background samples leads to more stable and consistent explanation scores across institutions. In summary, our contributions are:

1. We identify the absence of a representative background dataset as a key limitation for time-series explainability in non-IID federated learning and propose *TimeSHAP–FL* to address this challenge in a privacy-aware manner.
2. We analyze explanation quality for the task of sepsis onset prediction, showing that *TimeSHAP–FL* improves the consistency and fidelity of time-series

explanations compared to different baselines, particularly when combining synthetic and local background data.
3. We assess the similarity between GAN-generated and real sequences, demonstrating that the synthetic data does not reproduce individual patient trajectories while preserving sufficient diversity for explanation purposes.

2 Related Work

TimeSHAP–FL connects two research areas: federated XAI methods and the synthesis of time-series data in distributed settings. Existing work in federated XAI demonstrates that constructing a representative background dataset is essential for producing meaningful explanations. However, prior approaches are primarily designed for static tabular data and implicitly assume access to background distributions that do not capture temporal or causal dependencies. In parallel, research on synthetic time-series generation provides tools to model such dependencies in distributed settings, yet these methods have not been integrated with explanation frameworks. This disconnect leaves time-series explainability in federated learning largely unaddressed.

Federated XAI. Federated XAI is an emerging research topic with only a limited number of existing methods. Early approaches applied SHAP by averaging explanations across clients [4] or by computing feature attributions via integrated gradients [10]. These strategies depend on simple aggregation, which can distort explanations when client data distributions differ [12]. More advanced techniques focus on improving the representativeness of background datasets used during explanation generation. For example, federated clustering has been proposed to construct shared prototype samples [5]. Similarly, *SHAP–FL* [7] introduced a histogram based synthesis process that improves explanation quality while keeping data local. However, all existing federated XAI methods operate on static features and therefore ignore temporal dependencies that drive the behavior of time-series models. Methods such as *TimeSHAP* [2] and *WindowSHAP* [20] require background data that reflect temporal structure, which clients in FL do not possess. As a result, prior federated XAI approaches cannot be directly extended to time-series settings without explicitly addressing the absence of a temporally representative background distribution.

Synthesizing Time-Series Data. Given the need for temporally consistent background datasets, we turn to research on time-series data synthesis. Several models have been proposed to generate realistic sequences in federated settings. For instance, Singh et al. [23] introduced a federated *Long Short-Term Memory* (LSTM) for time-series generation. Similarly, Jiang et al. [15] developed a differentially private autoencoder for distributed temporal reconstruction [8]. More recently, algorithms such as *FedSyn* [1] applied powerful GANs to produce synthetic samples that preserve global temporal patterns without sharing raw data. These methods demonstrate that temporal structure can be learned and synthesized in FL. However, they were designed for data augmentation or privacy protection and have not been used to support XAI methods.

Overall, prior work either improves federated explainability without modeling temporal dependencies or synthesizes time-series data without connecting it to XAI. This creates a gap for a method that explicitly synthesizes temporally coherent sequences in FL and leverages them to enable reliable time-series explanations.

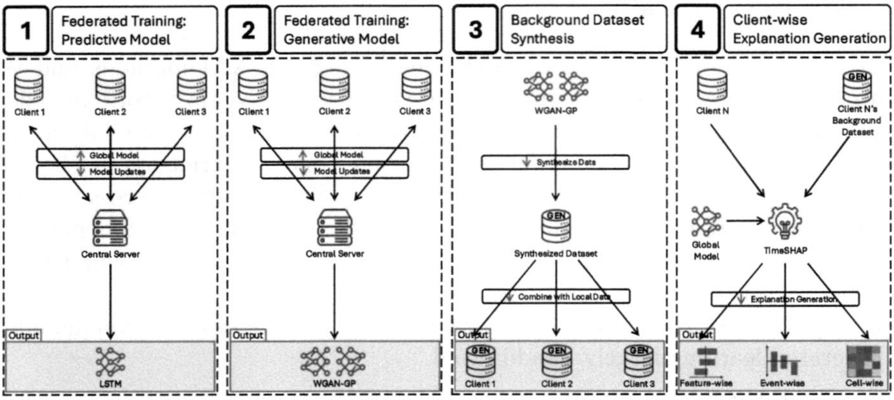

Fig. 1. *TimeSHAP–FL* Framework

3 Methodology

The proposed *TimeSHAP–FL* framework consists of four steps that together enable federated training, time-series data synthesis, and explanation generation. An overview is provided in Fig. 1, and we describe each step in detail below.

Step 1: Federated Training of the Predictive Model
Each client c holds a private dataset $\mathcal{D}_c$, which is split into a local training set $\mathcal{D}_c^{Train}$ and a test set $\mathcal{D}_c^{Test}$, both of which remain strictly on device to preserve privacy. The goal of this step is to collaboratively train a time-series prediction model across all clients without sharing raw data. This is achieved using FL, where clients train the model locally on $\mathcal{D}_c^{Train}$ and periodically transmit model updates to a central server. The server aggregates these updates into a global model [19] and redistributes the updated parameters for the next round. This iterative process continues until a pre-defined convergence criterion is met.

Much like *SHAP–FL* [7], *TimeSHAP–FL* is model and strategy agnostic. While our experiments employ an LSTM due to its suitability for sequential data, any differentiable temporal model (such as gated recurrent units or temporal convolutional networks) could be used within the same framework. The output of this step is a global time-series model trained collaboratively across all clients.

Step 2: Federated Training of the Generative Model
Next, clients jointly train a generative model that can synthesize realistic time-series samples representative of the federation without revealing local data. We employ a conditional *Wasserstein GAN with Gradient Penalty* (WGAN-GP) [11], where the generator G produces full temporal sequences conditioned on static patient features, and the critic D evaluates their realism. Following general design principles of Karras et al. [17], the generator processes latent vectors and static features through fully connected layers followed by a recurrent module, while the critic applies temporal convolutions to assess the joint static-dynamic structure of entire sequences.

The critic is optimized using the standard WGAN-GP loss, where $x \sim \mathcal{D}$ denotes a real time-series sample drawn from a client's local dataset, s represents static conditioning features, $\tilde{x} = G(z, s)$ is a generated sequence, and $\hat{x}$ is sampled uniformly along straight lines between pairs of real and generated samples. The critic $D(\cdot)$ outputs a scalar realism score, and λ controls the strength of the gradient penalty:

$$\mathcal{L}_D = \mathbb{E}_{\tilde{x} \sim G(z,s)}[D(\tilde{x})] - \mathbb{E}_{x \sim \mathcal{D}}[D(x)] + \lambda \mathbb{E}_{\hat{x}} \left[(\|\nabla_{\hat{x}} D(\hat{x})\|_2 - 1)^2 \right].$$

The generator is updated to minimize $-\mathbb{E}_{\tilde{x} \sim G(z,s)}[D(\tilde{x})]$, encouraging the generation of sequences that the critic assigns high realism scores. To reduce the risk of memorization, the critic is trained with *Differential Privacy* (DP) [8] by clipping per-sample gradients and adding Gaussian noise before sharing updates.

The federated GAN training follows the *FedSyn* procedure [1]: in each communication round, clients perform several local updates of G and D under DP, transmit their model updates, and the server aggregates them following an attention-based update rule [14]. The output of this step is a federated, privacy-preserving generative model capable of synthesizing time-series trajectories.

Step 3: Background Dataset Synthesis
After obtaining the generative model from *Step 2*, the central server synthesizes a shared background dataset that reflects the global temporal structure learned during federated training. Therefore, it broadcasts latent vectors $z \sim \mathcal{N}(0, I)$ to all clients. Each client locally draws static feature vectors from its own $\mathcal{D}_c^{Train}$, conditions the generator on these features, and produces synthetic sequences $G(z, s)$. Crucially, clients return only the generated samples to the server, never revealing real static or temporal data. The server aggregates all received samples into a unified synthetic dataset $\mathcal{D}^{Synth}$, which is safe to redistribute because the generator was trained under DP and does not reproduce individual trajectories.

However, relying exclusively on $\mathcal{D}^{Synth}$ may not yield optimal explanation performance for all clients. We therefore construct the final background dataset of each client c as a mixture of synthetic and local samples,

$$\mathcal{D}_c^{Background} = \alpha \, \mathcal{D}^{Synth} \cup (1 - \alpha) \, \mathcal{D}_c^{Train},$$

where α denotes the proportion of synthetic samples. Unless otherwise stated, we use $\alpha = 0.5$, and evaluate the effect of different ratios in Sect. 4.4. The output of this step is a globally consistent background dataset for each client.

Step 4: Client-wise Explanation Generation
In the final step, each client explains the predictions of the federated model trained in *Step 1* using *TimeSHAP*, employing its locally constructed background dataset $\mathcal{D}_c^{Background}$ from *Step 3*. *TimeSHAP* extends the *SHAP* framework to temporal settings by decomposing the prediction of a sequence model into contributions at multiple granularities [2]. Instead of sampling individual feature values independently, *TimeSHAP* perturbs entire temporal segments using background sequences, thereby preserving correlations and causal structure.

TimeSHAP produces three complementary types of explanations:

- **Feature-wise Explanations**, which attribute importance to individual features aggregated over the entire temporal trajectory. These values capture how much each variable (for example *Heart Rate* or *Blood Pressure* (BP)) contributes to the final prediction when considered across all events.
- **Event-wise Explanations**, which quantify the contribution of each time step to the model's prediction. This is implemented by recursively removing or replacing events with sequences drawn from $\mathcal{D}_c^{Background}$ and computing the corresponding SHAP value of each event.
- **Cell-wise Explanations**, which attribute importance to individual feature values at specific time steps. *TimeSHAP* estimates these contributions by perturbing single feature–time cells using values drawn from background sequences, while keeping all other cells fixed.

The output of this step is a set of feature-, event-, and cell-wise *TimeSHAP* explanations for each prediction made on $\mathcal{D}_c^{Test}$.

4 Empirical Evaluation

4.1 Experimental Setup

We evaluate *TimeSHAP–FL* on the task of early sepsis onset prediction, a clinically critical problem requiring transparent time-series models.

Data Setting. We use the *MIMIC-IV* dataset [16] and follow the preprocessing protocol of Düsing and Cimiano [6]. The dataset includes 28,610 patients admitted to one of seven ICUs (MICU, MICU/SICU, SICU, TSICU, CCU, CVICU, NSICU). Each patient is assigned to the ICU of first admission, which represents a federated client. We extract 26 clinically relevant features (see Fig. 2) and aggregate them into six hourly windows after ICU admission. Sepsis labels follow the *Sepsis-3* definition [22], marking patients as positive if sepsis occurs within the first 30 h. Missing values are imputed using linear interpolation.

Model Setting. For the predictive model, we use an LSTM-based architecture similar to prior work on the task [6], consisting of three LSTM layers (16 hidden units, batch normalization, dropout of 0.2), followed by two fully connected layers producing a binary sepsis prediction. The generative model is the federated

WGAN-GP described in Sect. 3, using the conditional recurrent generator and convolutional critic described earlier. Both models are trained for 50 federated rounds using *FedAtt* [14], chosen for its robustness to client imbalance.

Evaluation Setting. To assess explanation quality, we compute the *Spearman Correlation Score* between the ranking of *TimeSHAP* attributions and the ranking induced by the model's actual predictive behavior. This measures how consistently an explanation reflects the model's true decision structure. Spearman correlation focuses on rank consistency and is less sensitive to scale differences, making it suitable for comparing attribution structures. For two attribution vectors $a, b \in \mathbb{R}^n$, where a_i denotes the *TimeSHAP* attribution assigned to the i-th feature (or event) and b_i reflects the corresponding importance induced by the model's predictive behavior, we compute the Spearman Correlation Score. Let $\rho(a)$ and $\rho(b)$ denote the rank vectors obtained by sorting a and b, respectively. Spearman correlation is then defined as

$$\rho_s(a, b) = 1 - \frac{6\sum_i(\rho(a_i) - \rho(b_i))^2}{n(n^2 - 1)}.$$

Spearman correlation ranges from -1 to 1, where values close to 1 indicate a strong positive monotonic relationship. All results report mean and standard deviation across 5-fold cross-validation.

We compare four background dataset configurations: (1) each client's local training set D_c^{Train}, (2) the histogram-based background dataset of SHAP-FL [7], (3) our proposed *TimeSHAP–FL*, and (4) the aggregated training set D^{Train}, which serves as the centralized upper bound.

4.2 *TimeSHAP–FL* Explanations

This section illustrates the explanatory outputs produced by *TimeSHAP–FL* for the task of sepsis onset prediction. Figure 2 presents the aggregated feature-, event-, and cell-wise attributions for the *MICU*.

Figure 2 (a) shows the feature-wise explanations, where the features *SOFA*, *Heart Rate*, and *Creatinine* receive high absolute SHAP values. These findings are consistent with established clinical practice for sepsis diagnosis and risk stratification. In particular, the SOFA score is widely used to assess organ dysfunction in septic patients, elevated heart rate is a hallmark of systemic inflammatory response, and increased creatinine levels indicate impaired renal function, which is strongly associated with sepsis severity [9,22]. The prominence of these features suggests that the explanations produced by *TimeSHAP–FL* align well with clinically meaningful indicators routinely considered by medical practitioners.

The event-wise explanations in Fig. 2 (b) reveal that earlier time steps carry the greatest importance for predicting sepsis onset. This suggests that measurements obtained within the first hours after ICU admission provide the strongest discriminative signal for the model.

Figure 2 (c) provides the fine-grained cell-wise explanations. The highest SHAP values concentrate in the *Creatinine* feature across the earliest time steps,

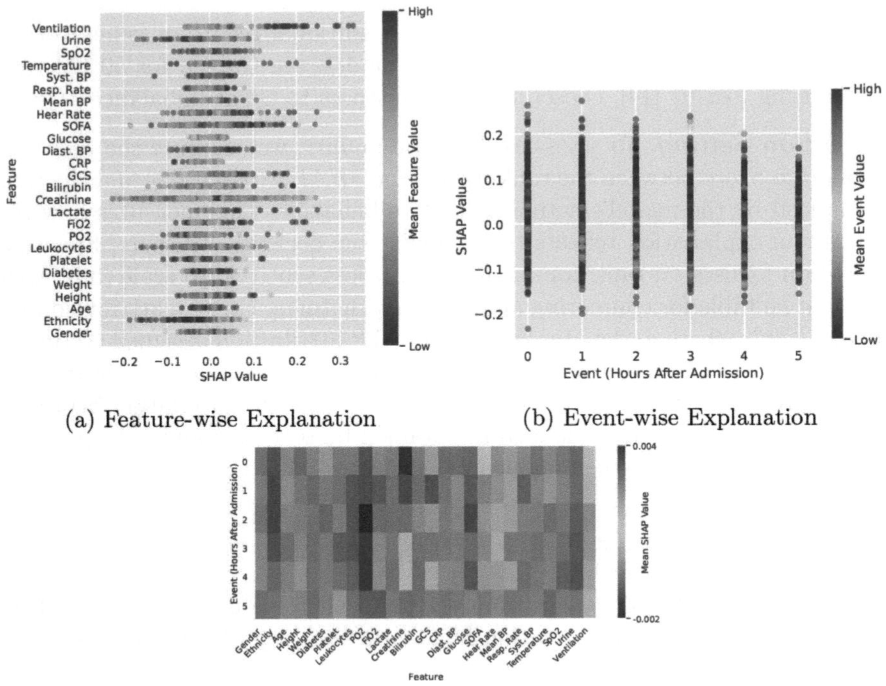

(a) Feature-wise Explanation (b) Event-wise Explanation

(c) Cell-wise Explanation

Fig. 2. Exemplary *TimeSHAP–FL* explanations for MICU.

reinforcing the clinical importance of early renal dysfunction for sepsis detection. Similarly, *Ventilation* status exhibits positive contributions, reflecting the increased sepsis risk commonly associated with mechanically ventilated patients in intensive care. In contrast, *PO2* shows comparatively low SHAP values during the early hours, suggesting that adequate oxygenation is associated with reduced risk. These temporal attribution patterns are consistent with how clinicians prioritize early organ function and supportive interventions when assessing sepsis risk in ICU settings.

Overall, the identified high-importance features and their temporal patterns are well aligned with established sepsis criteria and clinical assessment practices, indicating that *TimeSHAP–FL* produces explanations that are not only consistent across clients but also clinically plausible.

4.3 Explanation Quality

Table 1 reports Spearman Correlation Scores for feature-, event-, and cell-wise explanations across all ICUs and background dataset settings. Overall, we observe consistent improvements of *TimeSHAP–FL* over the local baseline and *SHAP–FL* for both feature-wise and event-wise explanations. These gains are expected: the synthetic background dataset captures the global training distribution, leading to better agreement with the actual model behavior, while the temporal coherence of the generated sequences allows *TimeSHAP–FL* to outperform *SHAP–FL*, which does not incorporate temporal dependencies during background construction.

Table 1. Spearman Correlation Scores per Background Datasets and Explanation Type (Best Performance per Dataset and Type is Bold)

	Background Data	MICU	MICU/ SICU	SICU	TSICU	CCU	CVICU	NSICU	Overall
Feature-wise Expl.	D_c^{Train}	0.6224 ±0.02	0.6292 ±0.03	0.6735 ±0.00	0.6239 ±0.02	0.6145 ±0.03	0.6203 ±0.01	0.6263 ±0.02	0.6314 ±0.01
	SHAP–FL	0.6013 ±0.00	0.6100 ±0.01	0.6793 ±0.02	0.6388 ±0.01	0.4664 ±0.02	0.6023 ±0.01	0.6197 ±0.01	0.6140 ±0.01
	TimeSHAP–FL	**0.6650** ±0.02	**0.6653** ±0.02	0.6997 ±0.02	**0.6727** ±0.02	**0.6476** ±0.04	**0.6656** ±0.02	**0.6933** ±0.02	**0.6727** ±0.01
	D^{Train}	0.6539 ±0.02	0.6586 ±0.02	**0.7029** ±0.01	0.6620 ±0.01	0.5641 ±0.08	0.6414 ±0.04	0.6571 ±0.01	0.6543 ±0.02
Event-wise Expl.	D_c^{Train}	0.6210 ±0.02	0.6246 ±0.03	0.6778 ±0.01	0.6683 ±0.01	0.6163 ±0.04	0.6245 ±0.01	0.6320 ±0.01	0.6349 ±0.00
	SHAP–FL	0.5526 ±0.03	0.5866 ±0.01	0.6646 ±0.01	0.6041 ±0.01	0.3221 ±0.03	0.5463 ±0.00	0.5933 ±0.05	0.5608 ±0.00
	TimeSHAP–FL	**0.6642** ±0.04	**0.6600** ±0.02	**0.7077** ±0.01	0.6717 ±0.02	**0.6590** ±0.06	**0.6611** ±0.05	**0.6549** ±0.03	**0.6674** ±0.01
	D^{Train}	0.6480 ±0.02	0.6467 ±0.01	0.6898 ±0.00	**0.6741** ±0.05	0.6077 ±0.02	0.6206 ±0.05	0.6708 ±0.01	0.6482 ±0.00
Cell-wise Expl.	D_c^{Train}	0.6538 ±0.01	0.6507 ±0.03	0.6742 ±0.01	0.6671 ±0.01	0.6475 ±0.04	0.6298 ±0.02	0.6522 ±0.01	0.6465 ±0.01
	SHAP–FL	0.6591 ±0.01	**0.6705** ±0.01	**0.7174** ±0.02	0.6673 ±0.00	0.6222 ±0.05	0.6485 ±0.03	0.6512 ±0.03	0.6630 ±0.01
	TimeSHAP–FL	**0.6635** ±0.01	**0.6705** ±0.02	0.7076 ±0.02	**0.6908** ±0.03	**0.6624** ±0.06	**0.6664** ±0.04	0.6541 ±0.02	**0.6708** ±0.02
	D^{Train}	0.6577 ±0.02	0.6413 ±0.01	0.7093 ±0.02	0.6691 ±0.01	0.6449 ±0.07	0.6536 ±0.04	**0.6802** ±0.02	0.6603 ±0.00

Interestingly, *TimeSHAP–FL* also surpasses the centralized gold standard (D^{Train}) for most clients in these two explanation types, although the improvements are smaller and not always statistically significant. Unlike previously, these results cannot be attributed to better representativeness or temporal structure, since the gold standard directly uses the full global dataset. Instead, as analyzed in Sect. 4.4, combining globally representative samples with a proportion of client-specific local samples can yield background datasets that are better aligned with the client's predictive characteristics than either source alone.

For cell-wise explanations, *TimeSHAP–FL* continues to achieve the highest scores overall and for most ICUs, but the margin over *SHAP–FL* and the gold standard is smaller. This is consistent with the mechanism by which cell-wise attributions are computed: they rely primarily on feature-level value distributions rather than temporally coherent sequences. Since *SHAP–FL* already provides representative feature distributions, the advantage of *TimeSHAP–FL* in capturing temporal structure plays a less central role.

Overall, *TimeSHAP–FL* achieves the highest Spearman scores across most settings, even outperforming the centralized gold standard for several clients, indicating that the mixed synthetic–local background dataset can produce better explanations under this metric.

4.4 Sanity Checks

Beyond the previous assessment of explanation quality, we conduct additional sanity checks to provide a deeper analysis of *TimeSHAP–FL*. Figure 3 summarizes two complementary investigations: panel (a) examines how different proportions of synthesized data in each client's background dataset $\mathcal{D}_c^{Background}$ affect explanation quality, while panel (b) evaluates potential privacy leakage arising from the generative model.

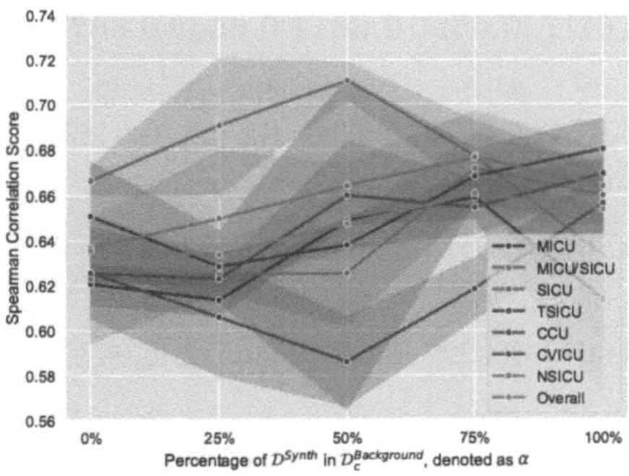

(a) $\mathcal{D}_c^{Background}$ Composition

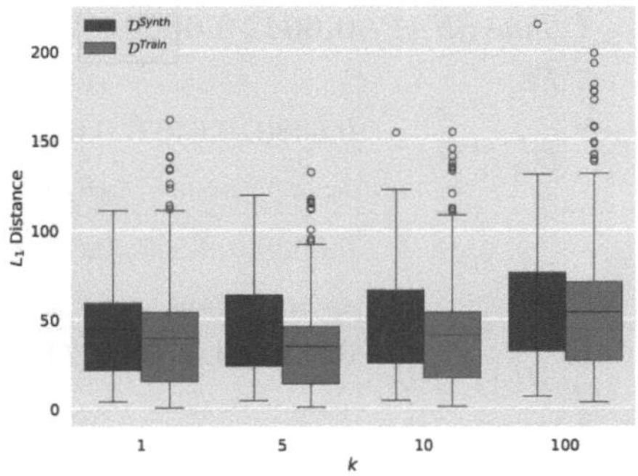

(b) Privacy Leakage Analysis

Fig. 3. Sanity Checks for *TimeSHAP–FL*.

Effect of Background Dataset Composition. In Sect. 3, we hypothesized that combining synthesized samples from $\mathcal{D}^{Synth}$ with local samples from $\mathcal{D}_c^{Train}$ improves the quality of client-specific explanations. Figure 3 (a) reports Spearman scores (averaged across feature-, event-, and cell-wise explanations) for background datasets containing between $\alpha = 0\%$ and $\alpha = 100\%$ synthetic data. The results confirm that the optimal setting is strongly ICU-dependent. Most ICUs, as well as the overall mean, achieve the best performance when $\alpha = 75\%$ of the background dataset is synthetic. However, there are notable exceptions: for example, *SICU* performs best at $\alpha = 50\%$ and also achieves the highest overall Spearman score among all clients. Conversely, *CCU* performs worst at this $\alpha = 50\%$ setting, instead favoring both smaller and larger proportions of synthetic samples. Overall, the observed variability in the optimal synthetic-to-local mixing ratio reflects genuine heterogeneity across clients rather than a limitation of the method, and suggests that adaptive or validation-based tuning strategies may be beneficial in practice rather than relying on a single universal setting.

Privacy Leakage Assessment. In addition to the theoretical privacy guarantees provided by DP during GAN training, we conduct an empirical test to ensure that *TimeSHAP-FL* does not replicate individual patients' trajectories. Figure 3 (b) compares, for $k \in \{1, 5, 10, 100\}$, the mean L_1 distance between (i) samples drawn from $\mathcal{D}^{Synth}$ and their k nearest neighbors in $\mathcal{D}^{Train}$, and (ii) samples drawn from $\mathcal{D}^{Train}$ and their k nearest neighbors in $\mathcal{D}^{Train}$. The underlying rationale is the following: if the GAN memorized and reproduced real patient trajectories, synthetic samples would appear unusually close to their nearest real neighbor, resulting in systematically lower distances. The results, however, show the opposite: for all values of k, synthetic samples exhibit slightly higher nearest-neighbor L_1 distances compared to real samples. This demonstrates that GAN-generated trajectories do not concentrate around individual training points and, on average, remain more distinct from the training distribution than real samples do from each other. Together with the theoretical guarantees from DP, these findings provide strong evidence that *TimeSHAP-FL* does not pose a meaningful risk of patient-level privacy leakage.

5 Conclusion

In this work, we introduced *TimeSHAP-FL*, a federated framework that enables fine-grained time-series explanations for models trained across multiple institutions while preserving data privacy. By combining a federated, differentially private WGAN-GP with the temporal attribution capabilities of *TimeSHAP*, our method provides feature-, event-, and cell-wise explanations that more accurately reflect the global model's behavior than existing federated XAI approaches. Applied to the high-stakes task of sepsis onset prediction, *TimeSHAP-FL* consistently improved explanation quality across ICUs and demonstrated that incorporating both synthesized and local samples into background datasets yields more stable and informative attributions. Beyond quantitative improvements, the resulting explanations aligned well with established clinical knowledge, highlighting the method's potential to support trustworthy model interpretation in real-world healthcare settings. Additional sanity checks further indicated that the synthesized samples do not replicate patient trajectories and therefore do not introduce additional privacy risks, reinforcing the suitability of the approach for sensitive medical data.

Limitations. Although our results are encouraging, several limitations remain. First, the computational and communication overhead associated with federated GAN training is non-negligible and was not quantified in detail. We focus on cross-silo FL scenarios, such as collaborations between hospitals, where institutions typically have sufficient computational resources and explanation quality is prioritized over strict efficiency constraints. A more systematic analysis of the trade-offs between explanation fidelity and resource consumption is therefore left for future work. Second, while the synthesized background dataset improves explanation quality, we did not study in depth how variations in the quality of GAN-generated samples affect downstream explanations. Third, our evaluation is based on a single clinical use case, which limits conclusions about generalizability across domains, data modalities, or FL configurations.

Future Work. Future research may address these limitations in several directions. Since *TimeSHAP–FL* does not explicitly model causal relationships in the data, integrating causal reasoning into federated explanations is a promising extension, for instance through causal SHAP [13]. Another direction is the development of an automated, client-specific strategy for selecting the optimal ratio α of synthetic and local samples in $\mathcal{D}_c^{Background}$, avoiding costly manual tuning. Finally, validating *TimeSHAP–FL* in real-world clinical environments together with domain experts will be essential to assess its usability, interpretability, and practical impact on clinical decision support.

Acknowledgements. This research was partially funded by the Ministry of Culture and Science of North Rhine-Westphalia under the grant no NW21-059A SAIL and the Deutsche Forschungsgemeinschaft as part of the TRR 318 "Constructing Explainability".

References

1. Behera, M.R., Upadhyay, S., Shetty, S., Priyadarshini, S., Patel, P., Lee, K.F.: Fedsyn: synthetic data generation using federated learning. arXiv preprint arXiv:2203.05931 (2022)
2. Bento, J., Saleiro, P., Cruz, A.F., Figueiredo, M.A., Bizarro, P.: Timeshap: explaining recurrent models through sequence perturbations. In: ACM SIGKDD Conference on Knowledge Discovery & Data Mining, pp. 2565–2573 (2021)
3. Caicedo-Torres, W., Gutierrez, J.: ISEEU: visually interpretable deep learning for mortality prediction inside the ICU. J. Biomed. Inf. **98** (2019)
4. Corbucci, L., Guidotti, R., Monreale, A.: Explaining black-boxes in federated learning. In: World Conference on XAI, pp. 151–163. Springer (2023)
5. Ducange, P., Marcelloni, F., Renda, A., Ruffini, F.: Consistent post-hoc explainability in federated learning through federated fuzzy clustering. In: 2024 IEEE International Conference on Fuzzy Systems, pp. 1–10. IEEE (2024)
6. Düsing, C., Cimiano, P.: Improving early sepsis onset prediction through federated learning. arXiv preprint arXiv:2509.20885 (2025)
7. Düsing, C., Cimiano, P.: Shap–FL: improving explainability for multicentric sepsis onset prediction through background dataset synthesis. In: International Conference on Artificial Intelligence in Medicine, pp. 120–129. Springer (2025)

8. Dwork, C., Roth, A., et al.: The algorithmic foundations of differential privacy. Found. Trends® Theor. Comput. Sci. **9**(3–4), 211–407 (2014)
9. Evans, L., et al.: Surviving sepsis campaign: international guidelines for management of sepsis and septic shock 2021. Crit. Care Med. **49**(11), e1063–e1143 (2021)
10. Fiosina, J.: Interpretable privacy-preserving collaborative deep learning for taxi trip duration forecasting. In: International Conference on Vehicle Technology and Intelligent Transport Systems, pp. 392–411. Springer (2021)
11. Gulrajani, I., Ahmed, F., Arjovsky, M., Dumoulin, V., Courville, A.C.: Improved training of wasserstein GANs. NeurIPS **30** (2017)
12. Herbinger, J., Bischl, B., Casalicchio, G.: Repid: regional effect plots with implicit interaction detection. In: International Conference on Artificial Intelligence and Statistics, pp. 10209–10233. PMLR (2022)
13. Heskes, T., Sijben, E., Bucur, I.G., Claassen, T.: Causal shapley values: exploiting causal knowledge to explain individual predictions of complex models. Adv. Neural. Inf. Process. Syst. **33**, 4778–4789 (2020)
14. Ji, S., Pan, S., Long, G., Li, X., Jiang, J., Huang, Z.: Learning private neural language modeling with attentive aggregation. In: 2019 International Joint Conference on Neural Networks (IJCNN), pp. 1–8. IEEE (2019)
15. Jiang, X., Zhou, X., Grossklags, J.: Distributed synthetic time-series data generation with local differentially private federated learning. IEEE Access (2024)
16. Johnson, A.E., et al.: Mimic-iv, a freely accessible electronic health record dataset. Sci. Data **10**(1), 1 (2023)
17. Karras, T., Aittala, M., Hellsten, J., Laine, S., Lehtinen, J., Aila, T.: Training generative adversarial networks with limited data. Adv. Neural. Inf. Process. Syst. **33**, 12104–12114 (2020)
18. Lundberg, S.M., Lee, S.I.: A unified approach to interpreting model predictions. Adv. Neural Inf. Process. Syst. **30** (2017)
19. McMahan, B., Moore, E., Ramage, D., Hampson, S., y Arcas, B.A.: Communication-efficient learning of deep networks from decentralized data. In: Artif. Intell. Stat., 1273–1282. PMLR (2017)
20. Nayebi, A., Tipirneni, S., Reddy, C.K., Foreman, B., Subbian, V.: Windowshap: an efficient framework for explaining time-series classifiers based on shapley values. J. Biomed. Inform. **144**, 104438 (2023)
21. Peracchio, L., Nicora, G., Buonocore, T.M., Bellazzi, R., Parimbelli, E.: Do you trust your model explanations? an analysis of XAI performance under dataset shift. In: International Conference on AI in Medicine, pp. 257–266. Springer (2024)
22. Singer, M., et al.: The third international consensus definitions for sepsis and septic shock (sepsis-3). JAMA **315**(8), 801–810 (2016)
23. Singh, G., Sharma, P., Bedi, J.: A novel integration of federated learning and LSTM for synthetic time series generation. In: Conference on Artificial Intelligence, pp. 15–26. Springer (2024)
24. Tonekaboni, S., Joshi, S., McCradden, M.D., Goldenberg, A.: What clinicians want: contextualizing explainable machine learning for clinical end use. In: Machine Learning For Healthcare Conference, pp. 359–380. PMLR (2019)

Grasynda: Graph-Based Synthetic Time Series Generation

Luis Amorim[1,2(✉)], Moisés Santos[2,3], Paulo J. Azevedo[1,5], Carlos Soares[2,3,4], and Vitor Cerqueira[2,3]

[1] Department of Informatics, University of Minho, Braga, Portugal
[2] Faculdade de Engenharia da Universidade do Porto, Porto, Portugal
up202402643@edu.fe.up.pt
[3] Laboratory for Artificial Intelligence and Computer Science (LIACC), Porto, Portugal
[4] Fraunhofer Portugal AICOS, Porto, Portugal
[5] HASLab, INESCTEC, Porto, Portugal

Abstract. Data augmentation is a crucial tool in time series forecasting, especially for deep learning architectures that require a large training sample size to generalize effectively. However, extensive datasets are not always available in real-world scenarios. Although many data augmentation methods exist, their limitations include the use of transformations that do not adequately preserve data properties. This paper introduces `Grasynda`, a novel graph-based approach for synthetic time series generation that: (1) converts univariate time series into a network structure using a graph representation, where each state is a node and each transition is represented as a directed edge; and (2) encodes their temporal dynamics in a transition probability matrix. We performed an extensive evaluation of `Grasynda` as a data augmentation method for time series forecasting. We use three neural network variations on six benchmark datasets. The results indicate that `Grasynda` consistently outperforms other time series data augmentation methods, including ones used in state-of-the-art time series foundation models. The method and all experiments are publicly available.

Keywords: Time series · Synthetic data · Forecasting

1 Introduction

In time series forecasting problems, deep learning approaches are increasingly used as effective alternatives to well-established methods such as ARIMA or exponential smoothing [1,4]. While neural networks, such as Chronos [1], NHITS [9] or PatchTST [7], have shown state-of-the-art forecasting performance in benchmark datasets, their effective application requires a sufficiently large training sample size [2,4]. To address this limitation, data augmentation techniques are often used to enrich the training set by generating artificial but realistic time series. Indeed, foundation models for time series, such as Amazon's

M. Baratchi et al. (Eds.): IDA 2026, LNCS 16513, pp. 398–410, 2026.
https://doi.org/10.1007/978-3-032-23833-7_29

Chronos [1,2], are typically trained on a mix of real and synthetic datasets to maximize their performance.

A variety of methods have been proposed for generating synthetic time series data, ranging from simple transformations such as jittering [23], to more advanced techniques such as pattern mixing (i.e., averaging multiple time series) [10] or generative models [24]. Despite their popularity, these methods face important limitations. Basic augmentations such as jittering or scaling [23] typically fail to either maintain temporal dependencies or to introduce sufficiently realistic variability. On the other hand, more sophisticated generative models, such as generative adversarial networks [24], require considerable computational resources and are prone to issues such as mode collapse or overfitting, especially when training data is limited.

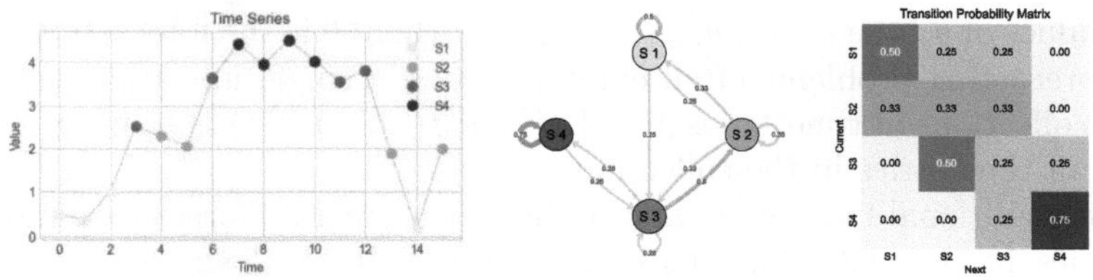

Fig. 1. Example plot of a time series in the time domain (left); its network representation (middle); and the corresponding transition matrix (right).

To address these limitations, we introduce `Grasynda` (GRAph-based SYNthetic time series DAta generation), a novel method that generates synthetic time series by transforming them into graph representation [22]. In this approach, the time series is discretized into a finite set of states, where each observed transition between consecutive states is represented as a directed edge, effectively capturing the temporal dynamics of the series. The working hypothesis is that the network conversion framework encodes the temporal dynamics into a transition probability matrix that effectively captures the underlying data generation process. The graph representation inherently captures both local patterns (immediate state transitions) and global structures (overall network connectivity patterns). `Grasynda` leverages this matrix to generate new synthetic series by sampling transitions according to the estimated probabilities, thus maintaining the statistical properties of the original data. Figure 1 illustrates this process, showing the original time series, its graph representation, and the corresponding transition matrix.

We validate the usefulness of the synthetic time series created by `Grasynda` based on a forecasting downstream predictive task. We conduct experiments using six benchmark datasets and three neural network architectures. The results suggest that augmenting time series datasets using `Grasynda` improves the forecasting accuracy of different neural networks. The proposed method also shows competitive forecast accuracy when compared with other data augmentation methods. The experiments and method are publicly available online[1].

In summary, the contributions of this paper are a novel graph-based method for synthetic time series generation, and an empirical evaluation of its effectiveness based on state-of-the-art neural network architectures applied to benchmark datasets.

2 Background

2.1 Time Series Forecasting

A univariate time series is a time-ordered set of observations $Y = \{y_1, y_2, \ldots, y_T\}$, where each value $y_t \in \mathbb{R}$ corresponds to the observed state at a specific time t, and T denotes the total length of the series. Forecasting refers to predicting future values of a time series, $y_{T+1}, \ldots, y_{T+h}$, where h is the forecasting horizon. Many forecasting problems often extend beyond a single univariate series, and involve collections of time series $\mathcal{Y} = \{Y^{(1)}, Y^{(2)}, \ldots, Y^{(N)}\}$ [12], where N is the number of time series in the collection.

Most traditional forecasting approaches train separate forecasting models for each series in the collection. This makes the capture of shared patterns and dependencies that may exist across the time series impossible. Global forecasting models [4] overcome this limitation by using the data from all available time series to build a forecasting model. Machine learning approaches to time series forecasting typically adopt an autoregressive modeling framework [4]. This approach models future values as a function of the l preceding observations (lags).

Several types of deep learning architectures have been developed for forecasting tasks, ranging from feedforward neural networks and recurrent-based architectures (e.g., LSTMs, GRUs) designed to capture temporal dependencies, to more recent Transformer-based models [15] which leverage attention mechanisms. Despite the development of these diverse architectures, models based on multi-layer perceptrons (MLPs) have consistently shown competitive forecasting performance [9].

One such state-of-the-art architecture based on MLPs is NHITS [9]. NHITS is based on stacked blocks of MLP layers connected with residual links. NHITS also combines two complementary techniques, multi-rate input sampling and hierarchical interpolation, leading to state-of-the-art forecasting performance.

Time series forecasting using Kolmogorov-Arnold Networks (KANs) recently emerged as a competitive alternative to MLPs [16]. KAN learns interpretable transformations inspired by the Kolmogorov-Arnold representation theorem. Contrary to MLPs, KANs place learnable activation functions directly on the connections of the network.

[1] https://github.com/Amorim009/Grasynda.

2.2 Data Augmentation

Machine learning algorithms, particularly neural networks which are known to excel in data-rich environments, generally achieve better performance with larger training sets [8]. The need for an adequate sample size is especially important when using deep learning models. Without sufficient data, they tend to memorize the training set rather than generalize to new inputs. Data augmentation aims to address these challenges by generating realistic variations of the original data, increasing diversity, and enabling models to generalize better to new, unseen conditions [4,14]. The augmentation process works by generating artificial but realistic time series that mimic the characteristics of the original data. Then, the synthetic data is added to the original dataset to increase the size of the training set.

Data generation techniques for time series forecasting have recently received increasing attention, leading to different approaches. Transformation methods [23], including time or magnitude warping (`T-Warp` and `M-Warp`), scaling (`Scaling`), or jittering (`Jitter`), are easily implementable due to their simplicity and computational efficiency [21,23]. Jittering involves adding random noise to each data point, aiming to improve model robustness to such variations. `Scaling` modifies the magnitude of the series by multiplying observations by a random factor, typically drawn from a standard Normal distribution. `T-Warp` distorts the temporal axis using smooth functions such as cubic splines, which are subsequently used to resample the series [20]. `M-Warp` is similar but applies these smooth distortions directly to the values of the time series, leading to smoother transformations relative to `Scaling` [20]. These methods are adaptable to domains such as healthcare [11]. However, they have limitations. Most importantly, they often fail to introduce sufficient variety or accurately replicate the original data generation process, losing inherent contextual integrity.

Some synthetic generation techniques focus on mixing patterns from multiple time series. These include methods such as Moving Block Bootstrap (`MBB`) combined with seasonal decomposition [5]; Dynamic Time Warping (DTW) Barycentric Averaging (`DBA`) [10] which averages series based on DTW; and `TSMixup` [1] which combines randomly sampled series segments using weighted averages. The practical impact of such methods is significant. For example, Chronos [1] is a state of the art foundation model by Amazon that was pre-trained with synthetic time series generated using `TSMixup`. Recently, more time series models are being trained with synthetic data. Although generative models address some challenges, they still require high computational cost and a significant hyperparameter tuning effort to generate data that is minimally useful [23].

2.3 Time Series Analysis Using Network Science

The application of network science to time series analysis has proven to be an effective approach that bridges two distinct domains: temporal data analysis and graph theory [6]. By transforming temporal data into network representations, this approach allows the use of graph theory to reveal complex patterns, nonlinear relationships, or multi-scale dynamics that conventional time and frequency-domain analyses may fail to detect. Furthermore, it offers more intuitive visualizations of dynamic behavior, which can reveal hidden structures in temporal data and improve interpretability [22].

Using network science for time series analysis involves transforming the temporal data into a graph representation. Several approaches have been proposed for this conversion. Quantile graphs [6], for example, discretize the time series and represent transitions between these states as edges in a graph. Another example is visibility graphs [22], which connect time series observations that can *see* each other based on a visibility criterion. Silva et al. [22] overview several other methods for time series analysis using network science.

3 The Grasynda Methodology

This proposed methodology, **Grasynda**, is illustrated in Fig. 2, consists of using a graph conversion of a time series. Sect. 3.1); Graph transitions are encoded in a matrix representation that is used as an engine for synthetic data generation. We build the probability matrix as a base for the generation process because it can be a useful tool to dynamically capture data point interactions (Sect. 3.1).

3.1 Time Series Graph

Discretization. The process begins by performing discretization on the original time series $Y = \{y_1, y_2, \ldots, y_T\}$. Each value $y_t \in Y$ is mapped to a discrete label $s_t \in \{1, 2, \ldots, k\}$ corresponding to the index of the state it belongs to. This results in a new discretized time series S:

$$S = \{s_1, s_2, \ldots, s_T\}, \quad S_t \in \{1, 2, \ldots, k\} \tag{1}$$

where s_t represents the corresponding label associated with the value y_t. Let R_j denote the set of original values assigned to state s_j:

$$R_j = \{y_t \in Y \mid s_t = j\}, \quad j \in \{1, \ldots, k\} \tag{2}$$

Graph Construction. After discretization, we create the graph structure based on the subsequent state transitions. This structure results in a directed weighted graph $G = (V, E, P)$, where:

- $V = \{1, 2, \ldots, k\}$ is the set of nodes, each representing a state s_i,
- $E \subseteq V \times V$ is the set of directed edges, where an edge $(s_i, s_j) \in E$ exists if there is at least one transition from s_i to s_j in the discretized time series S,
- $P = [p_{ij}]$ is the normalized edge weight matrix (transition matrix), where each weight p_{ij} corresponds to the empirical conditional probability of transitioning from state s_i to state s_j:

The elements of P satisfy:

$$p_{ij} = \frac{\text{count of transitions from } s_i \text{ to } s_j}{\text{total transitions from } s_i} \tag{3}$$

Since these are conditional probabilities, each row of P sums to 1: $\sum_{j=1}^{k} p_{ij} = 1$ for all $i \in \{1, \ldots, k\}$. Generally $p_{ij} \neq p_{ji}$, indicating the directed nature of the graph and the non-symmetric property of the transition matrix.

3.2 Data Generation

By sampling from the transition matrix P, we can generate synthetic sequences $\hat{S} = \{\hat{s}_1, \hat{s}_2, \ldots, \hat{s}_T\}$. We start by selecting an initial state $\hat{s}_1$, matching the first state from the original time series. The subsequent states are sampled iteratively based on the transition probabilities:

$$\hat{s}_{t+1} \sim P(\hat{s}_t) \tag{4}$$

This means that $\hat{s}_{t+1}$ is chosen according to the probability distribution given by the row of P corresponding to $\hat{s}_t$. In effect, the synthetic discretized synthetic sequence follows the same probabilistic transitions as the original discrete series. We can sample any number of states from the matrix to generate synthetic sequences of arbitrary length.

After obtaining a synthetic discrete time series $\hat{S}$, we convert it back to the original time series space. This is achieved by sampling from the value distribution of each corresponding state . Specifically, for each state $\hat{s}_i$ in $\hat{S}$, a synthetic value $\hat{y}_i$ is sampled from its distribution. $\hat{y}_t \sim \text{Uniform}(R_{\hat{s}_i})$

This results in a synthetic time series $\hat{Y} = \{\hat{y}_1, \hat{y}_2, \ldots, \hat{y}_T\}$. This sampling approach is expected to preserve the original temporal dynamics of the time series leveraging the probabilistic nature of the graph matrix transitions.

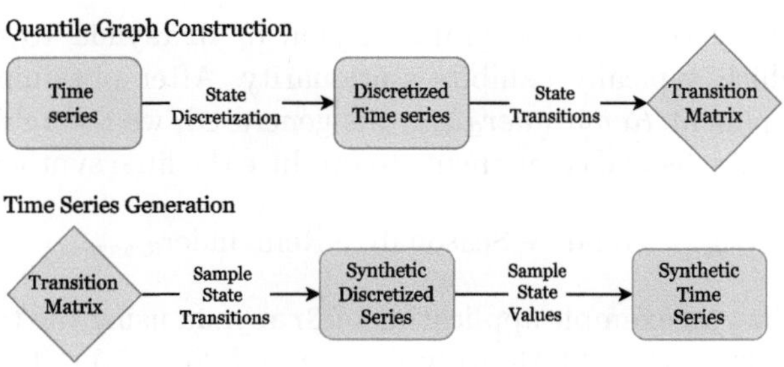

Fig. 2. Workflow behind `Grasynda` for building a graph from a time series and using it to create synthetic time series.

4 Quantile-Based Grasynda

In this section, we describe an instantiation of Grasynda using quantile graphs as the method for converting time series into network representations.

4.1 Time Series Quantile Graph

For the quantile-based version of our method, each value in the time series is assigned to a quantile using equal-frequency binning. This produces a discrete sequence with the same length as the original series, where each element corresponds to a quantile label. The sequence is defined by a directed graph in which each node represents a quantile level, and each edge captures the transitions observed between consecutive labels.

We count the transitions from one quantile to the others and normalise these counts to obtain the transition probability matrix of the quantile graph. The final synthetic time series is generated by following the general Grasynda methodology described in Sect. 3.

4.2 Data Preprocessing: Handling Non-stationarity

The adopted tool to convert time series into network structures, quantile graph conversion, focuses on transitions between consecutive observations. While this tool is adequate to capture local patterns through edge connections, it has limitations in preserving long-term temporal dependencies such as trend and seasonality that characterize non-stationary time series.

To address these limitations, we adopt a decomposition-based approach following Bergmeir et al. [5]. Our solution leverages Seasonal-Trend decomposition using LOESS (STL) to decompose each time series into three distinct components:

$$y_t = \text{Trend}_t + \text{Seasonal}_t + \text{Remainder}_t \tag{5}$$

We then apply the quantile graph version of `Grasynda` to the remainder component, which typically exhibits stationarity. After obtaining a synthetic remainder component $Remainder_{\texttt{Grasynda}}$ is generated, we recombine it with the original trend and seasonal components to produce the final synthetic time series:

$$\hat{y}_t = \text{Trend}_t + \text{Seasonal}_t + \text{Remainder}_{\texttt{Grasynda}} \tag{6}$$

Figure 3 shows an example application of `Grasynda` using the time series with identifier *ID90* from the M3 Monthly dataset (c.f. Sect. 5.1). The figure shows the time series in the standard time domain (a), the corresponding network representation using a quantile graph (b), and a synthetic time series generated from the graph using `Grasynda` (c). We note that this preprocessing operation may be useful for other instantiations of the methodology as well.

5 Experiments

To assess the effectiveness of the synthetic time series generated by **Grasynda**, we performed a series of experiments focused on a downstream forecasting task. Specifically, we evaluated **Grasynda** in comparison to several state-of-the-art synthetic time series generation methods, using six benchmark datasets and three neural network architectures. We address the following research question: *How effective is **Grasynda** as a data augmentation for time series forecasting?*

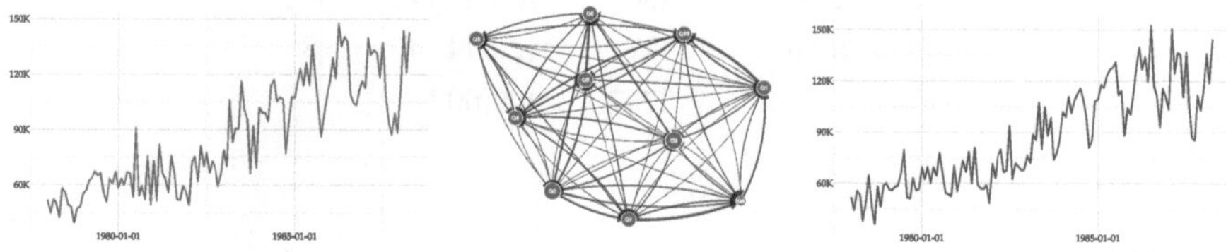

Fig. 3. Example plot of a time series in the time domain (left); its network representation using a quantile graph (middle); and a synthetic time series generated from the graph using **Grasynda** (right).

5.1 Experimental Design

Forecasting Problems. We use six publicly available benchmark datasets from three forecasting competitions: M1 [17], M3 [18], and Tourism [3]. These datasets span various application domains, such as industry, demography, and economics. Table 1 provides a detailed summary of their characteristics. In total, the datasets comprise 3,797 time series and 409,602 individual observations. For forecasting tasks, we set the horizon (h) to 12 for monthly time series and 8 for quarterly time series. The input window (number of lagged observations) is fixed at 24 for monthly and 8 for quarterly data, corresponding to two full seasonal cycles for each frequency.

Performance Estimation. To assess forecasting performance, we hold out the final h observations of each time series as a test set, while training models on all preceding data points. Forecast accuracy is evaluated using the Mean Absolute Scaled Error (MASE), a widely adopted, scale-independent metric. MASE is calculated as: $\text{MASE} = \frac{\text{MAE}}{\text{MAE}_{\text{naive}}}$, where $\text{MAE} = \frac{1}{T}\sum_{t=1}^{T}|y_t - \hat{y}_t|$ denotes the mean absolute error of the forecast, and $\text{MAE}_{\text{naive}} = \frac{1}{T-1}\sum_{t=2}^{T}|y_t - y_{t-1}|$ corresponds to the mean absolute error of a one-step naive forecast.

Given a training dataset $\mathcal{D} = \{Y^{(1)}, Y^{(2)}, \ldots, Y^{(M)}\}$, where each time series is $Y^{(i)} = \{y_1^{(i)}, y_2^{(i)}, \ldots, y_T^{(i)}\}$, we generate a synthetic series $\hat{Y}^{(i)} = \{\hat{y}_1^{(i)}, \hat{y}_2^{(i)}, \ldots, \hat{y}_T^{(i)}\}$ for every $Y^{(i)}$. The augmented training set, therefore, consists of both the original and synthetic time series:

$$\mathcal{D}_{\text{aug}} = \{Y^{(1)}, \ldots, Y^{(M)}, \hat{Y}^{(1)}, \ldots, \hat{Y}^{(M)}\} \tag{7}$$

Table 1. Summary of the datasets: average value, number of time series, number of observations, seasonal period, and forecasting horizon (h).

Dataset	Avg. Value	# Series	# Obs.	Period	h
M1-M	72.7	617	44,892	12	12
M1-Q	40.9	203	8,320	4	8
M3-M	117.3	1428	167,562	12	12
M3-Q	48.9	756	37,004	4	8
T-M	298.5	366	109,280	12	12
T-Q	99.6	427	42,544	4	8
Total	–	3797	409,602	–	–

5.2 Methods

We evaluate three state-of-the-art neural network forecasting models: NHITS [9], KAN [16], and MLP. These models were chosen for their competitive forecasting accuracy (see Sect. 2) and computational efficiency relative to other deep learning architectures, such as transformer-based models [25]. We include the seasonal naive method as a baseline, providing a reference point for the experimental results. Each neural network is implemented using the neuralforecast Python library and optimized via 10 iterations of random search for hyperparameter selection.

We benchmark the proposed method against several widely used data augmentation techniques: `Scaling` [21]; `TSMixup`, use by Amazon's time series foundational model Chronos [1,2]; `DBA` [10]; jittering (`Jitter`) [21]; magnitude warping (`M-Warp`) [13]; time warping (`T-Warp`) [19]; and seasonal moving block bootstrap (`MBB`) [5]. These methods are further described in Sect. 2. As a baseline, we also consider results using the original, non-augmented training data (`Original`). All augmentation methods are configured according to settings established in previous literature. For our method, `Grasynda`, we use 25 quantiles.

5.3 Results

Table 2 summarizes the experimental results, reporting the MASE scores for each synthetic time series generation method across different model architectures and datasets. The table also includes the average performance and average rank for each method (with lower ranks indicating better results). Asterisks ('*') indicate methods that achieve significantly better performance than `Original`, according to the Wilcoxon signed-rank test. The table confirms the validity of the experimental setup, as all the neural network models consistently outperform the seasonal naive baseline.

Overall, `Grasynda` shows consistently competitive forecasting performance compared to other synthetic time series generation approaches across all model architectures. It achieves the lowest (best) average MASE in two out of the

Table 2. MASE of each approach across different architectures and datasets. Best and second-best values are **bolded** and underlined, respectively. Asterisks (*) denote significant improvements over `Original`.

Model		Original	Grasynda	DBA	Jitter	M-Warp	MBB	Scaling	T-Warp	TSMixup	SNaive
NHITS	M1-M	0.977	**0.945***	0.948*	0.970	0.981	0.952	0.948*	0.977	0.978	1.221
	M1-Q	1.037	1.034	1.094	1.036	**0.974***	1.029	1.039	1.066	1.049	1.647
	M3-M	0.801	0.759*	0.779	0.786	0.773	0.764*	0.770	**0.757***	0.770	1.091
	M3-Q	1.199	1.220	1.136*	1.185	1.181	1.165*	**1.131***	1.193	1.193	1.417
	T-M	1.208	1.190*	1.193	1.200*	1.208	**1.184***	1.206	1.230	1.190*	1.345
	T-Q	1.635	**1.555***	1.565*	1.608*	1.628	1.663	1.603*	1.616	1.622	1.702
	Avg.	1.143	1.117	1.119	1.131	1.124	1.126	**1.112**	1.129	1.134	1.404
	Avg. Rank	7.3	**3.1**	4.4	5.2	5.8	3.7	3.8	6.0	5.8	10.0
MLP	M1-M	0.930	0.948	0.985	0.940	0.968	0.936	**0.926**	0.977	0.954	1.221
	M1-Q	1.070	1.075	1.064	1.034*	**0.996***	1.036*	1.040	1.076	1.062	1.647
	M3-M	0.774	0.768	0.777	0.760*	0.780	0.765*	0.767	**0.757**	0.761	1.091
	M3-Q	1.143	1.181	1.152	1.148	1.159	1.154	**1.112**	1.132	1.209	1.417
	T-M	1.198	1.195	2.545	1.213	1.216	1.196*	1.204	1.257	**1.183***	1.345
	T-Q	1.568	**1.517***	1.605	1.654	1.615	1.637	1.645	1.720	1.598	1.702
	Avg.	**1.114**	**1.114**	1.355	1.125	1.122	1.121	1.116	1.153	1.128	1.404
	Avg. Rank	4.2	5.0	7.0	4.3	6.0	4.2	**3.8**	6.3	4.5	9.7
KAN	M1-M	0.961	0.941	0.957	0.962	0.979	0.939	**0.936**	1.008	0.955	1.221
	M1-Q	1.013	1.022	1.043	1.013	**0.966***	1.016	1.030	1.071	1.044	1.647
	M3-M	0.784	0.779	0.777	0.783	0.796	**0.773***	0.797	0.798	0.775*	1.091
	M3-Q	1.223	1.207*	**1.146***	1.229	1.213	1.221	1.165*	1.206	1.271	1.417
	T-M	1.227	**1.190***	1.231	1.228	1.229	1.217	1.222	1.273	1.213*	1.345
	T-Q	1.571	**1.548**	1.549	1.631	1.591	1.600	1.623	1.670	1.642	1.702
	Avg.	1.130	**1.114**	1.117	1.141	1.129	1.128	1.129	1.171	1.150	1.404
	Avg. Rank	4.9	**3.0**	4.3	5.9	5.3	3.5	4.5	8.0	5.5	10.0
Effectiveness		–	**0.72**	0.56	0.50	0.39	0.67	0.67	0.33	0.56	0.0

three evaluated forecasting models, and ranks second in the remaining model. The average rank scores, calculated for each forecasting model, further indicate that `Grasynda` outperforms other data augmentation methods.

We also assessed the effectiveness of each approach in augmenting the training datasets. For each of the 18 evaluated experiments (6 datasets times 3 forecasting models), we counted the number of times each augmentation method outperformed the no-augmentation baseline (`Original`). `Grasynda` achieved better results than `Original` in 13 out of 18 cases (72%), representing the highest effectiveness among all augmentation methods. Additionally, in 7 of these 18 tests, `Grasynda` delivered a statistically significant improvement over `Original`. By contrast, most other augmentation approaches rarely showed significant improvements over the baseline, with the exception of `MBB`, which also achieved significant gains 7 times.

An important remark is the comparison between `Grasynda` and `TSMixup` [1]. Our model obtained a better average MASE and rank for all three architectures, suggesting that it outperforms `TSMixup`. This is particularly interesting because `TSMixup` is used to create data for training a state of the art time series foundation model, Chronos [2].

6 Discussion

The experimental results highlight the effectiveness of `Grasynda` as a data augmentation method for a time series forecasting downstream task. Across different benchmark datasets and forecasting architectures, `Grasynda` improved forecasting accuracy over the no-augmentation baseline (`Original`) in 72% of cases, and achieved the best or second-best average ranks among all compared methods. Its gains consistently surpassed those of state-of-the-art alternatives, including `TSMixup`, which has recently been leveraged to train state-of-the-art time series foundation models such as Chronos [1,2]. These findings underscore the potential of modeling time series structure using graph representations for generating realistic and useful synthetic data.

Despite these promising results, some limitations and opportunities for future research remain. First, the current version of `Grasynda` is specifically tailored to univariate time series and does not explicitly address multivariate data or long-range dependencies. Extending the graph construction and generation process to handle these cases is an open opportunity. Second, our approach relies on a straightforward quantile-based discretization scheme; more sophisticated discretization strategies (e.g. visibility graphs [22]) may further enhance its flexibility, or generalizability to diverse domains. Another consideration is that the quantile-based discretization scheme used in `Grasynda` does not directly model non-stationarity sources of variation, such as trend or seasonality. As a result, adequate preprocessing, such as STL decomposition, may be necessary to ensure the generated synthetic series are realistic. In effect, another promising direction is to integrate these aspects more directly into the graph-based generation framework.

Overall, the evidence shows the potential of employing graph-based approaches for synthetic time series generation, pointing toward interesting future research directions at the intersection of network science and time series modeling.

7 Conclusions

This paper introduced Grasynda, a novel synthetic time series generation method. By converting time series into graph representations and using transition probability matrices, Grasynda generates synthetic data.

Future work will pursue several promising directions. These include methodological improvements to Grasynda. Also exploring implementations with different graph construction or discretization techniques (e.g., visibility graphs) [22] and utilizing the generation process methodology to stress-test forecasting networks by introducing intentional variability.

Acknowledgements. This work was partily funded by the Agenda "Center for Responsible AI", nr. C645008882-00000055, investment project nr. 62, financed by the Recovery and Resilience Plan (PRR) and by European Union - NextGeneration EU. Funded by the European Union – NextGenerationEU. Views and opinions expressed are however those of the author(s) only and do not necessarily reflect those of the European Union or the European Commission. Neither the European Union nor the European Commission can be held responsible for them. Also funded by the European Union under the Horizon Europe Framework Programme Grant Agreement N°: 101095387. Views and opinions expressed are however those of the author(s) only and do not necessarily reflect those of the European Union or European Commission. Neither the European Union nor the European Health and Digital Executive Agency can be held responsible for them. This work was also funded by: UID/00027 of the LIACC - Artificial Intelligence and Computer Science Laboratory - funded by Fundação para a Ciência e a Tecnologia, I.P./ MCTES through the national funds.

This work is also funded by National Funds through the Portuguese funding agency, FCT - Fundação para a Ciência e a Tecnologia, within project LA/P/0063/2020. DOI 10.54499/LA/P/0063/2020 | https://doi.org/10.54499/LA/P/0063/2020.

Competing Interests. the authors are unaware of any competing interests.

References

1. Ansari, A.F., et al.: Chronos-2: from univariate to universal forecasting. arXiv preprint arXiv:2510.15821 (2025)
2. Ansari, A.F., et al.: Chronos: learning the language of time series (2024)
3. Athanasopoulos, G., Hyndman, R.J., Song, H., Wu, D.C.: The tourism forecasting competition. Int. J. Forecast. **27**(3), 822–844 (2011)
4. Bandara, K., Hewamalage, H., Liu, Y.H., Kang, Y., Bergmeir, C.: Improving the accuracy of global forecasting models using time series data augmentation (2020)
5. Bergmeir, C., Hyndman, R.J., Benítez, J.M.: Bagging exponential smoothing methods using stl decomposition and box-cox transformation. Int. J. Forecast. **32**(2), 303–312 (2016)
6. Campanharo, A.S.L.O., Sirer, M.I., Malmgren, R.D., Ramos, F.M., Amaral, L.A.N.: Duality between time series and networks. PLOS ONE **6** (08 2011)
7. Cerqueira, V., Roque, L., Soares, C.: Modelradar: aspect-based forecast evaluation. Mach. Learn. **114**(10), 229 (2025)

8. Cerqueira, V., Torgo, L., Soares, C.: A case study comparing machine learning with statistical methods for time series forecasting: size matters. J. Intell. Inf. Syst. **59**(2), 415–433 (2022)
9. Challu, C., Olivares, K.G., Oreshkin, B.N., Garza Ramirez, F., Mergenthaler Canseco, M., Dubrawski, A.: Nhits: neural hierarchical interpolation for time series forecasting. In: Proceedings of the AAAI Conference on Artificial Intelligence, vol. 37, no. 6, pp. 6989–6997 (2023)
10. Forestier, G., Petitjean, F., Dau, H.A., Webb, G.I., Keogh, E.: Generating synthetic time series to augment sparse datasets. In: 2017 IEEE International Conference on Data Mining (ICDM), pp. 865–870. IEEE (2017)
11. Hussain, Z., Gimenez, F., Yi, D., Rubin, D.: Differential data augmentation techniques for medical imaging classification tasks. In: AMIA Annual Symposium Proceedings. vol. 2017, p. 979. American Medical Informatics Association (2017)
12. Hyndman, R., Athanasopoulos, G.: Forecasting: principles and practice (2014)
13. Iwana, B.K., Uchida, S.: An empirical survey of data augmentation for time series classification with neural networks. PLOS ONE **16**(7), 1–32 (2021)
14. Khosla, C., Saini, B.S.: Enhancing performance of deep learning models with different data augmentation techniques: a survey. In: 2020 International Conference on Intelligent Engineering and Management (ICIEM), pp. 79–85 (2020)
15. Lim, B., Arık, S.Ö., Loeff, N., Pfister, T.: Temporal fusion transformers for interpretable multi-horizon time series forecasting. Int. J. Forecast., 1748–1764 (2021)
16. Liu, Z., et al.: Kan: Kolmogorov-arnold networks (2024)
17. Makridakis, S., et al.: The accuracy of extrapolation (time series) methods: results of a forecasting competition. J. Forecast. **1**(2) (1982)
18. Makridakis, S., Hibon, M.: The m3-competition: results, conclusions and implications. Int. J. Forecast. **16**(4), 451–476 (2000)
19. Rashid, K.M., Louis, J.: Time-warping: a time series data augmentation of IMU data for construction equipment activity identification. In: International Symposium on Automation and Robotics in Construction (ISARC) (2019)
20. Roque, L., Soares, C., Torgo, L.: Rhiots: a framework for evaluating hierarchical time series forecasting algorithms. In: Proceedings of the 30th ACM SIGKDD Conference on Knowledge Discovery and Data Mining, pp. 2491–2499 (2024)
21. Semenoglou, A.A., Spiliotis, E., Assimakopoulos, V.: Data augmentation for univariate time series forecasting with neural networks. Patt. Recogn. (2023)
22. Silva, V.F., Silva, M.E., Ribeiro, P., Silva, F.: Time series analysis via network science: concepts and algorithms. WIREs Data Mining Knowl. Discov. (2021)
23. Victor, A.O., Ali, M.I.: Enhancing time series data predictions: a survey of augmentation techniques and model performances. In: Proceedings of the 2024 Australasian Computer Science Week, pp. 1–13 (2024)
24. Wen, Q., Sun, L., Yang, F., Song, X., Gao, J., Wang, X., Xu, H.: Time series data augmentation for deep learning: a survey (2020)
25. Zeng, A., Chen, M., Zhang, L., Xu, Q.: Are transformers effective for time series forecasting?. In: Proceedings of the AAAI Conference on Artificial Intelligence, vol. 37, no. 9, pp. 11121–11128 (2023)

Section: Applications of AI/ML

Predicting and Interpolating Spatiotemporal Environmental Data: A Case Study of Groundwater Storage in Bangladesh

Anna Pazola[1,2](✉), Mohammad Shamsudduha[3], Richard G. Taylor[2], and Allan Tucker[1]

[1] Department of Computer Science, Brunel University of London, London, UK
anna.pazola@brunel.ac.uk
[2] Department of Geography, UCL, London, UK
[3] Department of Risk and Disaster Reduction, UCL, London, UK

Abstract. Geospatial observational datasets are often limited to point measurements, making temporal prediction and spatial interpolation essential for constructing continuous fields. This study evaluates two deep learning strategies for addressing this challenge: (1) a grid-to-grid approach, where gridded inputs are used to predict rasterised targets (aggregation before modelling), and (2) a grid-to-point approach, where gridded inputs model point targets, followed by kriging interpolation to fill the domain (aggregation after modelling). Using groundwater storage data from Bangladesh as a case study, we compare the efficacy of these approaches. Our findings indicate that spatial interpolation is substantially more difficult than temporal prediction. In particular, nearest neighbours are not always the most similar, and uncertainties in geology strongly influence point temporal behaviour. These insights motivate future work on advanced interpolation methods informed by clustering locations based on time series dynamics. Demonstrated on groundwater storage, the conclusions are applicable to other environmental variables governed by indirectly observable factors. Code is available at https://github.com/pazolka/interpolation-prediction-gwsa.

Keywords: Spatiotemporal data · U-net · Kriging · Groundwater storage · Bangladesh

1 Motivation

Spatial interpolation and temporal prediction are two modelling objectives that go hand-in-hand in environmental modelling and monitoring [7]. Because continuous data collection is costly and often infeasible, many datasets are only available at point locations, hindering various downstream tasks that rely on them, including planning, risk assessment and decision making. For example,

M. Baratchi et al. (Eds.): IDA 2026, LNCS 16513, pp. 413–426, 2026.
https://doi.org/10.1007/978-3-032-23833-7_30

estimating $PM_{2.5}$ concentrations is essential for protecting vulnerable populations, yet monitoring stations are sparse and purely physics-based forecasting can be constrained by uncertainties in emission inventories and meteorological inputs [15]. Similarly, quantifying groundwater storage change is critical for sustainable freshwater management - particularly in regions that depend heavily on groundwater for agriculture - to reduce the risk of aquifer over-exploitation [25]. However, widespread monitoring of groundwater levels is expensive, and numerical forecasting is often challenged by complex and site-specific interactions between anthropogenic pressures and natural variability.

This study explores how temporal prediction and spatial interpolation can be combined within data-driven models to pursue the "holy grail" of spatiotemporal modelling: estimating values at unobserved locations and future times. What are the existing approaches, especially deep learning-based solutions, for addressing this problem? Using an extensive spatiotemporal dataset on groundwater storage changes in Bangladesh, we identify key challenges and report performance of selected, well-established models, offering insights relevant for other environmental variables with complex and poorly understood governing physical processes. The key contribution of this paper is a comparison of selected models in addressing a coupled task of spatial interpolation and temporal prediction. Additionally, this is the first application of deep learning, specifically U-net and CNN-GRU, to groundwater storage dataset in Bangladesh.

As machine learning (ML) research and AI industry develops, primarily driven by large language models, the environmental cost of AI is widely recognised and discussed, e.g. [5]. In this work, we focus on lightweight models that contribute to the discourse on simpler, and more interpretable deep learning approaches, closely integrated with domain expertise, as an alternative to general-purpose, transformer-based models that are computationally expensive and environmentally costly [23].

2 Background

Estimation of continuous fields from observational datasets involves spatial interpolation and temporal gap filling. In geostatistics, Kriging is the best linear unbiased predictor (BLUP) and is one of the most commonly used interpolation techniques [6]. It takes into account not only distance but also correlation between neighbouring points. However, with its strong assumptions on Gaussianity and stationarity, and very high computational cost due to a creation of large distance matrix ($O(n^3)$ for time), it can be challenging to scale to large n without approximations [6]. Common approaches for temporal gap filling vary depending on the characteristics of time series, with autoregressive integrated moving average (ARIMA) and seasonal-trend decomposition-based methods, as well as ML models (e.g. random forest [4]) being popular choices. Merging these tasks might leverage the quality of estimations, by taking into consideration spatial and temporal dependencies simultaneously [6].

A variety of models are well-suited for both spatiotemporal interpolation and short-term extrapolation tasks. These can be broadly categorised into groups:

- *Statistical methods*: This includes space-time (3D) kriging [17], and its ML counterpart, spatiotemporal DeepKriging [19].
- *Deterministic generative models*: Examples include 2D/3D U-Net architectures [14] and stacks of convolutional neural networks (CNN) and recurrent layers (such as long short-term memory - LSTM, or gated recurrent unit - GRU). These models typically require prior rasterisation of the target data, or a subsequent interpolation step, for generating gridded outputs.
- *Probabilistic models*: These encompass spatial Gaussian processes (GP) - mathematically equivalent to kriging and similarly computationally intensive [28] - as well as neural processes [10].

These model families have been applied widely across environmental sciences. Most notably, U-net is highly flexible and has been successfully adapted to a wide range of tasks. Since its introduction by [22] for image segmentation, numerous U-net variants have been developed and applied in environmental sciences. These include regression for improved sea ice forecasting [20], and inpainting-based reconstruction of precipitation maps [18]. Spatiotemporal U-Net formulations have been used to produce space-time consistent gap-filled precipitation maps [14], and to generate $PM_{2.5}$ maps with high spatial and temporal resolution from multi-modal observations obtained at different times [21].

Despite these successes, U-net architecture performs image-to-image translation, which requires a gridded representation of the target variable. When observations are discrete and vector-based, rasterising them onto a regular grid (e.g., via binning) requires choosing an aggregation function and resolution, both of which can introduce information loss and artefacts. As an alternative, a simpler model consisting of convolutional and dense layers, and optionally augmented with temporally aware layers, can be used to model spatiotemporal dependencies between gridded inputs and point target, avoiding the need to rasterise the target. CNN-LSTM hybrids have been applied, for example, by [2] for predicting future streamflow at gauging stations from a 'weather video', and by [8] to forecast algal blooms in the coastal environments from the satellite data. This design also enables a clearer separation between the prediction step (learning temporal dynamics) and the interpolation step (mapping predictions to unobserved locations), allowing the interpolation stage to be tailored independently. It contrasts with spatiotemporal DeepKriging [19] that interpolates first, predicts later.

In this work, we study how different couplings of prediction and interpolation within selected model classes - U-Net variants, CNN-GRU models, and (Deep)Kriging - affect the estimation of a spatiotemporal target variable at unobserved locations. Although neural processes have become prominent in recent research on data-driven weather prediction [1], we exclude this model family from the scope of this analysis due to their comparatively high data requirements.

3 Methods

The method selection process prioritised computational efficiency and ability to handle gridded predictors and point targets with minimal pre- and postprocessing, aiming to generate spatially and temporally complete datasets for a specified area and time resolution.

Two strategies were used for handling the format of observed targets: rasterising point observations before modelling (grid-to-grid/image-to-image approach), or after modelling via spatial interpolation (grid-to-point/image-to-vector model with a subsequent interpolation step). In each case, we explored 2D (purely spatial) and 3D (spatio-temporal) versions of the models.

3.1 Grid-to-Grid Approach: U-Net

Point observations can be rasterised by aggregating all points that fall within the same grid cell. The resulting target grid contains spatial gaps, and, at the first glance, resembles an image inpainting problem, in which spatially complete gridded predictors are used to reconstruct missing pixels. However, as the target is unavailable at inference and gaps are fixed and identical across all samples, the problem is best viewed as supervised prediction on a fixed spatial domain.

We employ a variant of U-net, an image-to-image translation model [22], which has been successfully applied to similar problems, e.g. [18]. A purely spatial version of the U-net model uses 2D convolutions, pooling and upsampling, ahereas a spatiotemporal version replaces these with 3D counterparts (i.e. 3D convolutions, max-pooling and upsampling). In this way, the network can process temporal sequences, capturing time-dependent relationships (time lags) that are characteristic of environmental variables.

The number of encoder-decoder blocks is flexible and mostly independent from the input shape. However, the network depth should reflect the complexity and size of the inputs, i.e. shallower U-nets might be more appropriate for small inputs, reducing the number of parameters and avoiding overfitting.

Handling missing target pixels in U-net-style inpainting commonly involves using an explicit mask as one of the input channels [11], using partial or gated convolutions [18], and/or applying masked losses during training [21]. In our setting, because the gaps correspond to fixed, permanently missing regions, we opt for a masked loss that differs from the classic image inpainting objective. Instead of computing the loss only on missing pixels, our masked mean squared error (MSE) loss computes errors only over valid pixels. In the 2D U-net, masked loss averages per-pixel MSE over valid regions, ignoring masked areas. In the 3D U-net, masked loss is augmented with penalties for low-frequency (trend) and spatial-mean differences between predictions and targets:

$$Loss = w_{main} L_{MSE} + w_{trend} L_{lowpass} + w_{mean} L_{mean} \quad (1)$$

where L_{MSE}: standard masked per-pixel MSE, $L_{lowpass}$: masked MSE of low-frequency (trend) components obtained via normalized average pooling, L_{mean}: masked MSE between spatial means of true and predicted fields. This encourages the U-net to match both local details and large-scale trends, improving stability and physical consistency in the 3D (space-time) setting.

3.2 Grid-to-Point Approach: CNN/CNN-GRU Stack

Rather than rasterising point observations before model training, we explore the effects of decoupling the interpolation and prediction tasks, and building a predictive grid-to-point model, followed by a separate interpolation step. As demonstrated by [2], hybrid CNN-LSTM architectures are well suited to grid-to-point prediction. We therefore base our 3D grid-to-point model on their architecture, introducing several adaptations required by the case study, described in Sect. 4. The exact implementation is summarised in Fig. 2. The most important change is replacing the LSTM unit with a GRU that is more appropriate for a shorter input sequences. GRU has ~25% fewer parameters than LSTM for the same units and input dimensions, reducing overfitting risk in a smaller dataset. We also increased the feature dimension of the last CNN pair to prevent the model from shrinking the spatial information too aggressively. The 2D (purely spatial) variant mirrors the convolutional and max-pooling layout of the 3D model but excludes the temporal GRU component. The output of both grid-to-point models corresponds to the number of all point observations, therefore it predicts the target at all observed locations simultaneously.

Kriging. We convert point-based predictions into gridded fields using geostatistical interpolation implemented in Python with *skgstat*. For each time step t, we compute an empirical variogram from the station coordinates (expressed in km) and the predicted target values, and fit a spherical variogram model. The fitted variogram is then used to parameterise ordinary kriging, which is applied to interpolate predictions onto a regular spatial grid. During kriging, we limit the local neighborhood used for each grid cell to between 2 and 30 stations to balance robustness and locality. The resulting interpolated surface is evaluated at all grid points and reshaped to produce a 2D field for each timestep.

3.3 Spatiotemporal DeepKriging

We compare all above models to spatiotemporal DeepKriging, that conceptually echoes 3D kriging by treating time as an additional spatial dimension within a deep learning framework [19]. It separates interpolation (stage 1) from temporal prediction (stage 2), which opposes our CNN/CNN-GRU setup. Stage 1 performs interpolation at observation points using RBF-based embeddings of space and time coordinates, combined with scalar predictors. A feed-forward network trained with a quantile loss produces multiple predictive quantiles, enabling uncertainty estimation. Stage 2 then carries out grid-based forecasting with a ConvLSTM that ingests recent spatiotemporal sequences and full-grid predictors to predict future quantile fields, using a masked quantile loss to ignore invalid regions. Both stages rely on separate models that are trained independently.

4 Case Study: Groundwater Storage Changes in Bangladesh

Estimating changes in groundwater storage (ΔGWS) on a continuous grid is critical for informing sustainable water management from regional to local scales. Typically derived from a combination of land surface model outputs and satellite gravimetry, ΔGWS exhibit considerable uncertainty [3,12]. Many areas across the world, such as Bangladesh, rely heavily on groundwater for drinking water supplies and national food security through dry-season irrigation [25]. However, as documented in Bangladesh, excessive groundwater abstraction has led to localised depletion, most importantly in Dhaka and western regions [13]. Taken together with the existing water-quality challenges (salinity, arsenic), these pressures make robust groundwater risk management indispensable, especially as global warming puts considerable pressure on water resources [25].

Bangladesh has an extensive network of point observations of groundwater levels that can be used as a basis for data-driven modelling on the country scale, most recently by [9] using tree-based methods and shallow neural networks.

To create our predictive models, a set of gridded satellite-based predictors (precipitation, actual evapotranspiration, NDVI, elevation, terrestrial water storage changes) was selected based on the forcing variables used commonly in the literature, e.g. [9].

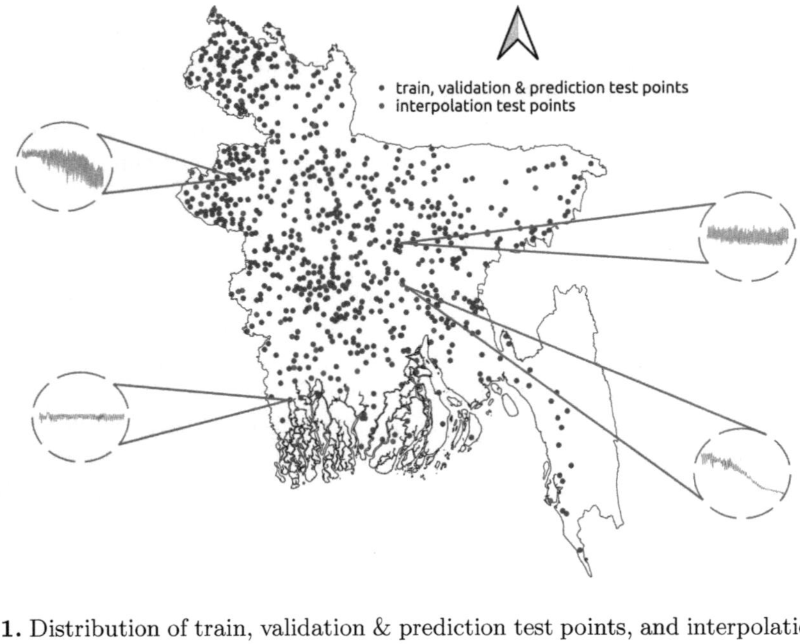

Fig. 1. Distribution of train, validation & prediction test points, and interpolation test (holdout) points. Groundwater dynamics in Bangladesh are characterised by a high spatial heterogeneity.

4.1 Input and Target Datasets

All geospatial datasets use the WGS84 geographic coordinates system. For kriging purposes, the coordinates were locally projected to WGS 84/UTM zone 46N.

Gridded Predictors. The gridded input data consists of the remotely sensed variables: MODIS' NDVI as a proxy for agricultural output and irrigation, TERRACLIMATE precipitation and actual evapotranspiration as key meteorological variables controlling the available amount of water, MERIT DEM elevation, and GRACE ΔTWS [24]. Input grids were reprojected to the common grid (0.25° x 0.25°), and the temporal resolution of the dataset was set to 1 month. The study period is dictated by the availability of GRACE data (04/2002–05/2017).

Groundwater Level Observations. The target dataset consists of point observations of groundwater levels (GWL) at 1023 locations in Bangladesh (Fig. 1). Original weekly monitored GWL time-series dataset can be obtained from the Bangladesh Water Development Board[1] through an online data request and payment. Processed monthly time-series dataset used in this work was compiled by [26].

GLDAS 2.2 Model Output. NASA's Global Land Data Assimilation System (GLDAS) data products provide a wide range of simulated hydrological variables, commonly used for estimating ΔGWS. In particular, GLDAS 2.2 Catchment Land Surface Model CLSM-F2.5 [16] simulates shallow groundwater storage explicitly. In this work, average monthly outputs at a native daily scale at 0.25° were evaluated against available in-situ observations, alongside the outputs of the data-driven models.

4.2 Preprocessing

The raw target dataset required careful curation, as it contained anomalous values linked to human error and unreported measuring well changes, i.e. replacement of the existing observational well with a different well in a close proximity that might have a different depth and tap into a different aquifer. Such well changes were evident in some GWL time-series as sudden changes in the magnitude of seasonality, range of reported values and trends, and identified subseries were treated independently.

As most of the employed methods were unable to handle temporal gaps, with the exception of DeepKriging, temporal gap filling based on random forest regression was performed on GWL time series using *R* package *missForest* [4, 27]. Subsequently, GWL values were converted to groundwater storage using an aquifer storage coefficient dataset compiled by [26], and normalised w.r.t. 2004–2009 for direct comparison with GRACE and GLDAS anomaly values. Thus, the

[1] http://www.hydrology.bwdb.gov.bd/index.php.

target values are denoted as ΔGWS and are expressed in meters. For grid-to-grid models, rasterisation was performed through averaging of point values to a 0.25° x 0.25° grid.

Table 1. Performance on ΔGWS dataset from Bangladesh, R^2 (mean and std) over 10-fold temporal CV. Train-val-test refer to the temporal splits. The interpolation holdout locations are independent from the train-val-test locations (Fig. 1) and span the whole study period. The division of interpolation set into train-val-test results aims to isolate the "interpolation + prediction" performance, where values are predicted in the future at unobserved locations. Raw grid-to-point (g2p) is also reported to demonstrate degradation of the predictive performance through a subsequent Kriging step in g2p models. All models were tested with 2D (spatial only) and 3D (spatiotemporal) inputs.

R^2	**prediction**						**interpolation**					
	train		**val**		**test**		**train**		**val**		**test**	
	mean	std	mean	std	mean	std	mean	std	mean	std	mean	std
DeepKrig	0.66	0.02	0.57	0.01	0.02	0.28	0.61	0.03	0.46	0.16	−0.02	0.31
GLDAS 2.2	0.08	–	−0.05	–	−0.11	–	0.08	–	−0.8	–	−0.44	–
2D Unet	0.60	0.04	0.44	0.12	0.48	0.08	0.58	0.04	0.38	0.17	0.45	0.12
3D Unet	0.58	0.07	0.45	0.10	0.45	0.06	0.57	0.07	0.41	0.14	0.44	0.10
2D g2p + Krig	0.64	0.02	0.49	0.13	0.50	0.08	0.60	0.03	0.38	0.17	0.45	0.11
3D g2p + Krig	0.64	0.03	0.48	0.15	0.50	0.08	0.61	0.03	0.39	0.18	0.45	0.13
2D g2p	0.85	0.01	0.67	0.14	0.67	0.06	–	–	–	–	–	–
3D g2p	0.85	0.04	0.65	0.15	0.64	0.07	–	–	–	–	–	–

4.3 Experiment Setup

To test both prediction and interpolation, we held out 8% of locations, selected at random - this was deemed suitable given the dense station network. Next, 10-fold temporal cross validation (CV) was performed on the remaining data, splitting the time series into train, validation and test sets, with a progressively growing training set, and validation and test sets of fixed length of 8 time steps. Unlike standard CV, successive training sets are supersets of those that come before them. Target and predictors were min-max scaled before modelling. Performance of the models were reported using the coefficient of determination R^2 and mean squared error MSE.

5 Results and Discussion

Detailed model implementation is summarised in Fig. 2. Table 1 shows the performance of employed data-driven models, and the alignment of the GLDAS 2.2

CLSM GWS simulations with the interpolation and prediction sets. Note that GLDAS 2.2 CLSM does not follow the train-validation-test CV, and the division is only retained for comparison with data-driven models. All data-driven models clearly outperform the physical GLDAS 2.2 CLSM model in simulating changes in groundwater storage in Bangladesh, as land surface models like GLDAS are known to overly generalise groundwater model components, leading to diverging predictions and considerable bias [12]. Nevertheless, data-driven models excel at learning spatiotemporal relationships whereas physics-driven models can generate physically consistent values and extremes beyond those seen in training data.

All models are light-weight, ranging from 118,561 (2D U-net) to 213,483 (3D CNN-GRU stack) parameters. However, the subsequent kriging step in the latter approach adds substantial computational complexity ($O(n^2)$ space, $O(n^3)$ time). The size of these models is comparable to the combined size of 2-stage DeepKriging ($\sim$ 200,000 parameters).

Poor performance of the DeepKriging model on the test sets might be explained by its simple shallow neural network architecture and its limited use of gridded predictors. The drop in performance on the test set indicates that the framework overfits training and validation data. Although other employed models rely on convolutional layers to extract spatial patterns from the gridded inputs, DeepKriging only takes into consideration scalar values. Groundwater levels often reflect changes in distant predictors, as recharge zones may lie far from where observations are made.

Upon a closer inspection of R^2 scores (Fig. 3), there is a clear distinction in performance of the individual tasks. The box plots of g2p models show a pure prediction spread, as these models don't involve interpolation. Adding kriging on top of them, and the alternative g2g approach, degrades these g2p estimates, which is marked by an increased spread in prediction in other models. At the same time, interpolation spread is consistently higher than the prediction spread of g2p models. This indicates that interpolation might be a more difficult problem, especially on the dataset that is as spatially heterogeneous as groundwater dynamics in Bangladesh. Even though the employed g2g and g2p models show minimal performance differences on the testing sets, we conclude that g2p CNN is the most suitable model for prediction, and that interpolation requires further research. This finding recognises that, as g2p CNN is one of the more sizeable models here, there is a trade-off between good performance and model size. MSE scores display a similar pattern, and can be examined in the code repository.

Visual inspection of the final CV fold (with a train-val-test ratio of 90%-5%-5%) reveals mixed performance for the interpolation holdout dataset (points excluded prior to the 10-fold CV). As illustrated in Fig. 4a, well *RJ092* situated near others with comparable seasonality and trend patterns shows modelled values that align closely with observed data. For well *RA-22* (Fig. 4b), the estimated values are primarily driven by its only nearby neighbour with very stable, regular seasonality. In contrast, well *DI030_1* (Fig. 4c) is surrounded by locations with stronger seasonal signals, experiencing consistent under- and overestimation of

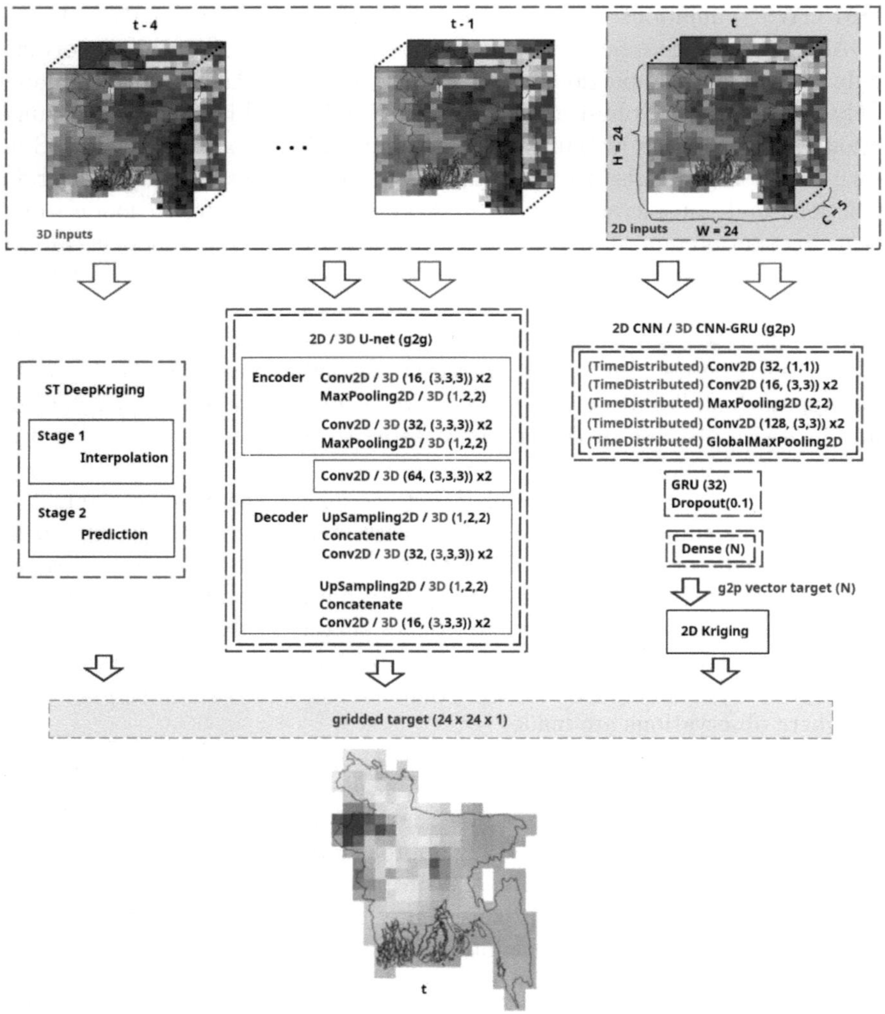

Fig. 2. Employed spatial (2D) and spatiotemporal (3D) models, as described in Sect. 3. 2D models are fed with gridded predictors ($C = 5$) from the same timestep; 3D models additionally use lagged predictors $t...t-4$. The optimal temporal lag was found empirically during preliminary analysis. Results of all approaches are reported in Fig. 3. Spatiotemporal DeepKriging was implemented according to [19]. Note that Stage 2 of DeepKriging only uses past predictors, as opposed to all other models.

ΔGWS values. Poor interpolation performance on wells *RA-22* and *DI030_1* might be caused by the fact that groundwater dynamics do not always follow the first geographical principle "everything is related to everything else, but near things are more related than distant things". In complex hydrogeological settings like Bangladesh, nearby points can tap into different hydrogeological conditions.

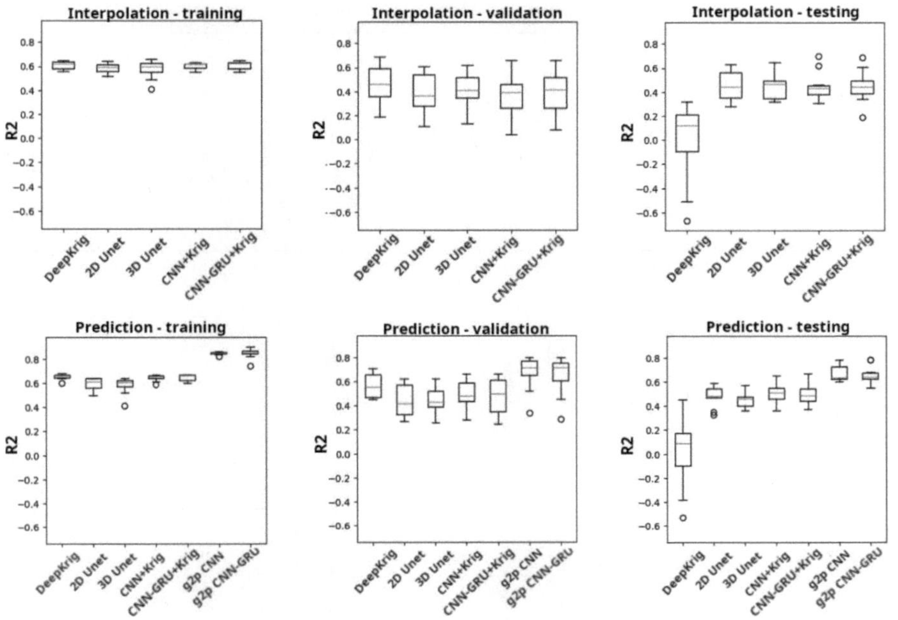

Fig. 3. Performance on ΔGWS dataset from Bangladesh on data-driven models, R^2 over 10-fold temporal CV. Refer to Table 1 for mean performance scores.

The heterogeneity of time-series dynamics, presented in Fig. 1, may occur in the neighbouring wells.

Figure 3 demonstrates that g2p models, mapping gridded predictors to point target, perform substantially better; interpolation of these points through kriging, as well as the alternative g2g modelling approach, degrade model fit at the observed points. Exploration of well groups reveals that the division into the regular grid cells often leads to the situation that wells that show similar patterns end up in separate grid cells, and the connection between them is lost. At the same time, wells that show very different patterns, are aggregated into the same grid cell, collectively producing a mean time series for that grid cell that is unrealistic. Therefore, image-like representation of the target variable, as required by the g2g models, may not be appropriate. The alternative g2p method that employs kriging on top of point predictions does not solve the problem described above. As kriging operates at the same grid of 0.25°, the raw g2p (point) performance is degraded to a similar extent, regardless of whether the target averaging takes place before DL modelling (g2g method) or after (g2p with interpolation).

This finding motivates a search for an alternative interpolation technique beyond a regular grid. As subsurface geology is often unknown, with surface geology being an (imperfect) proxy, interpolation informed by clustering of locations based on their time-series dynamics, is an interesting task to explore.

Based on the current dataset and study area, there appears to be little difference in performance of the spatial (2D) vs spatiotemporal (3D) versions of the

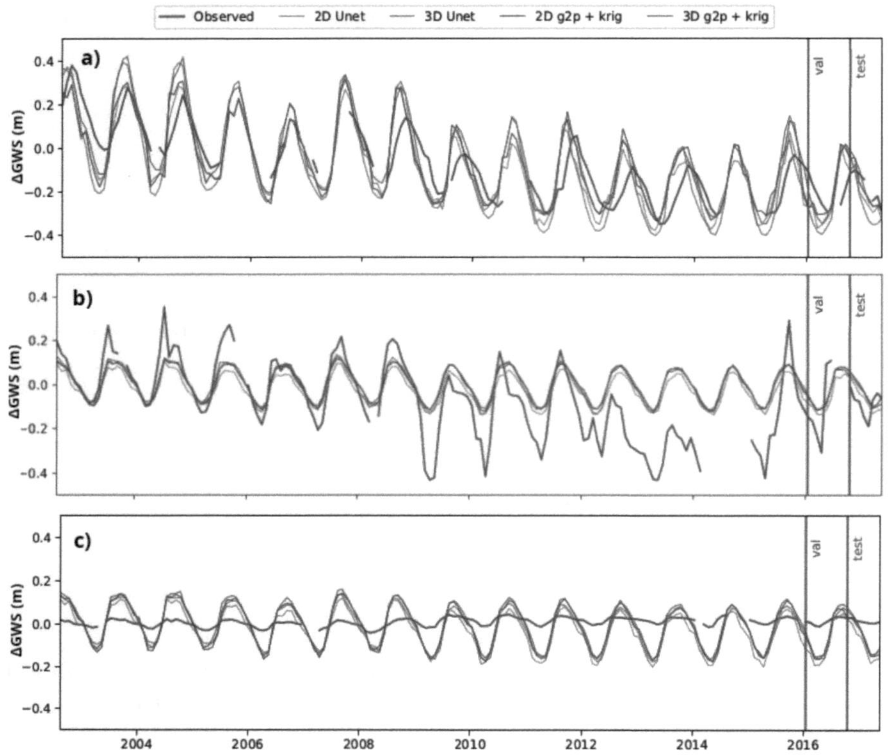

Fig. 4. Performance of the employed models at three example locations: (a) RJ092 (b) RA-22 (c) DI030 from the interpolation holdout set; last CV fold (95%-5%-5%).

g2p and g2g models. This outcome aligns with expectations: in humid environments like Bangladesh, groundwater responds rapidly to predictors, primarily precipitation. Future research will test whether this pattern persists with independent environmental variables.

Notably, we were unable to perform full 3D kriging on our dataset because of its prohibitive computational cost, even when employing localisation strategies. Since Gaussian processes can be viewed as a probabilistic generalisation of kriging and inherit a similarly unfavourable scaling with the number of observations, they are likewise impractical at this scale. Given this strong requirement for computational efficiency, a natural extension of this work is to explore neural processes, which operate in continuous domains while offering substantially improved scalability compared to Gaussian-process-based approaches.

6 Conclusions

This analysis explored the interplay of spatial interpolation and temporal prediction on an example dataset of groundwater storage changes in Bangladesh.

The interpolation task was shown to be more challenging than prediction, due to the complexity of subsurface processes and lack of spatial datasets on meaningful hydrogeological variables. Employed interpolation methods - both the target point rasterisation with subsequent grid filling using 2D/3D U-net, and point prediction with subsequent interpolation using Kriging, considerably degraded grid-to-point model fit at observed locations. Nevertheless, data-driven models substantially outperform the estimates of groundwater storage changes by the commonly employed process-based GLDAS 2.2 CLSM model in Bangladesh. Future work is motivated by the search for an advanced interpolation technique informed by clustering of locations based on their time-series dynamics, that is generalisable to other environmental variables with complex spatial patterns.

Acknowledgments. We thank the Bangladesh Water Development Board for providing groundwater-level dataset. This work was supported by a PhD studentship to AP from the London NERC Doctoral Training Programme (grant no. NE/S007229/1). RGT acknowledges support of a fellowship from the CIFAR Earth-4D program.

Disclosure of Interests. AT is a member of the International Symposium on Intelligent Data Analysis Council.

References

1. Allen, A., et al.: End-to-end data-driven weather prediction. Nature **641**(8065), 1172–1179 (2025). https://doi.org/10.1038/s41586-025-08897-0
2. Anderson, S., Radić, V.: Evaluation and interpretation of convolutional long short-term memory networks for regional hydrological modelling. Hydrol. Earth Syst. Sci. **26**(3), 795–825 (2022). https://doi.org/10.5194/hess-26-795-2022
3. Arifin, A., et al.: Plausibility criteria for GRACE-derived groundwater storage changes from aquifers globally. Geophys. Res. Lett. **52**(22), e2025GL118580 (2025). https://doi.org/10.1029/2025GL118580
4. Arriagada, P., Karelovic, B., Link, O.: Automatic gap-filling of daily streamflow time series in data-scarce regions using a machine learning algorithm. J. Hydrol. **598**, 126454 (2021). https://doi.org/10.1016/j.jhydrol.2021.126454
5. Bolón-Canedo, V., et al.: A review of green artificial intelligence: towards a more sustainable future. Neurocomputing **599**, 128096 (2024). https://doi.org/10.1016/j.neucom.2024.128096
6. Cressie, N., Johannesson, G.: Spatial prediction for massive datasets (2006)
7. Cressie, N., Wikle, C.K.: Statistics for spatio-temporal data (2011)
8. Ding, W., Li, C.: Algal blooms forecasting with hybrid deep learning models from satellite data in the Zhoushan fishery. Eco. Inf. **82**, 102664 (2024). https://doi.org/10.1016/j.ecoinf.2024.102664
9. Fahim, A.K.F., Kamal, A.S.M.M., Shahid, S.: Modeling spatial groundwater level patterns of Bangladesh using physio-climatic variables and machine learning algorithms. Groundw. Sustain. Dev. **25**, 101142 (2024). https://doi.org/10.1016/j.gsd.2024.101142
10. Garnelo, M., et al.: Neural processes (2018). https://doi.org/10.48550/arXiv.1807.01622

11. Geiss, A., Hardin, J.C.: Inpainting radar missing data regions with deep learning. Atmos. Meas. Techn. **14**(12), 7729–7747 (2021). https://doi.org/10.5194/amt-14-7729-2021
12. Gleeson, T., et al.: GMD perspective: the quest to improve the evaluation of groundwater representation in continental- to global-scale models. Geosci. Model Dev. **14**(12), 7545–7571 (2021). https://doi.org/10.5194/gmd-14-7545-2021
13. Janardhanan, S., et al.: Groundwater balance and long-term storage trends in the regional Indo-Gangetic aquifer in northwest Bangladesh. J. Hydrol. Reg. Stud. **49**, 101500 (2023). https://doi.org/10.1016/j.ejrh.2023.101500
14. Kishikawa, D., Muto, Y., Kotsuki, S.: Conditional Diffusion Models for Global Precipitation Map Inpainting (2025). https://doi.org/10.48550/arXiv.2507.20478
15. Koo, J.S., et al.: Development of PM2.5 forecast model combining ConvLSTM and DNN in Seoul. Atmosphere **15**(11) (2024). https://doi.org/10.3390/atmos15111276
16. Li, B., et al.: Global GRACE data assimilation for groundwater and drought monitoring: advances and challenges. Water Res. Res. **55**(9), 7564–7586 (2019). https://doi.org/10.1029/2018WR024618
17. Li, S., Griffith, D.A., Shu, H.: Temperature prediction based on a space-time regression-kriging model. J. Appl. Stat. **47**(7), 1168–1190 (2020). https://doi.org/10.1080/02664763.2019.1671962
18. Meuer, J., et al.: Infilling of missing rainfall radar data with a memory-assisted deep learning approach. Hydrol. Earth Syst. Sci. **29**(15), 3687–3701 (2025). https://doi.org/10.5194/hess-29-3687-2025
19. Nag, P., Sun, Y., Reich, B.J.: Spatio-temporal DeepKriging for interpolation and probabilistic forecasting. Spatial Stat. **57**, 100773 (2023). https://doi.org/10.1016/j.spasta.2023.100773
20. Palerme, C., et al.: Improving short-term sea ice concentration forecasts using deep learning. Cryosphere **18**(4), 2161–2176 (2024). https://doi.org/10.5194/tc-18-2161-2024
21. Porcheddu, A., et al.: Machine learning data fusion for high spatio-temporal resolution $PM_{2.5}$. Atmos. Meas. Techn. **18**(18), 4771–4789 (2025). https://doi.org/10.5194/amt-18-4771-2025
22. Ronneberger, O., Fischer, P., Brox, T.: U-net: convolutional networks for biomedical image segmentation (2015). https://doi.org/10.48550/arXiv.1505.04597
23. Rudin, C., et al.: Amazing things come from having many good models (2024). https://doi.org/10.48550/arXiv.2407.04846
24. Save, H.: CSR GRACE and GRACE-FO RL06 Mascon Solutions v02 (2020). https://doi.org/10.15781/cgq9-nh24
25. Shamsudduha, M.: Impacts of human development and climate change on groundwater resources in Bangladesh. In: Mukherjee, A. (ed.) Groundwater of South Asia, Singapore, pp. 523–544 (2018). https://doi.org/10.1007/978-981-10-3889-1_31
26. Shamsudduha, M., et al.: The Bengal water machine: quantified freshwater capture in Bangladesh. Science **377**(6612), 1315–1319 (2022). https://doi.org/10.1126/science.abm4730
27. Stekhoven, D.J., Bühlmann, P.: Missforest: nonparametric missing value imputation for mixed-type data. Bioinformatics **28**(1), 112–118 (2012). https://doi.org/10.1093/bioinformatics/btr597
28. Zhang, J., et al.: An efficient implementation for spatial-temporal gaussian process regression and its applications (2022). https://doi.org/10.48550/arXiv.2209.12565

Combining Dynamic Bayesian Networks with Population Dynamics Modelling to Predict Breeding Success in Seabirds

Alan Anderson[1(✉)], Neda Trifonova[1], Beth Scott[1], and Allan Tucker[2]

[1] School of Biological Sciences, University of Aberdeen, Aberdeen, UK
r09aa23@abdn.ac.uk

[2] Department of Computer Science, Brunel University, Uxbridge, UK

Abstract. Modelling dynamic relationships between marine animal species and their local environment is integral to evaluating the past and current state of any ecosystem. With significant changes in both climate and biodiversity occurring in recent decades, and with fluctuations expected to increase in frequency and severity, ecosystem models employed need to show high fidelity to recorded data to allow for accurate future predictions in response to long-term environmental instability. Holistic Dynamic Bayesian Network ecosystem models can identify the key relationships between environmental and species/biological variables, but do not necessarily capture crucial life-history characteristics relating to individual species population dynamics. In this study, we attempt to incorporate Leslie Matrix modelling of population dynamics into H-DBN models. Two 'testing' methods were investigated for our models, with the aim of applying realistic bounding to the rate of population increase for two contrasting species of seabird, measured through their breeding success. The first method used all available variables to both train and test the DBN model, while the second method used a reduced number of variables as 'evidence' to test the model. In early results, significant improvements in predictive accuracy of the models for both species was observed when population dynamics data was included. Specifically, after the addition of the Leslie Matrix output to the DBN ecosystem model, we were able to capture with high accuracy the inter-annual variability of the breeding success.

1 Introduction

The cumulative effects of climate change and increasing anthropogenic activity are a major issue of concern for the field of ecological study, with profound implications for all ecosystem types [9,15]. In the marine environment, a range of environmental changes are readily apparent and recognised within the scientific literature [3,7,10]. Marine ecosystems consist of complex dynamic interactions among species and the environment, the understanding of which has significant ecological and societal implications for predicting nature's response to changes in climate and biodiversity [6]. However, such relationships are not easily

M. Baratchi et al. (Eds.): IDA 2026, LNCS 16513, pp. 427–437, 2026.
https://doi.org/10.1007/978-3-032-23833-7_31

understood due to the cost of sampling such vast spatial volumes, and the complexity of gathering continuous data at multiple trophic levels over such large spatial regions. Though significant improvements have been made in coupled physics+biological ecosystem models such as NEMO-ERSEM [4], the impacts on species at the upper trophic levels remain particularly difficult to measure. Dynamic Bayesian Networks (DBNs) have been used as an effective method for capturing such complex relationships with limited available real-world/sampled data [13,14,16]. DBNs generalise Hidden Markov Models (HMMs) to model the dataset dynamics and incorporate a hidden variable (HV) linked to each node within an annual 'slice' (inter). This hidden variable serves to model missing data/unobserved variables and can infer some underlying 'state' of the timeseries when applied through an autoregressive link, potentially capturing relationships of a higher order [12].

In seabirds, when considering extreme events such as Avian flu which can rapidly kill off a high percentage of the total population, population recovery can only occur at a certain limited rate given the number of reproducing adults and the survival rate of both adults and juveniles. Understanding of these population dynamics is well reproduced by Leslie matrix models as described extensively by Caswell [2], which link individuals of a given species to their general population through their 'vital rates', which predict rates of population growth via age class specific fertility, and mortality. The use of these matrix models is particularly effective when classifying groups of individuals by their size or developmental stage, rather than by age, and this has been the case with 'life tables' used since the mid-1940s.

In this paper, we describe the methods used to combine the output from a simple population dynamics model, a Leslie Matrix based on real-world seabird data, with DBN models. These DBN models were informed by key physical and biogeochemical (BGC) drivers with previously identified inter-annual relationships [12]. Two methods are used for testing the DBN models to determine how much data is required for accurate predictions of breeding success for two species, Northern Gannets *(Morus bassanus)* and Common Guillemots *(Uria aalge)*.

Both methods use key drivers selected by the strength of their relationships to breeding success in the training stage, but the two methods differ in that method 1 uses all available drivers for both training and testing, while method 2 uses increasing numbers of drivers/variables as 'evidence' during the testing phase. This second method starts from using only the output of Leslie Matrix as evidence to determine if additional data is required to inform the model during testing. Comparisons of output from both methods are made to evaluate the effectiveness of each, particularly related to similar models which do not include Leslie Matrix population data. The main aims of this study were firstly to evaluate the effectiveness of combining DBN models with population dynamics methods, and secondly to determine which drivers are most pertinent in producing accurate predictions from such a combined approach.

2 Methods

2.1 Leslie Matrices and Vital Rates

Northern Gannets and Common Guillemots were chosen as test species for this analysis due to their contrasting foraging styles and the high data-scores associated with their survival rates from data provided by the Joint Nature Conservation Society (JNCC) [8]. Gannets are 'plunge divers' (diving into the water from altitude), and experienced long-term population growth over the last century, prior to severe outbreaks of avian flu in the early 2020s. By contrast, guillemots dive from the surface and hunt closer to the seabed for periods of one to two minutes. Guillemots have shown gradual decline in population counts over recent decades. Leslie Matrices for both seabird species were populated based on the basic life-cycle diagram depicted in Fig. 1. Vital rates for both species were derived from the JNCC report data which was already averaged over multiple UK colonies, as shown in Table 1.

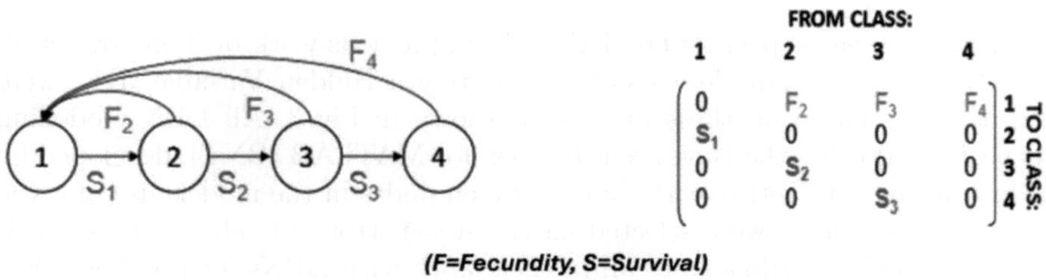

Fig. 1. Example template for life-cycle diagram and Leslie Matrix.

Table 1. Vital Rates for Gannets and Guillemots

	Survival						Fecundity	Dispersal	
	Juvenile	1–2yrs	2–3yrs	3–4yrs	4yrs-adult	Adult		Natal	Adult
Northern Gannet	0.424	0.829	0.891	0.895	0.895	0.919	0.700	0.100	-
Common Guillemot	0.560	0.792	0.917	0.917	0.917	0.939	0.672	0.580	0.070

Due to lack of real data for age-class distribution among multiple seabird colonies, both starting populations were modelled at a rate of 40% adults, with the juvenile and immature age-classes evenly distributed. After a maximum of 10 years, this age-class distribution was shown to come to equilibrium based on the vital rates. Example output for gannets from the Leslie Matrix model is shown as Fig. 2.

An initial population vector N_0 for each species was thus created and a population model for each species was run over a timespan of 40 years, with

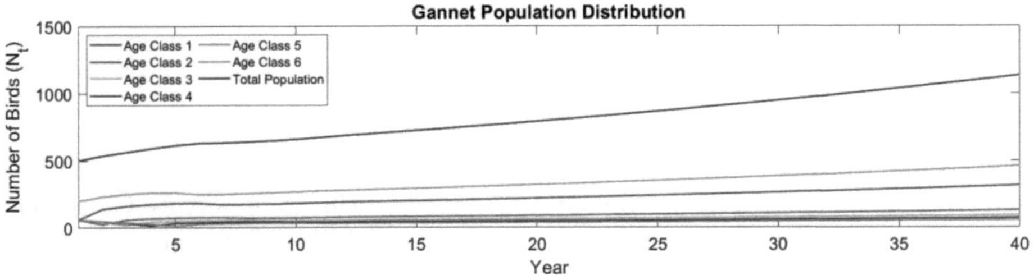

Fig. 2. Output from Leslie Matrix for Gannets.

the latter 30 years taken as output data. For the DBN modelling, the total population numbers N_t were taken as a separate node along with key drivers for each species as described in Sect. 2.2.

2.2 Building DBNs

Key drivers for each species were derived from previous work by Trifonova et al. [12]. Dynamic Bayesian Networks incorporating a Hidden Variable (HV) were then designed based on these drivers, as shown in Fig. 3. All DBN modelling was performed using the Bayes Net Toolbox for MATLAB (BNT), developed by Kevin Murphy [11]. Intra-yearly links between nodes in the models used for the following experiments were selected based on the strongest relationships previously identified by Trifonova et al. in their work with DBNs for the North Sea ecosystem. Variables which did not yet have associated drivers were connected between slices only through an auto-regressive (AR) link. The simplest networks tested used only the three primary drivers for each species, along with an additional three drivers for Demersal Biomass (DEM). Breeding Success (BS) was used as the proxy for species population status. To incorporate the Leslie Matrix data, yearly output for total population N_t of each seabird were included as an additional node, with the option to remove this node completely from the DBN for comparison purposes. The drivers for these networks are defined in Table 3.

Table 2. Main Drivers for Simple DBN Models

Gannet BS	Guillemot BS	Demersal Biomass
Horizontal Speed (HSpeed)	Bottom Temperature (BT)	Chlorophyll-A (Chla)
Sandeel Larvae abundance	Net Primary Production (netPP)	Net Primary Production (netPP)
Demersal Biomass	Demersal Biomass	Zooplankton Assemblage A6 (Zpl A6)

These simple networks relied heavily on auto-regressive (AR) links for several of the gannet and demersal biomass drivers. To mitigate this, additional physical and biological drivers were added to reduce direct reliance on AR links; primarily while analysing the gannet models. Secondary drivers were added as additional

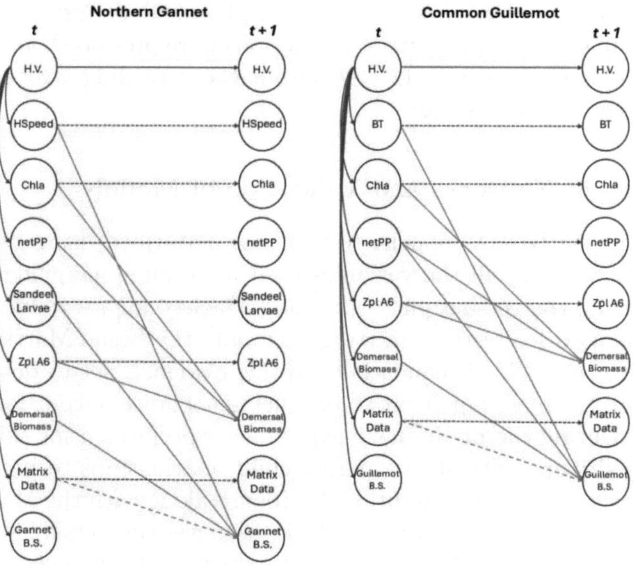

Fig. 3. Network Diagrams showing key inter-year relationships between all variables, for gannets and guillemots. Horizontal arrows represent AR links, while diagonal arrows show parent-child relationships between nodes.

nodes linked to the existing primary drivers depicted in Fig. 3, based on their previously identified 'parent - child' relationships. Firstly, bottom temperature (BT) was added as a direct driver for Chla & Sandeel larvae abundance, while Pelagic catch was added as an additional driver for sandeels. This allowed for removal of AR links from Chla and Sandeels, resulting in a 10-variable model for gannets. To test if adding a larger range of driver variables further improved the predictive accuracy of the models, additional physical variables were then added to the gannet model as drivers for netPP and ChlA. This allowed for the added benefit of trialling the 2nd 'testing' method using only the ocean physics and Leslie Matrix data as 'evidence', to reduce reliance on biological variables. A 13-variable DBN was thus built incorporating Sea Surface Temperature (SST) as an additional driver for ChlA, and Potential Energy Anomaly (PEA) along with Vertical current speed (VSpeed) added as drivers for Zpl A6. For each network configuration described in Sect. 2.2, models were trained using datasets from the deeper North Sea region defined by Trifonova et al. [12] with a timeseries from 1990 to 2019. All data were standardized before use. Non-parametric bootstrapping [5] was applied 250 times during the training of each model, with the Expectation Maximisation (EM) algorithm [1] used to infer the state of the Hidden Variable (HV). The junction tree algorithm was employed as an exact inference method. The standardised timeseries data was split into two-year pairs, and for each bootstrap iteration, a number of samples were taken from these pairs (using random sampling and replacement) to train the DBN model. Those pairs not used in training were identified as 'testing' pairs for each iteration. In all experiments, output predictions were produced only for the years 1991 to 2019, as for each year of test output the preceding year would be used as a start-point during testing.

2.3 Testing of Networks and Validation of Models

After training the HDBN models, testing was performed using two alternative methods; firstly, using all the variables used in training as 'evidence' from the test data during the testing phase (method 1). Secondly, a reduced number of variables was used for testing, starting with only the Leslie Matrix data. Other variables were then added sequentially to the 'evidence' set to determine which variables were the most pertinent to producing accurate output for each species (method 2). During the preliminary experiments performed for this paper, only the 'primary drivers' with strong parent-child relationships to breeding success for each species were used as additional evidence along with the Matrix data (see Fig. 3). For gannets, the additional evidence used were demersal biomass (DEM), sandeel larvae abundance (Sandeel), and horizontal current speed (HSpeed). For Guillemots, demersal biomass and bottom temperature (BT) were used. The sequences of variables used for training of DBN models for both species in the second method are listed in Table 3.

Table 3. Evidence Sets Used for Testing, Method 2

Species:	Gannet	Guillemot
Evidence Set 1	Matrix Data Only	Matrix Data Only
Evidence Set 2	Matrix Data + DEM	Matrix Data + DEM
Evidence Set 3	Matrix Data + DEM + Sandeel + Hspeed	Matrix Data + DEM + BT

Model output was initially compared through visual evaluation of the output plots and quantitatively by calculating the Sum of Squares Error (SSE) for each set of predictions. Confidence Intervals were also calculated for each yearly datapoint by calculating the mean value of all test runs and the standard deviation based on all test outputs. Multiple test runs were performed for each testing method to ensure repeatability of results, with 10 runs per network configuration. Error bars for each run of predictions were determined by calculating the standard deviation for each datapoint based on the predicted value of that datapoint as derived during the bootstrapping process.

3 Results and Discussion

Preliminary results showed significant improvement in the accuracy of predictions for both species. While the full reasoning for this is not known, further analysis will be performed in later work to determine how this improvement can be used to reduce the number of input variables required for training and testing HDBN models of this type. General analysis of the results for both methods is discussed in this section, followed by comparison between the 'testing' methods 1 and 2.

3.1 Comparison Between 'Testing' Evidence Methods

Sum-of-Squares Errors (SSEs) for all testing methods are shown in Table 4. Runs which showed both good predictive curves and low SSEs are highlighted in green. Though the mean SSEs for method 1 results are higher over 10 runs for both species, this can be partially explained by one poor fit in each set, which exhibited an anomalous high SSE, highlighted in red.

Table 4. SSE Results for All Experimental Methods

Run	1	2	3	4	5	6	7	8	9	10	Mean
Gannet Method 1, No Matrix	23.2	28.0	30.8	41.1	30.1	21.9	27.6	23.3	25.7	19.8	27.2
Gannet Method 1, With Matrix	26.6	159.7	23.0	28.7	22.4	20.6	38.1	26.0	22.2	23.4	39.1
Gannet Method 2	25.1	81.9	28.4	28.3	41.3	30.1	25.6	24.8	57.2	27.9	37.1
Guillemot Method 1, No Matrix	19.6	23.9	23.1	25.0	18.6	22.5	47.1	25.4	20.2	22.9	24.8
Guillemot Method 1, With Matrix	15.0	18.0	26.5	33.6	37.1	22.0	28.2	21.0	27.8	42.0	27.2
Guillemot Method 2	63.6	45.4	56.6	20.1	17.6	37.7	28.7	47.4	42.9	24.1	38.4

3.2 Projections for Gannets, Using Method 1

Results using the first HDBN method are shown in Fig. 4. Here, the fewest set of drivers were used as per Table 2, with three drivers for Gannet Breeding Success and three drivers for Demersal Biomass. The fit between the 'predicted' curve and the actual data is observed to be improved with the addition of Leslie Matrix data, also represented quantitatively with a lower SSE. The direction of each significant peak (whether increasing or decreasing) is much more accurately represented in Model 2, though the magnitudes of such peaks still show some deviation from the 'actual' data.

Additional Gannet Drivers: Additional drivers were gradually added to the HDBN network, as shown in Table 3. However, running the HDBN model using these additional drivers often did not show significant improvements in accuracy of fit, with some large errors occurring for several years and over multiple runs. Some (inconsistent) improvement was observed in the 12-variable model, while adding SST as a 13th variable ultimately reduced overall predictive accuracy for several of the experimental runs. Sample results for these latter two models are presented in Fig. 5.

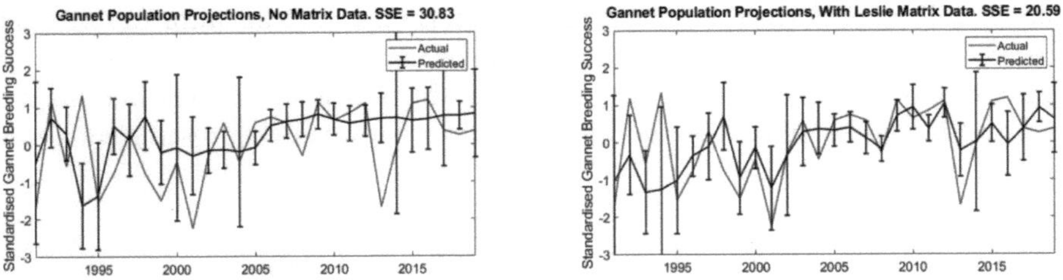

Fig. 4. Sample Results for 8-variable gannet model, without and with Leslie Matrix data.

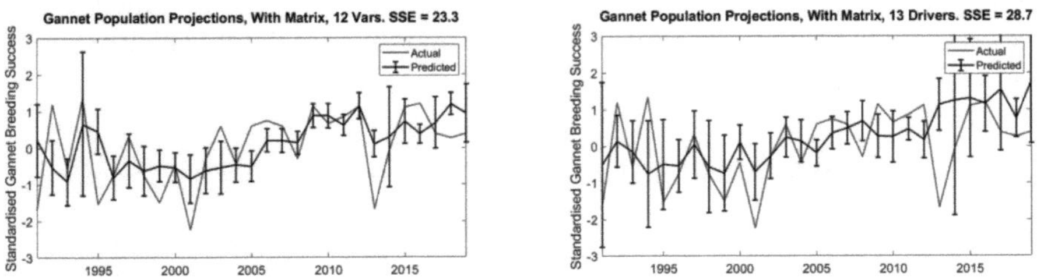

Fig. 5. Sample Results for 12 and 13-variable gannet models, with Leslie Matrix data.

3.3 Projections for Guillemots, Using Method 1

Results for the simplest Guillemot HDBN model both with and without using Leslie Matrix data are shown in Fig. 6. As was the case with the gannet model, the 'predicted' curve shows a significant improvement in matching the 'actual' data, in particular when observing the decline and sharp low peak at year 2007. This low peak remained largely uncaptured by the model when the Matrix data was not included during training.

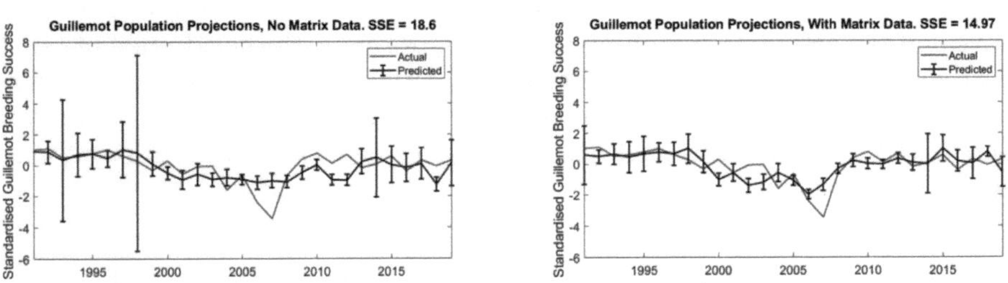

Fig. 6. Sample Results for 7-variable guillemot model, without and with Leslie Matrix data.

3.4 Projections for Both Species, Using Method 2

Results using the second 'testing' method for both species are shown in Fig. 7. It is clear from these plots that using only the Leslie Matrix data as 'evidence' for testing the model can capture a generalised trend over the timeseries, but yearly peaks are notably lost. This is particularly apparent for the guillemot model. Incorporating additional evidence during testing was shown to re-capture some of this acute accuracy, but did not show significant overall improvement in results when compared to method 1 where all variables were used for both phases. Results for both species using the matrix data in combination with demersal biomass and subsequently with additional primary drivers (sandeel larvae and HSpeed for gannets, bottom temperature for guillemots) as 'evidence' variables for the testing phase are depicted in Fig. 7c-f.

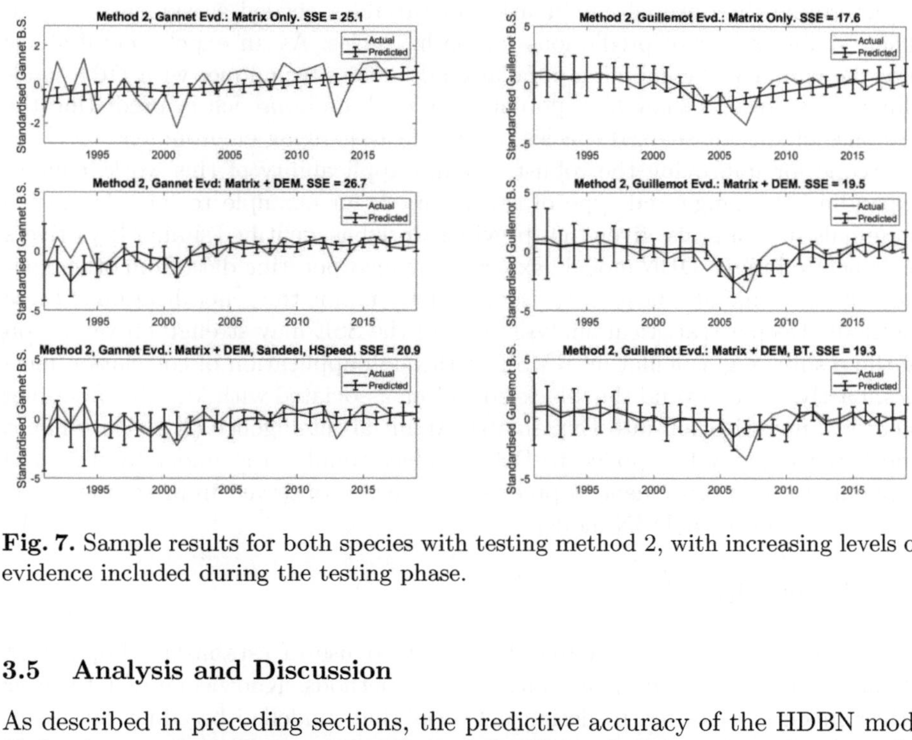

Fig. 7. Sample results for both species with testing method 2, with increasing levels of evidence included during the testing phase.

3.5 Analysis and Discussion

As described in preceding sections, the predictive accuracy of the HDBN models implemented for both species showed significant improvement when population predictions from the Leslie Matrix were included as an additional driver to breeding success. It is hypothesised that the population data serves as a form of 'rate-limiting' to better capture a realistic rate of population increase or decline in response to variation in other drivers, be they physical or biological.

Output plots for testing method 1 showed that even a small range of drivers for both species produce more accurate results when the population data is included. The guillemot models in particular showed very high improvement in goodness-of-fit, reflected by much lower SSEs in most runs in comparison to using the same drivers without the population data. For gannets, the inclusion of additional drivers to those described in Table 2 in the gannet networks did not significantly improve the accuracy of results. Incorporating these additional drivers into HDBN models to most accurately represent the full range of ecosystem dynamics was expected to improve predictive accuracy, so further investigation into the parent-child relationships implemented for the extra variables provides scope for further study.

Initial experiments using testing method 2 showed that, while the general trend of breeding success could be captured and represented using only the population data as evidence when testing the model, annual variation in magnitude were not reflected in the output plots, instead showing only a smoothed curve. It is therefore apparent that to capture the full annual variation in breeding success for either species, the model requires further data (physical and/or biological) during the testing phase. Additional experiments can later be performed to determine the optimal combination of variables needed as evidence to produce reliably accurate predictions for both species. As an experimental study into the viability of combining population dynamics techniques with DBN modelling, these results show high potential for such an approach in predicting the response of specific seabird species to their fast-changing environment.

Scope for improving the robustness and applicability of this work includes expanding the range and type of species used, for example to include marine mammals such as seals, given that previous work has identified similar key drivers on which to build a DBN model. Experiments incorporating data from other spatial regions would also be greatly useful in determining the generalization of these methods. Deeper statistical analysis beyond the SSE may strengthen validation of the results, e.g. calculating RMSE or through application of correlation metrics. Lastly, to go beyond the 'black-box' often associated with Machine Learning methods, investigation into Explainable Artificial Intelligence (XAI) techniques and how they may be applied to DBN models would aid in understanding and explaining the improvement in predictive accuracy observed from adding Leslie Matrix data into our DBN models.

4 Conclusion

In this paper, we have explored the combined use of established Population Dynamics techniques with innovative DBN methods. Knowledge of inter-year driver linkages was used to design networks that modelled breeding success of two seabird species, with significant improvements in predictive accuracy found to be achievable with the addition of Leslie Matrix population data as an additional DBN node. Inter-annual variability of seabird breeding success was predicted with high accuracy, particularly using the first 'testing' method, where all drivers were included as evidence. We also demonstrated that using only the Leslie Matrix data as evidence was effective in predicting the general trend of breeding success for each species, whether increasing or decreasing. Thus, even with limited recorded data, such a method may still provide adequate output data to determine the current and future direction of a species population growth or decline. Future work will include continued analysis of the key drivers required as evidence and the potential further optimisation of DBN models to reduce reliance on AR links.

References

1. Bilmes, J.: A gentle tutorial of the EM algorithm and its application to parameter estimation for gaussian mixture and hidden Markov models. Int. Comput. Sci. Inst. **4**(510), 126 (1998)
2. Caswell, H.: Matrix Population Models: Construction, Analysis, and Interpretation, 2nd edn. Sinauer Associates Inc. (2001)
3. Domonkos, P., Jan Kyselý, J., Piotrowicz, K., Petrovic, P., Likso, T.: Variability of extreme temperature events in south-central Europe during the 20th century and its relationship with large-scale circulation. Int. J. Climatol. **23**, 987–1010 (2003)
4. Edwards, K., Barciela, R., Butenschön, M.: Validation of the nemo-ersem operational ecosystem model for the north west European continental shelf. Ocean Sci. **8**, 983–1000 (2012)
5. Friedman, N., Goldszmidt, M., Wyner, A.: Data analysis with Bayesian networks: a bootstrap approach (2013)
6. Garcia Molinos, J., Halpern, B., Shoeman, D.: Climate velocity and the future global redistribution of marine biodiversity. Nat. Clim. Chang. **6**, 83–88 (2016)
7. Holliday, N., Bersch, N., Berx, B., Chafik, L., Cunningham, S.: Ocean circulation causes the largest freshening event for 120 years in eastern subpolar north Atlantic. Nat. Commun. **11** (2020)
8. Horswill, C., Robinson, R.: Review of seabird demographic rates and density dependence. JNCC Report No. 552 (2015)
9. Levin, S., Anderies, J., Adger, N., Barrett, S., Bennett, E.: Governance in the face of extreme events: lessons from evolutionary processes for structuring interventions, and the need to go beyond. Ecosystems **25**, 697–711 (2022)
10. Moore, P., Smale, D.: Impacts of climate change on shallow and shelf subtidal habitats, relevant to the coastal and marine environment around the UK. MCCIP Sci. Rev. **2020**, 272–292 (2020)
11. Murphy, K.: The bayes net toolbox for matlab (2001)
12. Trifonova, N.: Bayesian network modelling provides spatial and temporal understanding of ecosystem dynamics within shallow shelf seas. Ecol. Indicators **129** (2021)
13. Trifonova, N., Scott, B.: Ecosystem indicators: predicting population responses to combined climate and anthropogenic changes in shallow seas. Ecography (2023)
14. Tucker, A., Duplisea, D.: Bioinformatics tools in predictive ecology: applications to fisheries. Philos. Trans. R. Soc. B: Biol. Sci. **367**, 279–290 (2012)
15. Ummenhofer, C., Meehl, G.: Extreme weather and climate events with ecological relevance: a review. Philos. Trans. R. Soc. B: Biol. Sci. **372** (2017)
16. Uusitalo, L., Toczak, M., Muller-Karulis, B., Putnis, I., Trifonova, N., Tucker, A.: Hidden variables in a dynamic Bayesian network identify ecosystem level change. Eco. Inform. **45**, 9–15 (2018)

Bridging Forecast Accuracy and Inventory KPIs: A Simulation-Based Software Framework

So Fukuhara[1,2], Abdallah Alabdallah[1], Nuwan Gunasekara[1], and Sławomir Nowaczyk[1(✉)]

[1] Center for Applied Intelligent Systems Research, Halmstad University, Halmstad, Sweden
{sofuk25,abdallah.alabdallah,nuwan.gunasekara,slawomir.nowaczyk}@hh.se, s24d164@kagawa-u.ac.jp
[2] Kagawa University, Kagawa, Japan

Abstract. Efficient management of spare parts inventory is crucial in the automotive aftermarket, where demand is highly intermittent, and uncertainty drives substantial cost and service risks. Forecasting is therefore central, but the quality of a forecasting model should be judged not by statistical accuracy (e.g., MAE, RMSE, IAE) but rather by its impact on key operational performance indicators (KPIs), such as total cost and service level. Yet most existing work evaluates models exclusively using accuracy metrics, and the relationship between these metrics and operational KPIs remains poorly understood.

To address this gap, we propose a decision-centric simulation software framework that enables the systematic evaluation of forecasting models in realistic inventory management settings. The framework comprises: (i) a synthetic demand generator tailored to spare-parts demand characteristics, (ii) a flexible forecasting module that can host arbitrary predictive models, and (iii) an inventory control simulator that consumes the forecasts and computes, based on selected inventory control policy, operational KPIs. This closed-loop setup enables practitioners and researchers to evaluate models not only in terms of statistical error but also in terms of their downstream implications for inventory decisions.

Using a wide range of simulation scenarios, we show that improvements in conventional accuracy metrics do not necessarily translate into better operational performance, and that models with similar statistical error profiles can induce markedly different cost–service trade-offs. We analyze these discrepancies to characterize how specific aspects of forecast performance affect inventory outcomes and to derive actionable guidance for model selection. Overall, the framework operationalizes the link between demand forecasting and inventory management, shifting evaluation from purely predictive accuracy towards operational relevance in the automotive aftermarket and related domains.

An open-source implementation of the software, including all experimental results, is available at https://github.com/caisr-hh/TruckParts-Demand-Inventory-Simulator/releases/tag/IDA_2026.

M. Baratchi et al. (Eds.): IDA 2026, LNCS 16513, pp. 438–452, 2026.
https://doi.org/10.1007/978-3-032-23833-7_32

Keywords: Spare parts demand · Aftermarket logistics · Synthetic data generation · Demand forecasting · Inventory management · Simulation

1 Introduction

The automotive aftermarket plays a central role in the vehicle life cycle, ensuring the availability of spare parts for maintenance and repair over extended periods. Efficient inventory management is therefore crucial, both economically and environmentally. Overstocking ties up capital and leads to obsolescence-related scrapping, while understocking increases downtime, emergency shipments, and customer dissatisfaction. These trade-offs are particularly acute for rare and expensive components, where demand is sparse and highly uncertain, and for parts associated with emerging technologies such as electric-vehicle traction batteries. Limited supply, short shelf lives, rapid product evolution, and limited operational experience all contribute to the complexity of related logistics processes.

From a sustainability perspective, improving spare parts logistics offers substantial leverage. Better forecasts and more informed inventory decisions can reduce the number of components produced but never used, lower the frequency of rush orders, and shift transport away from fast but environmentally costly modes (e.g., air) towards slower but greener alternatives (e.g., sea or consolidated road shipments). At the same time, avoiding unnecessary back-and-forth movement of parts between warehouses, regional hubs, and dealers can substantially decrease the overall environmental footprint of vehicle operations. Achieving "the right part in the right place at the right time" is thus not only an economic objective, but also a key enabler for more resource-efficient and sustainable transport.

However, accurately predicting spare parts demand remains a major challenge. Demand is often intermittent or sporadic, with signals about upcoming needs being delayed or incomplete. Parts may be phased out or replaced with little notice, and supply chain capacity is often constrained. On a large scale, even determining the current life cycle stage of parts in the field is challenging. In practice, inventory policies at warehouses and dealers are still frequently based on coarse-grained rules, with manual and ad hoc adjustments made to account for local factors. This leads to suboptimal resource utilization, making it difficult to systematically assess the impact of new forecasting methods or data sources.

Recent advances in AI and ML offer new opportunities to address these issues. Richer data, including predictive maintenance signals, telematics, and warranty information, can provide earlier and more granular indications of future demand. At the same time, modern forecasting approaches, such as meta-learning and other data-driven methods that leverage information across large families of time series while adapting to local idiosyncrasies, are particularly well-suited to the heterogeneous and sparse nature of spare parts demand [17]. However, regardless of the data sources or modeling techniques employed, the evaluation of such models remains dominated by statistical accuracy metrics, with a limited understanding of how changes in these metrics translate into real-world operational performance.

This paper addresses the interface between demand forecasting and inventory management. We develop and study a decision-centric simulation framework that links forecasting models to downstream inventory decisions under realistic inventory management settings, including lead times and operational cost structures. The framework clarifies how different aspects of forecast performance shape cost–service trade-offs and highlights how common metrics can be misleading.

2 Spare Parts Forecasting Methods

This section briefly outlines commonly used spare parts forecasting approaches to provide context for the forecasting models considered in this study. Spare parts forecasting methods are broadly classified into two categories: time-series methods, which rely on historical demand data, and contextual methods, which incorporate additional information, such as expert judgment or equipment data [17]. Classical methods, such as Simple Exponential Smoothing (SES) and ARIMA, are often unsuitable for spare parts due to their erratic, lumpy, or intermittent demand patterns [17], leading many to use ML methods such as Random Forests, SVMs, or ANNs. In addition, DL methods, such as Long Short-Term Memory networks, have demonstrated strong performance in modeling complex temporal dependencies.

Time-series methods use historical demand data and can be parametric, nonparametric, or employ improvement strategies. Parametric approaches assume lead-time demand follows a known probability distribution. Croston's method separately forecasts demand intervals and sizes [3], later improved by the Syntetos-Boylan Approximation (SBA) with a bias correction term [19]. To address demand obsolescence, the TSB method incorporates demand probability forecasts that are updated every period [22]. Parametric bootstrapping methods simulate lead-time demand using estimated parameters but can underestimate variability, potentially leading to insufficient safety stock [9,18]. In contrast, nonparametric approaches reconstruct the empirical demand distribution without assuming a specific form, typically via bootstrapping. The method by [23] utilizes a Markov process for demand occurrence and performs well for highly lumpy demand, whereas neural networks excel at capturing nonlinear patterns, such as intermittence [8,14]. Forecasts can be further improved through strategies such as demand classification, which matches patterns with optimal methods [20], and data aggregation, which reduces variability from zero-demand periods to create more stable forecasts [16].

Contextual forecasting integrates historical data with external information. Judgmental forecasting adjusts statistical forecasts with expert knowledge, where systematic, particularly negative, adjustments can improve intermittent demand forecasts; however, the effectiveness of this approach varies [21]. Installed base forecasting uses information about the operational equipment portfolio,

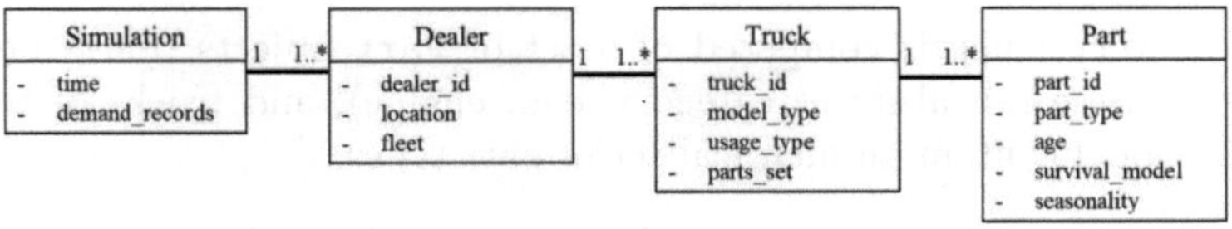

Fig. 1. Class Diagram.

incorporating data on quantity, age, and maintenance schedules, proving particularly effective for highly intermittent demand in B2B settings [4,11,24], despite challenges with data collection costs [1].

This paper is motivated and focuses on heavy-duty vehicle aftermarket logistics; however, our overall software framework is flexible and can also be applied to other domains by adapting relevant parameters.

3 Spare Parts Demand Data Generator

Synthetic data provides a practical alternative to real operational records, avoiding confidentiality constraints and offering a common evaluation environment. It can be generated at any scale or structure, enabling the testing of diverse scenarios. This study develops a synthetic demand generator for aftermarket truck parts. The simulator replicates truck operating structure, usage conditions, and seasonal factors, and generates demand time series by simulating part failures using a survival model.

We are not permitted to share real-world demand data from our industrial partner, but generating synthetic data with similar characteristics allows us to share our results with the scientific community and promotes reproducible research.

3.1 Concept and Structure

The simulator uses a Discrete Time Simulation (DTS) approach in which time advances in fixed increments (in this study, one day) and the system state is updated at each increment.

The class diagram of the simulator is shown in Fig. 1, employing a hierarchical structure `Dealer` → `Truck` → `Part`, as described below:

- The `Simulation` class manages the overall system, maintaining the current system time (simulation date) and creating a log of demand occurrences;
- Multiple `Dealer` objects are instantiated, with unique identifiers and location information. Each dealer manages a collection of trucks;
- Each `Truck` is identified by a model type and usage type;

- Finally, every truck is composed of a set of `Part` objects. Parts belong to a specific type (an abstract category, e.g., engine), and trucks of the same vehicle model contain an identical set of part types.

Part types have a specific "baseline" survival model and are affected by seasonality and usage in predetermined ways. A part instance, on the other hand, keeps track of its own age (a broken part is replaced with a new one). Part failures within a truck and usage patterns across trucks are assumed to be independent; relaxing this constraint is technically straightforward, but there is a lack of domain knowledge to allow a meaningful, realistic specification of more complex cases.

3.2 Survival Model

The simulator uses survival models to probabilistically simulate spare-parts demand. We have implemented survival models widely used in reliability engineering, treating time-to-event as a random variable and describing its distribution together with the effects of covariates [7,13].

In the simulator, part failure is the event of interest. For any part p, the failure probability is evaluated conditionally on the time elapsed since its last failure, denoted as t_p. Specifically, for a part that has survived until time t_p, the probability of failure within the time interval $[t_p,\, t_p + \Delta t]$ is given by

$$\Pr\left(t_p < T_p \leq t_p + \Delta t \mid T_p > t_p\right) = 1 - \exp\left(-\int_{t_p}^{t_p+\Delta t} h(u)\, du\right), \tag{1}$$

where Δt is the simulation time-step and T_p is the random variable for time to failure. $h(t)$ denotes the hazard function, defined as the instantaneous failure rate at t. Hazard functions have been extensively studied to capture a wide range of failure behaviors. Four parametric models, $\{$`Exponential`, `Weibull`, `Log-logistic`, `Gompertz`$\}$, are employed. These models exhibit distinct temporal hazard behaviors, enabling heterogeneous failure dynamics across parts. Their definitions and parameter settings, conditioned on the truck's usage style, are presented below. The `Hard` setting indicates severe operating conditions with higher and earlier failure risk, whereas `Normal` represents standard usage conditions.

Exponential Model. The hazard function is defined as

$$h(t) = \lambda, \qquad t \geq 0, \tag{2}$$

where λ is the model parameter. This constant rate signifies that the probability of failure remains the same at any point in time, reflecting the "memoryless" property, which means past events do not influence the likelihood of future events. It determines the median failure time t_m, at which the probability of failure reaches 50%, as follows:

$$t_m = \frac{\ln(2)}{\lambda}. \tag{3}$$

From a user perspective, t_m is a more interpretable parameter than λ. Accordingly, the simulator accepts a range of values for t_m as a configuration parameter, selects one uniformly at random for each part type, and computes λ accordingly.

Weibull Model. The hazard function is defined as

$$h(t) = \frac{k}{\lambda}\left(\frac{t}{\lambda}\right)^{k-1}, \qquad t \geq 0, \tag{4}$$

where λ and k are the scale and shape parameters, respectively, and t_m is:

$$t_m = \lambda\,(\ln 2)^{1/k}. \tag{5}$$

The simulator first assigns the value to k randomly (within a given range), and then (as above), calculates λ according to the user-specified desired median failure time t_m.

Log-logistic Model. The hazard function and median failure time t_m are defined as

$$h(t) = \frac{k\lambda\, t^{k-1}}{1 + (\lambda t)^k}, \qquad t > 0, \qquad t_m = \lambda^{-1}(\ln 2)^{1/k}. \tag{6}$$

Gompertz Model. The hazard function and median failure time t_m are defined as

$$h(t) = k\lambda e^{\lambda t}, \qquad t > 0, \qquad t_m = \frac{1}{\lambda}\ln\left(1 + \frac{\ln 2}{k}\right). \tag{7}$$

3.3 Seasonality

Seasonal conditions significantly impact failure probabilities, and the nature of this influence varies by component type and operating environment. For example, engine-related parts tend to exhibit higher failure rates during cold periods, driven by increased cold starts and additional load on auxiliary heating systems. In contrast, tire components are sensitive to road-surface conditions associated with freeze–thaw cycles in winter and to temperature-induced rubber deterioration during summer.

Moreover, regional climate further reinforces these effects: prolonged and harsher winters in cold regions amplify seasonal stresses on components, whereas high heat and humidity in hot regions accelerate mechanical degradation. We incorporate these seasonal effects into the hazard function, focusing on two seasons, summer and winter, and two regional types, southern and northern. Simulator introduces a seasonal coefficient $w(t)$ for each part, multiplying the baseline hazard function $h(t)$ as

$$h_{\text{seasonal}}(t) = w(t) \cdot h(t), \tag{8}$$

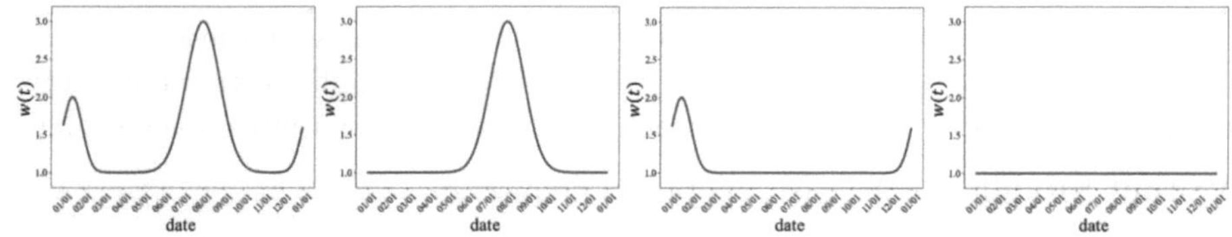

(a) both seasons (b) summer (c) winter (d) no seasonality

Fig. 2. Curves of seasonality coefficients $w(t)$ for the period of 12 months (Jan-Dec): a) both seasons, b) summer, c) winter, d) no seasonality.

and $w(t)$, capturing seasonal fluctuations, is modelled using a Radial Basis Function:

$$w(t) = \sum_{m=1}^{M} A_m \exp\left(-\frac{\|\max(|\mu_m - d(t)|,\, D - |\mu_m - d(t)|)\|^2}{\sigma_m^2}\right), \tag{9}$$

where M denotes the number of basis functions, and $m = 1, \ldots, M$ indexes each basis component; t denotes the cumulative simulation time and is mapped to the day-of-year via $d(t)$, with the length of one year D (typically $D = 365$). The parameters μ_m, σ_m, and A_m denote the center, width, and amplitude of the m-th basis function, controlling the timing, duration, and strength of seasonal effects. To ensure annual periodicity, the model employs the modular distance $\max(|\mu_m - d(t)|,\, D - |\mu_m - d(t)|)$ [5].

This formulation allows the seasonal coefficient $w(t)$ to flexibly represent region- and part-specific patterns. Examples of the patterns used in our experiments can be seen in Fig. 2.

3.4 Concept Drift

In practice, the number of trucks managed by a dealer often changes over time due to operational growth, demand fluctuations, or business decisions. The simulator models two types of concept drifts: sudden (the number of trucks for a dealer increases abruptly, for example, from 50 to 75 trucks, a dealer-specific random time point) and slow (the number of trucks increases gradually, for example, by one truck per week).

3.5 Simulation

Input Data. The main input parameters of the simulation are as follows:

- **Simulation start date**: denoted by T_0
- **Simulation horizon** (days): denoted by T.
- **Time-step size**: denoted by Δt.
- **Number of dealers**: denoted by N_d.

- **Number of trucks per dealer (range)**: $N_t \in [N_t^{(l)}, N_t^{(u)}]$.
- **Number of part types constituting a truck (range)**: $N_p \in [N_p^{(l)}, N_p^{(u)}]$.
- **Seasonality parameters**: the RBF parameters $(\boldsymbol{\mu}, \boldsymbol{\sigma}, \boldsymbol{A})$ in Eq. (9).
- **Parameters of survival models**: the range of the median time $t_m \in [t_m^{(l)}, t_m^{(u)}]$ in Eqs. 3, 5, 6 and 7.

Process Flow. The simulation proceeds according to the following steps:

1. **Initialization**: The simulation is initialized as follows:
   (a) The simulation time is initialized as `time` $\leftarrow 0$.
   (b) A total of N_d `Dealer` instances are created. For each dealer:
   - Assign a `location` from {`southern`, `northern`}.
   - Sample the beginning time of sudden drift, `sudden_time` $\in [T_0, T_0 + T]$.
   - Sample the beginning time of slow drift, `slow_time` $\in [T_0, T_0 + T]$.
   - Sample the number of trucks managed by the dealer from $N_t \in [N_t^{(l)}, N_t^{(u)}]$.
   - Assign a `survival_model` and `seasonality` category to each part type.

   (c) A total of N_t `Truck` instances are generated.
   - Each truck is assigned a `usage` from {`Normal`, `Hard`}.
   - Each part in a truck is assigned `age` $\leftarrow 0$.

2. **Demand Evaluation**: At any given time step, the simulator first evaluates which parts of the trucks failed, as follows:
   (a) *Failure probability:* The probability that part p fails within the interval $[t_p, t_p + \Delta t]$, given that it has survived up to time t_p (`age`), is computed as
   $$\Pr(t_p < T \leq t_p + \Delta t \mid T > t_p).$$
   (b) *Failure occurrence decision:* A failure is triggered when the probability exceeds a uniform random draw:
   $$\text{Demand} = \begin{cases} \text{True}, & \text{if } \Pr(t_p < T \leq t_p + \Delta t_p \mid T > t_p) \geq U(0,1), \\ \text{False}, & \text{otherwise.} \end{cases} \tag{10}$$
   (c) *Post-failure processing:* When a failure occurs, the simulator records the event (`time`, `dealer_id`, `truck_id`, `part_id`, `part_type`) and resets `age` $\leftarrow 0$.

3. **Concept Drift**: The simulator, for each dealer, evaluates:
   (a) *sudden drift*: When `time` reaches `sudden_time`, the dealer generates additional `Truck` instances: $N_t \leftarrow \alpha \cdot N_t, \quad \alpha \in \text{rand}(1.25, 1.5)$.
   (b) *slow drift*: If `time` $\geq$ `slow_time`, the dealer experiences slow drift and generates new `Truck` instance $N_t \leftarrow N_t + 1$.

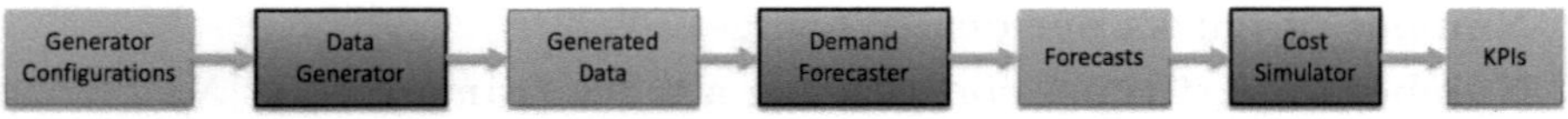

Fig. 3. Overall flow of the software framework

4. **Time Update and Termination**: The simulation time is incremented as $\texttt{time} \leftarrow \texttt{time} + \Delta t$, and the age of each part is updated accordingly, $\texttt{age} \leftarrow \texttt{age} + \Delta t$. If $\texttt{time} == T$, the simulation terminates; otherwise, it returns to Step 2.
5. **Output**: Upon termination, daily demand counts are aggregated for each part type within each dealer. The simulator outputs a set of daily demand time series data for all (`dealer_id`, `part_type`) combinations.

4 Cost Simulator

To evaluate the cost, we use the cost simulator [12], which employs Discrete-Event Simulation (DES) [2], a technique previously applied to supply chain planning [10]. The simulator models the inventory management process for individual parts at dealers, incorporating stochastic elements such as demand variability and fixed lead times. It operates through three primary event types: demand (customer orders), delivery (replenishment orders arriving), and inventory check (periodic reviews triggering reorders). The simulator [12] uses a "standard inventory policy" based on the assumption of normally distributed demand, which is commonly adopted as a baseline in inventory management; this assumption allows safety stock, reorder points, and order quantities to be calculated using established formulas [2,10]. It is common in industry to use these formulas even when demand is highly non-Gaussian, and the simulator presented in this paper allows us to evaluate the consequences of such choices.

The cost simulator incorporates multiple cost components to provide a comprehensive evaluation of inventory strategies, including holding costs, order costs, rush order costs, transportation costs, and badwill costs (penalties for service-level shortfalls). Inputs to the model include historical and forecasted demand data, lead times, and inventory policy parameters. In this work, we maintain constant inventory policy parameters for all parts across all dealers. Output metrics focus on KPIs such as *total cost* and *service level* (fill rate), enabling direct comparison of different forecasting approaches under controlled conditions. Figure 3 shows this setup as a flow diagram.

5 Experiments

We created 48 synthetic demand scenarios spanning three years of synthetic data, forecasted demand using several models, and evaluated corresponding business KPIs. Four concept-drift conditions were considered: without drift, sudden drift only, slow drift only, and combined drift. For each drift condition, we tested twelve parameter configurations constructed by combining four ranges of the number of trucks per dealer, $N_t \in [5, 10], [10, 30], [30, 50], [50, 100]$, with three ranges of median lifetime, $t_m \in [100, 150], [150, 365], [365, 730]$ days. The remaining settings were fixed, with the simulation horizon $T = 1095$ days, the simulation step $\Delta t = 1$ day, and four dealers ($N_d = 4$).

For forecasting, models were trained on the first two years of demand data and used to forecast daily demand over the subsequent one-year horizon. We employed three machine-learning models (XGBoost, Random Forest, and SVR) and four statistical time-series methods (ARIMA, Croston's method, TSB, and SBA). For the machine-learning models, input features included: (i) calendar-based features; (ii) categorical identifiers for part types and dealers; (iii) demand statistics (mean, median, and standard deviation) computed over the past 90 days; (iv) the number of consecutive zero-demand days prior to the current day; and (v) lagged demand values from the previous seven days. Hyperparameters were tuned to minimize MAE.

The Cost Simulator evaluates total inventory-management costs for each part type at each dealer over a one-year period. The following simplifying assumptions were made. All parts share the same price, unit cost, weight, initial inventory, and lead time. A single supplier with unlimited capacity. No partial shipments, cancellations, or modifications to orders. Orders are processed independently, and batching effects are ignored.

Demand Generation Results. We present two metrics for the generated demand: the Average Demand Interval (ADI) and the squared Coefficient of Variation (CV^2). ADI quantifies the demand intermittency by measuring the average interval between non-zero demands, whereas CV^2 captures demand variability. Figure 4 presents the ADI–CV^2 distribution of all generated demand series across all scenarios. The insets display representative demand series of selected parts: the first one has summer and winter seasonality with a sudden drift, while the second and third include only combined and slow drift, respectively. The majority of generated series fall within the intermittent region, confirming that the generator successfully produces real-world-like intermittent demand with substantial behavioral diversity.

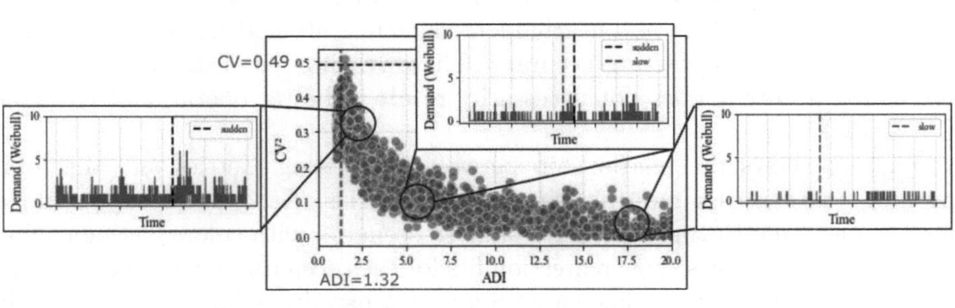

Fig. 4. ADI–CV^2 distribution for all demand series, with examples.

Forecasting Results. We evaluated the forecasting performance using four metrics: the Mean Absolute Error (MAE), the Root Mean Squared Error (RMSE), the coefficient of determination (R^2), and the Intermittent Accuracy Error (IAE) [6], which is designed to assess accuracy for highly intermittent demand (originally created for make-to-order manufacturing, which shares a lot of similarity with aftermarket logistics). Table 1 presents the mean and standard deviation of each metric across all scenarios, with the best value in each metric in bold. XGBoost and Random Forest achieved the best MAE, R^2, and IAE; Random Forest obtained the lowest MAE, RMSE, and IAE. In terms of R^2, all models show negative values, even when accounting for standard deviation, implying the demand series is highly irregular and challenging to predict.

Table 1. Prediction accuracy metrics and operational KPIs for each forecasting model (bold indicates the best performing model, per metric).

Model	MAE	RMSE	R^2	IAE	Total Cost	Service Level
XGBoost	**0.43** ± 0.36	0.60 ± 0.43	**0.02** ± 0.11	**0.80** ± 0.19	8.2×10^5	0.996
RandomForest	**0.43** ± 0.35	**0.59** ± 0.42	0.01 ± 0.10	**0.80** ± 0.19	7.5×10^5	**0.997**
SVR	0.50 ± 0.35	0.64 ± 0.46	−0.34 ± 1.05	0.84 ± 0.20	8.4×10^5	0.996
ARIMA	0.53 ± 0.57	0.68 ± 0.62	−0.12 ± 0.28	0.81 ± 0.16	7.5×10^5	**0.997**
CROSTON	0.52 ± 0.56	0.67 ± 0.62	−0.10 ± 0.25	**0.80** ± 0.16	$\mathbf{7.4 \times 10^5}$	**0.997**
SBA	0.51 ± 0.53	0.66 ± 0.59	−0.08 ± 0.20	**0.80** ± 0.16	7.8×10^5	0.996
TSB	0.57 ± 0.63	0.73 ± 0.68	−0.34 ± 0.71	0.83 ± 0.16	7.8×10^5	0.996

Cost Simulation Results. Table 1 also presents the total inventory-management cost and achieved service level for each forecasting model across all scenarios. The lowest total cost is achieved by Croston's method, followed by Random Forest, ARIMA, SBA, and TSB. In contrast, XGBoost and SVR incur substantially higher costs than the other models. This pattern stands in contrast to the forecasting accuracy in the same table, where Croston, SBA, TSB, and ARIMA perform markedly worse than XGBoost. Furthermore, the service levels remain consistently high and nearly identical across all models, despite substantial differences in forecasting accuracy. These results demonstrate that higher forecasting accuracy does not necessarily result in better operational KPIs, as seen in Fig. 5.

Figure 6 shows the relationship between forecasting metrics and the total inventory cost across multiple simulated scenarios. Each point represents a model's mean metric and cost under a given scenario, with colors distinguishing scenarios, and the lines show regression fits for each scenario. One can clearly see a counterintuitive pattern known as Simpson's paradox [15]. The per-scenario correlations, reflected in the slopes of the regression lines, vary widely and are predominantly negative, averaging −0.16 and −0.06 for MAE and RMSE, respectively, indicating that lower MAE or RMSE does not imply lower cost.

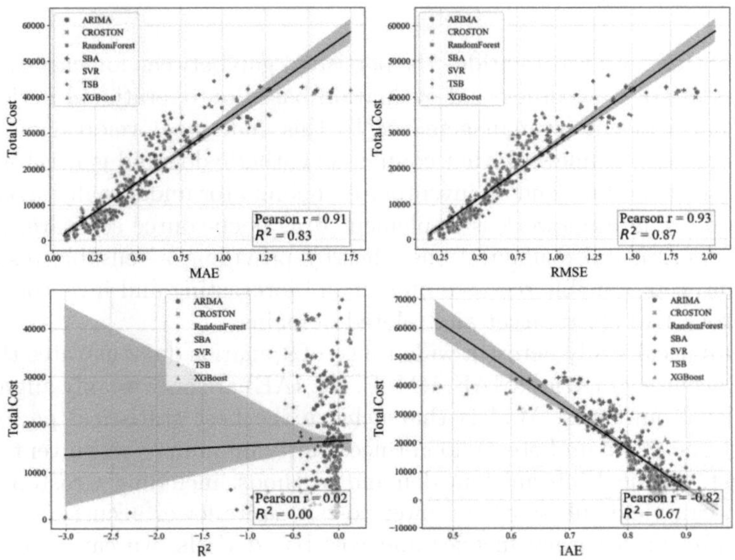

Fig. 5. Comparison between total cost and each evaluation metric.

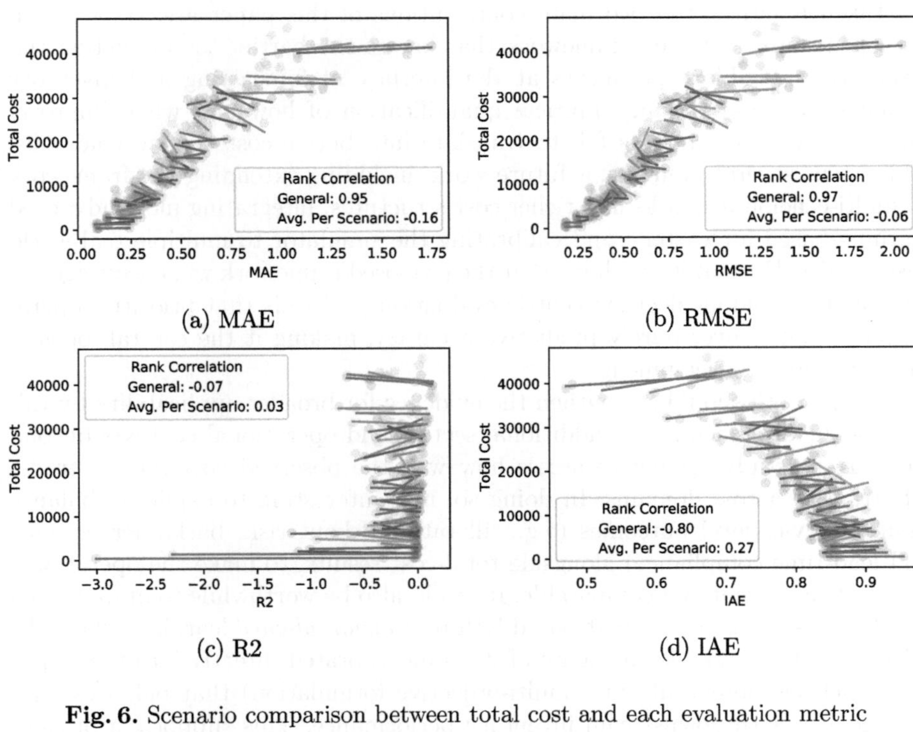

(a) MAE

(b) RMSE

(c) R2

(d) IAE

Fig. 6. Scenario comparison between total cost and each evaluation metric

6 Conclusions

The goal of this work was to bridge the persistent gap between forecast evaluation based on statistical accuracy and evaluation based on operational performance in spare parts inventory management. To this end, we developed a decision-centric simulation framework that couples a synthetic demand generator, a flexible forecasting module, and an inventory cost simulator under realistic operating assumptions. The framework is implemented as open-source software, together with all experimental configurations, thereby providing a reusable testbed for transparent, reproducible research on demand forecasting and inventory control in the automotive aftermarket and related domains.

Our empirical study across a wide range of scenarios demonstrates that conventional accuracy metrics (MAE, RMSE, R2, IAE) are only weakly aligned with business-level outcomes. Models that achieve the best statistical performance (XGBoost and Random Forest) do not necessarily minimize total inventory cost. In contrast, classical intermittent-demand methods, including Croston's, SBA, and TSB, can yield substantially lower costs despite lower accuracy scores. By jointly analyzing accuracy metrics and cost-based KPIs, we can quantify this discrepancy and show that model rankings can change dramatically once service levels, rush orders, and other cost components are taken into account. These findings highlight the risk of relying solely on statistical error measures when selecting forecasting models for real inventory systems.

Taken together, the two main contributions of this paper are: (i) a modular, open-source software framework that lowers the barrier to conducting rigorous, reproducible experiments at the interface of forecasting and inventory management; and (ii) an empirical quantification of how and when improvements in accuracy metrics fail to translate into better cost–service trade-offs. This opens several avenues for future work, including extending the framework to multi-echelon networks and richer cost structures, integrating more advanced learning-based forecasters, and calibrating the simulator to multiple real-world case studies. Ultimately, we hope that the proposed framework will encourage the community to adopt decision-centric evaluation protocols that prioritize operational relevance over purely predictive accuracy, making it the central focus of methodological development.

Future work should strengthen the evidence for broader applicability by validating the approach across additional sectors and operational contexts beyond the current study setting to assess how well the observed cost–service trade-offs transfer across domains. In doing so, it is interesting to explicitly balance desired service-level outcomes (e.g., fill rate/stockout risk, backorder volume, and lead-time compliance) alongside total-cost results, to make the operational impact transparent and comparable. It would also be worthwhile to move from a purely post-hoc use of the cost-based KPI to *decision-aligned* learning, where the KPI (or a differentiable surrogate of it) is incorporated directly into the objective function (potentially in a multi-objective formulation) that balances forecast fidelity with downstream inventory performance. This supports a broader agenda for decision support in efficient industrial operations and for the practical integration of ML objectives with business-relevant KPIs.

Acknowledgments. The work was carried out with support from The Knowledge Foundation and from Vinnova (Sweden's innovation agency) through the Vehicle Strategic Research and Innovation Programme, FFI.

Disclosure of Interests. The authors have no competing interests to declare that are relevant to the content of this article.

References

1. Andersson, J., Jonsson, P.: Big data in spare parts supply chains: the potential of using product-in-use data in aftermarket demand planning. Int. J. Phys. Distrib. Logistics Manag. **48**(5), 524–544 (2018)
2. Banks, J., Carson, J.S., Nelson, B.L., Nicol, D.M.: Discrete-event system simulation. Pearson Prentice Hall, 4th edn. (2005)
3. Croston, J.D.: Forecasting and stock control for intermittent demands. Oper. Res. Q. **23**(3), 289–303 (1972)
4. Dekker, R., Pinçe, Ç., Zuidwijk, R., Jalil, M.N.: On the use of installed base information for spare parts logistics: a review of ideas and industry practice. Int. J. Prod. Econ. **143**(2), 536–545 (2013)
5. Duvenaud, D.: The kernel cookbook: advice on covariance functions. https://www.cs.toronto.edu/duvenaud/cookbook (2014)
6. Flyckt, J.: Decision support through global demand forecasting: challenges and directions in make-to-order manufacturing organisations (2025)
7. Gorjian, N., Ma, L., Mittinty, M., Yarlagadda, P., Sun, Y.: A review on reliability models with covariates. In: Kiritsis, D., Emmanouilidis, C., Koronios, A., Mathew, J. (eds.) Engineering Asset Lifecycle Management, pp. 385–397. Springer, London (2010)
8. Gutierrez, R.S., Solis, A.O., Mukhopadhyay, S.: Lumpy demand forecasting using neural networks. Int. J. Prod. Econ. **111**(2), 409–420 (2008)
9. Hasni, M., Aguir, M.S., et al.: On the performance of adjusted bootstrapping methods for intermittent demand forecasting. Int. J. Prod. Econ. **216**, 145–153 (2019)
10. Hellström, D., Johnsson, M.: Simulation-based inventory optimization. Int. J. Phys. Distribut. Logistics Manag. **50**(8), 861–879 (2020)
11. Hua, Z., Zhang, B., Yang, J., Tan, D.: A new approach of forecasting intermittent demand for spare parts inventories in the process industries. J. Opera. Res. Soc. **58**(1), 52–61 (2007)
12. Kamil, J., Amer, M.: Prediction Intervals for ML-driven Automotive Service Market Logistics. Master's thesis, Halmstad University, Halmstad, Sweden (2025). MSc Programme
13. Karim, M.R., Islam, M.A.: Reliability and Survival Analysis. Springer Singapore (2019). https://doi.org/10.1007/978-981-13-9776-9
14. Kourentzes, N.: Intermittent demand forecasts with neural networks. Int. J. Prod. Econ. **143**(1), 198–206 (2013)
15. Neufeld, E.: Simpson's paradox in artificial intelligence and in real life. Comput. Intell. **11**(1), 1–10 (1995). https://doi.org/10.1111/j.1467-8640.1995.tb00021.x

16. Nikolopoulos, K., Syntetos, A.A., Boylan, J.E., Petropoulos, F., Assimakopoulos, V.: An aggregate-disaggregate intermittent demand approach (adida) to forecasting: an empirical proposition and analysis. J. Oper. Res. Soc. **62**(3), 544–554 (2011)
17. Pinçe, Ç., Turrini, L., Meissner, J.: Intermittent demand forecasting for spare parts: a critical review. Omega **105**, 102513 (2021)
18. Snyder, R.: Forecasting sales of slow and fast moving inventories. Eur. J. Oper. Res. **140**(3), 684–699 (2002)
19. Syntetos, A.A., Boylan, J.E.: The accuracy of intermittent demand estimates. Int. J. Forecast. **21**(2), 303–314 (2005)
20. Syntetos, A.A., Boylan, J.E., Croston, J.: On the categorization of demand patterns. J. Oper. Res. Soc. **56**(5), 495–503 (2005)
21. Syntetos, A.A., Nikolopoulos, K., Boylan, J.E., Fildes, R., Goodwin, P.: The effects of integrating management judgement into intermittent demand forecasts. Int. J. Prod. Econ. **118**(1), 72–81 (2009)
22. Teunter, R.H., Syntetos, A.A., Babai, M.Z.: Intermittent demand: linking forecasting to inventory obsolescence. European J. Oper. Res. **214**(3), 606–615 (2011)
23. Willemain, T.R., Smart, C.N., Schwarz, H.F.: A new approach to forecasting intermittent demand for service parts inventories. Int. J. Forecast. **20**(3), 375–387 (2004)
24. Zhu, S., van Jaarsveld, W., Dekker, R.: Spare parts inventory control based on maintenance planning. Reliabil. Eng. Syst. Safety **193**, 106600 (2020)

LeTMEMo: Leveraging Topic Modeling for Evaluating (Closed-Vocabulary) Models

Vu Minh Hoang Dang(✉) and Rakesh M. Verma

University of Houston, Houston, USA
vdang9@uh.edu, rmverma2@central.uh.edu

Abstract. This study introduces LeTMEMo, a framework that leverages unsupervised topic modeling to evaluate and enhance closed-vocabulary models and questionnaires. We use SERVQUAL as a case study, a service quality theoretical framework and dictionary that is widely used in social science and marketing research to quantify service quality concepts, including reliability, empathy, assurance, tangibles, and responsiveness. However, its fixed word lists and domain-general design limit its ability to capture the semantic patterns present in dynamic and context-specific corpora. Using LeTMEMo with BERTopic on the YelpCHI dataset, we examine how the discovered topics align with SERVQUAL dimensions across a broad set of embedding and clustering configurations. To assess lexical coverage and potential gaps, we compare two pipelines. One pipeline keeps only SERVQUAL terms, and the other removes only SERVQUAL terms while keeping the rest of the vocabulary intact. The results show that BERTopic, especially when paired with pretrained embeddings and topic-controlled clustering, generates coherent topics that align with SERVQUAL dimensions. The pipeline that excludes SERVQUAL terms reveals themes that fall outside the dictionary, suggesting opportunities for extensions and enhancements to SERVQUAL. LeTMEMo can potentially be applied to other closed-vocabulary models, such as the popular LIWC.

Keywords: BERTopic · Dictionary · SERVQUAL · Lexical Evaluation

1 Introduction

Closed-vocabulary lexicons such as SERVQUAL [20] and LIWC [21] provide interpretable mappings between words and predefined conceptual categories. These dictionaries are widely used in service research and social sciences, but their fixed vocabularies and domain-general design often struggle to capture the fluid and contextual language found in user-generated reviews. Prior studies show that manually constructed lexicons can miss informal expressions, emerging terminology, and domain-specific semantics [7,10,15], which raises questions about their reliability in modern applications.

M. Baratchi et al. (Eds.): IDA 2026, LNCS 16513, pp. 453–465, 2026.
https://doi.org/10.1007/978-3-032-23833-7_33

We introduce LeTMEMo, a framework that uses unsupervised topic modeling to evaluate how well closed vocabularies represent the semantic space of domain-specific corpora. Instead of relying on predictive performance or lexical overlap, LeTMEMo examines how latent topics discovered from text align with the predefined dimensions of a dictionary, which reveals both coverage and conceptual gaps.

We use the YelpCHI corpus [17] as a case study because it contains two service domains, hotel and restaurant, and exhibits diverse service-related language across these contexts. Although topic modeling often begins with methods such as LDA [3], our analysis focuses on neural topic models, specifically BERTopic [9], which provides flexible control over embeddings and clustering. We evaluate 91 combinations of embedding models and clustering algorithms under two vocabulary modes: one that keeps only SERVQUAL terms and one that excludes them. This design allows us to assess how different modeling choices influence dictionary alignment.

Our contributions are as follows:

- We introduce LeTMEMo as a systematic framework for evaluating closed vocabularies using topic and dictionary alignment, addressing the lack of methods that assess dictionary coverage through unsupervised latent semantic structure.
- We compare a wide range of embedding models and clustering algorithms and identify those that best recover SERVQUAL-related semantics.
- We show how the dictionary-included and dictionary-excluded modes reveal both the strengths and the boundaries of SERVQUAL in real consumer reviews.

2 Related Work

Closed-vocabulary lexicons such as SERVQUAL [20], LIWC [21], and the NRC Emotion Lexicon [16] map words to psychologically meaningful categories and are widely used in marketing, psychology, and social science. These dictionaries offer interpretability and are grounded in established theoretical frameworks. However, their fixed vocabularies and domain-independent design often limit their ability to adapt to emerging expressions, domain-specific jargon, or evolving user-generated text. Several reviews have also noted that SERVQUAL has conceptual and methodological limitations, including unstable dimensions and limited domain coverage [5,6,12].

Prior work on dictionary-based text analysis falls into several strands. One strand examines lexical coverage and representational validity. For example, Rocklage et al. introduce the Evaluative Lexicon and show how large-scale ratings can extend emotional meaning beyond LIWC categories [22]. Lu et al. evaluate established sentiment lexicons in clinical field notes and report substantial domain mismatch when lexicons are transferred to specialized contexts [14]. Another strand uses computational methods to refine or expand lexicons. Ojeda

Hernández et al. apply formal concept analysis to construct and validate domain-specific sentiment lexicons [19], demonstrating how structured concept lattices can reveal gaps and redundancies in manually constructed dictionaries. Tausczik and Pennebaker provide a foundational review of LIWC and document how its categories behave across diverse genres of text [24].

A separate line of research explores unsupervised topic modeling for discovering latent thematic structure. Brookes and McEnery evaluate its use in discourse studies, highlighting strengths and pitfalls [4]. Korenčić et al. introduce a framework assessing how models recover reference concept sets [11]. Lind et al. use topic modeling for comparative multilingual analysis [13]. Recent work examines automated evaluation using large language models [23].

However, little research uses topic modeling to evaluate closed vocabularies directly. Studies typically assess lexicons through predictive utility, expert evaluation, or embedding similarity, using topic models for exploration rather than evaluating fixed dictionaries. Thus, how dictionary terms relate to latent semantic structure in domain corpora remains unclear.

Our work bridges these research strands by introducing a topic modeling based framework for evaluating closed vocabularies. Instead of relying on predictive performance or lexical overlap alone, we examine how dictionary terms align with topics discovered from data. By applying a wide range of topic modeling configurations to the YelpCHI corpus and comparing discovered topics to SERVQUAL dimensions, we provide a scalable and unsupervised method for assessing the coverage, representational adequacy, and conceptual limits of closed vocabularies in domain-specific contexts.

3 Methodology

Our goal is to evaluate the representational adequacy of closed-vocabulary dictionaries by examining how well their predefined categories align with latent semantic structure discovered through topic modeling. To achieve this, we introduce LeTMEMo, a dual-mode evaluation framework illustrated in Fig. 1. LeTMEMo systematically explores a large configuration space of embedding models and clustering algorithms under two vocabulary conditions. The subsections below describe each component of the framework, including the topic modeling setup, the vocabulary modes, the alignment procedure, and the discovery of new dimensions.

SERVQUAL [20] is a widely used framework in marketing and service research that defines five core dimensions of perceived service performance: Tangibles, Reliability, Responsiveness, Assurance, and Empathy. It provides a structured vocabulary that links linguistic cues to well-established service constructs, which makes it a popular choice for survey design, customer experience analysis, and applied text-based research. Its broad adoption and clearly defined categories make it a suitable reference dictionary for evaluating how well data-driven topic models capture service-related semantics.

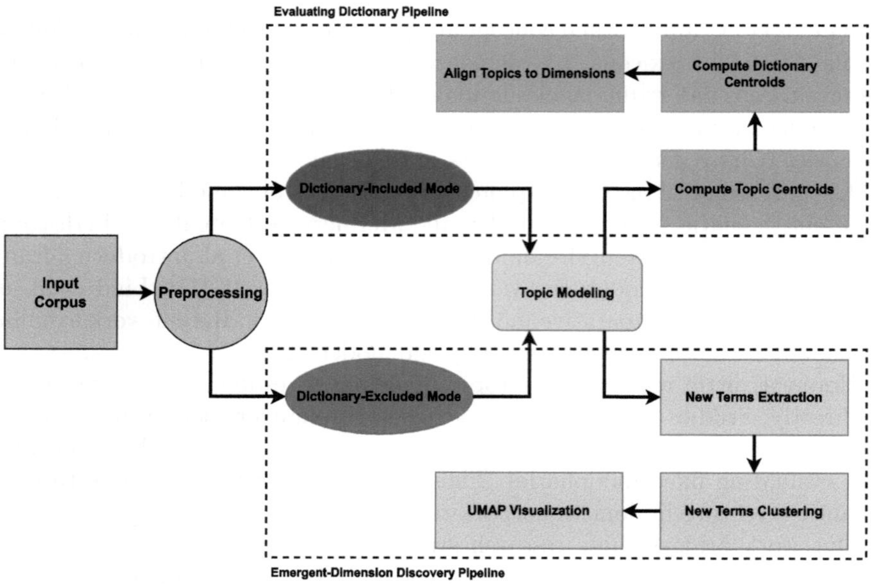

Fig. 1. An overview of the proposed LeTMEMo framework.

3.1 Topic Modeling Framework

We compare several topic modeling families that differ in how they infer latent semantic structure from text. Latent Dirichlet Allocation (LDA) [3] and hierarchical LDA (hLDA) [8] rely on bag-of-words assumptions and provide limited contextual sensitivity, which restricts their effectiveness on short or informal review texts. Top2Vec [1] and its contextual extension C-Top2Vec [2] improve semantic coherence by embedding documents and words in a shared vector space, yet they offer limited control over modeling components and do not allow users to specify the number of topics.

We select BERTopic [9] as the primary modeling framework due to its modular design. It allows independent configuration of embedding models, dimensionality reduction, and clustering algorithms, enabling a systematic evaluation of modeling choices. Crucially, BERTopic supports clustering methods that permit explicit control over the number of topics, allowing us to align the number of discovered topics with SERVQUAL's five conceptual dimensions when needed.

3.2 Embedding and Clustering Configuration Space

We evaluate 13 pretrained embedding models from the SentenceTransformers library[1], including `all-mpnet-base-v2`, `all-distilroberta-v1`, and multilingual variants such as `multi-qa-mpnet-base-dot-v1`. The full list of embedding

[1] https://sbert.net/docs/sentence_transformer/pretrained_models.html.

models can be found at Fig. 2. These models differ in architecture, training objectives, and multilingual capabilities.

For clustering, we evaluate seven algorithms: (1) Agglomerative Clustering, (2) BIRCH, (3) DBSCAN, (4) HDBSCAN, (5) KMeans, (6) MiniBatchKMeans, and (7) Spectral Clustering. In this study, we give priority to clustering methods that allow direct control over the number of output clusters, since this capability is crucial for aligning discovered topics with SERVQUAL's five conceptual dimensions. For algorithms that support explicit topic count specification, we set the number of clusters to five. For others, the number of clusters is inferred from the data based on density or connectivity thresholds.

3.3 Dictionary-Included and Dictionary-Excluded Modes

To evaluate both the scope and the boundaries of the SERVQUAL dictionary, we construct two vocabulary conditions. In the *dictionary-included* mode, all words not present in the dictionary are removed from the corpus. This isolates the expressive capacity of the dictionary itself and tests whether its vocabulary is sufficient to generate coherent service-quality themes. In the *dictionary-excluded* mode, all dictionary terms are removed. This forces the topic model to organize the remaining vocabulary and exposes semantic structures not represented in the dictionary. Each BERTopic configuration is applied under both modes, enabling direct comparison of dictionary-driven and open-vocabulary topic structures. We also tested the model on the original text without any dictionary-based filtering. However, the generated topics were noisy and lacked interpretability, so we excluded them from this study.

3.4 Semantic Alignment with SERVQUAL

To measure how well the discovered topics relate to the SERVQUAL dimensions, we embed each dimension as a centroid vector by averaging the embeddings of all dictionary terms for that dimension using the same sentence encoder used in topic modeling. Each topic is represented by a weighted centroid computed from its top representative words. We then use cosine similarity to compare each topic centroid with each dimension centroid.

We also evaluate alignment at the configuration level. A configuration refers to one complete run of BERTopic, defined by a specific choice of embedding model and clustering algorithm. For each configuration, we build a matrix of pairwise cosine distances between all topic centroids and all dimension centroids. We then apply the Hungarian algorithm to this matrix to find the best possible one-to-one assignment between topics and dimensions. The resulting score summarizes how closely the configuration reproduces the conceptual structure represented by the dictionary.

3.5 Emergent-Dimension Discovery

To investigate semantic coverage beyond the dictionary, we analyze the best-performing BERTopic configuration under the dictionary-excluded mode. For each topic, we extract all representative terms that do not appear in the dictionary. These new terms are embedded using the same encoder and clustered via agglomerative clustering with cosine distance. The number of clusters is selected from a small range ($k \in [4, 8]$) using silhouette scores and manual inspection. Each resulting cluster represents a candidate emergent dimension. We embed each cluster as a centroid and compute its cosine similarity to the SERVQUAL centroids; clusters that remain distant from all dimensions are interpreted as genuinely new constructs not represented by the dictionary.

4 Experimental Setup

4.1 Dataset and Dictionary

All experiments use the YelpCHI corpus [17], which contains reviews from two domains (hotels and restaurants) annotated as truthful or deceptive. After removing duplicates and excluding entries with incomplete metadata, the corpus contains 67,395 reviews. The dataset spans a wide range of service-related expressions, making it suitable for evaluating whether the dictionary captures the semantic variation present in real consumer feedback. In particular, hotel and restaurant reviews naturally foreground service delivery, staff interactions, and physical environment, which correspond closely to the SERVQUAL dimensions. This alignment makes YelpCHI a practical setting for testing dictionary coverage under realistic consumer language.

For the lexicon, we use the dictionary compiled by Nguyen et al. [18]. This version standardizes SERVQUAL terms for NLP applications and provides the specific term set used in all experiments.

4.2 Preprocessing

We use spaCy as the preprocessing toolkit for tokenization and normalization. Specifically, we tokenize the text, lowercase tokens, and remove punctuation and whitespace tokens. We also mask named entities and common identifiers (emails, URLs, and numbers) to reduce spurious cues from rare strings. To preserve multi-word SERVQUAL entries, we match them with spaCy's `PhraseMatcher` and merge them into single underscore-joined tokens so that dictionary phrases are treated as single lexical units.

After this common normalization, we apply two distinct preprocessing pipelines that differ in how they handle the vocabulary. In the *dictionary-included* pipeline, we remove English stopwords using the NLTK stopword list and then retain only SERVQUAL-matched tokens. This design intentionally isolates the dictionary signal by retaining only SERVQUAL terms, and we assess coherence and coverage via embedding-based topic centroids rather than document-level context or co-occurrence.

In the *dictionary-excluded* pipeline, we remove all SERVQUAL-matched tokens while keeping the remaining vocabulary intact (including stopwords). This complementary setting forces the topic model to organize non-dictionary language and surfaces prominent themes that lie outside SERVQUAL, enabling a direct comparison between dictionary-driven structure and out-of-dictionary semantic regions.

4.3 Implementation

Implementation is done in Python using BERTopic, HuggingFace Transformers, SentenceTransformers, scikit-learn, and UMAP. Each BERTopic run produces topic–word lists, document–topic assignments with soft probabilities, topic centroid embeddings, and UMAP projections for visualization. We compute clustering with Euclidean distance on L2-normalized embeddings to align with BERTopic's defaults (effectively consistent with cosine structure). We also use the default configurations for the clustering algorithms (e.g., k-means++ initialization for K-Means), as exhaustive fine-tuning across all embedders and clusters is not computationally feasible.

5 Results

5.1 Alignment Across Embedding and Clustering Configurations

Figure 2 reports the average Hungarian alignment scores for all combinations of embedding models and clustering algorithms. The strongest overall performance is achieved by `multi-qa-mpnet-base-dot-v1` paired with Spectral Clustering, which reaches an alignment score of 0.54. Across the embedding space, MPNet-based models consistently produce higher alignment scores than lighter encoders, indicating that deeper contextual representations capture service-related semantics more effectively. In contrast, smaller models such as the MiniLM and distilroberta variants yield weaker alignment across nearly all clustering methods. The multilingual distiluse models show the lowest scores in the heatmap, suggesting limited suitability for this domain. Among multilingual alternatives, `paraphrase-multilingual-MiniLM-L12-v2` performs notably better than other lightweight encoders. Overall, the results highlight substantial variation across configurations and demonstrate that both encoder capacity and clustering choice influence how closely the discovered topics align with SERVQUAL dimensions.

5.2 Evaluating Clustering Algorithms with the Best Embedder

To identify the most effective clustering algorithm given the best-performing embedder, we analyze the cosine similarity between topic centroids and dimension centroids for `multi-qa-mpnet-base-dot-v1` across all clustering methods (Fig. 3). Spectral Clustering produces the strongest and most consistent alignment patterns. In particular, several Spectral topics exhibit high similarity across multiple dimensions, with Topic 1 showing clear correspondence to *Assurance*, *Tangibles*, *Empathy*, and especially *Responsiveness*. This indicates that Spectral forms coherent topic representations that capture the structure of the SERVQUAL framework more effectively than other clustering methods.

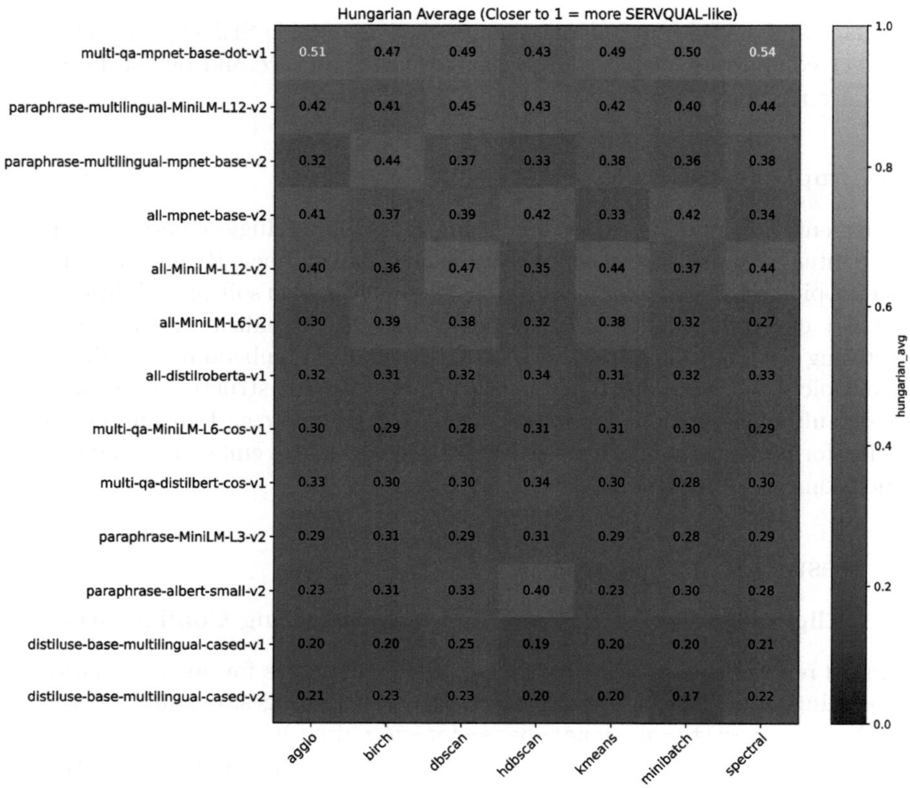

	agglo	birch	dbscan	hdbscan	kmeans	minibatch	spectral
multi-qa-mpnet-base-dot-v1	0.51	0.47	0.49	0.43	0.49	0.50	0.54
paraphrase-multilingual-MiniLM-L12-v2	0.42	0.41	0.45	0.43	0.42	0.40	0.44
paraphrase-multilingual-mpnet-base-v2	0.32	0.44	0.37	0.33	0.38	0.36	0.38
all-mpnet-base-v2	0.41	0.37	0.39	0.42	0.33	0.42	0.34
all-MiniLM-L12-v2	0.40	0.36	0.47	0.35	0.44	0.37	0.44
all-MiniLM-L6-v2	0.30	0.39	0.38	0.32	0.38	0.32	0.27
all-distilroberta-v1	0.32	0.31	0.32	0.34	0.31	0.32	0.33
multi-qa-MiniLM-L6-cos-v1	0.30	0.29	0.28	0.31	0.31	0.30	0.29
multi-qa-distilbert-cos-v1	0.33	0.30	0.30	0.34	0.30	0.28	0.30
paraphrase-MiniLM-L3-v2	0.29	0.31	0.29	0.31	0.29	0.28	0.29
paraphrase-albert-small-v2	0.23	0.31	0.33	0.40	0.23	0.30	0.28
distiluse-base-multilingual-cased-v1	0.20	0.20	0.25	0.19	0.20	0.20	0.21
distiluse-base-multilingual-cased-v2	0.21	0.23	0.23	0.20	0.20	0.17	0.22

Fig. 2. Hungarian alignment scores across embedding and clustering configurations.

KMeans and MiniBatchKMeans also generate reasonably aligned clusters, especially for concrete dimensions such as *Tangibles* and *Assurance*. In contrast, density-based methods like DBSCAN and HDBSCAN yield fewer than five topics, which prevents a full one-to-one mapping to SERVQUAL's five dimensions and results in more diffuse alignments. Agglomerative and BIRCH offer moderate alignment but exhibit less stability across dimensions compared to Spectral. Overall, these results identify Spectral Clustering as the most suitable algorithm when combined with `multi-qa-mpnet-base-dot-v1`, and we use this configuration as the basis for our subsequent analyses.

5.3 Topic-SERVQUAL Alignment and Effect of Deduplication

Using the best configuration (`multi-qa-mpnet-base-dot-v1` with Spectral), we first examine the detailed alignment between the five discovered topics and the five SERVQUAL dimensions. As shown in Fig. 4, Topic 1 exhibits the strongest structure, with high similarity to *Assurance*, *Tangibles*, and *Responsiveness*. Topic 2 also aligns well with *Tangibles*, whereas Topics 3 and 4 show weaker

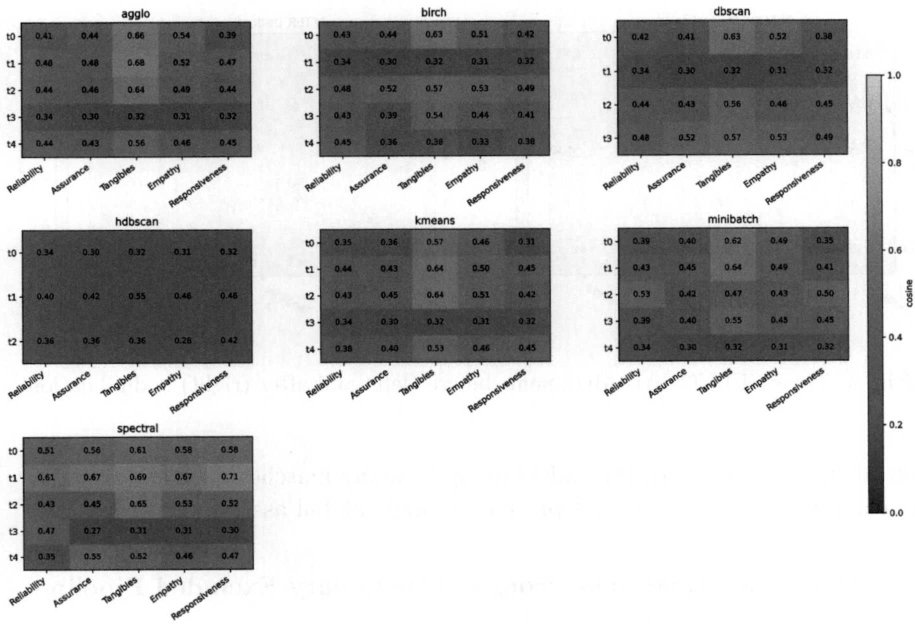

Fig. 3. Topic-to-SERVQUAL cosine similarity matrices for each clustering algorithm using the best-performing embedder (`multi-qa-mpnet-base-dot-v1`).

patterns. These results clarify which dimensions are well captured by the model and which ones remain less supported in the corpus.

We then apply a light deduplication step to ensure that each word appears in only one topic. When a word occurs in multiple topics, we keep the topic with the highest cTFIDF weight. If two or more topics are assigned the same weight, we break ties using the cosine similarity between the word and each topic centroid. Most duplicated words have low weights, so deduplication produces only minor changes. Adjustments occur only when a high-weight term is reassigned. For example, the word *error-free* appears in both Topic 0 and Topic 3 before deduplication, and assigning it to Topic 3 slightly reduces the similarity between Topic 0 and *Reliability*. Aside from these targeted shifts, the overall alignment pattern remains stable, which shows that the configuration is robust.

We use the Hungarian algorithm to obtain a strict one-to-one alignment between topics and SERVQUAL dimensions, which reflects how well a configuration can reconstruct the dictionary's conceptual structure. This choice is appropriate here because we set the number of clusters to five for topic-controlled clustering algorithms, making the assignment comparable across runs. As a sanity check, we also considered an exclusive greedy one-to-one baseline that iteratively selects the highest-similarity topic–dimension pair while preventing reuse of any topic or dimension. This yields the same qualitative conclusions about which configurations align best and which dimensions are less supported. More flexible

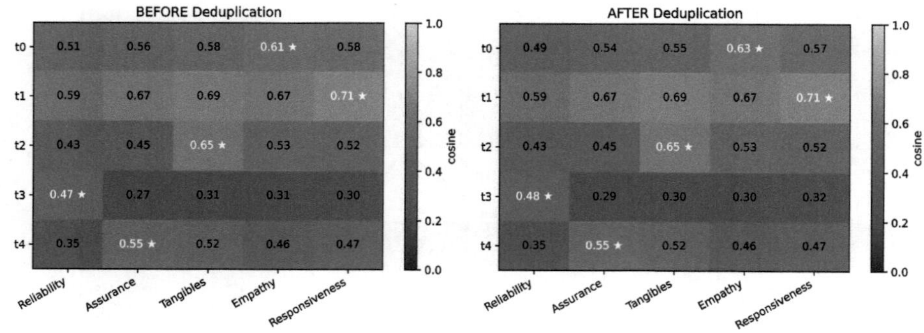

Fig. 4. Topic-SERVQUAL Alignment, before (left) and after (right) deduplication.

soft alignment strategies that allow many-to-many matches may better capture conceptual overlap, but do not produce a single global assignment score.

5.4 Emergent Dimensions from the Dictionary-Excluded Pipeline

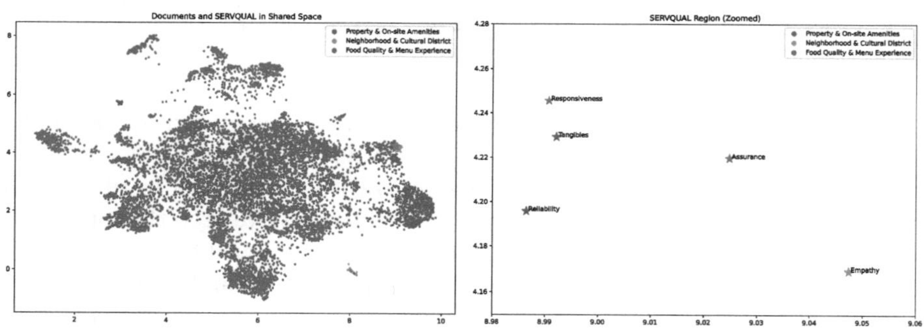

Fig. 5. UMAP projection of hotel and restaurant review embeddings generated by the dictionary-excluded pipeline. Colors indicate emergent clusters, and the five SERVQUAL centroids (stars) are shown for reference.

To evaluate semantic structure beyond the predefined dictionary, we apply the dictionary-excluded pipeline using the best configuration. As shown in Fig. 5, the UMAP projection reveals several broad and coherent regions of consumer language. Restaurant reviews dominate the corpus and form a large cluster centered on food quality and menu experience, while a smaller cluster reflects property features and on-site amenities.

The centroids appear only in a narrow overlap between these regions. Although this placement is sensible since SERVQUAL was designed to capture general evaluative language, it occupies only a small portion of the overall embedding space. Most review content lies far from this area, which indicates that the dictionary covers only a limited semantic range. This finding aligns with longstanding critiques that SERVQUAL does not fully capture experiential, contextual, or domain-specific aspects of service encounters [5,6,12].

The zoomed view further shows that all five dimensions are tightly grouped, reinforcing that the dictionary encodes a narrow band of service-quality language. In contrast, the dictionary-excluded pipeline uncovers several prominent regions outside SERVQUAL's coverage. These include experiential and intangible aspects such as ambience, atmosphere, or neighborhood context, as well as domain-specific areas like menu variety and property amenities.

Overall, this analysis identifies a concrete drawback of SERVQUAL: its fixed dimensions do not span the broader semantic space present in hospitality and restaurant reviews. At the same time, the emergent clusters recovered by our framework offer a data-driven path for extending SERVQUAL by suggesting potential new dimensions that capture experiential and domain-specific themes.

5.5 Summary

Taken together, the results show that BERTopic can recover themes that relate to the SERVQUAL framework when it uses high-capacity multilingual embeddings and clustering methods that form stable centroid structures. Several topics show clear correspondence to individual dimensions, while others capture only partial or diffuse relationships. These patterns suggest that topic modeling offers a practical way to examine how well a closed-vocabulary dictionary represents the semantic space of real consumer reviews. At the same time, the presence of large clusters that do not align with any predefined dimension highlights the limitations of the dictionary and points to areas where extensions or refinements are needed.

6 Limitations

This study has several limitations. First, topic modeling is an unsupervised method, so the discovered topics can vary across runs, parameter settings, and clustering choices. This variability affects how consistently topics align with dictionary dimensions. Second, the alignment procedure depends on embedding-based similarity between topic centroids and dictionary centroids. Although this approach captures broad semantic trends, it may overestimate similarity when words appear in different contexts or underestimate it when related concepts use varied surface forms. Third, the evaluation lacks an external gold standard. High similarity to dictionary centroids is treated as evidence of coverage, but meaningful themes that fall outside the predefined categories may be undervalued or overlooked. Finally, although the framework is designed to be model-agnostic, the implementation relies on BERTopic due to its modular structure and practical performance. Results may differ for other topic modeling pipelines that do not offer comparable control over embeddings and clustering.

7 Conclusion

This study presents LeTMEMo, a topic modeling based framework for evaluating and extending closed-vocabulary dictionaries. Using SERVQUAL as a case study, we show that carefully configured BERTopic models can recover latent themes that align with predefined dictionary dimensions, while also revealing substantial portions of consumer language that lie outside the dictionary's scope. The dictionary-included and dictionary-excluded pipelines jointly expose both the representational strengths and the limitations of fixed vocabularies when applied to real-world review data.

LeTMEMo is readily applicable to other closed-vocabulary resources such as LIWC. By comparing topic centroids with category centroids, the framework provides a principled way to assess category coverage, overlap, and omission in domain-specific corpora, and to identify candidate extensions grounded in observed language use. Future work will apply LeTMEMo across additional lexicons and domains and explore automated methods for recommending new terms and dimensions based on discovered topics.

Acknowledgments. Research partly supported by NSF grant 2244279, ARO grant W911NF-23-1-0191 and a US Department of Transportation grant for CyberCare. Verma is the founder of Everest Cyber Security and Analytics, Inc.

References

1. Angelov, D.: Top2Vec: distributed representations of topics. arXiv preprint arXiv:2008.09470 (2020)
2. Angelov, D.: Contextualized Top2Vec: enhancing topic representations using pre-trained language models. In: Findings of the Association for Computational Linguistics: EMNLP 2024, pp. 11418–11432. Association for Computational Linguistics (2024). https://aclanthology.org/2024.findings-emnlp.790
3. Blei, D.M., Ng, A.Y., Jordan, M.I.: Latent Dirichlet allocation. J. Mach. Learn. Res. **3**, 993–1022 (2003)
4. Brookes, G., McEnery, T.: The utility of topic modelling for discourse studies: a critical review. Discourse Stud. **21**(1), 3–21 (2019). https://doi.org/10.1177/1461445618814032
5. Buttle, F.: SERVQUAL: review, critique, research agenda. Eur. J. Mark. **30**(1), 8–32 (1996). https://doi.org/10.1108/03090569610105762
6. Coulthard, M.: Measuring service quality: a review and critique of research using SERVQUAL. Int. J. Mark. Res. **46**(4), 479–497 (2004)
7. Eichstaedt, J., et al.: Closed- and open-vocabulary approaches to text analysis: a review, quantitative comparison, and recommendations. Psychol. Methods **26**(4), 398–421 (2021). https://doi.org/10.1037/met0000335
8. Griffiths, T.L., Steyvers, M.: Hierarchical topic models and the nested Chinese restaurant process. In: Advances in Neural Information Processing Systems, vol. 16 (2004)
9. Grootendorst, M.: BERTopic: neural topic modeling with a class-based TF–IDF procedure. In: Proceedings of the 2022 Conference on Empirical Methods in Natural Language Processing: System Demonstrations, pp. 14–23 (2022)

10. Hutto, C.J., Gilbert, E.: VADER: a parsimonious rule-based model for sentiment analysis of social media text. In: Proceedings of the Eighth International AAAI Conference on Weblogs and Social Media (ICWSM), pp. 216–225 (2014)
11. Korenčić, D., Ristov, S., Repar, J., Šnajder, J.: A topic coverage approach to evaluation of topic models. arXiv preprint arXiv:2012.06274 (2020)
12. Ladhari, R.: A review of twenty years of SERVQUAL research. Int. J. Qual. Serv. Sci. **1**(2), 172–198 (2009). https://doi.org/10.1108/17566690910971445
13. Lind, F., Rönngren, D., Björklund, J.: Building the bridge: topic modeling for comparative text analysis in multilingual contexts. J. Multiling. Multicult. Dev. **43**(2), 180–196 (2022). https://doi.org/10.1080/01434632.2021.1965973
14. Lu, K.J.Q., Meaney, C., Guo, E., Leung, F.H.: Evaluating the applicability of existing lexicon based sentiment analysis tools to medical resident field notes. JMIR Med. Educ. **9**(1), e41953 (2023). https://doi.org/10.2196/41953
15. Mohammad, S.M.: Sentiment analysis: detecting valence, emotions, and other affectual states from text. Emot. Rev. **13**(2), 104–115 (2021). https://doi.org/10.1177/17540739211005734
16. Mohammad, S.M., Turney, P.D.: Crowdsourcing a word emotion association lexicon. Comput. Intell. **29**(3), 436–465 (2013)
17. Mukherjee, A., Venkataraman, V., Liu, B., Glance, N.S.: What Yelp fake review filter might be doing? In: Proceedings of the Seventh International AAAI Conference on Weblogs and Social Media (ICWSM), pp. 409–418. AAAI Press (2013)
18. Nguyen, N.T., Dang, V.M.H., Hancer, M., Verma, R.M.: Leveraging service quality model as a feature set with machine learning for fake hotel review detection. In: Proceedings of the 30th Annual Graduate Education and Graduate Student Research Conference in Hospitality and Tourism, Tempe, Arizona (2025). https://www.gradconfht.com/wp-content/uploads/2025/01/2025-Graduate-Conference-Proceedings_Final.pdf
19. Ojeda-Hernández, M., Lagunas, C., Zepeda-Mendoza, M.L.: Lexicon based sentiment analysis in texts using formal concept analysis. Int. J. Approximate Reasoning **155**, 104–112 (2023). https://doi.org/10.1016/j.ijar.2023.01.015
20. Parasuraman, A., Zeithaml, V.A., Berry, L.L.: SERVQUAL: a multiple item scale for measuring consumer perceptions of service quality. J. Retail. **64**(1), 12–40 (1988)
21. Pennebaker, J.W., Francis, M.E., Booth, R.J.: Linguistic Inquiry and Word Count: LIWC 2001. Lawrence Erlbaum Associates, Mahwah (2001)
22. Rocklage, M.D., Rucker, D.D., Nordgren, L.F.: The evaluative lexicon 2.0: the measurement of emotionality, extremity, and valence in language. Behav. Res. Methods **50**(4), 1327–1344 (2018). https://doi.org/10.3758/s13428-017-0975-6
23. Stammbach, D., Lapesa, G., Padó, S.: Revisiting automated topic model evaluation with large language models. arXiv preprint arXiv:2305.12152 (2023)
24. Tausczik, Y.R., Pennebaker, J.W.: The psychological meaning of words: LIWC and computerized text analysis methods. J. Lang. Soc. Psychol. **29**(1), 24–54 (2010). https://doi.org/10.1177/0261927X09351676

Author Index

M. Baratchi et al. (Eds.): IDA 2026, LNCS 16513, pp. 467–468, 2026.
https://doi.org/10.1007/978-3-032-23833-7

MIX
Papier aus verantwortungsvollen Quellen
Paper from responsible sources
FSC® C105338

If you have any concerns about our products,
you can contact us on
ProductSafety@springernature.com

In case Publisher is established outside the EU,
the EU authorized representative is:
Springer Nature Customer Service Center GmbH
Europaplatz 3, 69115 Heidelberg, Germany

Printed by Libri Plureos GmbH
in Hamburg, Germany